The Genera of Lactic Acid Bacteria

The Lactic Acid Bacteria
Volume 2

The Genera of Lactic Acid Bacteria

Edited by

B.J.B. WOOD
Department of Bioscience and Biotechnology
University of Strathclyde
Glasgow
UK

and

W.H. HOLZAPFEL
Federal Research Centre for Nutrition
Institute of Hygiene and Toxicology
Karlsruhe
Germany

SPRINGER-SCIENCE+BUSINESS MEDIA, B.V.

First edition 1995

Originally published by Chapman & Hall in 1995
Softcover reprint of the hardcover 1st edition 1995

Typeset in 10/12pt Times by Cambrian Typesetters, Frimley, Surrey

Cover photograph is a scanning electron micrograph of *Lactococcus lactis* subsp. *lactis* Bu2-60. (Courtesy of Horst Neve, Kiel, Germany.)

ISBN 978-1-4613-7666-8 ISBN 978-1-4615-5817-0 (eBook)
DOI 10.1007/978-1-4615-5817-0

A catalogue record for this book is available from the British Library

Library of Congress Catalog Card Number: 94–73502

∞ Printed on permanent acid-free text paper, manufactured in accordance with ANSI/NISO Z39.48-1992 (Permanence of Paper).

Series preface

The Lactic Acid Bacteria is planned as a series in a number of volumes, and the interest shown in it appears to justify a cautious optimism that a series comprising at least five volumes will appear in the fullness of time. This being so, I feel that it is desirable to introduce the series by providing a little of the history of the events which culminated in the decision to produce such a series. I also wish to indicate the boundaries of the group 'The Lactic Acid Bacteria' as I have defined them for the present purposes, and to outline my hopes for future topics in the series.

Historical background

I owe my interest in the lactic acid bacteria (LAB) to the late Dr Cyril Rainbow, who introduced me to their fascinating world when he offered me a place with him to work for a PhD on the carbohydrate metabolism of some lactic rods isolated from English beer breweries by himself and others, notably Dr Dora Kulka. He was particularly interested in their preference for maltose over glucose as a source of carbohydrate for growth, expressed in most cases as a more rapid growth on the disaccharide; but one isolate would grow only on maltose. Eventually we showed that maltose was being utilised by 'direct fermentation' as the older texts called it, specifically by the phosphorolysis which had first been demonstrated for maltose by Doudoroff and his associates in their work on maltose metabolism by a strain of *Neisseria meningitidis*.

I began work on food fermentations when I came to Strathclyde University, and I soon found myself involved again with the bacteria which I had not touched since completing my doctoral thesis. In 1973 J.G. Carr, C.V. Cutting and G.C. Whiting organised the 4th Long Ashton Symposium *Lactic Acid Bacteria in Beverages and Food*, and from my participation in that excellent conference arose a friendship with Geoff Carr. The growing importance of these bacteria was subsequently confirmed by the holding, a decade later, of the first of the Wageningen Conferences on the LAB. Discussions about the LAB, and the fact that they were unusual and important enough to have entire conferences devoted to them, with Mr George Olley of Applied Science Publishers Ltd (who had proposed the idea which resulted in the production of *Microbiology of Fermented Foods*) gave rise to the idea that there might be scope for a book on them.

Subsequent discussions with Geoff Carr refined 'a book' into 'a multi-volume series' and rather slowly Geoff and I began to lay rough plans for such a series, a process greatly helped by our preference for planning sessions in pubs. Sadly, Geoff died before the plans reached fruition, but his thinking contributed much to my development of the ideas which we hatched together, and I think that he would have approved of the final form which the series is taking. I have tried to achieve a multi-national spread among the authors, which would certainly have met with his approval. I have also tried to achieve a direct and clear style of writing through my directions to authors, and my editing, as both he and Cyril Rainbow valued clarity very highly, and deplored the sort of scientific writing which seemed to have been generated under the delusion that proper science should demonstrate its importance by being as obscure as possible.

The lactic acid bacteria

Most of us think that we know what we mean when we use the expression 'lactic acid bacteria'. We tend to think of them as a reasonably coherent group, and this is convenient for a great many purposes. In fact, however, the LAB remind me of a cloud in the sky, which can look rather solid and well-bounded when viewed from a distance but becomes more and more fuzzy and intangible the closer that one approaches to it. The LAB are all Gram-positive; anaerobic, micro-aerophilic or aero-tolerant; catalase negative; rods or cocci; most importantly they all produce lactic acid as the sole, major or an important product from the energy-yielding fermentation of sugars. It used to be thought that all LAB were non-motile and non-sporing, although we now have the Sporolactobacilli, and motile organisms are reported which would otherwise fit with the LAB. *Listeria* species produce lactic acid as a major metabolic product, but I doubt if many people working with the LAB as normally understood, would wish to admit *Listeria* to membership of the club. The LAB as a group had a 'squeaky clean' image, with such adverse effects as have been reported being, in general, ascribed to unusual circumstances such as abnormal immune status. This despite the Streptococci, as originally defined, including both beneficial and harmful organisms. Yet *Listeria* is closely related (according to modern classifications) to undoubted LAB, and *Streptococcus* as now defined seems to consist almost entirely of rather disagreeable specimens, so it seems a little irrational to include one of these genera and exclude the other; I can offer no real defence of my decision to do so despite this evidently logical argument, other than the hoary old one of 'accepted practice'. After much discussion, and with due acknowledgement of the rapidly evolving situation which is leading to the establishment of numerous new genera, the genera which receive chapters in Volume 2 will be regarded as the principal ones for the purposes of this series. This whilst acknowledging the right of individual authors to define the LAB in the manner which seems most appropriate to the story which they have to tell, and recognising that there will be well-established additional genera of LAB before the series has reached its conclusion.

Future plans

Future plans will inevitably be modified in the light of experience gained as the series develops, not least the feedback from readers and the comments and suggestions of reviewers. At present the following volumes are envisaged (not necessarily in the order listed:

Genetics of the LAB.
Physiology and Biochemistry.
Uses. This volume will cover their role in food fermentations, lactic acid production, vitamin assays, etc.

It is hoped and believed that together these volumes will form a lasting primary source and, as the computer experts would say, 'benchmark' against which future progress in our understanding of this diverse group of bacteria can be measured.

Brian J.B. Wood

Preface

There is very little to say about this volume which can usefully add to the material in chapter 1. I would, however, like to take the opportunity to thank various people, not least my co-editor Professor Wilhelm Holzapfel. While I recognise the paramount importance of taxonomy as the map by which we find our way around the confusing worlds of biology, I must admit that my grasp of the finer points of this discipline is not always secure. Wilhelm's hand on the tiller was always more confident than mine could ever be, but perhaps even more important was his remarkably extensive network of contacts. I am certain that his standing contributed significantly to our success in persuading busy people to give up their time to write difficult reviews. He also contributed significantly to my hitherto disgracefully limited understanding of the newer methods which are so changing both the methods employed in taxonomical investigations and (as a consequence thereof) our appreciation of the subtle relationships between, and even the evolution of, the kingdom Prokaryota.

Preparation of this book began with Elsevier Applied Science, and I was very pleased that it was one of the titles selected by Chapman and Hall when the transfer to them took place. I remain very grateful to the Elsevier staff for prodding me into getting started on this volume, and to the Blackie Academic and Professional staff for applying the necessary pressure when I seemed to be flagging along the way. Gratitude is also due to the authors for their excellent chapters, which have made this dry-seeming topic come vividly alive to at least one reader. I do not suppose that many users of this book will read every chapter of it in the way that Wilhelm Holzapel and I have needed to, but I hope that some will make the effort, as I am sure that they will find it rewarding. I hope also that it can stimulate thinking which will lead to new research areas; for my part, I am by no means satisfied that we know all that there is to know about the biochemical mechanisms which underly the very traditional, but still useful 'sugar assimilation and fermentation tests'. I also remain unclear as to the reasons why some LAB metabolise arginine to ornithine as they do; the immediate objective is to obtain ATP of course, but I find it difficult to believe that the average LAB is often in a natural environment which has a level of arginine sufficient to repay the expenditure required to put the necessary enzymes in place. I venture to suggest that much weightier speculations than these can be sparked by the material contained in this book.

Brian J.B. Wood

Contributors

B. Biavati	Facolta di Scienze Agrarie, Istituto di Microbiologia Agraria e Technica, Università degli Studi di Bologna, via Filippo Re 6, I-40126 Bologna, Italy
D. Claus	DSM-Deutsche Sammlung von Mikroorganismen und Zellkulturen GmbH, D-38124 Braunschweig, Germany
F. Dellaglio	Dipartimento di Scienze e Technologie Agro-Alimentari e Microbiologiche, Università degli Studi del Molise, I-86100 Campobasso, Italy
L.A. Devriese	Laboratorium voor Bacteriologie, en Mycologie van de huisdieren, Universiteit Gent, Salisburylaan 133, B-9820 Merelbeke, Belgium
L.M.T. Dicks	Department of Microbiology, Private Bag X5018, ZA-5900 University of Stellenbosch, Stellenbosch, South Africa
D. Fritze	DSM-Deutsche Sammlung von Mikroorganismen und Zellkulturen GmbH, D-38124 Braunschweig, Germany
W.P. Hammes	Institut für Lebensmitteltechnologie, Fachgebiet Allgemeine Lebensmitteltechnologie und -mikrobiologie, Universität Hohenheim, Garbenstraße 25, D-70559 Stuttgart 70, Germany
J.M. Hardie	Department of Oral Microbiology, The London Hospital Medical College, Turner Street, London E1 2AD, UK
W.H. Holzapfel	Institut für Hygiene und Toxikologie, Bundesforschungsanstalt für Ernährung, Engesserstr. 20, D-76131 Karlsruhe, Germany
W. Ludwig	Lehrstuhl für Mikrobiologie, Technische Universität München, Arcisstr 21, D-80333 München, Germany
D. Palenzona	Facolta di Scienze Agrarie, Istituto di Microbiologia Agraria e Technica, Università degli Studi di Bologna, via Filippo Re 6, I-40126 Bologna, Italy

B. Pot LMG-Cultuur Collectie, Laboratorium voor Microbiologie, Universiteit Gent, K.L. Ledeganckstraat 35, B-9000 Gent, Belgium

U. Schillinger Institut für Hygiene und Toxikologie, Bundesforschungsanstalt für Ernährung, Engesserstr. 20, D-76131 Karlsruhe, Germany

K.H. Schleifer Lehrstuhl für Mikrobiologie, Technische Universität München, Arcisstr 21, D-80333 München, Germany

B. Sgorbati Facolta di Scienze Agrarie, Istituto di Microbiologia Agraria e Technica, Università degli Studi di Bologna, via Filippo Re 6, I-40126 Bologna, Italy

W.J. Simpson BRF International, Lyttel Hall, Coopers Hill Road, Nutfield, Redhill, Surrey RH1 4HY, UK

H. Taguchi Kirin Brewery Company Limited, Brewing Research Laboratories, Beer Division, Technical Center, 1-17-1, Namamugi, Tsurumi-ku, Yokahama, 230, Japan

M. Teuber Eidgenössische Technische Hochschule Zürich, Schmelzbergstrasse 9, ETH Zentrum, CH-8092 Zürich, Switzerland

S. Torriani Dipartimento di Scienze e Technologie Agro-Alimentari e Microbiologiche, Università degli Studi del Molise, I-86100 Campobasso, Italy

R.F. Vogel Lehrstuhl für Technische Mikrobiologie, Technische Universität München, 85350 Freising-Weihenstephan, Germany

R.A. Whiley Department of Oral Microbiology, The London Hospital Medical College, Turner Street, London E1 2AD, UK

B.J.B. Wood Department of Bioscience and Biotechnology, Royal College Building, University of Strathclyde, George Street, Glasgow G1 1XW, UK

Contents

1 Lactic acid bacteria in contemporary perspective

W.H. HOLZAPFEL and B.J.B. WOOD

1.1 Introduction

Defining the scope of this text was a difficult task in practice, although it seemed simple when it was first proposed. Twenty years ago, when the first Lactic Acid Bacteria (LAB) Conference was held at Long Ashton Research Centre, Bristol, UK, it would have comprised chapters on the lactobacilli, streptococci, pediococci, leuconostocs and (with some dissent) bifidobacteria. Since then, much has happened, as this text shows. The genus *Streptococcus* has been stripped of its dairy and enteric species, which have been organized into new genera. The genus *Sporolactobacillus* has been recognized; however, the chapter on spore-forming organisms also includes discussion of the genus *Bacillus* with respect to species resembling the LAB in key physiological features such as the fermentative production of lactic acid as main end-product and relation to oxygen. Historically, the LAB have had a rather clean and wholesome image, with the genus *Streptococcus* as originally conceived presenting the seemingly anomalous picture of a genus which encompassed benign dairy organisms, animal commensals, true pathogens and commensals with a potential for pathogenicity. This complex situation has been resolved to some extent with the removal of the dairy species to the new genus *Lactococcus* and one motile species to *Vagococcus*. The other genus, *Enterococcus*, has some phenotypic features, different from that of the lactococci. It is also of interest that some close phylogenetic relationship exists between the genera *Enterococcus*, *Vagococcus* and *Carnobacterium*, in addition to 'common' physiological features such as their ability to grow at pH values around 9.5, much higher than the lactobacilli do. On the other hand, recent reports indicate that even members of the 'food-grade' lactobacilli can manifest in an apparently pathogenic role under opportunistic circumstances. Within the wider context, however, the question of situation and association may dictate the potential of any microorganism for pathogenic behaviour; with respect to the LAB this issue should be treated with greatest caution and hesitation.

Traditionally, the LAB are defined by the formation of lactic acid as a sole or main end-product from carbohydrate metabolism. The fact that bacilli may meet this criterion has long been recognized, although they have never been considered as belonging to the LAB. We have moved some way toward rectifying this anomaly with the broad sweep of chapter 11 – Spore-Forming Lactic Acid Producing Bacteria – but other anomalies remain. For example, *Actinomyces israelii* will, under appropriate anaerobic conditions, obtain energy by homofermentative conversion of carbohydrates to lactic acid. The modern recognition that the bifidobacteria are properly classified close to the actinomycetes reduces, but does not remove the dilemma. The 'conventional' phenotypic approach in LAB taxonomy will continue to have its rightful place in applied (food) microbiology laboratories. This will simplify the practical handling of, for example, 'food-grade' strains and may be supportive of considerations and legislative procedures towards their approval for application. On the other hand, phylogenetic evidence has clearly separated the bifidobacteria (with >50 mol% G+C in the DNA) from other LAB, all containing <50 mol% G+C.

Another difficulty is presented by the increasingly important food-borne pathogen *Listeria monocytogenes* and its relatives. Like the staphylococci they are catalase-positive, they ferment sugars with the production of lactic acid and seem to meet many of the other criteria typical of this rather loosely defined group called LAB, which still has no formal taxonomic status. Since bacteriocins are typically active against related organisms, it is interesting that quite a number of bacteriocinogenic LAB show antimicrobial activities also against *Listeria* and some *Staphylococcus* spp.

Currently, the work of Collins and others is generating persuasive arguments for creating several new but (as yet) small genera. It seems reasonable to suggest that at least some of these new genera will, as understanding of their significance spreads through the microbiological community, become as large and important as any of the genera which could currently be described as 'established'. From this, one might conclude that this text is already outdated, but this is surely the fate of any text on taxonomy at a time when the discovery of new organisms, revision of existing taxa, and the development of new and ever more subtle molecular biological techniques for studying the biochemical basis for taxonomy and phylogeny are all developing with their current rapidity.

1.2 Carbohydrate metabolism

Essentially the metabolism of carbohydrates by LAB can be classified as homolactic or heterolactic, although it is doubtful if any organism could be said to metabolise carbohydrates to lactic acid to the exclusion of all other products. Homolactic fermentation follows the familiar Embden–Meyerhof–Parnas (EMP) pathway for glycolysis. Unlike animal muscle, however, LAB may either form D(−)- or L(+)-lactic acid, or a racemic mixture of the two isomers. This has diagnostic significance in some phenotypic identification schemes.

During preparation of their chapters a number of authors raised questions about the presentation of metabolic pathways, including the EMP path. After careful consideration, it was decided that, while it is intended that a future volume will be devoted entirely to the biochemical activities of LAB, a limited amount of biochemistry is essential in this volume, for it to stand as a complete work in its own right. In order to save repetition authors were asked to limit their individual biochemical discussions to matters unique to their particular genus, or so central to it that they were unable to present their chapter without including the material in question. This is especially the case for the genus *Lactococcus*, which probably has been the object of the most intensive and fundamental metabolic studies among the LAB. We undertook to include the biochemical pathways which are common throughout the LAB in this prefatory chapter. The authors of the *Leuconostoc* chapter (chapter 7) had originally proposed to include three metabolic diagrams drawn from the excellent review by Kandler (1983), and we are grateful to them for agreeing to the transfer of those diagrams from their chapter to here and to the copyright holders for allowing us to reproduce the diagrams. Figure 1.1 shows the three main pathways associated with hexose metabolism in LAB

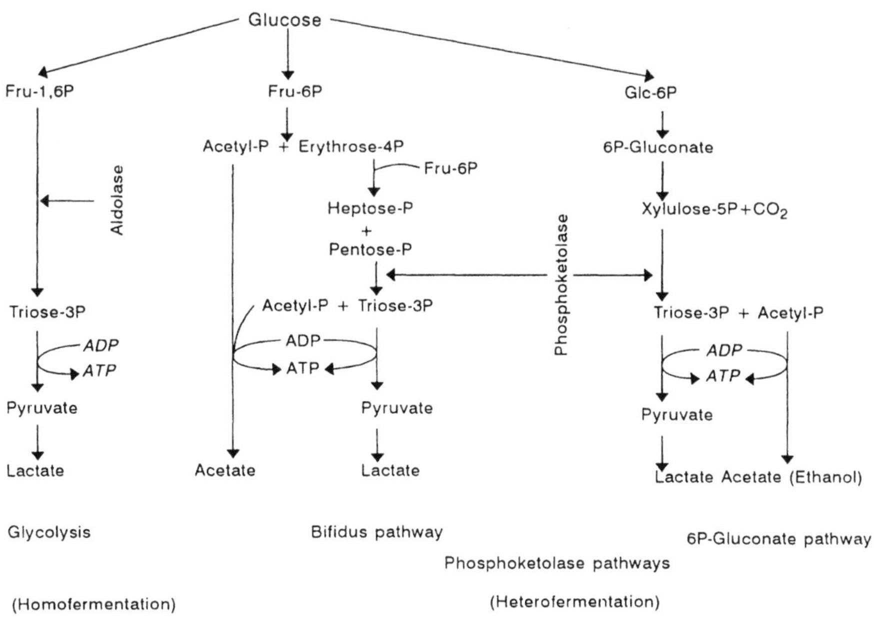

Figure 1.1 Schematic presentation of the main pathways of hexose fermentation in lactic acid bacteria (Reproduced from Kandler, 1983, by permission of Kluwer Academic Publishers.)

in a simplified form, allowing easy recognition of the principal similarities and differences. In the diagram glucose is shown as the starting point, although any metabolisable hexose may feed into these pathways with the appropriate emendments. The EMP route of glycolytic homofermentation is characterized by the key role of aldolase, as compared to phosphoketolase serving as key enzyme in the other two pathways. The 6-phosphogluconate pathway which yields carbon dioxide, lactate, acetate, and (in some cases) ethanol is found in some form in organisms exhibiting the heterolactic type of fermentation, except for the bifidobacteria, which utilize the 'Bifidus' pathway.

The chapter on the bifidobacteria offers a rather more detailed version of this latter route for hexose metabolism. Figure 1.2 illustrates the differences and relationships between galactose and lactose metabolism consequent upon uptake by the permease system or the phosphoenolpyruvate phosphotransferase system. Figure 1.3 illustrates the ingenious way in which a variety of pentoses and pentose alcohols are routed through the phosphoketolase pathway via conversion to D-xylulose-5-phosphate, eventually yielding equimolar amounts of lactate and acetate.

The chapter on lactococci (chapter 6) offers valuable additional information on energy production in LAB. Although the presentation naturally reflects the interests of that particular group of authors, it must be appreciated that the metabolic considerations therein are not confined

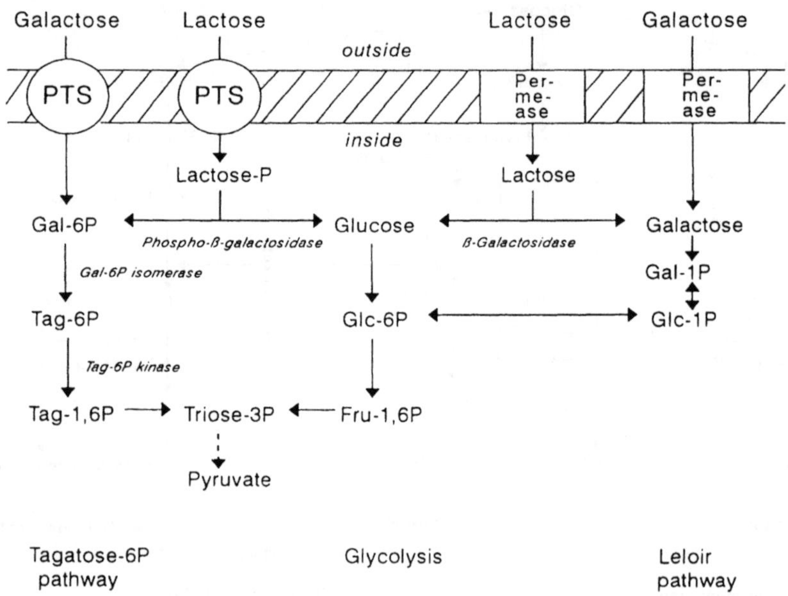

Figure 1.2 Lactose and galactose uptake and dissimilation in some lactic acid bacteria (Reproduced from Kandler, 1983, by permission of Kluwer Academic Publishers.)

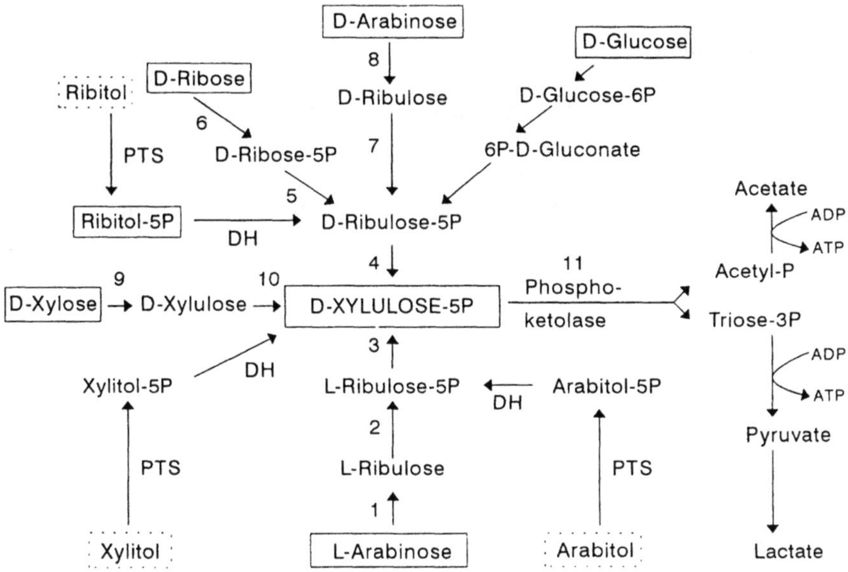

Figure 1.3 Pentose fermentations in lactic acid bacteria (Reproduced from Kandler, 1983, by permission of Kluwer Academic Publishers.) 1, L-Arabinose ketol-isomerase; 2, ATP: L-ribulose 5-phosphotransferase; 3, L-ribulose 5-phosphate 4-epimerase; 4, D-ribulose 5-phosphate 3-epimerase; 5, D-ribose 5-phosphate ketol-isomerase; 6, ATP: D-ribose 5-phosphotransferase; 7, ATP: D-ribulose 5-phosphotransferase; 8, D-aribinose ketol-isomerase; 9, D-xylose ketol-isomerase; 10, ATP: D-xylulose 5-phosphotransferase; 11, D-xylulose 5-phosphate D-glyceraldehyde 3-phosphate lyase; DH, dehydrogenase.

to that one genus. The solute transport mechanisms shown in their Figure 6.4 are of particularly general applicability, and ATP production by the arginine deiminase pathway (Figure 6.5) is also found in organisms from other genera. The malolactic fermentation (Figure 6.6) is perhaps associated in many workers' minds more with *Leuconostoc* than *Lactococcus*, but is significant in at least these two genera. Figure 6.7 directs attention to the lac operon as source of enzymes significant in lactose utilization by industrially important *Lactococcus lactis* strains. Citrate metabolism in dairy *Lc. lactis* is valued as the source of diacetyl, but in other situations the same product represents an unacceptable spoilage of the product, e.g. as produced by *Pediococcus damnosus* in beer.

We hope that this short introduction will be of help in reading the ensuing chapters, and again emphasize that the biochemical presentation here is deliberately limited to the minimum we consider to be essential for present purposes. The intention is that a future volume will be devoted to a very detailed consideration of biochemical topics.

Acknowledgement

One of the authors (B.J.B.W.) wishes to acknowledge his gratitude to colleagues in the Dept. of Food Science & Technology, University of California, Davis, CA 95616, USA, where he was on sabbatical leave during the later parts of this text's assembly into its final form.

Reference

Kandler, O. (1983) Carbohydrate metabolism in lactic acid bacteria. *Antonie van Leeuwenhoek Journal of Microbiology*, **49**, 209–224.

2 Phylogenetic relationships of lactic acid bacteria

K.H. SCHLEIFER and W. LUDWIG

2.1 Introduction

Prokaryotes possess a simple morphology, reveal no ontogeny and lack in general fossil records. Therefore, phylogenetic relationships can only be deduced by comparative sequence analyses of conserved, homologous and ubiquitously distributed macromolecules. The current knowledge on the phylogenetic relatedness of bacteria is mainly based upon comparative sequence analysis of 16S ribosomal ribonucleic acid (16S rRNA; Woese, 1987). Based on these data bacteria can be divided into at least 12 major lines of descent, so-called phyla (Figure 2.1). Comparative sequence analyses of other conserved macromolecules such as 23S rRNA, elongation factor Tu or β-subunit of ATPase support the 16S rRNA data (Schleifer and Ludwig, 1989; Ludwig *et al.*, 1993).

Lactic acid bacteria comprise a diverse group of Gram-positive, non-spore-forming bacteria. They occur as cocci or rods and generally lack catalase, although pseudo-catalase can be found in rare cases. They are chemo-organotrophic and grow only in complex media. Fermentable

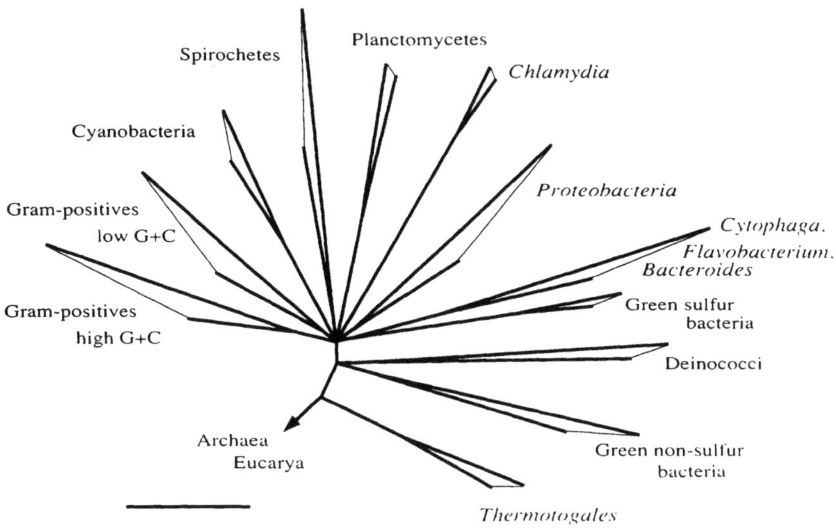

Figure 2.1 Phylogenetic tree of bacteria based upon 16S rRNA sequence comparison. The bar indicates 10% expected sequence divergence.

carbohydrates are used as energy source. Hexoses are degraded mainly to lactate (homofermentatives) or to lactate and additional products such as acetate, ethanol, CO_2, formate or succinate (heterofermentatives). Lactic acid bacteria are found in foods (dairy products, fermented meat, sour dough, fermented vegetables, silage, beverages), on plants, in sewage, but also in the genital, intestinal and respiratory tracts of man and animals (Hammes *et al.*, 1991).

Based on 16S and 23S rRNA sequence data, the Gram-positive bacteria form two lines of descent (Figure 2.2). One phylum consists of Gram-positive bacteria with a DNA base composition of less than 50 mol% guanine plus cytosine (G+C), the so-called *Clostridium* branch, whereas the other branch (actinomycetes) comprises organisms with a G+C content that is higher than 50 mol%. The typical lactic acid bacteria, such as *Carnobacterium*, *Lactobacillus*, *Lactococcus*, *Leuconostoc*, *Pediococcus* and *Streptococcus*, have a G+C content of less than 50 mol% and belong to the *Clostridium* branch. Originally, the genus *Bifidobacterium* was considered to be a member of the lactic acid bacteria, but based on the high DNA G+C content and from 16S rRNA data it is now quite clear that bifidobacteria belong to the actinomycetes branch (Figure 2.2). There are also other Gram-positive bacteria that are important for food or feed production and are members of the Actinomycetes branch, e.g. *Brevibacterium*, *Corynebacterium*, *Microbacterium* and *Propionibacterium*. In

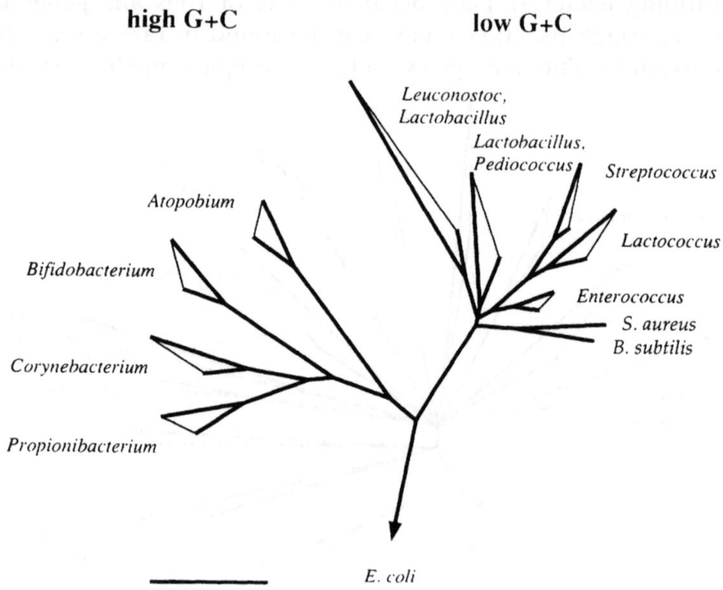

Figure 2.2 Phylogenetic trees of Gram-positive bacteria. The bar indicates 10% expected sequence divergence.

the present overview only the genuine lactic acid bacteria as well as their relatives with a low DNA G+C content and the genus *Bifidobacterium* will be considered.

2.2 The genera *Lactobacillus*, *Leuconostoc* and *Pediococcus*

The three genera are traditionally treated separately because of their different morphology and/or fermentation patterns. However, phylogenetically they are intermixed (Figure 2.3). The generic description does not exactly reflect the phylogenetic relationships. Even the internal structure of the genus *Lactobacillus* does not correlate well with the original subdivisions *Thermobacterium*, *Streptobacterium* and *Betabacterium* or with metabolic traits such as homo- and heterofermentative pathways. Based on 16S rRNA studies it was suggested to subdivide the genus *Lactobacillus* and related genera into three groups (Collins *et al.*, 1991). From a careful sequence analysis of the available 16S rRNA data of lactobacilli and related organisms (Olsen, G.J. *et al.*, 1991; de Rijk *et al.*, 1992) a phylogenetic tree was reconstructed by using different methods (distance matrix, parsimony and maximum likelihood analyses). These analyses showed that the *Leuconostoc* group is clearly separated from the other lactic acid bacteria, whereas the other two groups, the *Lb. delbrueckii* group and the *Lb. casei* group, are more closely related and hardly possible to distinguish (Figure 2.3). Despite this uncertainty we shall discuss the three groups separately.

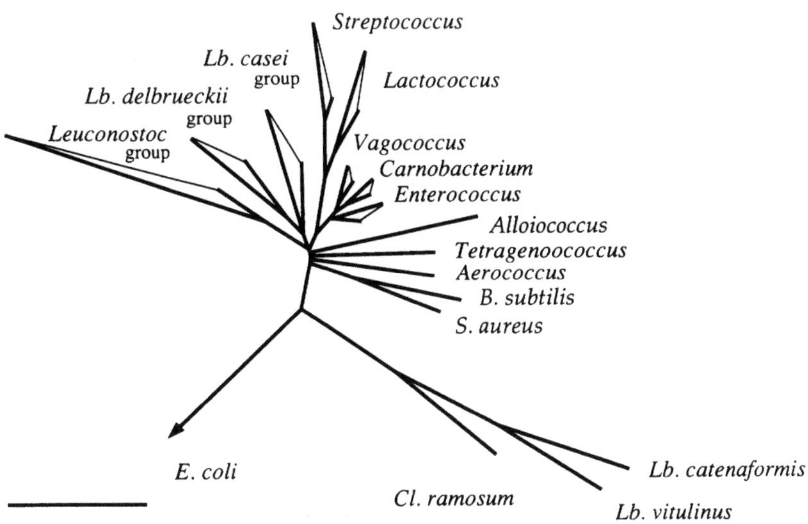

Figure 2.3 Phylogenetic tree of lactic acid bacteria and related bacteria. The bar indicates 10% expected sequence divergence.

(1) The first group comprises *Lb. delbrueckii* and other obligately homofermentative lactobacilli such as *Lb. acidophilus*, *Lb. amylophilus*, *Lb. amylovorus*, *Lb. crispatus*, *Lb. gallinarum*, *Lb. gasseri*, *Lb. helveticus*, *Lb. jensenii*, *Lb. johnsonii* and *Lb. kefiranofaciens* (Figure 2.4; Collins *et al.*, 1991). There are two exceptions, *Lb. acetotolerans* and *Lb. hamsteri*, which are facultative heterofermenters. The three subspecies of *Lb. delbrueckii*, namely *Lb. delbrueckii* subsp. *delbrueckii*, *Lb. delbrueckii* subsp. *bulgaricus* and *Lb. delbrueckii* subsp. *lactis* can neither be distinguished by differences in their 16S rRNA sequences nor by DNA–DNA hybridization since they share over 80% DNA similarity. *Lb. acidophilus* and related lactobacilli were previously divided into genotypic groups A and B (Johnson *et al.*, 1980), each consisting of several subgroups. Later it could be shown that the type strain of *Lb. acidophilus* belongs to DNA group A1 (Lauer and Kandler, 1980), subgroup A2 consists of *Lb. crispatus* (Cato *et al.*, 1983), subgroup A3 was assigned to *Lb. amylovorus* and subgroup A4 to *Lb. gallinarum* (Fujisawa *et al.*, 1992). Subgroup B1 was described as *Lb. gasseri* (Lauer and Kandler, 1980) and subgroup B2 as *Lb. johnsonii* (Fujisawa *et al.*, 1992). The *Lb. delbrueckii* group embraces most of the obligately homofermentative organisms. However, there are also obligately homofermentative organisms which phylogenetically belong to the *Lb. casei–Pediococcus* subgroup, e.g. *Lb. animalis*, *Lb. mali*, *Lb. ruminis*, *Lb. salivarius* and *Lb. sharpeae*.

(2) The second group, the so-called *Lb. casei–Pediococcus* group is the largest of the three groups (Figure 2.5). It comprises more than 30 *Lactobacillus* species and five pediococcal species (Collins *et al.*, 1991; Schleifer *et al.*, 1992). Most of the members are facultatively heterofermentative organisms. However, some are obligately hetero- or homofermentative. A list of species belonging to the *Lb. casei–Pediococcus*

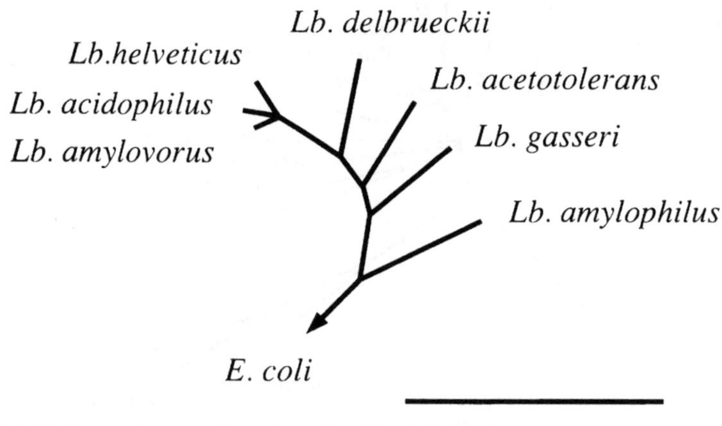

Figure 2.4 Phylogenetic tree of the *Lactobacillus delbrueckii* group. The bar indicates 10% expected sequence divergence.

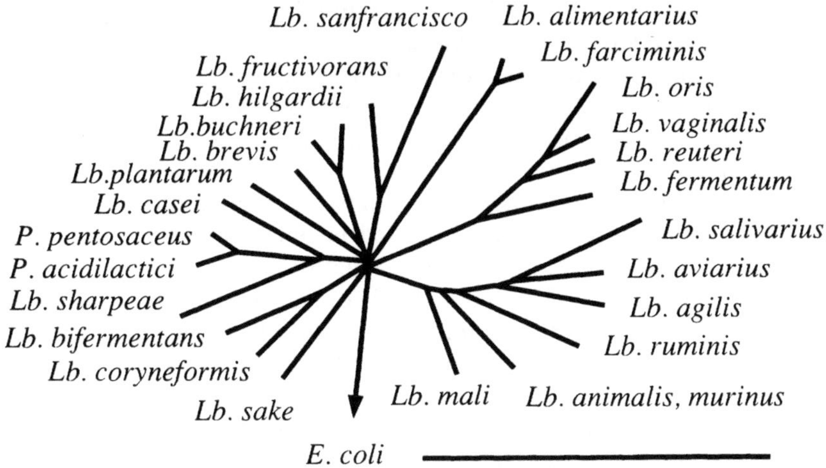

Figure 2.5 Phylogenetic tree of the *Lactobacillus casei–Pediococcus* group. The bar indicates 10% expected sequence divergence.

Table 2.1 List of lactic acid bacteria belonging to the *Lactobacillus casei–Pediococcus* group

Fermentation type	Species
Obligate homofermenters	*Lb. animalis*, *Lb. aviarius*, *Lb. ruminis*, *Lb. salivarius*, *Lb. sharpeae*, *Lb. yamanashiensis* (*Lb. mali*) *P. damnosus*, *P. dextrinicus*, *P. parvulus*
Facultative heterofermenters	*Lb. agilis*, *Lb. alimentarius*, *Lb. bifermentans*, *Lb. casei*, *Lb. coryniformis*, *Lb. curvatus*, *Lb. graminis*, *Lb. homohiochii*, *Lb. paracasei*, *Lb. pentosus*, *Lb. plantarum*, *Lb. rhamnosus*, *Lb. sake* *P. acidilactici*, *P. pentosaceus*
Obligate heterofermenters	*Lb. brevis*, *Lb. buchneri*, *Lb. fermentum*, *Lb. fructivorans*, *Lb. hilgardii*, *Lb. kefir*, *Lb. oris*, *Lb. parabuchneri*, Lb. reuteri, *Lb. sanfrancisco*, *Lb. suebicus*, *Lb. vaccinostercus*, *Lb. vaginalis*

group is compiled in Table 2.1. The best known species are *Lb. plantarum* and *Lb. casei*. DNA–DNA hybridization studies have shown that strains of *Lb. plantarum* are genetically not homogeneous and can be divided into different genotypic groups (Dellaglio *et al.*, 1975). A close relative of *Lb. plantarum* is *Lb. pentosus*. *Lactobacillus casei sensu lato* also comprises several genotypic groups. DNA–DNA hybridization studies showed that the majority of the strains designated *Lb. casei* subsp. *casei* and members of other subspecies (*alactosus*, *pseudoplantarum* and *tolerans*) were closely related to each other but not to the type strain of *Lb. casei* subsp. *casei* (Collins *et al.*, 1989a). The former strains were described as a new species, *Lb. paracasei*. *Lactobacillus casei* subsp. *rhamnosus* was elevated to the species rank as *Lb. rhamnosus* since only a low DNA similarity was found with other members of the *Lb. casei* group (Collins *et al.*, 1989a; Dellaglio *et al.*, 1991). *Lactobacillus murinus* and *Lb. animalis* show an identical 16S rRNA sequence and at least some of the strains are also physiologically closely related (Kandler and Weiss, 1986). Based on 16S rRNA studies four of the five pediococcal species form a distinct clade, whereas *P. dextrinicus* is more closely related to certain lactobacilli, e.g. *Lb. coryneformis* and *Lb. bifermentans*, than to the other pediococci (Collins *et al.*, 1991). The two pediococci included in our analyses are most closely related to *Lb. sharpeae* and *Lb. casei* (Figure 2.5).

(3) The third group is composed of all members of the genus *Leuconostoc* and obligately heterofermentative lactobacilli. The so-called *Leuconostoc* group (Yang and Woese, 1989) can be clearly subdivided into two subgroups (Figure 2.6). One subgroup comprises all leuconostocs (except *Leuc. paramesenteroides*) and *Lb. fructosus*, whereas the other subgroup consists of obligately heterofermentative lactobacilli and *Leuc. paramesenteroides*.

(4) There are several bacteria which were erroneously placed in the genus *Lactobacillus*. *Lactobacillus catenaformis* is more closely related to certain clostridia (*Cl. ramosus*, *Cl. innocuum*) and *Erysipelothrix rhusiopathiae* than to any member of the genus *Lactobacillus* (Stackebrandt and Woese, 1979). The same is true for *Lb. vitulinus* (Figure 2.3). However, *Lb. catenaformis* and *Lb. vitulinus* are not so closely related that they could be allocated to the same genus. *Lactobacillus minutus*, *Lb. rimae*, *Lb. uli* and *Streptococcus parvulus* are more closely related with high DNA G+C than with the low DNA G+C Gram-positive bacteria and were recently reclassified in a new genus, *Atopobium* (Collins and Wallbanks, 1992; Figure 2.2). *Lactobacillus xylosus* and *Lb. hordniae* were also misclassified and were reclassified as *Lactococcus lactis* (Schleifer *et al.*, 1985).

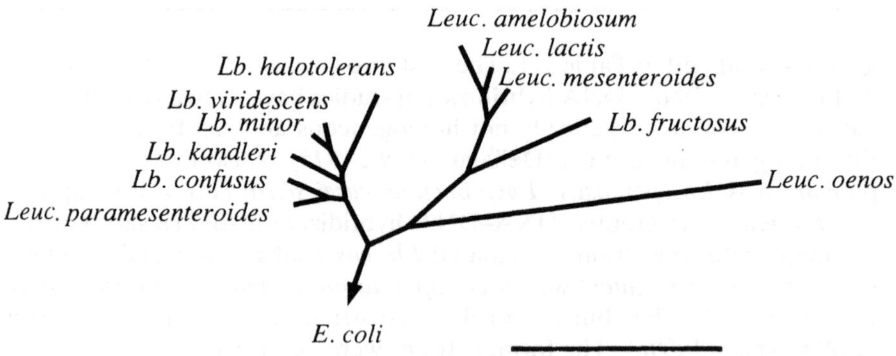

Figure 2.6 Phylogenetic tree of the *Leuconostoc* group. The bar indicates 10% expected sequence divergence.

2.3 The genus *Carnobacterium*

Atypical lactobacilli isolated from chicken meat, vacuum-packaged meat or fish are physiologically and biochemically very similar and were placed in a new genus, *Carnobacterium* (Collins *et al.*, 1987). In contrast to typical lactobacilli they cannot grow on acetate agar but grow at pH 8.5–9.0. Later studies have revealed that carnobacteria form a phylogenetically coherent group, quite distinct from all other lactobacilli (Figure 2.7; Wallbanks *et al.*, 1990). The genus *Carnobacterium* consists of five species, namely *C. divergens*, *C. funditum*, *C. gallinarum*, *C. mobile* and *C. piscicola*. Comparative 16S rRNA sequence analyses have shown that *Lb. maltaromicus* has a very similar sequence to that of *C. piscicola* (Collins *et al.*, 1991). However, the identity of the two species has to be confirmed by DNA–DNA hybridization studies.

2.4 The genera *Streptococcus*, *Lactococcus*, *Enterococcus* and *Vagococcus*

The genus *Streptococcus sensu lato* is not a phylogenetically coherent unit but consists of four subgroups that have been conferred genus status (Schleifer and Kilpper-Bälz, 1987; Collins *et al.*, 1989b; Figure 2.3).

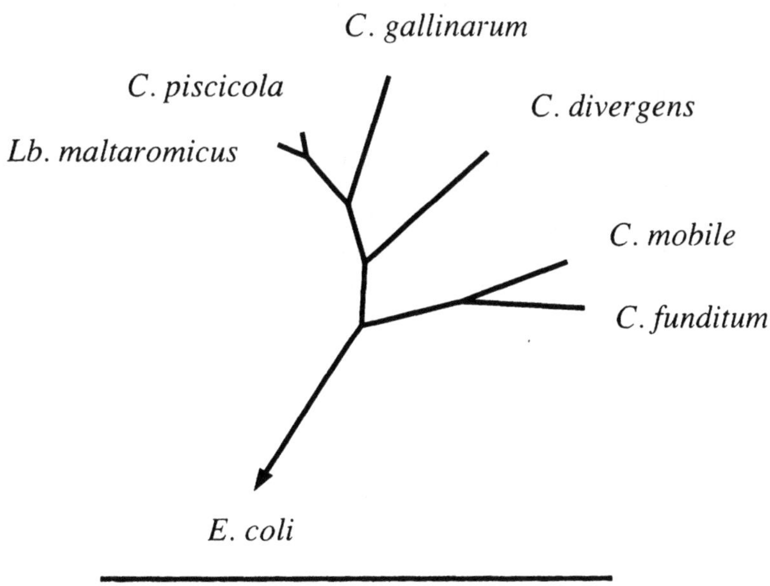

Figure 2.7 Phylogenetic tree of carnobacteria. The bar indicates 10% expected sequence divergence.

(1) The genus *Streptococcus sensu stricto* can be divided in the pyogenic and the oral or viridans streptococci. The intrageneric relationships of the genus *Streptococcus sensu stricto* were studied by DNA–rRNA hybridization studies (Kilpper-Bälz and Schleifer, 1984; Kilpper-Bälz *et al.*, 1985) and partial 16S rRNA sequence analysis (Bentley *et al.*, 1991). The results obtained with both methods are in good agreement. However, the 16S rRNA sequence data indicate that besides *S. pyogenes*, *S. canis*, *S. equi*, *S. dysgalactiae* and *S. iniae* also *S. agalactiae*, *S. parauberis*, *S. porcinus* and *S. uberis* belong to the pyogenic streptococci. The oral streptococci can be subdivided into four groups. The *S. mutans* group with seven species (*S. criscetus*, *S. downei*, *S. ferus*, *S. macacae*, *S. mutans*, *S. rattus*, *S. sobrinus*), the *S. salivarius* group with *S. salivarius*, *S. thermophilus* and *S. vestibularis*, the *S. anginosus* group consisting of *S. anginosus*, *S. constellatus* and *S. intermedius* and finally the *S. oralis* group with six species (*S. gordonii*, *S. mitis*, *S. oralis*, *S. parasanguis*, *S. pneumoniae*, *S. sanguis*). *Streptococcus bovis* forms, together with *S. equinus* and *S. alactolyticus*, a distinct cluster well separated from both the pyogenic and oral groups (Bentley *et al.*, 1991). Two species, namely *S. acidominimus* and *S. suis* are genetically distinct and show no specific relationship to any of the other streptococcal species.

(2) The genus *Lactococcus* was proposed by Schleifer *et al.* (1985) to accommodate the non-motile, mesophilic streptococci carrying a group N antigen. Four species (*Lc. garviae*, *Lc. lactis*, *Lc. plantarum*, *Lc. raffinolactis*) were mentioned in the original description of the genus *Lactococcus*. One new species, *Lc. piscium*, has been described since then (Williams *et al.*, 1990).

(3) The genus *Enterococcus* was revived by Schleifer and Kilpper-Bälz (1984) to accommodate the species '*Streptococcus*' *faecalis* and '*Streptococcus*' *faecium*. An extended study led to the transfer of five other group D streptococci to the genus *Enterococcus* (Collins *et al.*, 1984). Currently, the genus *Enterococcus* consists of 19 species. Comparative 16S rRNA sequence studies indicate that the enterococci are somewhat closer related to *Carnobacterium* and *Vagococcus* than to *Streptococcus* and *Lactococcus* (Figure 2.3). Based on comparative 16S rRNA sequence analysis several species groups can be distinguished within the genus *Enterococcus*. The four species *Ent. durans*, *Ent. faecium*, *Ent. hirae* and *Ent. mundtii* form a distinct cluster as do *Ent. avium*, *Ent. raffinosus*, *Ent. malodoratus* and *Ent. pseudoavium* and the two species *Ent. casseliflavus* and *Ent. gallinarum*. A possible relationship was also found between *Ent. columbae* and *Ent. cecorum*. All the other enterococcal species analysed so far form individual lines of descent. *Enterococcus faecalis* and *Ent. solitarius* represent the deepest branching within the phylogenetic tree of the genus *Enterococcus*. The species *Ent. solitarius* reveals even a closer relationship to *Tetragenococcus halophilus* than to other enterococcal species.

(4) Motile group N streptococci which were isolated from chicken faeces and river water and were quite distinct from lactococci were described by Collins *et al.* (1989c) as *Vagococcus fluvialis*. A second species, *V. salmoninarum*, was described by Wallbanks *et al.* (1990) for a group of lactic acid bacteria isolated from diseased salmonid fish.

2.5 The genera *Aerococcus*, *Alloiococcus*, *Tetragenococcus* and *Atopobium*

The genus *Aerococcus* currently contains two species, *A. urinae* and *A. viridans*. They can be distinguished from other catalase-negative Gram-positive cocci by their tendency to divide in two planes at right angles to form tetrads. For a long time *A. viridans* has been the only species of the genus, but recently *Aerococcus*-like organisms were classified as *A. urinae* (Aguirre and Collins, 1992). It could be also shown that *Pediococcus urinaeequi* is very closely related to *A. viridans* and should be transferred to the genus *Aerococcus* (Collins *et al.*, 1990). The genus *Aerococcus* represents a distinct line of descent within the Gram-positive bacteria with a low DNA G+C content (Figure 2.3).

Large Gram-positive cocci, often present as diplococci or tetrads, and phenotypically most closely resembling aerococci and streptococci were isolated from human middle ear fluid. However, in contrast to aerococci and streptococci the isolated bacteria were catalase positive. Comparative 16S rRNA sequence analysis revealed that the unknown bacterium formed a long, distinct line of descent within the Gram-positive bacteria with low G+C content (Aguirre and Collins, 1992; Figure 2.3). The name *Alloiococcus otitis* was proposed for this unique bacterium (Aguirre and Collins, 1992).

Pediococcus halophilus shows no close relationship to other pediococci. 16S rRNA sequence analysis exhibited that this species forms a distinct line of descent quite separate from pediococci and aerococci. Therefore, it was proposed that *P. halophilus* be reclassified in a new genus *Tetragenococcus* (Collins *et al.*, 1990; Figure 2.3).

Lactobacillus minutus does not belong to the Gram-positive bacteria with low G+C content but is more closely related to the high G+C content Gram-positive bacteria. A number of oral strains assigned to this species were transferred to a new species, *Lb. uli* (Olsen, I. *et al.*, 1991). In the same study the new species *Lb. rimae* was described for isolates from human gingival crevices. Comparative 16S rRNA sequence analysis of *Lb. minutus* and *Lb. rimae* exhibited that these species are related to each other and to *Streptococcus parvulus* (Collins and Wallbanks, 1992). Based on this study it was proposed to reclassify these species in a new genus, *Atopobium* (Figure 2.2).

2.6 The genus *Bifidobacterium*

Members of the genus *Bifidobacterium* exhibit a high G+C content of DNA (55–67 mol%). Based on 16S rRNA sequence analysis they belong to the Gram-positive bacteria with a high G+C content (Figure 2.2). Physiologically they resemble genuine lactic acid bacteria since they are also saccharoclastic and produce lactate and acetate as major fermentation end-products (Biavatti *et al.*, 1991). However, they possess a special pathway of hexose fermentation, the so-called Bifidus pathway or fructose 6-phosphate shunt. The complete pathway employs enzymes of the glycolytic and pentose phosphate pathways and contains a second phosphoketolase that cleaves fructose 6-phosphate to acetyl phosphate and erythrose 4-phosphate. The presence of this enzyme, fructose 6-phosphate phosphoketolase, is a reliable characteristic for the identification of a bacterium as a member of the genus *Bifidobacterium*. The 16S rRNA sequences of a limited number of bifidobacteria are known. From these data it is obvious that bifidobacteria form a phylogenetically coherent unit within the Gram-positive bacteria with high G+C content (Figure 2.8). There is only one other organism within this cluster, namely *Gardnerella vaginalis*.

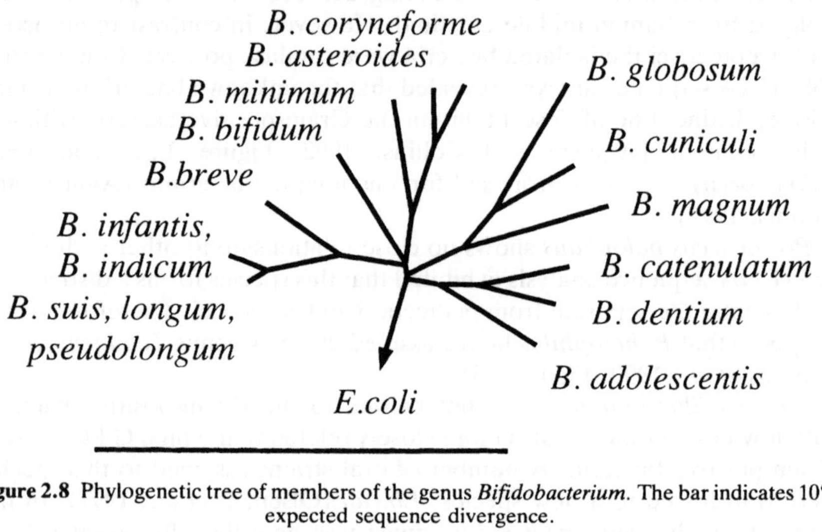

Figure 2.8 Phylogenetic tree of members of the genus *Bifidobacterium*. The bar indicates 10% expected sequence divergence.

Acknowledgement

This work was supported by grants from the Bundesministerium für Forschung und Technologie (0319274A) and the European Community (BIO-CT-910263).

References

Aguirre, M. and Collins, M.D. (1992) Phylogenetic analysis of *Alloiococcus otitis* gen. nov., spec. nov., an organism from human middle ear fluid. *International Journal of Systematic Bacteriology*, **42**, 79–83.

Bentley, R.W., Leigh, J.A. and Collins, M.D. (1991) Intrageneric structure of *Streptococcus* based on comparative analysis of small-subunit rRNA sequences. *International Journal of Systematic Bacteriology*, **41**, 487–491.

Biavatti, B., Sgorbati, B. and Scardovi, V. (1991) The genus *Bifidobacterium*. In *The Prokaryotes. A Handbook on the Biology of Bacteria*, 2nd edn (eds Balows, A., Trüper, H.G., Dworkin, M., Harder, W. and Schleifer, K.H.). Springer, New York, USA, pp. 816–833.

Cato, E.P., Moore, W.E.C. and Johnson, J.L. (1983) Synonymy of strains of '*Lactobacillus acidophilus*' group A2 (Johnson *et al.*, 1980) with the type strain of *Lactobacillus crispatus* (Brygoo and Aladame, 1953) Moore and Holdeman 1970. *International Journal of Systematic Bacteriology*, **33**, 426–428.

Collins, M.D. and Wallbanks, S. (1992) Comparative sequence analysis of the 16S rRNA genes of *Lactobacillus minutus*, *Lactobacillus rimae* and *Streptococcus parvulus*: proposal for the creation of a new genus *Atopobium*. *FEMS Microbiology Letters*, **95**, 235–240.

Collins, M.D., Jones, D., Farrow, J.A.E., Kilpper-Bälz, R. and Schleifer, K.H. (1984) *Enterococcus avium* nom. rev., comb. nov.; *E. casseliflavus* nom. rev. comb. nov.; *E. durans* nom. rev., comb. nov.; *E. gallinarum* comb. nov.; and *E. malodoratus*sp. nov. *International Journal of Systematic Bacteriology*, **34**, 220–223.

Collins, M.D., Farrow, J.A.E., Phillips, B.A., Ferusu, S. and Jones, D. (1987) Classification of *Lactobacillus divergens*, *Lactobacillus piscicola*, and some catalase-negative, asporogeneous, rod-shaped bacteria from poultry in a new genus *Carnobacterium*. *International Journal of Systematic Bacteriology*, **37**, 310–316.

Collins, M.D., Phillips, B.A. and Zanoni, P. (1989a) Deoxyribonucleic acid homology studies of *Lactobacillus casei*, *Lactobacillus paracasei* sp. nov., subsp. *paracasei* and subsp. *tolerans*, and *Lactobacillus rhamnosus* sp. nov., comb. nov. *International Journal of Systematic Bacteriology*, **39**, 105–108.

Collins, M.D., Facklam, R.R., Farrow, J.A.E. and Williamson, R. (1989b) *Enterococcus raffinosus* sp. nov., *Enterococcus solitarius* sp. nov. and *Enterococcus pseudoavium* sp. nov. *FEMS Microbiology Letters*, **57**, 283–286.

Collins, M.D., Ash, C., Farrow, J.A.E., Wallbanks, S. and Williams, A.M. (1989c) 16S Ribosomal ribonucleic acid sequence analyses of lactococci and related taxa. Description of *Vagococcus fluvialis* gen. nov., sp. nov. *Journal of Applied Bacteriology*, **67**, 453–460.

Collins, M.D., Williams, A.M. and Wallbanks, S. (1990) The phylogeny of *Aerococcus* and *Pediococcus* as determined by 16S rRNA sequence analysis: description of *Tetragenococcus* gen nov. *FEMS Microbiology Letters*, **70**, 255–262.

Collins, M.D., Rodrigues, U.M., Ash, C., Aguirre, M., Farrow, J.A.E., Martinez-Murica, A., Phillips, B.A., Williams, A.M. and Wallbanks, S. (1991) Phylogenetic analysis of the genus *Lactobacillus* and related lactic acid bacteria as determined by reverse transcriptase sequencing of 16S rRNA. *FEMS Microbiology Letters*, **77**, 5–12.

Dellaglio, F., Bottazzi, V. and Vescovo, M. (1975) Deoxyribonucleic acid homology among *Lactobacillus* species of the subgenus *Streptobacterium* Orla-Jensen. *International Journal of Systematic Bacteriology*, **25**, 160–172.

Dellaglio, F., Dicks, L.M.T., Du Toit, M. and Torriani, S. (1991) Designation of ATCC 334 in place of ATCC 393 (NCDO 161) as the neotype strain of *Lactobacillus casei* subsp. *casei* and rejection of the name *Lactobacillus paracasei* (Collins *et al.*, 1989) Request for an opinion. *International Journal of Systematic Bacteriology*, **41**, 340–342.

De Rijk, P., Neefs, J.-M., Van de Peer, Y. and De Wachter, R. (1992) Compilation of small ribosomal subunit RNA sequences. *Nucleic Acids Research*, **20** (Supplement), 2175–2189.

Fujisawa, T., Benno, Y., Yaeshima, T. and Mitsuoka, T. (1992) Taxonomic study of the *Lactobacillus acidophilus* group, with recognition of *Lactobacillus gallinarum* sp. nov. and *Lactobacillus johnsonii* sp. nov. and synonymy of *Lactobacillus acidophilus* group A3 (Johnson *et al.*, 1980) with the type strain of *Lactobacillus amylovorus* (Nakamura, 1981). *International Journal of Systematic Bacteriology*, **42**, 487–491.

Hammes, W.P., Weiss, N. and Holzapfel, W.P. (1991) The genera *Lactobacillus* and *Carnobacterium*. In *The Prokaryotes. A Handbook on the Biology of Bacteria: Ecophysiology, Isolation, Identification, Applications* (eds Balows, A., Trüper, H.G., Dworkin, M., Harder, W. and Schleifer, K.H.). Springer, New York, USA, pp. 1535–1594.

Johnson, J.L., Phelps, C.F., Cummins, C.S., London, J. and Gasser, F. (1980) Taxonomy of the *Lactobacillus acidophilus* group. *International Journal of Systematic Bacteriology*, **30**, 53–68.

Kandler, O. and Weiss, N. (1986) Regular, nonspring Gram-positive bacteria. In *Bergey's Manual of Systematic Bacteriology* (eds Sneath, P.H.A., Mair, N., Sharpe, M.E. and Holt, J.G.). Williams and Wilkins, Baltimore, USA, pp. 1208–1234.

Kilpper-Bälz, R. and Schleifer, K.H. (1984) Nucleic acid hybridization and cell wall composition studies of pyogenic streptococci. *FEMS Microbiology Letters*, **24**, 355–364.

Kilpper-Bälz, R., Wenzig, P. and Schleifer, K.H. (1985) Molecular relationship and classification of some viridans streptococci as *Streptococcus oralis* and emended description *Streptococcus oralis* (Bridge and Sneath, 1982) *International Journal of Systematic Bacteriology*, **35**, 482–488.

Lauer, E. and Kandler, O. (1980) *Lactobacillus gasseri* sp. nov., a new species of the subgenus *Thermobacterium*. *Zentralblatt für Bakteriologie, Parasitenkunde, Infektionskrankheiten und Hygiene 1. Abteilung Originale*, **C1**, 75–78.

Ludwig, W., Neumann, J., Klugbauer, N., Brockmann, E., Roller, C., Jilg, S., Reetz, K., Schachtner, I., Ludvigsen, A., Wallner, G., Bachleitner, M., Fischer, U. and Schleifer, K.H. (1993) Phylogenetic relationships of bacteria based on comparative sequence analysis of elongation factor Tu and ATP-synthase β-subunit genes. *Antonie van Leeuwenhoek International Journal of General and Molecular Microbiology*, **64**, 285–305.

Olsen, G.J., Larsen, N. and Woese, C.R. (1991) The ribosomal RNA data base project. *Nucleic Acids Research*, **19** (Supplement), 2017–2021.

Olsen, I., Johnson, J.L., Moore, L.V.H. and Moore, W.E.C. (1991) *Lactobacillus uli* sp. nov. and *Lactobacillus rimae* sp. nov. from human gingival crevice and emended descriptions of *Lactobacillus minutus* and *Streptococcus parvulus*. *International Journal of Systematic Bacteriology*, **41**, 261–266.

Schleifer, K.H. and Kilpper-Bälz, R. (1984) Transfer of *Streptococcus faecalis* and *Streptococcus faecium* to the genus *Enterococcus* nom. rev. as *Enterococcus faecalis* comb. nov. and *Enterococcus faecium* comb. nov. *International Journal of Systematic Bacteriology*, **34**, 31–34.

Schleifer, K.H. and Kilpper-Bälz, R. (1987) Molecular and chemotaxonomic approaches to the classification of streptococci, enterococci and lactococci: a review. *Systematic Applied Microbiology*, **6**, 1–19.

Schleifer, K.H. and Ludwig, W. (1989) Phylogenetic relationships among bacteria. In *Hierarchy of Life* (eds Fernholm, B., Bremer, K. and Jörnvall, H.). Excerpta Medica, Elsevier Science Publishers BV, Amsterdam, The Netherlands, pp. 103–117.

Schleifer, K.H., Kraus, J., Dvorak, C., Kilpper-Bälz, R., Collins, M.D. and Fischer, W. (1985) Transfer of *Streptococcus lactis* and related streptococci to the genus *Lactococcus* gen. nov. *Systematic and Applied Microbiology*, **6** 183–195.

Schleifer, K.H., Ludwig, W., Amann, R., Hertel, C., Ehrmann, M., Köhler, W. and Krause, A. (1992) Phylogenetic relationships of lactic acid bacteria and their identification with nucleic acid probes. In *Les Bacteries Lactiques. Actes du Colloque LACTIC 91* (eds Novel, G. and Le Querler, J.-F.). Centre du Publications de l'Université de Caen, France, pp. 23–32.

Stackebrandt, E. and Woese, C.R. (1979) A phylogenetic dissection of the family Micrococcaceae. *Current Microbiology*, **2**, 317–322.

Wallbanks, S., Martinez-Murcia, A.J., Fryer, J.L. Philips, B.A. and Collins, M.D. (1990) 16S rRNA sequence determination for members of the genus *Carnobacterium* and related lactic acid bacteria and description of *Vagococcus salmoninarum* sp. nov. *International Journal of Systematic Bacteriology*, **40**, 224–230.

Williams, A.M., Fryer, J.L. and Collins, M.D. (1990) *Lactococcus piscium* sp. nov. a new *Lactococcus* species from salmonid fish. *FEMS Microbiology Letters*, **68**, 109–114.

Woese, C.R. (1987) Bacterial evolution. *Microbiological Reviews*, **51**, 221–271.

Yang, D. and Woese, C.R. (1989) Phylogenetic structure of the leuconostocs: an interesting case of a rapidly evolving organism. *Systematic and Applied Microbiology*, **12**, 145–149.

3 The genus *Lactobacillus*

W.P. HAMMES and R.F. VOGEL

3.1 Introduction

Lactobacilli are Gram-positive, non-spore-forming, rods or coccobacilli with a G+C content of DNA usually below 50 mol%. They are strictly fermentative, aero-tolerant or anaerobic, aciduric or acidophilic and have complex nutritional requirements (e.g. for carbohydrates, amino acids, peptides, fatty acid esters, salts, nucleic acid derivatives, and vitamins). They do not synthesize porphyrinoids and thus, are devoid of heme-dependent activities. Strains of some species can use porphorinoids from the environment and exhibit activities of catalase, nitrite reduction or even cytochromes (Meisel, 1991). Pseudo-catalase is formed in strains of *Lb. mali*. With glucose as a carbon source lactobacilli may be either homofermentative, producing more than 85% lactic acid, or hetero-fermentative, producing lactic acid, CO_2, ethanol (and/or acetic acid) in equimolar amounts. In the presence of oxygen or other oxidants increased amounts of acetate may be produced at the expense of lactate or ethanol, whereby one additional mole of ATP is gained via the acetate kinase reaction. Thus, variations in the metabolic end products may occur. Various compounds (e.g. citrate, malate, tartrate, quinolate, nitrate, nitrite, etc.) may be metabolized, and used as energy source (e.g. via building up a proton motive force) or electron acceptors.

Lactobacilli are found where rich, carbohydrate-containing substrates are available, and thus, in a variety of habitats, such as mucosal membranes of man and animal (oral cavity, intestine and vagina), on plants or material of plant origin, in manure and man-made habitats such as sewage and fermenting or spoiling food.

Recently, lactobacilli were reviewed by Hammes *et al.* (1991) and Pot *et al.* (1994). In the former publication a thorough survey of the isolation, ecophysiology, identification and application of the organisms was presented. For the role of lactobacilli in health and disease, Volume I of *The Lactic Acid Bacteria* (Wood, 1992) provides an excellent survey. This chapter focuses on the taxonomy, phylogeny and evolution of the genus *Lactobacillus*.

It is remarkable that for an unequivocal identification of a species of the genus *Lactobacillus* it is not always sufficient to use the classical methods of physiological testing. In this regard the paper of Pot *et al.* (1994) provides an excellent review dealing with the methods for identifying lactobacilli and presenting a critical evaluation of their limitations. Clearly, the analyses of DNA or RNA composition (hybridization or sequencing) is the basis of systematic taxonomy. rRNA targeted DNA probes can be applied for species identification and alternatively, SDS-PAGE of whole cell proteins provides an accurate tool when performed under well-standardized conditions.

Progress in the understanding of the conservative nature of specific macromolecules (chapter 2) allows one to treat the lactobacilli under the aspects of a systematic grouping. This enables one to combine the well-known physiological and biochemical with phylogenetic data in a polyphasic approach for a grouping of the hitherto confusing multitude of the species of the genus *Lactobacillus*.

3.2 Grouping of lactobacilli

In Table 3.1 those 65 recognized species of the genus *Lactobacillus* are compiled that are either included in the approved list (AP) of Skerman *et al.* (1980) or validly published (VP). As pointed out in chapter 2, nine species have been transferred to other genera or species leaving 56 species which can readily be allotted to three evolutionary related groups named *Lb. delbrueckii* group, the *Lb. casei–Pediococcus* group and *Leuconostoc* group (Yang and Woese, 1989; Collins *et al.*, 1991; Hammes *et al.*, 1991). Based on the type of fermentation performed by the lactobacilli another three groups arrangement can be achieved in analogy to the original treatment of the lactobacilli by Orla-Jensen (1919). The author allotted the rod-shaped, catalase-negative species of the lactic acid bacteria to the genera *Thermobacterium*, *Streptobacterium* and *Betabacterium* which grouping was later on the subject of several re-definitions. The history of this grouping was reviewed by Pot *et al.* (1994).

The primary interest of Orla-Jensen's (1919) early description of the lactic acid bacteria was directed to identify those bacteria useful in the dairy industry, with the particular interest in the study of those bacteria occurring in Danish 'dairy cheese'. Following the chain of contamination of the raw milk, the author consistently studied the lactic acid bacteria of plants, animals and human beings, adults and children. He finally included into his thorough investigation fermenting food substrates such as beets, potatoes, mash, sauerkraut and sourdough. Thus, Orla-Jensen dealt with all important habitats of these organisms already at this early time. In his study he recognized 10 species. This number increased only slowly to 15 and 25 species, described in the 7th and 8th editions, respectively, of Bergey's manual and finally to 44 recognized species in the latest, 9th

Table 3.1 List of species of the genus *Lactobacillus*

Grouping*	Species number†	Species‡
Ba	18	*Lb. acetotolerans*[VP] (Etani *et al.*, 1986)
Aa	1	*Lb. acidophilus*[AL]
Bb	32	*Lb. agilis*[VP] (Weiss *et al.*, 1981)
Bb	20	*Lb. alimentarius*[VP] (Reuter, 1983)
Aa	2	*Lb. amylophilus*[VP] (Nakamura and Cromwell, 1979)
Aa	3	*Lb. amylovorus*[VP] (Nakamura, 1981)
		[*Lb. animalis*[VP] (Dent and Williams, 1982)] → *Lb. murinus*
Ab	12b	*Lb. aviarius* subsp. *araffinosus*[VP] (Fujisawa *et al.*, 1984)
Ab	12a	*Lb. aviarius* subsp. *aviarius*[VP] (Fujisawa *et al.*, 1984)
Bb		[*Lb. bavaricus*[VP] (Stetter and Stetter, 1980)] → *Lb. sake, Lb. curvatus* (Anon, 1987)
Bb	21	*Lb. bifermentans*[VP] (Kandler *et al.*, 1983a)
Cb	35	*Lb. brevis*[AL]
Cb	36	*Lb. buchneri*[AL]
Bb	22	*Lb. casei* subsp. *casei*[AL]
		[*Lb. catenaformis*[AL]] → no longer regarded as *Lactobacillus* (Stackebrandt and Teuber, 1988; Kandler and Weiss, 1986)
Cb	37	*Lb. collinoides*[AL]
Cc	48	*Lb. confusus*[AL]
Bb	23a	*Lb. coryniformis* subsp. *coryniformis*[AL]
Bb	23b	*Lb. coryniformis* subsp. *torquens*[AL]
Aa	4	*Lb. crispatus*[AL]
Bb	24	*Lb. curvatus*[AL]
Aa	5a	*Lb. delbrueckii* subsp. *bulgaricus*[VP] (Weiss *et al.*, 1983b)
Aa	5b	*Lb. delbrueckii* subsp. *delbrueckii*[AL]
Aa	5c	*Lb. delbrueckii* subsp. *lactis*[VP] (Weiss *et al.*, 1983b)
Ab	13	*Lb. farciminis*[VP] (Reuter, 1983)
Cb	38	*Lb. fermentum*[AL]
Cb	39	*Lb. fructivorans*[AL]
Cc	52	*Lb. fructosus*[AL]
Aa	6	*Lb. gallinarum*[VP] (Fujisawa *et al.*, 1992)
Aa	7	*Lb. gasseri*[VP] (Lauer and Kandler, 1980)
Bb	25	*Lb. graminis*[VP] (Beck *et al.*, 1988)
Cc	53	*Lb. halotolerans*[VP] (Kandler *et al.*, 1983b)
Ba	19	*Lb. hamsteri*[VP] (Mitsuoka and Fujisawa, 1987)
Aa	8	*Lb. helveticus*[AL]
Cb	40	*Lb. hilgardii*[AL]
Bb	26	*Lb. homohiochii*[AL]
Bb	27	*Lb. intestinalis*[VP] (Fujisawa *et al.*, 1990)
Aa	9	*Lb. jensenii*[AL]
Aa	10	*Lb. johnsoni*[VP] (Fujisawa *et al.*, 1992)
Cc	55	*Lb. kandleri*[VP] (Holzapfel and van Wyk, 1982)
Cb	41	*Lb. kefir*[VP] (Kandler and Kunath, 1983)
Aa	11	*Lb. kefiranofaciens*[VP] (Fujisawa *et al.*, 1988)
Cb	42	*Lb. malefermentans*[VP] (Farrow *et al.*, 1988)
Ab	15	*Lb. mali*[AL]
		[*Lb. maltaromicus*[AL]] → *Carnobacterium piscicola* (Collins *et al.*, 1991)
Cc	56	*Lb. minor*[VP] (Kandler *et al.*, 1983b)
		[*Lb. minutus*[AL]] → *Atopobium* (Collins and Wallbanks, 1992)
Bb	28	*Lb. murinus*[VP] (Hemme *et al.*, 1980)
Cb	43	*Lb. oris*[VP] (Farrow and Collins, 1988)
Cb	44	*Lb. parabuchneri*[VP] (Farrow *et al.*, 1988)

Table 3.1 *continued*

Grouping*	Species number†	Species‡
Bb	29a	*Lb. paracasei* subsp. *paracasei*VP (Collins *et al.*, 1989)
Bb	29b	*Lb. paracasei* subsp. *tolerans*VP (Collins *et al.*, 1989)
Bb	33	*Lb. pentosus*VP (Zanoni *et al.*, 1987)
Cb	46	*Lb. pontis*VP (Vogel *et al.*, 1994)
Bb	34	*Lb. plantarum*AL
Cb	45	*Lb. reuteri*VP (Kandler *et al.*, 1980)
Bb	30	*Lb. rhamnosus*VP (Collins *et al.*, 1989)
		[*Lb. rimae*VP] → *Atopobium* (Collins and Wallbanks, 1992)
		[*Lb. rogosae*AL] → no longer regarded as *Lactobacillus* (Stackebrandt and Teuber, 1988; Kandler and Weiss, 1986)
Ab	16	*Lb. ruminis*AL
Bb	31	*Lb. sake*AL
Ab	14a	*Lb. salivarius* subsp. *salicinius*AL
Ab	14b	*Lb. salivarius* subsp. *salivarius*AL
Cb	50	*Lb. sanfrancisco*VP (Weiss and Schillinger, 1984)
Ab	17	*Lb. sharpeae*VP (Weiss *et al.*, 1981)
Cb	48	*Lb. suebicus*VP (Kleynmans *et al.*, 1989)
		[*Lb. uli*AL] → *Atopobium* (Olsen *et al.*, 1991; Collins and Wallbanks, 1992)
Cb	49	*Lb. vaccinostercus*VP (Okada *et al.*, 1979)
Cb	47	*Lb. vaginalis*VP (Embley *et al.*, 1989)
Cc	54	*Lb. viridescens*AL
		[*Lb. vitulinus*AL → no longer regarded as *Lactobacillus* (see chapter 2)

*See Tables 3.2–3.4.
†See section 3.3.
‡VP, Validly published; AL, approved list.

edition. The still ongoing increase in the number of new species results from the emerging new taxonomic methods which allow a more precise identification of strains isolated some time ago and, to some extent, from the continued investigation of habitats.

The latest grouping of lactobacilli by Kandler and Weiss (1986) relies on biochemical–physiological criteria and neglects classical criteria of Orla-Jensen such as morphology and growth temperature since many of the recently described species did not fit into the traditional classification scheme. In an attempt to combine the group definitions of Kandler and Weiss (1986) with the grouping based on the phylogenetic relationship (*vide supra*), the species have been rearranged as shown in Tables 3.2–3.4. Groups **A**, **B**, **C** correspond to the respective groups I, II, III of Kandler and Weiss (1986).

It is remarkable that the terms 'obligately homofermentative' and 'facultatively heterofermentative' lactobacilli used for groups I and II, respectively, may be misleading. For example, Barre (1978) described

Table 3.2 Phylogenetic grouping and key characteristics of Group **A** lactobacilli (obligately homofermentative)

	Species*	Phylogenetic grouping	Peptidoglycan type	G+C content (mol(%))	Lactic acid isomer(s)	Growth (°C) 15/45	NH_3 from arginine	Carbohydrates fermented† Amygdalin	Cellobiose	Galactose	Lactose	Maltose	Mannitol	Mannose	Melibiose	Raffinose	Salicin	Sucrose	Trehalose
1	*Lb. acidophilus*	**a**	Lys-DAsp	34–37	DL	−/+	−	+	+	+	+	+	−	+	d	d	+	+	d
2	*Lb. amylophilus*	**a**	Lys-DAsp	44–46	L	+/−	ND	−	−	+	−	+	−	+	−	−	−	−	−
3	*Lb. amylovorus*	**a**	Lys-DAsp	40–41	DL	−/+	ND	+	+	+	−	+	−	+	−	−	+	+	+
4	*Lb. crispatus*	**a**	Lys-DAsp	35–38	DL	−/+	−	+	+	+	+	+	−	+	−	−	+	+	−
5a	*Lb. debrueckii* subsp. *bulgaricus*	**a**	Lys-DAsp	49–51	D	−/+	−	−	−	−	+	−	−	−	−	−	−	−	−
5b	*Lb. delbrueckii* subsp. *delbrueckii*	**a**	Lys-DAsp	49–51	D	−/+	d	−	d	−	−	d	−	+	−	−	−	+	d
5c	*Lb. delbrueckii* subsp. *lactis*	**a**	Lys-DAsp	49–51	D	−/+	d	+	d	d	+	+	−	+	−	−	+	+	+
6	*Lb. gallinarum*	**a**	Lys-DAsp	36–37	DL	+/+	ND	+	+	+	d	+	−	+	+	+	+	+	−
7	*Lb. gasseri*	**a**	Lys-DAsp	33–35	DL	−/+	−	+	+	+	d	d	−	+	d	d	+	+	d
8	*Lb. helveticus*	**a**	Lys-DAsp	38–40	DL	−/+	−	−	−	+	+	d	−	d	−	−	−	−	d
9	*Lb. jensenii*	**a**	Lys-DAsp	35–37	D	−/+	+	+	+	+	−	d	d	+	−	−	+	+	+
10	*Lb. johnsonii*	**a**	Lys-DAsp	33–35	DL	+/+	ND	+	+	+	d	+	−	+	d	d	+	+	d
11	*Lb. kefirano-faciens*	**a**	ND	34–35	D(L)	−/−	ND	−	−	+	+	+	−	ND	+	+	−	+	−

Table 3.2 *continued*

Species*	Phylogenetic grouping	Peptidoglycan type	G+C content (mol(%))	Lactic acid isomer(s)	Growth (°C) 15/45	NH_3 from arginine	Carbohydrates fermented† Amygdalin	Cellobiose	Galactose	Lactose	Maltose	Mannitol	Mannose	Melibiose	Raffinose	Salicin	Sucrose	Trehalose
12a *Lb. aviarius* subsp. *araffinosus*	**b**	Lys-DAsp	39–43	L(D)	−/ND	ND	d	d	−	−	+	−	+	−	−	d	+	+
12b *Lb. aviarius* subsp. *aviarius*	**b**	Lys-DAsp	39–43	DL	−/ND	ND	d	+	d	d	+	−	+	d	+	+	+	+
13 *Lb. farciminis*	**b**	Lys-DAsp	34–36	L(D)	+/−	+	+	+	+	+	+	−	+	−	−	+	+	+
14a *Lb. salivarius* subsp. *salicinus*	**b**	Lys-DAsp	34–36	L	−/+	−	−	−	+	+	+	+	−	+	+	+	+	+
14b *Lb. salivarius* subsp. *salivarius*	**b**	Lys-DAsp	34–36	L	−/+	−	−	−	+	+	+	+	−	+	+	−	+	+
15 *Lb. mali*	**b**	DAP	32–34	L	+/−	−	+	d	d	−	−	−	+	−	−	+	+	+
16 *Lb. ruminis*	**b**	DAP	44–47	L	−/d	−	+	+	+	d	+	−	+	+	+	+	+	−
17 *Lb. sharpeae*	**b**	DAP	53	L	+/−	−	+	+	+	+	+	−	+	−	−	+	−	−

*Group **a** species belong to the *Lb. delbrueckii* group; Group **b** organisms to the *Lb. casei–Pediococcus* group.
†Symbols: +, 90% or more of strains are positive; −, 90% or more of strains are negative; d, 11–89% of strains are positive; ND, no data available; DAP, diaminopimelic acid. Parenthesized isomers indicate <15% of total lactic acid.

Table 3.3 Phylogenetic grouping and key characteristics of Group **B** lactobacilli (facultatively heterofermentative)

	Species*	Phylogenetic grouping	Peptidoglycan type	G+C content (mol(%))	Lactic acid isomer(s)	Growth (°C) 15/45	Carbohydrates fermented† Amygdalin	Arabinose	Cellobiose	Esculin	Gluconate	Mannitol	Melezitose	Melibiose	Raffinose	Ribose	Sorbitol	Sucrose	Xylose
18	*Lb. acetotolerans*	**a**	Lys-DAsp	35–36.5	DL	–/–	–	–	d	+	–	d	–	–	–	d	–	–	–
19	*Lb. hamsteri*	**a**	Lys-DAsp	33–35	DL	–/ND	ND	+	+	+	+	+	–	+	+	+	+	+	d
20	*Lb. alimentarius*	**b**	Lys-DAsp	36–37	L(D)	+/–	ND	d	+	+	+	–	–	–	–	+	–	+	–
21	*Lb. bifermentans*	**b**	Lys-DAsp	45	DL	+/–	–	–	–	–	–	+	–	–	–	+	+	–	–
22	*Lb. casei*	**b**	Lys-DAsp	45–47	L	+/–	+	–	+	+	+	+	+	–	–	+	+	+	–
23a	*Lb. coryniformis* subsp. *coryniformis*	**b**	Lys-DAsp	45	D(L)	+/–	–	–	–	d	+	+	–	d	d	–	d	+	–
23b	*Lb. coryniformis* subsp. *torquens*	**b**	Lys-DAsp	45	D	+/–	–	–	–	–	+	+	–	–	–	–	–	+	–
24	*Lb. curvatus*	**b**	Lys-DAsp	42–44	DL	+/–	–	–	+	+	+	–	–	–	–	+	–	d	–
25	*Lb. graminis*	**b**	Lys-DAsp	41–43	DL	+/–	+	–	+	+	–	–	–	–	–	–	–	+	+
26	*Lb. homohiochii*	**b**	Lys-DAsp	35–38	DL	+/–	–	–	d	ND	–	d	–	–	–	d	–	–	–
27	*Lb. intestinalis*	**b**	Lys-DAsp	33–35	DL	–/+	–	–	d	–	ND	+	–	d	d	d	–	+	–
28	*Lb. murinus*	**b**	Lys-DAsp	43–44	L	–/+	d	+	+	+	–	d	–	+	+	+	–	+	–
29a	*Lb. paracasei* subsp. *paracasei*	**b**	Lys-DAsp	45–47	L/DL‡	+/d	+	–	+	+	+	+	+	–	–	+	d	+	–
29b	*Lb. paracasei* subsp. *tolerans*	**b**	Lys-DAsp	45–47	L	+/–	–	–	–	–	w	–	–	–	–	–	–	–	–
30	*Lb. rhamnosus*	**b**	Lys-DAsp	45–47	L	+/+	+	d	+	+	+	+	+	–	–	+	+	+	–
31	*Lb. sake*	**b**	Lys-DAsp	42–44	DL	+/–	+	+	+	+	+	–	–	+	–	+	–	+	–
32	*Lb. agilis*	**b**	DAP	43–44	L	–/+	+	–	+	+	–	+	+	+	+	+	d	+	–
33	*Lb. pentosus*	**b**	DAP	46–47	DL	+/–	+	+	+	ND	+	+	d	+	+	+	+	+	+
34	*Lb. plantarum*	**b**	DAP	44–46	DL	+/–	+	d	+	+	+	+	+	+	+	+	+	+	d

*Group **a** species belong to the *Lb. delbrueckii* group; Group **b** organisms to the *Lb. casei–Pediococcus* group.
†For symbols, see Table 3.2; w, weak positive reaction
‡Strains formerly designated *Lb. casei* subsp. *pseudoplantarum* produce DL-lactic acid.

Table 3.4 Phylogenetic grouping and key characteristics of Group **C** lactobacilli (obligately heterofermentative)*

Species†	Phylogenetic grouping	Peptidoglycan type	G+C content (mol(%))	Growth (°C) 15/45	NH_3 from arginine	Carbohydrates fermented‡												
						Arabinose	Cellobiose	Esculin	Galactose	Maltose	Mannose	Melezitose	Melibiose	Raffinose	Ribose	Sucrose	Trehalose	Xylose
35 *Lb. brevis*	**b**	Lys-DAsp	44–47	+/–	+	+	–	d	d	+	–	–	+	d	+	d	–	d
36 *Lb. buchneri*	**b**	Lys-DAsp	44–46	+/–	+	+	–	d	d	+	–	+	+	d	+	d	–	d
37 *Lb. collinoides*§	**b**	Lys-DAsp	46	+/–	+	+	–	+	+	+	–	+	+	–	+	–	–	+
38 *Lb. fermentum*	**b**	Orb-DAsp	52–54	–/+	+	d	d	–	+	+	w	–	+	+	+	+	d	d
39 *Lb. fructivorans*	**b**	Lys-DAsp	38–41	+/–	+	–	–	–	–	d	–	–	–	–	w	d	–	–
40 *Lb. hilgardii*	**b**	Lys-DAsp	39–41	+/–	+	–	–	–	d	+	–	d	–	–	+	d	–	+
41 *Lb. kefir*	**b**	Lys-DAsp	41–42	+/–	+	d	–	–	–	+	–	–	+	–	+	–	–	–
42 *Lb. male-fermentans*§	**b**	Lys-DAsp	41–42	+/–	+	–	–	–	–	+	–	–	–	–	+	–	–	–
43 *Lb. oris*	**b**	Lys-DAsp	49–51	–/d	–	+	d	d	+	+	d	–	+	+	+	+	d	+
44 *Lb. parabuchneri*	**b**	Lys-DAsp	44	+/ND	+	+	–	–	+	+	ND	+	+	+	+	+	–	–
45 *Lb. reuteri*	**b**	Lys-DAsp	40–42	–/+	+	+	–	ND	+	+	–	–	+	+	+	+	–	–
46 *Lb. pontis*	**b**	Orn-DAsp	53–55	+/+	+	–	–	–	d	+	–	–	d	d	+	+		–
47 *Lb. vaginalis*	**b**	Orn-DAsp	38–41	–/+	ND	–	–	d	+	+	+	–	+	+	d	+	–	–
48 *Lb. suebicus*	**b**	DAP	40	+/d	ND	+	d	–	+	+	ND	–	d	–	+	d	–	+
49 *Lb. vaccinostercus*	**b**	DAP	36	–/–	–	+	w	–	w	+	–	–	–	–	+	–	–	+
50 *Lb. sanfrancisco*	**b**	Lys-Ala	36–38	+/–	–	–	–	ND	d	+	–	–	–	–	d	d	–	–

51 *Lb. confusus*	**c**	Lys-Ala	45–47	+/+	+	−	+	+	+	+	+	−	−	−	+	+	−	+
52 *Lb. fructosus*	**c**	Lys-Ala	47	+/−	−	−	−	ND	−	−	−	−	−	−	−	−	−	−
53 *Lb. halotolerans*	**c**	Lys-Ala-Ser	45	+/−	+	−	−	−	−	+	+	−	−	−	+	−	+	−
54 *Lb. viridescens*	**c**	Lys-Ala-Ser	41–44	+/−	−	−	−	−	−	+	+	−	−	−	−	d	d	−
55 *Lb. kandleri*	**c**	Lys-Ala-Gly-Ala$_2$	39	+/−	+	−	−	−	+	−	−	−	−	−	+	−	−	−
56 *Lb. minor*	**c**	Lys-Ser-Ala$_2$	44	+/−	+	−	+	+	−	+	+	+	−	−	+	+	+	−

*The following sugars are generally fermented: fructose (exceptions: *Lb. malefermentans*, *Lb. sanfrancisco*, (some strains) and *Lb. vaccinostercus*) and glucose. The following sugars are generally not fermented: amygdalin (exceptions: *Lb. confusus* and *Lb. oris*), mannitol (exception: *Lb. kandleri*), rhamnose and sorbitol.

†*Group **b** species belong to the *Lb. casei–Pediococcus* group; Group **c** organisms to the *Leuconostoc* group.

‡For symbols, see Tables 3.2 and 3.3.

§Phylogenetic grouping due to peptidoglycan type.

thermophilic lactobacilli which ferment pentoses homofermentatively, and also strains related to *Lb. salivarius* (Raibaud *et al.*, 1973) further described as *Lb. murinus* (Hemme *et al.*, 1980) exhibited the same property. Unfortunately, the description of new species usually does not include the analysis of the end products derived from the fermentation of pentoses, and therefore, the enzymes of the pentose phosphate pathway may be present permitting a homofermentative metabolism of pentoses in lactobacilli. Nevertheless, maintaining the traditional terms is justified with regard to hexose utilization. However, at low substrate concentration and under strictly anaerobic conditions, some facultatively heterofermentative species may produce acetate, ethanol and formate instead of lactate from pyruvate. Thus, the definitions have to be used in awareness of their limitations and are given as follows.

- **Group A**: Obligately homofermentative lactobacilli. Hexoses are almost exclusively (>85%) fermented to lactic acid by the Embden–Meyerhof–Parnas (EMP) pathway. The organisms possess fructose-1,6-bisphosphate-aldolase but lack phosphoketolase, and therefore, neither gluconate nor pentoses are fermented.
- **Group B**: Facultatively heterofermentative lactobacilli. Hexoses are almost exclusively fermented to lactic acid by the EMP pathway. The organisms possess both aldolase and phosphoketolase, and therefore, not only ferment hexose but also pentoses (and often gluconate). In the presence of glucose, the enzymes of the phosphogluconate pathway are repressed.
- **Group C**: Obligately heterofermentative lactobacilli. Hexoses are fermented by the phosphogluconate pathway yielding lactate, ethanol (acetic acid) and CO_2 in equimolar amounts. Pentoses enter this pathway and may be fermented.

Within these three groups the species are arranged according to their phylogenetic relationship. The letter **a** indicates the affiliation to the *Lb. delbrueckii* group, **b** to the *Lb. casei–Pediococcus* group and **c** to the *Leuconostoc* group. Thus, the combination of the letters **Aa** defines a species as belonging to the obligately homofermentative lactobacilli affiliated in the *Lb. delbrueckii* group, whereas **Cb** means that the species is obligately heterofermentative phylogenetically belonging to the *Lb. casei–Pediococcus* group, etc.

3.2.1 Obligately homofermentative lactobacilli (Group A)

As shown in Table 3.2, the group of the obligately homofermentative lactobacilli comprises 17 species, 11 belonging to group **Aa** and six to group **Ab**. All members of the *Lb. delbrueckii* group (**Aa**) are characterized by the Lys-DAsp type of their peptidoglycan, whereas those of the *Lb. casei–Pediococcus* group (**Ab**) have either the Lys-DAsp type (four species) or the DAP-direct type (three species).

Group **Aa** is named after the type strain of *Lb. delbrueckii* of the genus *Lactobacillus*. This species contains three subspecies, *Lb. delbrueckii* subsp. *delbrueckii*, *Lb. delbrueckii* subsp. *bulgaricus*, and *Lb. delbrueckii* subsp. *lactis*. Very closely related to this species is *Lb. jensenii* which can be distinguished by its G+C content of the DNA. A second cluster in the *Lb. delbrueckii* group comprises *Lb. acidophilus* and physiologically closely related species with similar physiological properties. This group appeared to be quite heterogeneous in DNA–DNA hybridization studies (Lauer and Kandler, 1980) and was divided in two main genotypic subgroups referred to as A and B (Johnson *et al.*, 1980) which shared less than 25% DNA–DNA homology whereas strains within each subgroup shared a similarity of 75–100%. Recent studies on the systematics of *Lb. acidophilus* employing electrophoresis of soluble cellular proteins or lactate dehydrogenase and DNA–DNA reassociation indicated that *Lb. acidophilus* strains include six genomospecies (Fujisawa *et al.*, 1992). This finding was confirmed by the results of highly standardized SDS-PAGE of whole cell proteins (Pot *et al.*, 1994) and rRNA targeted oligonucleotide probes. These new techniques allow clear differentiation between the *Lb. acidophilus* strains of the six subgroups which are designated *Lb. acidophilus* (A1), *Lb. crispatus* (A2), *Lb. amylovorus* (A3), *Lb. gallinarum* (A4), *Lb. gasseri* (B1) and *Lb. johnsonii* (B2). The parentheses refer to the original grouping proposed by Johnson *et al.* (1980). These species can otherwise not be differentiated by simple phenotypic assays. *Lb. helveticus* is closely related to *Lb. acidophilus* with respect to a DNA–DNA homology, biochemical features and 16S rRNA sequence. Other obligate homofermenters which fall into the *Lb. delbrueckii* group are *Lb. amylophilus*, *Lb. kefiranofaciens*, *Lb. acetotolerans* and *Lb. hamsteri*. The latter two species can ferment pentoses and therefore were transferred to group **Ba**, *vide infra*.

Lb. aviarius, *Lb. mali*, *Lb. ruminis*, *Lb. salivarius*, *Lb. sharpeae* (Collins *et al.*, 1991) and *Lb. farciminis* (see chapter 2) belong to group **Ab**, and thus, are members of the *Lb. casei-Pediococcus* group. Kandler and Weiss (1986) used the epithet *Lb. yamanashiensis* instead of *Lb. mali.* That changing use of these epithets is the cause of some confusion. *Lb. yamanashiensis* was first described as an isolate from grape must by Nonomura *et al.* (1965). Carr and Davies (1970) thereafter described *Lb. mali* which had been isolated from apple juice and must. This species was included into the Approved List of Bacterial Names. A comparison of the biochemical and physiological properties of these strains with those described by Nonomura *et al.* (*Lb. yamanashiensis*) was performed by Carr and Davies (1977). Because of the high congruence of the results observed for all strains, the authors proposed to give *Lb. yamanashiensis* the priority and to treat their isolates from apple substrates as *Lb. yamanashiensis* subsp. *mali*. In a recent investigation Kaneuchi *et al.* (1988) concluded from the high physiological and genotypical relationship that a separation of subspecies is not justified. Because of these findings and since the species description of *Lb. yamanashiensis* (Nonomura *et al.*, 1965; Nonomura and Ohara, 1967; Nonomura, 1983) is rather insufficient, Kaneuchi *et al.* (1988) proposed to give *Lb. mali* the priority over *Lb. yamanashiensis*.

The organisms of groups **Aa** and **Ab** are found in all habitats that are described for lactobacilli. However, animal- and man-associated species predominate in these groups. This is consistent with their higher optimum growth temperature that gave rise to their traditional grouping as 'Thermobacteria'. *Lactobacillus delbrueckii*, *Lb. helveticus* and *Lb. kefiranofaciens* are important for food fermentations.

A close phylogenetic relationship between *Lb. farciminis* (Group **Ab**) and *Lb. alimentarius* (Group **Bb**) is shown in Figure 2.5. This finding suggests that the species of group **A** may have developed from metabolically more potent organisms of the **b** group type (compare section 3.2.4 with Figure 3.1). This may have occurred by a simple loss of phosphoketolase activity. Within group **Ab** this development may have at least two independent origins as is suggested by the occurrence of the Lys-DAsp and DAP-direct types of peptidoglycan in groups **Ab** and **Bb**.

*3.2.2 Facultatively heterofermentative lactobacilli (Group **B**)*

The species of lactobacilli belonging to the group of facultatively heterofermentative organisms are compiled in Table 3.3. Two species, *Lb. acetotolerans* and *Lb. hamsteri* constitute group **Ba**, meaning that phylogenetically these organisms fall into the *Lb. delbrueckii* group. The presence of the Lys-DAsp type of peptidoglycan is consistent with this grouping, and these species may be regarded as the more conservative representative of the *Lb. delbrueckii* group in terms of a greater metabolic potential as indicated by the presence of both aldolase and phosphoketolase (*vide supra*). In fact, for *Lb. acetotolerans* two types of strains were described by Etani *et al.* (1986), group I ferments ribose and group II does not. Strains of group II might therefore be placed into group **A**. These strain differences clearly show the dilemma of grouping lactobacilli according to physiological characteristics.

Group **Bb** contains 15 species, 12 of which contain Lys-DAsp and three DAP in their peptidoglycan. In contrast to Kandler and Weiss (1986) we have included into group **Bb** *Lb. bifermentans* since, in agreement with the group definition, this organism possesses both key enzymes, aldolase and phosphoketolase.

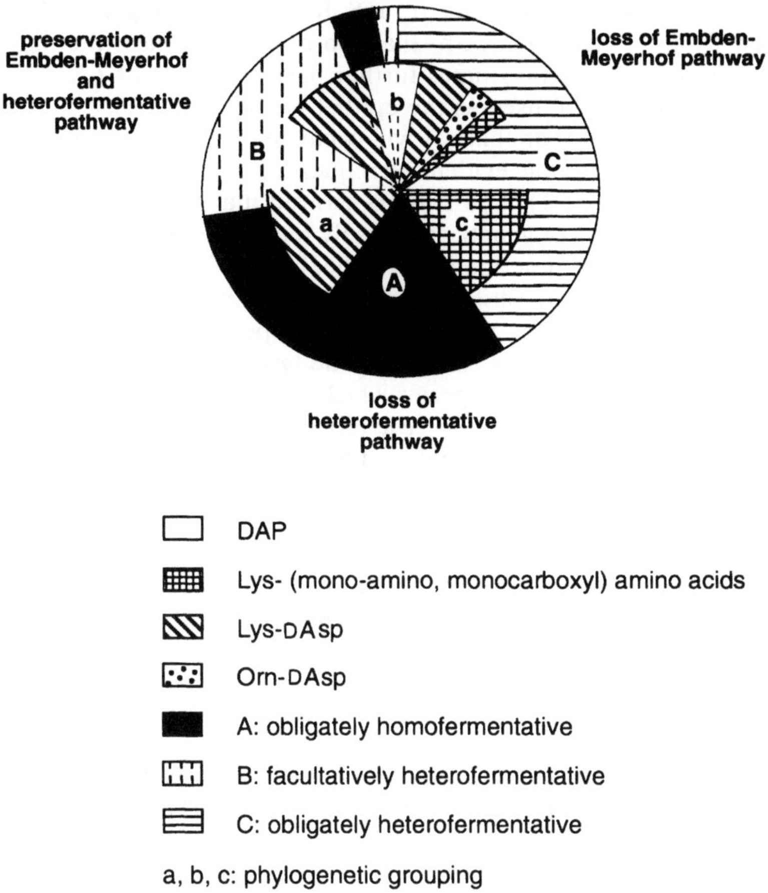

Figure 3.1 Grouping of lactobacilli and their possible evolution from common ancestors.

Lb. bifermentans is characterized by fermenting glucose homofermentatively. However, dependent on the pH, lactate can be metabolized to ethanol, acetic acid and CO_2 and H_2. The utilization of lactate (and/or pyruvate) is rather common for group **Bb**-organisms. For example, formate and acetate are formed by *Lb. pentosus* when grown anaerobically, in the presence of citrate (Czelovsky *et al.*, 1992), by *Lb. casei* (Vries *et al.*, 1970) and by *Lb. bulgaricus* (Rhee and Pack, 1980), or CO_2 and acetate are formed in the presence of oxidants (Kandler, 1983). Thus, depending on the composition of the growth medium, pH or redox potential quite different end products can be observed, especially with the species of the *Lb. casei–Pediococcus* group (**b**). *Lactobacillus animalis* has not been included into this group following the arguments of Kandler and Weiss (1986) who emphasized the high relatedness of *Lb. murinus* and *Lb.*

animalis, and in fact, the data of 16S RNA sequencing introduced in chapter 2 are consistent with that observation. Since the effective publication of *Lb. murinus* dates further back than that of *Lb. animalis*, *Lb. murinus* should be used as the epithet for this species.

It can be foreseen that changes will occur for the species *Lb. casei* and *Lb. paracasei*. With ample evidence a request for an opinion was presented by Dellaglio *et al.* (1991). The authors showed that the type strain of *Lb. casei* (ATCC 393) is not genetically closely related to several subspecies of *Lb. casei* as they were described by Kandler and Weiss (1986). This had led Collins *et al.* (1989) to describe *Lb. paracasei* sp. nov. that included these types of strains. It appears, however, that the species *Lb. paracasei* should be rejected and changed for *Lb. casei*. The type strain of *Lb. casei* would have to be allotted to a new species which includes also one strain of *Lb. rhamnosus* and '*Lactobacterium*' *zeae* (Kuznetsov, 1959).

Another change might be expected for *Lb. pentosus* since a 16S rRNA sequence similarity of >99% was shown for *Lb. pentosus* and *Lb. plantarum* (Collins *et al.*, 1991). On the other hand, it was indicated by DNA–DNA hybridization experiments performed with 28 strains of *Lb. plantarum* that some genotypic variation is still hidden among strains classically identified as *Lb. plantarum* (Dellaglio *et al.*, 1975). *Lactobacillus plantarum* and *Lb. pentosus* characteristically have DAP in their peptidoglycan and can clearly be differentiated from *Lb. agilis*, which is the other DAP-type species in group **Bb**, since the configuration of lactate is L(+) for *Lb. agilis* and DL for *Lb. plantarum*.

Lactobacillus casei and *Lb. plantarum* have been isolated from many different habitats and the same is true for *Lb. sake*. The two latter species are known for their unusual metabolic potential. For example, heme-dependent catalase and nitrite reduction occurs in both species. *Lactobacillus plantarum* strains may, in addition, reduce nitrate and exhibit pseudo-catalase activity (Hammes *et al.*, 1990), and *Lb. sake* may form slime and utilize arginine for ATP generation (Montel and Champomier, 1987). *Lactobacillus curvatus* is phylogenetically related to *Lb. sake* (Collins *et al.*, 1991) but does not possess its unusual properties.

The majority of group **B** species are often found associated with foods and perform either controlled fermentation (*Lb. casei*, *Lb. curvatus*, *Lb. sake*, *Lb. plantarum*) (*Lb. graminis* in fermenting silage) or spoilage, preferentially of refrigerated and packaged goods (same organisms plus *Lb. alimentarius*, *Lb. bifermentans*, *Lb. homohiochii*).

3.2.3 Obligately heterofermentative lactobacilli (Group **C***)*

The group of the obligately heterofermentative lactobacilli comprises 22 species, 16 (group **Cb**) of which belong to the *Lb. casei–Pediococcus* group and six (group **Cc**) to the *Leuconostoc* group. Like all species of the *Lb.*

casei–Pediococcus group the species of group **Cb** possess either a Lys-DAsp (in most cases) or DAP type of peptidoglycan. The presence of ornithine instead of lysine in *Lb. fermentum* and *Lb. vaginalis* is a reliable taxonomic marker to identify these species. Because of the chemical similarity of lysine and ornithine the Orn-DAsp type is not considered as major variation that would otherwise have suggested the treatment of these species outside the group of species with Lys-DAsp. The formerly separately listed *Lb. cellobiosus* was a third ornithine-containing species, but DNA–DNA hybridization revealed that this species is identical to *Lb. fermentum* (Vescovo *et al.*, 1979). A unique exception is *Lb. sanfrancisco* found exclusively in fermenting sourdough and its peptidoglycan of the Lys-Ala type. The study of the sourdough microflora in the authors' laboratory revealed the presence of *Lb. pontis* sp. nov. (Vogel *et al.*, 1994) with the Orn-DAsp type peptidoglycan. On the basis of 16S rRNA sequence comparison and protein pattern analysis this species was identified as a member of group **Cb**.

The species of group **Cb** comprise the organisms of Orla Jensen's genus '*Betabacterium*'. They may be regarded as derived from the metabolically more potent species of the **Bb** group, as it was discussed with the organisms of the **A** group. In this case however, the change consisted in a loss of the key enzymes of the EMP pathway aldolase and triose phosphate isomerase.

As with the species of groups **A** and **B** it is quite difficult and in some cases virtually impossible to identify group **C** species solely on physiological criteria. Therefore, strains were often misidentified and, after the more modern methods of taxonomy were established, it was necessary to allot these strains to either recognized old species or to new species that had to be described. For example, Vescovo *et al.* (1979) investigated the DNA homology of 24 strains of '*Lb. brevis*' and observed that 13 strains had to be allotted to *Lb. confusus*, *Lb. collinoides*, *Lb. hilgardii* or *Lb. kefir*, and strains of '*Lb. brevis*' isolated from the oral cavity or saliva (Davies 1955; Hayward and Davies, 1956; Hayward, 1957) were allotted by Farrow and Collins (1988) to the new species *Lb. oris*.

Lactobacillus kefir was isolated from kefir grains by Kandler and Kunath (1983). The properties of this species indicate that it is identical with the former type species of the genus *Lactobacillus* – '*Lb. caucasicum*' (Beijerinck, 1901; Orla-Jensen, 1919). Since no strain of '*Lb. caucasicum*' was kept available, this species was rejected by opinion 38, Jud. Comm. 1971, 99 (Lapage *et al.*, 1975).

Species characterized by high tolerance to ethanol, a limited spectrum of fermentable carbohydrates and pronounced acidophily were treated as a separate group within the 'Betabacteria' by Rogosa (1970) and Sharpe (1979). Species of this group are *Lb. fructivorans* and *Lb. hilgardii*. Formerly included into this group were '*Lb. trichodes*', '*Lb. heterohiochii*' and '*Lb. desidiosus*'. The former two species were shown to be synonyms of *Lb. fructivorans* (Weiss *et al.*, 1983a) and '*Lb. desidiosus*' appears to be heterogeneous, with some strains belonging to *Lb. hilgardii* (Vescovo *et al.*, 1979). *Lactobacillus suebicus* might be the species most tolerant to ethanol and low pH. This species ferments, however, a rather broad spectrum of carbohydrates. Genotypically, it is closely related to *Lb. vaccinostercus* (Collins *et al.*, 1991) and the presence of the DAP-type peptidoglycan in both species is consistent with this observation. The genotypic relationship does not include an ecological relationship, since *Lb. vaccinostercus* exhibits only poor tolerance to ethanol and low pH and occurs in cowdung whereas *Lb. suebicus* is found in fermenting mashes of apples and pears. Thus, a grouping according to neither the physiological properties nor the ecology of the species allows any conclusions of relationships among lactobacilli in general, and 'Betabacteria' (group **C**), specifically.

The classical 'Betabacteria' *Lb. brevis*, *Lb. buchneri* and *Lb. fermentum* were isolated from quite different habitats such as dairy products, fermenting plant materials, sourdough and the intestinal tracts of man and animals. Some habitats may be questioned since (see above) the identification methods were not adequate for a long time. In general, it appears that these species preferentially occur in plant association in the broadest sense, e.g. in grain, mashes, fermenting vegetable, etc. On the other hand, *Lb. vaginalis*, *Lb. oris* and *Lb. reuteri* appear to be preferentially associated with man and animals.

A similar preferential association can also be seen for species of the *Leuconostoc* group (**Cc**). *Lactobacillus confusus*, *Lb. fructosus* and *Lb. kandleri* are species mainly isolated from plant origin and *Lb. halotolerans*, *Lb. viridescens* and *Lb. minor* from animal sources. *Lactobacillus viridescens* was originally considered to consist of two subspecies namely subsp. *haloterans* (Reuter, 1970) and subsp. *minor* (Abo-Elnaga and Kandler, 1965b) but these were eventually described as separate species by Kandler *et al.* (1983a,b).

3.2.4 Conclusions

In Figure 3.1 the grouping of the species of the genus *Lactobacillus* is summarized. Components are the phylogenetic branches **a**, **b** and **c**, the peptidoglycan types found in these branches and the type of fermentation **A**, **B** and **C**. It can be seen that group **b** organisms (*Lb. casei–Pediococcus* group) are most heterogeneous with regard to both fermentation type and peptidoglycan types. In fact, all variations that can be found in the lactobacilli *in toto* are present in group **b**. On the other hand, group **c** organisms are most homogenous in the basic peptidoglycan type (characteristic for the genus *Leuconostoc*) and additionally, in containing exclusively obligately heterofermentative species. Group **a** is also homogenous with regard to the peptidoglycan type (Lys-DAsp) but is heterogeneous in so far as two species are facultatively heterofermentative among the body of obligately homofermentative species. This figure may also be conceived as a way of evolution, wherein from some ancestral species with diverse peptidoglycan types, and being facultatively heterofermentative, a straight development led to the majority of group **b** species. Group **a** and **c** species are characterized by a loss of one metabolic pathway connected with the development of species with just one peptidoglycan type. The loss of the enzymes of the heterofermentative and homofermentative pathway, respectively, occurred for a second (later) time within group **b**.

3.3 Description of the species

All cells are Gram-positive and non-spore-forming, usually catalase-negative, non-motile, and facultatively anaerobic unless otherwise stated. The grouping of the species together with the patterns of sugar fermentation and important physiological properties are given in Tables 3.2–3.4. The numbering of the species is in accordance with that used in these tables. The numbers given after the reference are the page numbers on which the descriptions of the species are provided.

Aa

1 *Lactobacillus acidophilus* (Moro, 1900; Hansen and Mocquot, 1970, p. 326) (*Bacillus acidophilus* Moro, 1900, p. 115). a.ci.do_phi.lus. M.L.n. *acidum* acid; Gr.adj. *philus* loving; M.L.adj. *acidophilus* acid-loving. Group **Aa**. Cells are rods with rounded ends, 0.6–0.9 by 1.5–6 µm, occurring singly, in pairs, and in short chains. Riboflavin, pantothenic acid, folic acid, and niacin are required for growth. Pyridoxal, thiamine, thymidine, and vitamin B_{12} are not required. Strains were isolated from the intestinal tract of man and animals, human mouth and vagina. The type strain is ATCC 4356.

2 *Lactobacillus amylophilus* (Nakamura and Cromwell, 1979, p. 539). a.my.lo_phi.lus. Gr.n. *amylum* starch; Gr.adj. *philus* loving; M.L.adj. *amylophilus* starch-loving. Group **Aa**. Cells are thin rods of 0.5–0.7 by 2.0–3.0 µm, occurring singly and in short chains. Actively ferments starch and displays extracellular amylolytic enzyme activity. Riboflavin, pyridoxal, pantothenic acid, folic acid, and niacin are required for growth. Thiamine is not required. Strains were isolated from swine waste-corn fermentations. The type strain is NRRL-B-4437 (Nakamura and Cromwell, 1979, p. 539).

3 *Lactobacillus amylovorus* (Nakamura, 1981, p. 61). a.my.lo'vo.rus. Gr.n. *amylum* starch; L.v. *vorare* to devour; M.L.adj. *amylovorus* starch destroying. Group **Aa**. Cells are rods of 1.0 by 3.0–5.0 μm, occurring singly and in short chains. Actively ferments starch and displays extracellular amylolytic enzyme activity. Riboflavin, pantothenic acid, folic acid, and niacin are required for growth. Thiamine is not required. Strains were isolated from cattle waste-corn fermentations. The type strain is NRRL-B-4437 (Nakamura and Cromwell, 1979). Synonymous with *Lb. acidophilus* group A3 of Johnson *et al.* (1980) (Fujisawa *et al.*, 1992).

4 *Lactobacillus crispatus* (Brygoo and Aladame, 1953; Moore and Holdeman, 1970, p. 15) (*Eubacterium crispatum* Brygoo and Aladame 1953, p. 641). cris.pa'tus. L.part.adj. *crispatus* curled, crisped, referring to the morphology originally observed in broth media. Group **Aa**. Cells are straight to slightly curved rods of 0.8–1.6 by 2.3–11 μm, occurring singly or in short chains. Some strains grow at 48–53°C. Strains were isolated from human faeces, vagina and buccal cavities, crops and caeca from chicken; also found in patients with purulent pleurisy, leucorrhoea and urinary tract infection. The type strain is ATCC 33820. Synonymous with *Lb. acidophilus* group A2 of Johnson *et al.* (1980) (Cato *et al.*, 1983).

5 *Lactobacillus delbrueckii* (Leichmann, 1896; Beijerinck, 1901, p. 229) (*Bacillus delbrücki* (sic) Leichmann, 1896, p. 284). del.bruec'ki.i. M.L.gen.n. *delbrueckii* of Delbrück; named for M. Delbrück, a German bacteriologist. Group **Aa**. Cells are rods with rounded ends, of 0.5–0.8 by 2.0–9.0 μm, occurring singly and in short chains. Pantothenic acid and niacin are generally required for growth. Some strains require riboflavin, folic acid, vitamin B_{12} and thymidine. Thiamine, pyridoxine, biotin and *p*-amino benzoic acid are not required. Three subspecies are currently recognized.

5a *Lactobacillus delbrueckii* subsp. *bulgaricus* (Orla-Jensen, 1919; Weiss, *et al.*, 1984, p. 270) (*Thermobacterium bulgaricum* Orla-Jensen, 1919, p. 556). bul.ga'ri.cus. M.L.adj. *bulgaricus* Bulgarian. Strains were isolated from yoghurt and cheese. Ferments only few carbohydrates. The type strain is ATCC 11842 (Weiss *et al.*, 1983b).

5b *Lactobacillus delbrueckii* subsp. *delbrueckii* (Leichmann, 1896; Weiss *et al.*, 1983b, p. 556). Strains were isolated from plant material fermented at high temperatures (40–53°C). The type strain is ATCC 9649 (Weiss *et al.*, 1983b).

5c *Lactobacillus delbrueckii* subsp. *lactis* (Orla-Jensen, 1919; Weiss *et al.*, 1984, p. 270) (*Thermobacterium lactis* Orla-Jensen, 1919, p. 164). lac'tis.L.n. *lac* milk; L.gen.n. *lactis* from milk). Strains were isolated from milk, cheese, compressed yeast and grain mash. The type strain is ATCC 12312 (Weiss *et al.*, 1983b).

6 *Lactobacillus gallinarum* (Fujisawa *et al.*, 1992, p. 489). gallin.ar'um. L.n. *gallina* the hen. L.gen.pl. *gallinarum* of hens. Group **Aa**. Cells from BL agar plate cultures are short to long rods of 0.5–1.5 by 1.5–10 μm and occur singly, in pairs, and sometimes in short chains. Colonies on BL agar are 0.5–2.0 mm in diameter, circular to slightly irregular, entire, grayish brown to reddish brown, and rough. Tolerant to 4% NaCl. The strains were isolated from chicken intestine. The type strain is ATCC 33199. Synonymous with *Lb. acidophilus* group A4 of Johnson *et al.* (1980) (Fujisawa *et al.*, 1992).

7 *Lactobacillus gasseri* (Lauer and Kandler, 1980, p. 77). gas.ser'i. M.L.gen.n. *gasseri* of F. Gasser who pioneered studies on lactate dehydrogenases of *Lactobacillus* species. Group **Aa**. Cells form rods with rounded ends, 0.6–0.8 by 3.0–5.0 μm occurring singly and in chains. Frequent formation of 'mini cells' and snakes. Colonies are usually rough and flat. Growth is greatly enhanced by anaerobiosis and 5% CO_2. Milk is coagulated with rare exceptions. In contrast with other lactobacilli the DAla-DAla termini of the peptide subunits in the peptidoglycan are preserved due to the lack of D,D-carboxypeptidase action. Strains were isolated from the mouth, intestine, faeces, and vagina. The type strain is DSM 20243. Synonymous with *Lb. acidophilus* group B1 of Johnson *et al.* (1980) (Lauer and Kandler, 1980).

8 *Lactobacillus helveticus* (Orla-Jensen, 1919; Bergey *et al.*, 1925, p. 184) (*Thermobacterium helveticum* Orla-Jensen, 1919, p. 164). hel.ve'ti.cus. L.adj. Suisse. Group **Aa**. Cells are rods, 0.7–0.9 by 2.0–6.0 μm, occurring singly or in chains. Some strains grow at 50–52°C. Riboflavin, pantothenic acid, and pyridoxal are required for growth. Thiamine, folic acid, biotin, vitamin B_{12} and thymidine are not required. Strains were isolated from sour milk, cheese starter cultures and cheese, particularly Emmental and Gruyère cheese. The type strain is ATCC 15009.

9 *Lactobacillus jensenii* (Gasser *et al.*, 1970, p. 221). jen.se'ni.i. M.L. gen.n. *jensenii* of Jensen; named for Orla-Jensen, a Danish microbiologist. Group **Aa**. Cells form rods with rounded ends. 0.6–0.8 by 2.0–4.0 μm, occurring singly and in short chains. Strains were isolated from human vaginal discharge and blood clot. The type strain is ATCC 25285.

10 *Lactobacillus johnsonii* (Fujisawa *et al.*, 1992, p. 489). john.so'ni.i. M.L.gen.n. *johnsonii* of Johnson; named for Johnson, an American microbiologist. Group **Aa**. Cells from BL agar plate cultures are short to long rods of 0.5–1.5 by 1.5–10 μm and occur singly, in pairs, and sometimes in short chains. Colonies on BL agar are 0.5–2.0 mm in diameter, circular to slightly irregular, entire, greyish brown to reddish brown, and rough. Tolerant to 4% NaCl. The strains were isolated from the faeces of chicken, mice, calves and pigs. The type strain is ATCC

33200. Synonymous with *Lb. acidophilus* group B2 of Johnson *et al.* (1980) (Fujisawa *et al.*, 1992).

11 *Lactobacillus kefiranofaciens* (Fujisawa *et al.*, 1988, p. 13). ke.fi.rano. fa'ci.ens. L.n. *kefiran*, a polysaccharide of kefir grain; L.v. *facere* to produce; M.L.part.adj. *kefiranofaciens*, kefiran producing. Group **Aa**. Cells form rods, 0.8–1.2 by 3.0–20 μm, occurring singly, in pairs, or occasionally in short chains. Grows on modified KPL agar (pH 5.5) at 30°C in 10 days producing colonies that are circular or irregular, 0.5–3.0 μm in diameter, convex, transparent to translucent, white, smooth to rough, and ropey. Strains were isolated from kefir grains. The type strain is ATCC 43761.

Ab

12 *Lactobacillus aviarius* (Fujisawa *et al.*, 1984, p. 419). a.vi.a'ri.us. L.adj. *aviarius* pertaining to birds. Group **Ab**. Cells are short to coccoid rods of 0.5–1.0 by 0.5–1.6 μm, with rounded ends, occurring singly or in short chains. Surface colonies on BL agar after 2 days of anaerobic incubation are 0.3–1.2 μm in diameter, round, globular, yellowish-white to reddish-brown, with a smooth surface and entire edge. Strictly anaerobic. The final pH of glucose broth is 3.9–4.0. Two subspecies are currently recognized.

12a *Lactobacillus aviarius* subsp. *aviarius* (Fujisawa *et al.*, 1984, p. 419). The strains were isolated from chicken and duck alimentary tract and faeces. The type strain is DSM 20655.

12b *Lactobacillus aviarius* subsp. *araffinosus* (Fujisawa *et al.*, 1984, p. 419). a.raffi.no'sus. Gr.pref. *a* not; L.M.adj. *raffinosus* of raffinose; M.L.adj. *araffinosus* not fermenting raffinose. The strains were isolated from chicken duodenum. The type strain is DSM 20653.

13 *Lactobacillus farciminis* (Reuter 1983, p. 672). far.ci'mi.nis. L.n. *farcimen* sausage; L.gen.n. *farciminis* of sausage. Group **Ab**. Cells are slender rods of 0.6–0.8 by 2.0–6.0 μm, occurring singly or in short chains. Grows in the presence of 10% NaCl and occasionally 12% NaCl. Strains were isolated from fermented dry sausage and sourdough. The type strain is DSM 20184.

14 *Lactobacillus salivarius* (Rogosa *et al.*, 1953, p. 691). sa.li.va'ri.us. L.adj. *salivarius* salivary. Group **Ab**. Cells are rods with rounded ends, of 0.6–0.9 by 1.5–5.0 μm, occurring singly or in chains of varying length. Strains were isolated from the mouth and intestinal tract of man and hamster and the intestinal tract of chicken. Two subspecies are recognized.

14a *Lactobacillus salivarius* subsp. *salicinius* (Rogosa *et al.*, 1953, p. 691).

sa.li.ci'ni.us. M.L.adj. *salicinius* pertaining to salicin, a glycoside. Ferments salicin and esculin but not rhamnose. The type strain is ATCC 11742.

14b *Lactobacillus salivarius* subsp. *salivarius* (Rogosa *et al.*, 1953, p. 691). Ferments rhamnose but not salicin and esculin. The type strain is ATCC 11741.

15 *Lactobacillus mali* (Carr and Davies, 1970, p. 769) (*Lactobacillus yamanashiensis* Nonomura *et al.*, 1965). ma'li. L.n. *malus* apple; L.gen.n. *mali* of the apple. Group **Ab**. Cells are slender rods of 0.6 by 1.8–4.2 μm, occurring singly, in pairs, palisades and irregular clumps. Motile with a few peritrichous flagella. Most strains exhibit pseudocatalase activity when grown on MRS agar containing (w/v) 0.1% glucose. Significant amounts of menaquinones, predominantly with eight or nine isoprene units were found in *Lb. yamanashiensis* (Collins and Jones, 1981). Strains were isolated from freshly pressed apple juice, cider and wine must. The type strain is ATCC 27053 (Kaneuchi *et al.*, 1988).

16 *Lactobacillus ruminis* (Sharpe *et al.*, 1973, p. 47). ru'mi.nis. L.n. *rumen* throat; L. gen.n. *ruminis* of rumen. Group **Ab**. Cells are rods of 0.6–0.8 by 3.0–5.0 μm, occurring singly or in short chains. Motile by peritrichous flagella; motility not always easy to demonstrate and often sluggish. Surface growth is obtained only under reduced oxygen pressure; growth in liquid media is improved by the addition of cysteine-HCl. Unlike the strains isolated from the rumen many strains from sewage were nonmotile and failed to grow at 45°C. Strains were isolated from the bovine rumen and from sewage. The type strain is ATCC 277780.

17 *Lactobacillus sharpeae* (Weiss *et al.*, 1981, p. 251). shar.pe.ae. M.L. gen.n. *sharpeae* of Sharpe; named for Sharpe, an English bacteriologist. Group **Ab**. Cells are rods with rounded ends, 0.6–0.8 by 3.0–8.0 μm, with a pronounced tendency to form 'snakes' and, after prolonged incubation, long characteristically wrinkled chains. In broth cultures, a flocculent sediment is observed. Strains were isolated from municipal sewage. The type strain is DSM 20505 (Weiss *et al.*, 1981).

Ba

18 *Lactobacillus acetotolerans* (Etani *et al.*, 1986, p. 547). a.ce.to. tole.rans. L. n. *acetum* vinegar; L. pres. part. *tolerans* tolerating, enduring; M.L.part.adj. *acetotolerans* vinegar tolerating. Group **Ba**. Produces D,L-lactic acid homofermentatively from glucose. Ribose may be fermented, and therefore this strain belongs to group **B** while its phylogenetic position is **a**. Cells are rods of 0.4–0.5 by 1.1–3.4 μm occurring singly, in pairs, or occasionally in short chains. The colonies are 0.3–1.5 mm in diameter, circular to irregular, convex, opaque, yellowish white, rough, and undulate when grown at 30°C for 14 days on Briggs agar (pH 5.0). Growth is generally observed at pH 3.3–6.6 at 23–40°C. Cells are resistant to 4–5% and to 9–11% of acetic acid at pH 3.5 and 5.0, respectively. Riboflavin, pantothenic acid, folic acid, uracil and Tween 80 are required for growth. Thiamine, *p*-aminobenzoic acid, biotin, adenine, guanine, xanthine, mevalonic acid, acetic acid and ethanol are not required. The habitat of this species is fermented vinegar broth. The type strain is JCM 3825.

19 *Lactobacillus hamsteri* (Mitsuoka and Fujisawa, 1987, p. 272). hams'ter.i. M.L.gen.n. *hamsteri* of the hamster from which the original isolate was derived. Group **Ba**. Produces D,L-lactic acid homofermentatively from glucose. Ribose is fermented, and therefore this strain belongs to group **B** while its phylogenetic position is **a**. Cells form long stout rods with rounded ends, 1.0–1.3 by 5–10 μm, arranged singly, in pairs and in short chains. Surface colonies on BL agar after two days of anaerobic incubation are 0.7–3 μm in diameter, round, umbonate, brown, with a rough surface and erosed edge. Strictly anaerobic. The final pH on glucose broth is 3.7. Isolated from the intestine and faeces of hamster. The type strain is JCM 6256.

Bb

20 *Lactobacillus alimentarius* (Reuter, 1983, p. 672). a.li.men.ta'ri.us. L.adj. pertaining to food. Group **Bb**. Cells are short, slender rods of 0.6–0.8 by 1.5–2.5 μm. Growth in the presence of 10% NaCl. Acetoin is produced from glucose. Strains were isolated from marinated fish products, meat products (fermented sausages and sliced prepacked sausages) and sourdough. The type strain is DSM 20249.

21 *Lactobacillus bifermentans* (Kandler *et al.*, 1983b, p. 896). bi.fer.men'tans. L.pref. *bis* twice; L.part. *fermentans* leavening; M.L.part.adj. *bifermentans* doubly fermenting. Group **Bb**. Cells are irregular rods with rounded or often tapered ends, 0.5–1.0 by 1.5–2.0 μm, occurring singly, in pairs or irregular short chains, often forming clumps. Homofermentative with production of D,L-lactic acid in media containing more than 1% fermentable hexoses. The organism therefore belongs to group **B**. At pH >4.0 lactic acid is fermented to acetic acid, ethanol, CO_2 and H_2. Strains were isolated from spoiled Edam and Gouda cheese where it forms undesired small cracks. The type strain is DSM 20003.

22 *Lactobacillus casei* (Orla-Jensen, 1919; Hansen and Lessel, 1971, p. 71) (*Streptobacterium casei* Orla-Jensen, 1919, p. 166). ca'sei. L.n. *caseus* cheese; L.gen.n. *casei* of cheese. Group **Bb**. Cells are rods of 0.7–1.1 by

2.0–4.0 μm, often with square ends and tending to form chains. Riboflavin, folic acid, calcium pantothenate and niacin are required for growth. Pyridoxal or pyridoxamine is essential or stimulatory. Thiamine, vitamin B_{12} and thymidine are not required. Strains were isolated from milk and cheese, dairy products and dairy environments, sourdough, cow dung, silage, human intestinal tract, mouth and vagina, sewage. The type strain is ATCC 393. ATCC 334 was suggested as neo-type strain by Dellaglio *et al.* (1991).

23 *Lactobacillus coryniformis* (Abo-Elnaga and Kandler, 1965a, p. 18). co.ry'ni.for'mis. Gr.n. *coryne* a club; L.adj. *formis* shaped; M.L.adj. *coryniformis* club-shaped. Group **Bb.** Cells short, often coccoid rods, 0.8–1.1 by 1.0–3.0 μm, occurring singly, in pairs or in short chains. Pantothenic acid, niacin, riboflavin, biotin and *p*-aminobenzoic acid are required for growth of the majority of strains. Folic acid, pyridoxine, thiamine and vitamin B_{12} are not required. Strains were isolated from silage, cow dung, dairy barn air and sewage. Two subspecies are recognized.

23a *Lactobacillus coryniformis* subsp. *coryniformis* (Abo-Elnaga and Kandler, 1965a, p. 18). The lactic acid produced from glucose contains 15–20% of the L-isomer. The type strain is DSM 20001.

23b *Lactobacillus coryniformis* subsp. *torquens* (Abo-Elnaga and Kandler, 1965a, p. 18), tor'quens L.pres.part. *torquens* twisting. Exclusively D(−)-lactic acid is produced. The type strain is ATCC 25600.

24 *Lactobacillus curvatus* (Troili-Petersson, 1903; Abo-Elnaga and Kandler, 1965a, p. 19). (*Betabacterium curvatum* Troili-Petersson, 1903, p. 137). cur.va'tus. L.past.part. *curvatus* curved. Group **Bb**. Cells are bean-shaped rods with rounded ends, 0.7–0.9 by 1.0–2.0 μm, occurring in pairs and short chains, closed rings of usually four cells, or horseshoe forms are frequently observed. Some strains are at first motile, motility is lost on subculture. Some strains are able to grow even at 2–4°C. L-LDH is activated by FDP and Mn^{2+}. Possesses lactic acid racemase the biosynthesis of which is induced by L(+)-lactic acid. Racemase induction is generally not repressed by acetate. Strains were isolated from cow dung, milk, silage, sauerkraut, prepacked finished dough and meat products. Some of the atypical streptobacteria from herbage, silage, fermented meat products and vacuum-packed meat reported in the past belong to *Lb. curvatus*. The type strain is ATCC 25601.

25 *Lactobacillus graminis* (Beck *et al.*, 1988, p. 282). gra'mi.nis. L.n. *gramen* grass, N.L.gen. *graminis* of grass. Group **Bb**. Cells are slightly curved rods with rounded ends, 0.7–1.0 by 1.5–2.0 μm, occurring singly, in pairs or short chains, slightly curved. Colonies are smooth and round.

Flocculent sediment after three days of growth in MRS broth. Strains were isolated from grass silage. The type strain is DMS 20719.

26 *Lactobacillus homohiochii* (Kitahara *et al.*, 1957, p. 118). ho'mo.hi.o'chi.i. Gr.adj. *homos* like, equal; Japanese n. *hiochi* spoiled sake; M.L.gen.n. *homohiochii* intended to mean homofermentative *Lactobacillus* of hiochi. Group **Bb**. Cells are rods with rounded ends, 0.7–0.8 by 2.0–4.0 μm, occasionally 6 μm. Does not grow in MRS broth. In Rogosa SL broth supplemented with DL-mevalonic acid (30 mg/litre) and ethanol (40 ml/litre) copious growth is obtained at 30°C after a marked lag phase of 4–7 days. No growth at 45°C at an initial pH >5.5. Resistant to 13–16% ethanol. D-Mevalonic acid is essential; ethanol is promotive. Strains were isolated from spoiled sake. The type strain is ATCC 15434.

27 *Lactobacillus intestinalis* (Hemme, 1984; Fujisawa *et al.*, Mitsuoka, 1990, p. 303). in.tes.tin.al'is. L.n.adj. *intestinalis* pertaining to the intestine. Group **Bb**. First described as Thermobacterium Group Th4 by Raibaud *et al.* (1973) and as *Lb. intestinalis* by Hemme (1974) but not validly published. Ferments D-ribose, better in the presence of small amounts of glucose and L-arabinose. Malate is decarboxylated. Cells form rods with rounded ends, 0.6–0.8 by 2.0–6.0 μm, occurring singly, in pairs or occasionally in short chains. Surface colonies on glucose–blood–liver agar are 0.7–2.5 mm in diameter, round, flat, and light brown with rough surfaces and erose edges after 2 days of anaerobic incubation at 37°C. Strains were isolated from the intestine of rats and mice. The type strain of the Th4 Group and thus of *Lb. intestinalis* is CNRZ 219 = ATCC 49335.

28 *Lactobacillus murinus* (Hemme *et al.*, 1980, p. 306). mu.ri'nus. L.adj. *murinae*, of mice. Group **Bb**. First described as *Thermobacterium* Group Th5 by Raibaud *et al.* (1973). Cells are rods with rounded ends of 0.8–1.0 by 2.0–4.0 μm, occurring frequently in chains. D-ribose and L-arabinose are fermented, but not D-arabinose and D-xylose. Malate is decarboxylated. Riboflavin is required for growth. Thiamine and vitamin B_{12} are not required. Strains represent the dominant autochthoneous *Lactobacillus* flora of the digestive tract of Murinae (mice and rats). The type strain of the Th5 Group and thus of *Lb. murinus* is CNRZ 220.

29 *Lactobacillus paracasei* (Collins *et al.*, 1989, p. 108). pa.ra.ca'se.i. Gr.prep. *para*, resembling; L.gen.n. *casei*, a specific epithet; M.L.adj. *paracasei*, resembling *L. casei*. Group **Bb**. Cells are rod shaped, 0.8–1.0 by 2.0–4.0 μm, often with square ends, and occur singly or in chains. Growth at 10 and 40°C. Some strains grow at 5 and 45°C. A few strains (formerly *Lb. casei* subsp. *pseudoplantarum*) produce inactive lactic acid due to the activity of L-lactic acid racemase. Two subspecies are validly published.

29a *Lactobacillus paracasei* subsp. *paracasei* (Collins *et al.*, 1989, p. 108).

Strains were isolated from dairy products, sewage, silage, humans, and clinical sources. The type strain is NCDO 151 (Collins *et al.*, 1989).

29b *Lactobacillus paracasei* subsp. *tolerans* (Collins *et al.*, 1989, p. 108). to.le'rans. L.pres.part. *tolerans* tolerating, enduring; means survival during the pasteurization of milk. Group **Bb**. Cells are rod shaped, 0.8–1.0 by 2.0–4.0 μm, often with square ends, and occur singly or in chains. Survives heating at 72°C for 40 s. The strains are isolated from dairy products. The type strain is ATCC 25599 (Collins *et al.*, 1989).

30 *Lactobacillus rhamnosus* (Collins *et al.*, 1989, p. 108) (*Lactobacillus casei* subsp. *rhamnosus* Hansen, 1968, p. 76). rham.no'sus. M.L.adj. *rhamnosus*, pertaining to rhamnose. Group **Bb**. Cells are rod shaped, 0.8–1.0 by 2.0–4.0 μm, often with square ends, and occur singly or in chains. Some strains grow at 48°C. The strains were isolated from dairy products, sewage, humans, and clinical sources. The type strain is ATCC 7469 (Collins *et al.*, 1989).

31 *Lactobacillus sake* (Katagiri *et al.*, 1934, p. 157). sa'ke. Japanese n. *sake* rice wine. Group **Bb**. Cells are rods with rounded ends, 0.6–0.8 by 2.0–3.0 μm, occurring singly and in short chains, frequently slightly curved and irregular, especially during the stationary growth phase. Some strains grow at 2–4°C. L-LDH is activated by FDP and Mn^{2+}. Possesses lactic acid racemase; induction of racemase in most strains is repressed by acetate. Therefore, the majority of the strains produce L(+)-lactic acid in MRS broth whereas DL-lactic acid is produced in cabbage press juice. A few strains, however, produce inactive lactic acid also in MRS broth. Some strains produce L(+)-lactate exclusively and were described as *Lb. bavaricus* (Stetter and Stetter, 1980, p. 601). Strains were isolated from sake starter, sauerkraut and other fermented plant material, meat products and prepacked finished dough. Some of the atypical streptobacteria from herbage, silage, fermented meat products and vacuum packed meat reported in the past probably belong to *Lb. sake*. The type strain is ATCC 15521.

32 *Lactobacillus agilis* (Weiss *et al.*, 1981, p. 252). a'gi.lis. L.adj. *agilis* agile, motile. Group **Bb**. Cells are rods with rounded ends, 0.7–1.0 by 3.0–6.0 μm, occurring singly, in pairs or in short chains. Motile with peritrichous flagella. Colonies are usually rough and flat. Surface growth is enhanced by anaerobiosis. Produces L(+)-lactic acid with about 10% D(−)-lactic acid from glucose. The type strain is DSM 20509.

33 *Lactobacillus pentosus* (Fred *et al.*, 1921, p. 410; Zanoni *et al.*, 1987, p. 339). pen.to'sus. M.L.adj. *pentosus* of pentose, pertaining to pentoses. Group **Bb**. Growth at 40°C. Glycerol is fermented. Cells are straight rods with rounded ends, 1.0–1.2 by 2.0–5.0 μm, occurring singly, in pairs or in short chains. Strains were isolated from corn silage, fermenting olives, and sewage. The type strain in ATCC 8041.

34 *Lactobacillus plantarum* (Orla-Jensen, 1919; Bergey *et al.*, 1923, p. 250) (*Streptobacterium plantarum* Orla-Jensen, 1919, p. 174). plan. ta'rum. L.fem.n. *planta* a sprout; M.L.n.gen.pl.n. *plantarum* of plants. Group **Bb**. Cells are straight rods with rounded ends, 0.9–1.2 by 3.0–8.0 μm, occurring singly, in pairs or in short chains. Some strains are able to reduce nitrate. Occasionally, strains exhibit pseudocatalase activity especially if grown under glucose limitation. Cell walls contain either ribitol or glycerol teichoic acid. Calcium pantothenate and niacin are required for growth. Thiamine, pyridoxal or pyridoxamine, folic acid, vitamin B_{12}, thymidine or deoxyribosides are not required. Riboflavin is generally not required. Strains were isolated from dairy products and environments, silage, sauerkraut, pickled vegetables, sourdough, cow-dung, and the human mouth, intestinal tract and stools, and from sewage. The type strain is ATCC 14917.

Cb

35 *Lactobacillus brevis* (Orla-Jensen, 1919; Bergey *et al.*, 1934, p. 312). (*Betabacterium breve* Orla-Jensen, 1919, p. 175). bre'vis. L.adj. short. Group **Cb**. Cells are rods with rounded ends, 0.7–1.0 by 2.0–4.0 μm, occurring singly and in short chains. Calcium pantothenate, niacin, thiamine and folic acid are required for growth. Riboflavin, pyridoxal and vitamin B_{12} are not required. Strains were isolated from milk, cheese, sauerkraut, sourdough, silage, cow dung, faeces and intestinal tract of humans and rats. The type strain is ATCC 14869.

36 *Lactobacillus buchneri* (Henneberg, 1903; Bergey *et al.*, 1923, p. 251) (*Bacillus buchneri* (sic) Henneberg, 1903, p. 163). buch.ne'ri. M.L.gen.n. *buchneri* of Buchner; named for Buchner, a German bacteriologist. Group **Cb**. Cells are rods with rounded ends, 0.7–1.0 by 2.0–4.0 μm, occurring singly and in short chains. Strains were isolated from milk, cheese, fermenting plant material and human mouth. The type strain is ATCC 4005.

37 *Lactobacillus collinoides* (Carr and Davies, 1972, p. 470). col.li. no.i'des. L.adj. *collinus* hilly; Gr.n. *idus* form, shape; M.L.adj. *collinoides* hill-shaped, pertaining to colony form. Group **Cb**. Cells are rods with rounded ends, 0.6–0.8 by 3.0–5.0 μm, with a tendency to form long filaments, occurring singly, in palisades and irregular clumps. Growth is distinctly improved by the addition of 20% tomato juice and by replacement of glucose by maltose. Strains were isolated from cider. The type strain is ATCC 27612.

38 *Lactobacillus fermentum* (Beijerinck, 1901) (*Lactobacillus fermenti* Beijerinck, 1901, p. 233). fer.men'tum. L.n. *fermentum* ferment, yeast. Group **Cb**. Cells are rods, 0.5–0.9 μm and highly variable in length, mostly occurring singly or in pairs. Calcium pantothenate, niacin and thiamine are required for growth. Riboflavin pyridoxal and folic acid are not required. Strains were isolated from yeast, milk products, sourdough, fermenting plant material, manure, sewage and human mouth and faeces. The type strain is ATCC 14931.

39 *Lactobacillus fructivorans* (Charlton *et al.*, 1934, p. 1). fruc.ti.vo'rans. L.n. *fructus* fruit; L.v. *vorare* to eat; M.L.pres.part. *fructivorans* fruit-eating, intending to mean fructose-devouring. Group **Cb**. Cells are rods with rounded ends, 0.5–0.8 by 1.5–4.0 μm, occasionally as long as 20 μm, occurring singly, in pairs and in chains, very long, more or less curved or coiled filaments often observed. Surface growth is enhanced by reduced O_2 pressure or under anaerobic conditions. Fructose is used as electron acceptor resulting in the formation of mannitol. Acidophilic; favourable pH is 5.0–5.5; no growth at an initial pH >6.0. Nutritionally fastidious, at least on primary isolation. Depending on the source of isolation, mevalonic acid, tomato juice and/or ethanol are required for growth. Some strains, especially those from non-alcohol-containing source, often become less fastidious during laboratory transfers and grow well on MRS broth. Strains were isolated from spoiled mayonnaise, salad dressings and vinegar preserves, from spoiled sake, dessert wines and aperitifs. The type strain is ATCC 8288. Synonymous with *Lb. heterohiochii* (Kitahara *et al.*, 1957) and *Lb. trichodes* (Fornachon *et al.*, 1949).

40 *Lactobacillus hilgardii* (Douglas and Cruess, 1936, p. 115). hil.gar'di.i. M.L.gen.n. *hilgardii* named for Hilgard, an American enologist. Group **Cb**. Cells are rods with rounded ends, 0.5–0.8 by 2.0–4.0 μm, occurring singly, in short chains, and frequently in long filaments. The optimum pH for growth and carbohydrate fermentation is in the range of 4.5–5.5. Grows in the presence of 15–18% ethanol. Originally isolated from California table wines but obviously widely distributed in wines of different origin. The type strain is ATCC 8290.

41 *Lactobacillus kefir* (Kandler and Kunath, 1983, p. 672). ke'fir. Turkish. n. *kefir*, a Caucasian sour milk. Group **Cb**. Cells are rods with rounded ends, 0.6–0.8 by 3.0–15 μm, with a tendency to form chains of short rods or long filaments. Strains were isolated from kefir grains and drink kefir. The type strain is DSM 20587.

42 *Lactobacillus malefermentans* (Russel and Walker, 1953; Farrow *et al.*, 1988, p. 165). ma.le.fer.men'tans. L.adj. *malum* bad. M.L.v. *fermentare* to ferment. L.part.adj. *fermentans* fermenting. *malefermentans* badly fermenting, referring to spoiled beer. Group **Cb**. Cells are rods, 0.8–1.0 by

1.5–2.5 μm, occurring singly, in pairs or in chains. Strains were isolated from beer. The type strain is NCDO 1410.

43 *Lactobacillus oris* (Farrow and Collins, 1988, p. 116). or′is. L.n. *os* mouth. L.gen.n. *oris* of the mouth. Group **Cb**. Cells are rods of 0.8–1.0 by 2.0–4.0 μm with rounded ends, occurring singly or in pairs, seldom in short chains. Colonies are small, raised, semi-rough, or rough. Strains were isolated from human saliva. The type strain is NCDO 2160.

44 *Lactobacillus parabuchneri* (Farrow *et al.*, 1988, p. 165). pa.ra.buch.ne′ri. Gr.prep. *para*, resembling; *buchneri* specific epithet; M.L.adj. *parabuchneri* resembling *Lb. buchneri*. Group **Cb**. Cells are rods, 0.8–1.0 by 2.0–4.0 μm, occurring singly, in pairs or in chains. Growth at 10 and 40°C. Can be distinguished from *Lb. buchneri* in not producing acid from D-xylose. These species share a DNA–DNA homology of 30% in hybridizations. Strains were isolated from human saliva, cheese and contaminated brewery yeasts. The type strain is NCDO 2748.

45 *Lactobacillus reuteri* (Lerche and Reuter, 1962; Kandler *et al.*, 1980, p. 266) (*Lactobacillus fermentun* Type II Lerche and Reuter, 1962, p. 462). reu′te.ri. M.L.n. *reuteri* named for Reuter, a German bacteriologist. Group **Cb**. Cells are slightly irregular, bent rods with rounded ends, 0.7–1.0 by 2.0–5.0 μm with rounded ends, occurring singly, in pairs, or in small clusters. The strains were isolated from faeces of humans and animals, from meat products and sourdough. The type strain is DSM 20016.

46 *Lactobacillus pontis* (Vogel *et al.*, 1994). pon′tis L.gen.n. *pons*, the bridge, referring to BRIDGE which is the abbreviation for the Commission of the European Communities Research Programme entitled *B*iotechnology *R*esearch, for *I*nnovation *D*evelopment and *G*rowth in *E*urope. During this programme the organism was isolated and characterized by the joined efforts of three laboratories. Group **Cb**. Cells are non-motile, non-sporing, Gram-positive slender rods that occur singly, in pairs and chains (diameter, 0.3–0.6 μm; length 4–6 μm). Some strains have a strong tendency to form chains of long bent rods or even resembling a vine tendril. Colonies on sanfrancisco agar are 1–2 mm in diameter, rough circular plateaux with irregular border and a smooth convex centre, translucent and greyish after 2–5 days of anaerobic incubation at 30°C. All strains grow at 15 and 45°C. The main fermentation products from maltose or fructose are lactate, acetate, ethanol, glycerol and CO_2. Catalase activity is not detected.

The terminal pH in sanfrancisco-medium ranges from 3.9–4.2. All strains ferment ribose and fructose. Few strains also ferment maltose. All strains cleave arginine. The DNA G+C content is 53–55 mol%. Strains were originally isolated from rye sourdough and can be predominant in some batches. The type strain is LTH 2587 (DSM 8475, LMG 14187).

47 *Lactobacillus vaginalis* (Embley *et al.*, 1989). va'gi.na.lis. L.gen.n. *vaginalis*, of the vagina. Group **Cb**. Cells are rods of 0.5–1.0 by 2.0–3.0 μm with rounded ends, occurring singly or in pairs, seldom in short chains. Colonies are white to grey, small to large (1–5 mm in diameter), and semirough, often with raised areas. Strains were isolated from the vagina of patients suffering from trichomoniasis. The type strain is NCTC 12197.

48 *Lactobacillus suebicus* (Kleynmans *et al.*, 1989). suè.bi.cus, L.adj. *suebicus* suebian; derived from Suebia, a county in southern Germany. Group **Cb**. Cells are rods of 0.5–1.0 by 2.0–3.0 mm with rounded ends, occurring singly or in pairs, seldom in short chains. Colonies are white, round, convex, smooth, slimy, 2 mm in diameter on Homohiochii agar after incubation for 4 days at 30°C in an atmosphere of 95% N_2 and 5% CO_2. The lowest limits for growth are 10°C and pH 2.8. Cells tolerate 14% ethanol at pH 3.3. Habitats are stored apple and pear mashes. The type strain is DSM 5007.

49 *Lactobacillus vaccinostercus* (Okada *et al.*, 1979, p. 217). vac.ci.no.ster'cus. L.adj. *vaccinus* from cows, L.n. *stercus* dung; M.L.adj. *vaccinostercus* from cow dung. Group **Cb**. Cells are rods of 0.9–1.0 by 1.5–2.5 μm with rounded ends, occurring mostly in pairs. Thiamine, pantothenic acid, niacin and biotin are essential. Pyridoxal, *p*-aminobenzoic acid and folic acid are not required. Strains were isolated from cow dung. The type strain is ATCC 33310.

50 *Lactobacillus sanfrancisco* (Kline and Sugihara, 1971; Weiss and Schillinger, 1984, p. 503). san.fran.cis'co. M.L. *sanfrancisco* San Francisco, named after the city where the sourdough from which the organism was first isolated had been propagated for more than 100 years. Group **Cb**. Cells are rods with rounded ends, 0.6–0.8 by 2.0–4.0 μm, occurring singly and in pairs. Does not grow reasonably well in MRS broth unless freshly prepared yeast extract is added and the initial pH is lowered to 5.6. Fresh isolates will adapt and exhibit good growth on MRS containing maltose instead of glucose. Strains were isolated from wheat and rye sourdough. The type strain is ATCC 27651. Synonymous with *Lb. brevis* var. *lindneri* (Spicher and Schröder, 1978). In our laboratory the type strain did not ferment galactose but fermented ribose.

Cc

51 *Lactobacillus confusus* (Holzapfel and Kandler, 1969; Sharpe *et al.*, 1972, p. 396) (*Lactobacillus coprophilus* subsp. *confusus* Holzapfel and

Kandler, 1969, p. 665). con.fu'sus. L.v. *confudere* to confuse; L.past.part. *confusus* confused, an allusion to its original confusion with *Leuconostoc*. Group **Cc**. Cells are short rods of 0.8–1.0 by 1.3–3.0 μm with a tendency to thicken at one end, occurring singly, rarely in short chains. Dextran is produced from sucrose. Strains were isolated from sugarcane and carrot juice, occasionally found in raw milk and sewage. The type strain is ATCC 10881.

52 *Lactobacillus fructosus* (Kodama, 1956, p. 705). fruc.to'sus. M.L.adj. *fructosus* of fructose, pertaining to fructose. Group **Cc**. Cells are rods of 0.5–0.8 by 2.0–4.0 μm, occurring singly or in pairs, seldom in short chains. Growth in MRS broth is markedly improved if glucose is replaced by fructose. Strains were isolated from flowers. The type strain is ATCC 13162.

53 *Lactobacillus halotolerans* (Kandler *et al.*, 1983a, p. 672). ha.lo.to'le.rans. Gr.n. *hals*, *halos* salt; L.pres.part. *tolerans* tolerating, enduring; M.L.part.adj. *halotolerans* salt-tolerating. Group **Cc**. Cells are irregular, short, even coccoid rods with round tapered ends, 0.5–0.7 by 1.0–3.0 μm, sometimes longer, with a tendency to forming coiling chains, clumping together. Colonies on MRS agar are greyish white, smooth, 2–3 mm in diameter. Growth is greatly enhanced by reduced O_2 pressure or anaerobiosis. Good growth occurs up to 12, very weak growth up to 14% NaCl. Strains were isolated from meat products. The type strain is DSM 20190 (Kandler *et al.*, 1983a).

54 *Lactobacillus viridescens* (Kandler and Abo-Elnaga 1966; Niven and Evans, 1957, p. 758) (*Lactobacillus corynoides* subsp. *corynoides* Kandler and Abo-Elnaga, 1966, p. 573). *Lactobacillus viridescens* is incorrectly cited in the Approved Lists of Bacterial Names as *Lactobacillus viridescens* (Kandler and Abo-Elnaga, 1966, p. 573). vi.ri.des'cens. M.L.pres.part. *viridescens* growing green, greening. Group **Cc**. Cells are small, often slightly irregular rods with rounded to tapered ends, 0.7–0.9 by 2.0–5.0 μm, occurring singly or in pairs. Pantothenate, niacin, thiamin, riboflavin, and biotin are required for growth. Folic acid and pyridoxal may be stimulatory. Strains were isolated from discoloured meat products and pasteurized milk. The type strain is ATCC 12706.

55 *Lactobacillus kandleri* (Holzapfel and van Wyk, 1982, p. 439). kand'le.ri. M.L.gen.n. *kandleri* of Kandler; named for Kandler, a German microbiologist. Group **Cc**. Cells are partly irregular rods of 0.7–0.8 by 1.0–5.0 μm, occurring singly or in pairs, seldom in short chains. Slime is produced from sucrose. Strains were isolated from a desert spring. Original habitat not known, but may probably be plants. The type strain is DSM 20593.

56 *Lactobacillus minor* (Kandler *et al.*, 1983a, p. 672). mi'nor. L.dim.adj.

minor smaller. Group **Cc**. Cells are irregular, short rods with rounded or tapered ends, 0.6–0.8 by 1.5–4.0 μm, sometimes longer, often bent with unilateral swellings. Often in pairs or short chains with a tendency to form loose clusters. Colonies on MRS agar are greyish white, smooth, 2–4 mm in diameter. Growth is greatly enhanced by reduced O_2 pressure or anaerobiosis. Good growth occurs up to 8, very weak growth up to 10% NaCl. Strains were isolated from the sludge of milking machines. The type strain is DSM 20014 (Kandler *et al.*, 1983a).

Note added at proof

The species of group **Cc** were recently allotted to the new genus *Weissella* by Collins *et al.* (1993). The request for an opinion replacing the type strain of *Lactobacillus casei* and rejecting the name *L. paracasei* by Dellaglio *et al.* (1991) was denied (Wayne, 1994).

Acknowledgement

The authors thank Dr D. Hemme INRA, France for critical reading of the manuscript and clarifying information erroneously given in literature on *L. intestinalis* and *L. murinus*.

References

Abo-Elnaga, I.G. and Kandler, O. (1965a) Zur Taxonomie der Gattung *Lactobacillus* Beijerinck. I. Das Subgenus *Streptobacterium* Orla-Jensen. *Zentralblatt für Bakteriologie, Parasitenkunde, Infektionskrankheiten und Hygiene*, **119**, 1–36.

Abo-Elnaga, I.G. and Kandler, O. (1965b) Zur Taxonomie der Gattung *Lactobacillus* Beijerinck. II. Das Subgenus *Betabacterium* Orla-Jensen. *Zentralblatt für Bakteriologie, Parasitenkunde, Infektionskrankheiten und Hygiene*, **119**, 117–129.

Anon (1987) International Committee on Systematic Bacteriology, Subcommittee on the Taxonomy of the lactobacilli, bifidobacteria, and related organisms, Minute 6 of the Meeting September 1986. *International Journal of Systematic Bacteriology*, **37**, 469–470.

Barre, P. (1978) Identification of thermobacteria and homofermentative, thermophilic pentose-utilizing lactobacilli from high temperature fermenting grape musts. *Journal of Applied Bacteriology*, **44**, 125–129.

Beck, R., Weiss, N. and Winter, J. (1988) *Lactobacillus graminis* sp. nov., a new species of facultatively heterofermentative lactobacilli surviving at low pH in grass silage. *Systematic Applied Microbiology*, **10**, 279–283.

Beijerinck, M.W. (1901) Sur les ferments lactiques de l'industrie. *Archives Néerlandaises des Sciences Extractes et Naturelles* (Sect. 2) **6**, 212–243.

Bergey, D.H., Harrison, F.C., Breed, R.S., Hammer, B.W. and Huntoon, F.M. (1923) In *Bergey's Manual of Determinative Bacteriology*, 1st edn. The Williams and Wilkins Co., Baltimore, USA.

Bergey, D.H., Harrison, F.C., Breed, R.S., Hammer, B.W. and Huntoon, F.M. (1925) In *Bergey's Manual of Determinative Bacteriology*, 2nd edn. The Williams and Wilkins Co., Baltimore, USA.

Bergey, D.H., Breed, R.S., Hammer, B.W., Huntoon, F.M., Murray, E.G.D. and Harrison, F.C. (1934) In *Bergey's Manual of Determinative Bacteriology*, 4th edn. The Williams and Wilkins Co., Baltimore, USA, pp. 1–664.

Brygoo, E.R. and Aladame, N. (1953) Etude d'une espèce nouvelle anaérobie stricte du genre *Eubacterium: E. crispatum* n. sp. *Annales de l'Institut Pasteur*, **84**, 640–651.

Carr, J.G. and Davies, P.A. (1970) Homofermentative lactobacilli of ciders including *Lactobacillis mali* sp. nov. *Journal of Applied Bacteriology*, **33**, 768–774.

Carr, J.G. and Davies, P.A. (1972) The ecology and classification of strains of *Lactobacillus collinoides* nov. spec.: A bacterium commonly found in fermenting apple juice. *Journal of Applied Bacteriology*, **35**, 463–471.

Carr, J.G. and Davies, P.A. (1977) The relationship between *Lactobacillus mali* from cider and *Lactobacillis yamanashiensis* from wine. *Journal of Applied Bacteriology*, **42**, 219–228.

Cato, E.P., Moore, W.E.C. and Johnson, J.L. (1983) Synonymy of strains of '*Lactobacillus acidophilus*' group A2 (Johnson *et al.*, 1980) with the type strain of *Lactobacillus cripatus* (Brygoo & Aladame, 1953) Moore & Holdeman, 1970. *International Journal of Systematic Bacteriology*, **33**, 426–428.

Charlton, D.B., Nelson, M.E. and Werkman, C.H. (1934) Physiology of *Lactobacillus fructivorans* sp. nov. isolated from spoiled salad dressing. *Iowa State Journal of Science*, **9**, 1–11.

Collins, M.D. and Jones, D. (1981) The distribution of isoprenoid quinone structural types in bacteria and their taxonomic implications. *Microbiological Reviews*, **45**, 316–354.

Collins, M.D. and Wallbanks, S. (1992) Comparative sequence analysis of the 16S rRNA genes of *Lactobacillis minutus*, *Lactobacillus rimae*, and *Streptococcus parvulus*: proposal for the creation of a new genus *Atopobium*. *FEMS Microbiology Letters*, **95**, 235–240.

Collins, M.D., Phillips, B.A. and Zanoni, P. (1989) Deoxyribonucleic acid homology studies of *Lactobacillus casei*, *Lactobacillus paracasei* sp. nov., subsp. *paracasei* and subsp. *tolerans* and *Lactobacillus rhamnosus* sp. nov., comb. nov. *International Journal of Systematic Bacteriology*, **39**, 105–108.

Collins, M.D., Rodrigues, U.M., Ash, C., Aguirre, M., Farrow, J.A.E., Martinez-Murcia, A., Phillips, B.A., Williams, A.M. and Wallbanks, S. (1991) Phylogenetic analysis of the genus *Lactobacillus* and related lactic acid bacteria as determined by reverse transcriptase sequencing of 16S rRNA. *FEMS Microbiology Letters*, **77**, 5–12.

Collins, M.D., Samelis, J., Metaxopoulos, J. and Wallbanks, S. (1993) Taxonomic studies on some leuconostoc-like organisms from fermented sausages: description of a new genus *Weissella* for the *Leuconostoc paramesenteoides* group of species. *Journal of Applied Bacteriology*, **75**, 595–603.

Czelovsky, J., Wolf, G. and Hammes, W.P. (1992) Production of formate, acetate, and succinate by anaerobic fermentation of *Lactobacillus pentosus* in the presence of citrate. *Applied Microbiology and Biotechnology*, **37**, 94–97.

Davies, G.H.G. (1955) The classification of lactobacilli from the human mouth. *Journal of General Microbiology*, **13**, 481–493.

Dellaglio, F., Bottazzi, V. and Vescovo, M. (1975) Deoxyribonucleic acid homology among *Lactobacillus* species of the subgenus *Streptobacterium* Orla-Jensen. *International Journal of Systematic Bacteriology*, **25**, 160–172.

Dellaglio, F., Dicks. L.M.T., Du Tolt M. and Torriani, S. (1991) Designation of ATCC 334 in place of ATCC 393 (NCDO 161) as the neotype strain of *Lactobacillus casei* subsp. *casei* and rejection of the name *Lactobacillus paracasei* (Collins *et al.*, 1989) Request for an opinion. *International Journal of Systematic Bacteriology*, **41**, 340–342.

Dent, V.E. and Williams, R.A.D. (1982) *Lactobacillus animalis* sp. nov., a new species of *Lactobacillus* from the alimentary canal of animals. *Zentralblatt für Bakteriologie, Parasitenkunde, Infektionskrankheiten und Hygiene*, **C3**, 377–386.

Douglas, H.C. and Cruess, W.V. (1936) A *Lactobacillus* from California wine: *Lactobacillus hilgardii. Food Research*, **1**, 113–119.

Embley, T.M., Faquir, N., Bossart, W. and Collins, M.D. (1989) *Lactobacillus vaginalis* sp. nov. from the human vagina. *International Journal of Systematic Bacteriology*, **39**, 368–370.

Etani, E., Masai, H. and Suzuki, K.-I. (1986) *Lactobacillus acetotolerans*, a new species from fermented vinegar broth. *International Journal of Systematic Bacteriology*, **36**, 544–549.

Farrow, J.A.E. and Collins, M.D. (1988) *Lactobacillus oris* sp. nov. from the human oral cavity. *International Journal of Systematic Bacteriology*, **38**, 116–118.

Farrow, J.A.E., Phillips, B.A. and Collins, M.D. (1988) Nucleic acid studies on some heterofermentative lactobacilli: description of *Lactobacillus malefermentans* sp. nov. and *Lactobacillus parabuchneri* sp. nov. *FEMS Microbiology Letters*, **55**, 163–168.
Fornachon, J.C.M., Douglas, H.C. and Vaughn, R.H. (1949) *Lactobacillus trichodes* nov. sp., a bacterium causing spoilage in appetizer and dessert wines. *Hilgardia*, **19**, 119–132.
Fred, E.B., Peterson, W.H. and Anderson, J.A. (1921) The characteristics of certain pentose destroying bacteria, especially as concerns their action on arabinose. *Journal of Biological Chemistry*, **48**, 385–412.
Fujisawa, T., Shirasaka, S., Watabe, J. and Mitsuoka, T. (1984) *Lactobacillus aviarius* sp. nov.: a new species isolated from the intestine of chickens. *Systematic Applied Microbiology*, **5**, 414–420.
Fujisawa, T., Adachi, S., Toba, T., Arihara, K. and Mitsuoka, T. (1988) *Lactobacillus kefiranofaciens* sp. nov. isolated from kefir grains. *International Journal of Systematic Bacteriology*, **38**, 12–14.
Fujisawa, T., Itoh, K., Benno, Y. and Mitsuoka, T. (1990) *Lactobacillus intestinalis* (ex Hemme 1974) sp. nov., nom. rev., isolated from the intestines of mice and rats. *International Journal of Systematic Bacteriology*, **40**, 302–304.
Fujisawa, T., Benno, Y., Yaeshima, T. and Mitsuoka, T. (1992) Taxonomic study of the *Lactobacillus acidophilus* group, with recognition of *Lactobacillus gallinarum* sp. nov. and *Lactobacillus johnsonii* sp. nov. and synonymy of *Lactobacillus acidophilus* group A3 (Johnson *et al.*, 1980) with the type strain of *Lactobacillus amylovorus* (Nakamura, 1981). *International Journal of Systematic Bacteriology*, **42**, 487–491.
Gasser, F., Mandel, M. and Rogosa, M. (1970) *Lactobacillus jensenii* sp. nov., a new representative of the subgenus *Thermobacterium. Journal of General Microbiology*, **62**, 219–222.
Hammes, W.P., Bantleon, A. and Min, S. (1990) Lactic acid bacteria in meat fermentation. *FEMS Microbiology Letters*, **87**, 165–173.
Hammes, W.P., Weiss, N. and Holzapfel, W.H. (1991) The genera *Lactobacillus* and *Carnobacterium*. In *The Prokaryotes. Handbook on the Biology of Bacteria: Ecophysiology, Isolation, Identification, Applications* (eds Balows, A., Trüper, H.G., Dworkin, M., Harder, W. and Schleifer, K.H.). Springer, New York, USA, pp. 1535–1594.
Hansen, P.A. (1968) Type strains of *Lactobacillus* species. A report by the taxonomic subcommittee on lactobacilli and closely related organisms. International committee on nomenclature of bacteria of the international association of microbiological societies. American Type Culture Collection, Rockville, Maryland, USA.
Hansen, P.A. and Lessel, E.F. (1971) *Lactobacillus casei* (Orla-Jensen) comb. nov. *International Journal of Systematic Bacteriology*, **21**, 69–71.
Hansen, P.A. and Mocquot, G. (1970) *Lactobacillus acidophilus* (Moro) comb. nov. *International Journal of Systematic Bacteriology*, **20**, 325–327.
Hayward, A.C. (1957) A comparison of *Lactobacillus* species from human saliva with those from other natural sources. *British Dental Journal*, **102**, 450–451.
Hayward, A.C. and Davies G.H.G. (1956) The isolation and classification of *Lactobacillus* strains from human saliva samples. *British Dental Journal*, **101**, 43–46.
Hemme, D. (1974) Taxonomie des lactobacilles homofermentaires du tube digestif du rat. Description de deux nouvelles espèces *Lactobacillus intestinalis*, *Lactobacillus murini*. Thèse, University of Paris VII, France.
Hemme, D., Raibaud, P., Ducluzeau, R., Galpin, J.-V., Sicard, P. and van Heijenoort, J. (1980) *Lactobacillus murinus* n. sp., une nouvelle espèche de la flore dominante autochtone du tube digestif du rat et de la souris. *Annales de l'Institut Pasteur*, **131**, 297–308.
Henneberg, W. (1903) Zur Kenntnis der Milchsäurebakterien der Brennereimaische, der Milch, des Bieres, der Presshefe, der Melasse, des Sauerkohls, der sauren Gurken und des Sauerteiges, sowie einige Bemerkungen uber die Milchsaurebakterien des menschlichen Magens, *Z. Spiritusind.*, **26**, 22–31; see: *Zentralblatt für Bakteriologie, Parasitenkunde, Infektionskrankheiten und Hygiene*, **11**, 163.
Holzapfel, W.H. and Kandler, O. (1969) Zur Taxonomie der Gattung *Lactobacillus* Beijerinck. VI. *Lactobacillus coprophilus* subsp. *confusus* nov. subsp., eine neue Unterart der Untergattung *Betabacterium. Zentralblatt für Bakteriologie, Parasitenkunde, Infektionskrankheiten und Hygiene*, **123**, 657–666.
Holzapfel, W.H. and van Wyk, E.P. (1982) *Lactobacillus kandleri* sp. nov., a new species of

the subgenus *Betabacterium* with glycine in the peptidoglycan. *Zentralblatt für Bakteriologie, Parasitenkunde, Infektionskrankheiten und Hygiene*, **C3**, 495–502.

Johnson, J.L., Phelps. C.F., Cummins, C.S., London, J. and Gasser, F. (1980) Taxonomy of the *Lactobacillus acidophilus* group. *International Journal of Systematic Bacteriology*, **30**, 53–68.

Kandler, O. (1983) Carbohydrate metabolism in lactic acid bacteria. *Antonie van Leeuwenhoek*, **49**, 209–224.

Kandler, O. and Abo-Elnaga, I.G. (1966) Zur Taxonomie der Gattung *Lactobacillus* Beijerinck. IV. *L. corynoides* ein Synonym von *L. viridescens*. *Zentralblatt für Bakteriologie, Parasitenkunde, Infektionskrankheiten und Hygiene*, **120**, 753–759.

Kandler, O. and Kunath, P. (1983) *Lactobacillus kefir* sp. nov., a component of the microflora of kefir. *Systematic Applied Microbiology*, **4**, 286–294.

Kandler, O. and Weiss, N. (1986) Regular, non-sporing Gram-positive rods. In: *Bergey's Manual of Systematic Bacteriology*, Vol. 2 (eds. Sneath, P.H.A., Mair, N., Sharpe, M.E. and Holt, J.G.). Williams and Wilkins, Baltimore, pp. 1208–1234.

Kandler, O., Stetter, K.O. and Köhl, R. (1980) *Lactobacillus reuteri* sp. nov., a new species of heterofermentative lactobacilli. *Zentralblatt für Bakteriologie, Parasitenkunde, Infektioniskrankheiten und Hygiene*, **1**, 264–269.

Kandler, O., Schillinger, U. and Weiss, N. (1983a) *Lactobacillus halotolerans* sp. nov., nom. rev. and *Lactobacillus minorsp.* nov., nom. rev. *Systematic Applied Microbiology*, **4**, 280–285.

Kandler, O., Schillinger, U. and Weiss, N. (1983b) *Lactobacillus bifermentans* sp. nov., nom. rev., an organism forming CO_2 and H_2 from lactic acid. *Systematic Applied Microbiology*, **4**, 408–412.

Kaneuchi, C., Seki, M. and Komagate, K. (1988) Taxonomic study of *Lactobacillus mali* Carr and Davies 1970 and related strains: validation of *Lactobacillus mali* Carr and Davies 1970 over *Lactobacillus yamanashiensis* Nonomura 1983. *International Journal of Systematic Bacteriology*, **38**, 269–272.

Katagiri, H., Kitahara, K. and Fukami, K. (1934) The characteristics of the lactic acid bacteria isolated from moto, yeast mashes for sake manufacture. IV. Classification of the lactic acid bacteria. *Bulletin of the Agricultural Chemical Society of Japan*, **10**, 156–157.

Kitahara, K., Kaneko, T. and Goto, O. (1957) Taxonomic studies on the hiochibacteria, specific saprophytes of sake. II. Identification and classification of the hiochi-bacteria. *Journal of General Microbiology*, **3**, 111–120.

Kleynmans, U., Heinzl, H. and Hammes, W.P. (1989) *Lactobacillus suebicus* sp. nov., an obligately heterofermentative *Lactobacillus* species isolated from fruit mashes. *Systematic Applied Microbiology*, **11**, 267–271.

Kline, L. and Sugihara, T.F. (1971) Microorganisms of the San Francisco sourdough process. II. Isolation and characterization of undescribed bacterial species responsible for the souring activity. *Applied Microbiology*, **21**, 102–110.

Kodama, R. (1956) Studies on the nutrition of lactic acid bacteria. Part 1. *Lactobacillus fructosus* nov. sp., a new species of lactic acid bacteria. *Journal of the Agricultural Chemical Society of Japan*, **30**, 705–708.

Kuznetzov, V.D. (1959) A new species of lactic acid bacteria. *Microbiology*, **28**, 248–351.

Lapage, S.P., Sneath, P.H., Lessel, E.F., Skerman, V.B.D., Seeliger, H.P.R. and Clark, W.A. (1975) *International Code of Nomenclature of Bacteria*. American Society of Microbiology, Washington, DC, USA.

Lauer, E. and Kandler, O. (1980) *Lactobacillus gasseri* sp. nov., a new species of the subgenus *Thermobacterium*. *Zentralblatt für Bakteriologie, Parasitenkunde, Infektionskrankheiten und Hygiene*, **C1**, 7578.

Leichmann, G. (1986) Über die freiwillige Säurung der Milch. *Zentralblatt für Bakteriologie, Parasitenkunde, Infektionskrankheiten und Hygiene*, **2**, 777–780.

Lerche, M. and Reuter, G. (1962) Das Vorkommen aerob wachsender Gram-positiver Stäbchen des Genus *Lactobacillus* Beijerinck im Darminhalt erwachsener Menschen. *Zentralblatt für Bakteriologie, Parasitenkunde, Infektionskrankheiten und Hygiene*, **185**, 446–481.

Meisel, J. (1991) Zum Einfluß von Hämatin auf den Stoffwechsel von Milchsäurebakterien. Thesis, Universität Hohenheim.

Mitsuoka, T. and Fujisawa, T. (1987) *Lactobacillus hamsteri*, a new species from the intestine of hamsters. *Proceedings of the Japanese Academy*, **63**, 269–272.

Montel, M.-C. and Champomier, M.-C. (1987) Arginine catabolism in *Lactobacillus saké* isolated from meat. *Applied and Environmental Microbiology*, **53** (11), 2683–2685.

Moore, W.E.C. and Holdeman, L.V. (1970) *Propionibacterium, Arachnia, Actinomyces, Lactobacillus* and *Bifidobacterium*. In *Outline of Clinical Methods in Anaerobic Bacteriology*, 2nd rev. edn. (eds Cato, E.P., Cummins, C.S., Holdeman, L.V., Johnson, J.L., Moore, W.E.C., Smibert, R.M. and Smith L.D.S.). Virginia Polytechnic Institute Anaerobe Laboratory, Blacksburg, VA, USA, pp. 15–22.

Moro, E. (1900) Über den *Bacillus acidophilus* n. sp. *Jahrbuch der Kinderheilkunde*, **52**, 38–55.

Nakamura, L.K. (1981) *Lactobacillus amylovorus*, a new starch-hydrolyzing species from cattle waste-corn fermentations. *International Journal of Systematic Bacteriology*, **31**, 56–63.

Nakamura, L.K. and Cromwell, C.D. (1979) *Lactobacillus amylophilus*, a new starch-hydrolyzing species from swine waste-corn fermentation. *Developments in Industrial Microbiology*, **20**, 531–540.

Niven, J.C.F. and Evans, J.B. (1957) *Lactobacillus viridescens* nov. spec. a heterofermentative species that produces a green discoloration of cured meat pigments. *Journal of Bacteriology*, **73**, 758–759.

Nonomura, H. (1983) *Lactobacillus yamanashiensis* subsp. *yamanashiensis* and *Lactobacillus yamanashiensis* subsp. *mali* sp. and subsp. nov., nom. rev. *International Journal of Systematic Bacteriology*, **33**, 406–407.

Nonomura, H. and Ohara, Y. (1967) Die Klassifikation der Äpfelsäure-Milchsäure-Bakterien. *Mitteilungen aus Klosterneuburg*, **17A**, 449–466.

Nonomura, H., Yamazaki, T. and Ohara, Y. (1965) Die Äpfelsäure-Milchsäure-Bakterien, welche aus japanischen Weinen isoliert wurden. *Mitteilungen aus Klosterneuburg*, **15A**, 241–254.

Okada, S., Suzuki, Y. and Kozaki, M. (1979) A new heterofermentative *Lactobacillus* species with meso-diaminopimelic acid in peptidoglycan, *Lactobacillus vaccinostercus* Kozaki and Okada sp. nov. *Journal of General and Applied Biology*, **25**, 215–221.

Olsen, I., Johnson, J.L., Moore, L.V.H. and Moore, W.E.C. (1991) *Lactobacillus uli* sp. nov. and *Lactobacillus rimae* sp. nov. from human gingival crevice and emended descriptions of *Lactobacillus minutus* and *Streptococcus parvulus*. *International Journal of Systematic Bacteriology*, **41**, 261–266.

Orla-Jensen, S. (1919) The lactic acid bacteria. A.F. Høst and Son, Koeniglicher Hof-Boghandel, Copenhagen.

Pot, B., Ludwig, W., Kersters, K. and Schleifer, K.H. (1994) Taxonomy of lactic acid bacteria. In: *Bacteriocins of lactic acid bacteria: Genetics and Applications* (eds De Vuyst, L. & Vandamme, E.J.). Chapman and Hall, Glasgow, UK, pp. 13–89.

Raibaud, P., Galpin, J.V., Ducluzeau, R., Mocquot, G. and Oliver, G. (1973) La genre *Lactobacillus* dans le tube digestif du rat. I. Caractères des souches homofermentaires isolées de rats holo- et gnotoxéniques. *Annales de l'Institut Pasteur*, **124A**, 83–109.

Reuter, G. (1970) Laktobazillen und eng verwandte Mikroorganismen in Fleisch und Fleischerwaren. 2. Die Charakterisierung der isolierten Laktobazillenstämme. *Fleischwirtschaft*, **50**, 954–962.

Reuter, G. (1983) *Lactobacillus alimentarius* sp. nov. and *Lactobacillus farciminis* sp. nov., nom. rev. *Systematic Applied Microbiology*, **4**, 277–279.

Rhee, S.K. and Pack, M.H. (1980) Effect of environmental pH on fermentation balance of *Lactobacillus bulgaricus*. *Journal of Bacteriology*, **144**, 217–221.

Rogosa, M. (1970) Characters used in the classification of lactobacilli. *International Journal of Systematic Bacteriology*, **20**, 519–533.

Rogosa, M., Wiseman, R.F,, Mitchell, J.A., Disraely, M.N. and Beaman, A.J. (1953) Species differentiation of oral lactobacilli from man including descriptions of *Lactobacillus salivarius* nov. spec. and *Lactobacillus cellobiosus* nov. spec. *Journal of Bacteriology*, **65**, 681–699.

Russel, C. and Walker, T.K. (1953) *Lactobacillus malefermentans* n. sp. isolated from beer. *Journal of General Microbiology*, **8**, 160–162.

Sharpe, M.E. (1979) Lactic acid bacteria in the dairy industry. *Journal of the Society of Dairy Technology*, **32**, 9–18.

Sharpe, M.E., Garvie, E.I. and Tilbury, R.H. (1972) Some slime forming heterofermentative species of the genus *Lactobacillus. Applied Microbiology*, **23**, 389–397.

Sharpe, M.E., Latham, M.J., Garvie, E.I., Zirngibl, J. and Kandler, O. (1973) Two new species of *Lactobacillus* isolated from the bovine rumen, *Lactobacillus ruminis* sp. nov. and *Lactobacillus vitulinus* sp. nov. *Journal of General Microbiology*, **77**, 37–49.

Skerman, V.B.D., McGowan, V. and Sneath, P.H.A. (1980) Approved list of bacterial names. *International Journal of Systematic Bacteriology*, **30**, 225–420.

Spicher, G. and Schröder, R. (1978) Die Mikroflora des Sauerteiges. IV. Mitteilung Untersuchungen uber die art in 'Reinzuchtsauern' anzutreffenden stäbchenförmigen Milchsäurebakterien (Genus *Lactobacillus* Beijerinck). *Zeitschrift für Lebensmitteluntersuchung und forschung*, **167**, 342–354.

Stackebrandt, E. and Teuber, M. (1988) Molecular taxonomy and phylogenetic position of lactic acid bacteria. *Biochimie*, **70**, 317–324.

Stetter, H. and Stetter, K.O. (1980) *Lactobacillus bavaricus* sp. nov., a new species of the subgenus *Streptobacterium. Zentralblatt für Bakteriologie, Parasitenkunde, Infektionskrankheiten und Hygiene*, **C1**, 70–74.

Troili-Petersson, G. (1903) Studien über die Mikroorganismen des schwedischen Guterkäses. *Zentralblatt für Bakteriologie, Parasitenkunde, Infektionskrankheiten und Hygiene*, **11**, 120–143.

Vescovo, M., Dellaglio, F., Botazzi, V. and Sarra, P.G. (1979) Deoxyribonucleic acid homology among *Lactobacillus* species of the subgenus *Betabacterium* Orla-Jensen. *Microbiologica*, **2**, 317–330.

Vogel, R.F., Böcker, G., Stolz, P., Ehrmann, M., Fanta, D., Ludwig, W., Pot, B., Kersters, K., Schleifer, K.H. and Hammes, W.P. (1994) Identification of lactobacilli from sourdough and description of *Lactobacillus pontis* sp. nov. *International Journal of Systematic Bacteriology*, **44**, 223–229.

Vries, De W., Kapteijn, W.M.C., Beek, E.G. and van der Stouthamer, A.H. (1970) Molar growth yields and fermentation balances of *Lactobacillus casei* L3 in batch cultures and in continuous cultures. *Journal of General Microbiology*, **63**, 333–345.

Wayne, L.G. (1994) Actions of the Judicial Commission of the International Committee on Systematic Bacteriology on Requests for Opinions Published Between January 1985 and July 1993. *International Journal of Systemic Bacteriology*, **44**, 177–178.

Weiss, N. and Schillinger, U. (1984) *Lactobacillus sanfrancisco* nov. spec., nom. rev. *Systematic Applied Microbiology*, **5**, 230–232.

Weiss, N., Schillinger, U., Laternser, M. and Kandler, O. (1981) *Lactobacillus sharpeae* sp. nov. and *Lactobacillus agilis* sp. nov., two new species of homofermentative, mesodiaminopimelic acid-containing lactobacilli isolated from sewage. *Zentralblatt für Bakteriologie, Parasitenkunde, Infektionskrankheiten und Hygiene*, **C2**, 242–253.

Weiss, N., Schillinger, U. and Kandler, O. (1983a) *Lactobacillus trichodes*, and *Lactobacillus heterohiochii*, subjective synonym of *Lactobacillus fructivorans. Systematic Applied Microbiology*, **4**, 507–511.

Weiss, N., Schillinger, U. and Kandler, O. (1983b) *Lactobacillus lactis, Lactobacillus leichmanii* and *Lactobacillus bulgaricus*, subjective synonyms of *Lactobacillus delbrueckii* subsp. *lactis* comb. nov. and *Lactobacillus delbrueckii* subsp. *bulgaricus* comb. nov. *Systematic Applied Microbiology*, **4**, 552–557.

Weiss, N., Schillinger, U. and Kandler, O. (1984) In: Validation of the publication of new names and new combinations previously published outside the IJSB. List No. 14. *International Journal of Systematic Bacteriology*, **34**, 270–271.

Wood, B.J.B. (1992) *The Lactic Acid Bacteria*, Vol. 1, The Lactic Acid Bacteria in Health and Disease. Elsevier Applied Science, London, USA.

Yang, D. and Woese, C.R. (1989) Phylogenetic structure of the 'Leuconostocs': an interesting case of a rapidly evolving organism. *Systematic Applied Microbiology*, **12**, 145–149.

Zanoni, P., Farrow, J.A.E., Phillips, B.A. and Collins, M.D. (1987) *Lactobacillus pentosus* (Fred, Peterson and Anderson) sp. nov., nom. rev. *International Journal of Systematic Bacteriology*, **37**, 339–341.

4 The genus *Streptococcus*

J.M. HARDIE and R.A. WHILEY

4.1 Introduction

The genus *Streptococcus* consists of Gram-positive, spherical or ovoid cells that are typically arranged in chains or pairs. These cocci are facultatively anaerobic, non-sporing, catalase-negative, homofermentative, and have complex nutritional requirements. Many of the known species are parasitic in man or other animals and some are important pathogens (Jones, 1978; Hardie, 1986; Colman, 1990). Chain-forming cocci were observed in wounds by Billroth (1874) and he applied the term 'streptococcos' to such organisms to designate their morphological arrangement (Jones, 1978). A few years later, Rosenbach (1884) first used the word *Streptococcus* in the generic sense and described the species *Streptococcus pyogenes* which is now the type species of the genus. This species was originally isolated from suppurative lesions in humans. Subsequently, in early studies by Nocard and Mollereau (1887), Schütz (1887, 1888) and Talamon (1883) – cited by Colman (1990) – several other varieties of streptococci were isolated from different sources, including *S. agalactiae* from cows with mastitis and streptococci from both equine and human cases of pneumonia.

During the period around the turn of the last century, the association between streptococci and a variety of human and animal diseases was established. Around this time the importance of morphologically similar bacteria, then classified as streptococci, in the dairy industry was also recognized. Thus, by the 1930s a large number of *Streptococcus*-like bacteria had been described and a multiplicity of names existed in the published literature. Despite numerous attempts over the years, as outlined below, it is only comparatively recently that the classification of these clinically and industrially important bacteria has been brought to a more generally acceptable level, although uncertainties remain in a few areas.

4.2 Classification

Although many distinct taxa have been recognized amongst the streptococci, their classification and nomenclature have caused considerable confusion over the years. One of the first characters to be recognized and used for distinguishing between isolates was the ability of certain clinically important streptococci to cause complete (β-)haemolysis around colonies grown on blood-containing culture media (Schottmüller, 1903). Other species were found to produce greening or α-haemolysis under certain conditions, whilst some caused no change to the red blood cells (Brown, 1919). Although these haemolytic changes are still useful for descriptive purposes, they have never provided a reliable basis for taxonomic subdivision of the genus.

An excellent review of the early attempts at classification of the streptococci has been provided by Jones (1978) and only a few of the major studies will be referred to in this section. Andrewes and Horder (1906) produced a classification based on a combination of biochemical, physiological and morphological characteristics which were applied to a large number of human, animal, milk and environmental isolates. They described eight groups of streptococci, designated *S. pyogenes* (identical to that of Rosenbach, 1884), *S. equinus*, *S. mitis*, *S. salivarius*, *S. anginosus*, *S. faecalis* and the pneumococci (the latter were not given a species name by these workers although they were recognized as streptococci). In a major study on the lactic acid bacteria isolated mainly from dairy products, published a few years later, Orla-Jensen (1919) extended the range of tests applied to include growth under different conditions, such as varying temperatures and salt concentrations, in addition to fermentation reactions and morphological features. He described nine groups of streptococci, some of which are now recognized as belonging to the genera (Schleifer and Kilpper-Bälz, 1987) *Lactococcus* (*S. lactis* and *S. cremoris*) or *Enterococcus* (*S. faecium*, *S. liquefaciens*).

An important development in streptococcal classification occurred with the introduction of serological methods for recognizing a series of specific cell wall antigens referred to as 'group antigens'. Lancefield (1933) first demonstrated the presence of a particular carbohydrate antigen in *S. pyogenes* which was designated Group A, and this led to the extension of the Lancefield grouping scheme to other streptococci designated B, C, G, etc. The immunochemical properties of several of these antigens were subsequently studied in great detail (e.g. Krause and McCarty, 1962), some of which, for example, Group D, were found to be teichoic acids (Krause, 1972). In some cases, including Group A (*S. pyogenes*) and Group B (*S. agalactiae*), further serological subdivisions have been made based on other antigenic components (such as M, T and R proteins in Lancefield Group A) and these have proved extremely useful for typing purposes in epidemiological investigations (Maxted, 1978).

Although detection of the Lancefield group antigens has been of immense value for identification of some of the major human and animal pathogens, the application of serological methods to the genus as a whole has been considerably less successful. As had been shown in more recent investigations, not all streptococcal species possess a unique group antigen, whilst several of the recognized antigens are not confined to a single species. Thus, in most cases, the mere presence of a particular group antigen does not allow a streptococcal isolate to be identified to species level, unless supported by other evidence.

A significant, and to some extent prophetic, contribution to streptococcal taxonomy was made by Sherman (1937) who divided the genus into four main groups which were named 'pyogenic', 'viridans', 'lactic' and '*Enterococcus*'. These subdivisions were based on the ability to grow at 10°C and 45°C, to survive at 60°C for 30 min, to grow at pH 9.6, in 0.1% methylene blue, and at different concentrations of sodium chloride. As described below, only the first two of the groups currently remain in the genus *Streptococcus*, the 'lactic' and '*Enterococcus*' groups having been designated as separate genera.

The 'viridans' or 'oral' group of streptococci, some members of which were described at the beginning of this century by Andrewes and Horder (1906), have been a source of considerable confusion over the years (Hardie and Marsh, 1978a; Jones, 1978; Hardie & Whiley, 1992). The definition of species within this group was greatly improved during the 1960s and 1970s, thanks to the contributions of Colman and others who started to apply more modern numerical and chemotaxonomic methods to the study of these streptococci (Colman and Williams, 1965, 1972; Colman, 1968). In more recent times, the use of molecular methods such as DNA–DNA hybridization and nucleic acid sequencing, in addition to phenotypic characters, has clarified the situation still further, confirming the validity of many previously described species and enabling the description of others (Coykendall, 1989; Kilian *et al.*, 1989a; Bentley *et al.*, 1991; Hardie and Whiley, 1992).

At about the time when Vol. 2 of *Bergey's Manual of Systematic Bacteriology* (Sneath *et al.*, 1986) was under preparation, major changes to the composition of the genus *Streptococcus* were being proposed as a result of extensive molecular and chemotaxonomic studies. These led to the creation of two new genera, *Enterococcus* and *Lactococcus*, to encompass species that were formerly included in Sherman's '*Enterococcus*' and 'lactic' groups, such as *S. faecalis* (now *E. faecalis*) and *S. lactis* (now *L. lactis*) (Schleifer *et al.*, 1985; Schleifer and Kilpper-Bälz, 1984). Thus, the definition of the genus *Streptococcus sensu stricto* is now more restricted although it still includes a large number of species (currently 39), as described later in detail.

4.3 Morphology

Cells of streptococci are normally spherical or ovoid in shape, but some species may appear as short rods under certain cultural conditions. They are typically arranged in chains or pairs, chain formation being seen best in broth cultures. Individual cells are usually 0.8–1.2 μm in diameter and chains lengths vary from a few cells to over 50, depending on the strain and the growth conditions. It is not unusual for cells in older cultures to appear Gram-variable, whilst some strains may be highly pleomorphic on initial isolation.

Some species produce capsules, either of hyaluronic acid in the case of *S. pyogenes* or a variety of type-specific polysaccharides in *S. pneumoniae*, but this is not a regular feature throughout the genus as a whole. Several species produce extracellular polysaccharides when grown in the presence of sucrose, including both glucans and fructans (Hardie and Marsh, 1978a). A variety of surface structures and appendages have been described in different streptococcal species, including fimbriae and fibrils, which may be responsible for adhesion of the organisms to various surfaces (Handley, 1990; Hogg, 1992).

4.4 Cultural characteristics

Growth on solid media generally requires enrichment with blood, serum or glucose. Colonies of most species rarely exceed 1 mm in diameter after 24 h incubation at 37°C on blood agar and are usually non-pigmented, often appearing slightly translucent. On sucrose-containing media, extracellular polysaccharide producers display a variety of colonial forms which may facilitate recognition of species such as *S. mutans* and *S. salivarius* (Hardie and Marsh, 1978b). Although streptococci are facultatively anaerobic, many strains grow optimally under microaerophilic or anaerobic conditions (with CO_2) rather than in air, and some have an absolute requirement for CO_2, particularly on initial isolation. Putative obligately anaerobic streptococci such as *S. morbillorum* (now *Gemella morbillorum*), *S. parvulus* (now *Atopobium parvulus*) *S. hansenii* and *S. pleomorphus* (both more closely related to clostridia), have now been reclassified into other genera (Schleifer and Kilpper-Bälz, 1987; Kilpper-Bälz and Schleifer, 1988; Collins and Wallbanks, 1992; Hardie and Whiley, 1994).

Growth in liquid media is enhanced by addition of glucose or some other fermentable carbohydrate, but unless the medium is well buffered (as in Todd–Hewitt broth) the fall in pH will soon become inhibitory. The appearance of broth cultures varies from diffuse turbidity to granular growth with a clear supernatant, depending on the particular species and strains.

As mentioned previously, one of the best-known characteristics of streptococci is their ability to produce different types of haemolysis on blood-containing media (Brown, 1919). Originally, these changes were described from streptococcal cultures in pour plates, but haemolysis can also be observed around surface growth on layered blood plates or in stab plates (Ruoff, 1992). The different types of haemolysis seen, namely complete (β), partial or greening (α) or none (gamma), are dependent upon the organisms concerned, the type of blood used (horse, sheep, human, etc.), the composition of the basal medium, and the atmospheric conditions. In some species, the appearance of green, α-haemolytic zones around colonies grown aerobically may be due to the production of hydrogen peroxide (Colman, 1990; Ruoff, 1992).

4.5 Biochemistry/physiology

4.5.1 Carbohydrate metabolism

Streptococci ferment glucose and other carbohydrates, yielding L-lactate as the main end product when growing rapidly under conditions of carbohydrate excess. Under glucose-limited conditions, and at low dilution rates in continuous culture, other end products are detected, such as formate, acetate and ethanol, as a result of a switch to different metabolic pathways (Ellwood, 1976). The wide and variable range of carbohydrates that can be utilized by different species of streptococci forms the basis of many of the commonly used phenotypic tests that have been employed in diagnostic identification schemes.

4.5.2 Other requirements

The nutritional requirements of streptococci are generally complex, although they have not been determined in detail for all species. For those streptococci that have been examined, they include amino acids, peptides, purines, pyrimidines and vitamins, in addition to a source of energy. In most cases such nutrients are provided by using complex culture media which often contain meat extract, peptone and blood or serum. However, some strains of 'nutritionally variant' streptococci require the addition of pyridoxal hydrochloride in order to allow growth (Bouvet *et al.*, 1981).

Some streptococcal species are able to break down arginine and this is also another energy-yielding mechanism. Under experimental conditions in the chemostat, an increased yield of glucose-limited cells can be obtained by adding arginine to the system.

4.5.3 *Temperature and salt tolerance*

Before the separation of enterococci and lactococci into distinct genera, determination of the range of temperatures at which isolates could grow and their ability to withstand different concentrations of sodium chloride, bile, and other chemicals, were important differential criteria. Those species which remain in *Streptococcus* generally grow within the range 20–42°C, with 37°C or thereabouts as the optimum temperature in most cases.

4.5.4 *Oxygen*

As mentioned previously, streptococci are facultatively anaerobic and are usually not markedly affected by the presence of oxygen. They are catalase negative, cannot synthesize haem compounds, and some species produce hydrogen peroxide when grown aerobically. For routine purposes, almost all strains will grow satisfactorily in atmospheres of air+10% CO_2, or anaerobically in a mixture of nitrogen (70–80%), hydrogen (10–20%) and CO_2 (10–20%).

4.6 Cell wall composition

As with other Gram-positive bacteria, the main structural component of the cell wall of streptococci is peptidoglycan (murein), together with various other associated polysaccharides, some of which form the basis of the Lancefield serological grouping system. Peptidoglycan consists of glucan chains that are cross-linked by short peptides and which contain alternating units of β-1,4-linked *N*-acetylglucosamine and *N*-acetylmuramic acid (Schleifer and Kandler, 1972). Different types of peptidoglycan structure have been described, depending on the chemical nature of the cross-linking of the adjacent stem peptides, and these types have been shown to have considerable value as taxonomic markers (Schleifer and Seidl, 1985). Cell wall polysaccharides have also been used as chemotaxonomic markers within the streptococci (Colman and Williams, 1965). Most species characteristically contain rhamnose as one of the sugar components, together with various combinations of glucose and galactose. However, rhamnose is absent from *S. oralis* and *S. pneumoniae*, both of which contain a ribitol teichoic acid. The chemical composition of the polysaccharide antigens of several species of streptococci have been determined, including Lancefield groups A, A-variant, B, C, E and G (Schleifer and Kilpper-Bälz, 1987; Colman, 1990), as well as the type-specific antigens within the *S. mutans* group (Hamada and Slade, 1980) and some other species. Some of the known chemical and serological characteristics of streptococci are summarized in Table 4.1.

Table 4.1 Serological markers and cell compositions of streptococci (species names are arranged in currently recognized species groups)

Species groups	Serological markers	Murein type	Characteristic cell wall polysaccharide components*
Oral streptococci			
S. mutans	Serotype c, e or f	Lys-Ala$_{2-3}$	Rha, Gluc
S. sobrinus	Serotype d, or h, g (or –)	Lys-Thr-Ala	Rha, Gluc, Gal
S. cricetus	Serotype a	Lys-Thr-Ala	Rha, Gluc, Gal
S. rattus	Serotype b	Lys-Ala$_{2-3}$	Rha, Gal, Glyc
S. macacae	Serotype c	ND	ND
S. downei	Serotype h	Lys-Thr-Ala	ND
S. ferus	Serotype c	Lys-Ala$_{2-3}$	ND
S. salivarius	Lancefield K, –	Lys-Ala$_{2-3}$ Lys-Thr-Ala	Rha, Gluc, Gal, GalNAc
S. vestibularis	–		
S. thermophilus	–	Lys-Ala$_{2-3}$	ND
S. intermedius	–	Lys-Ala$_{1-3}$	Rha, Gluc
S. constellatus	– or Lancefield F,† A or C	Lys-Ala$_{1-3}$	Rha, Gluc, Gal,
S. anginosus	– or Lancefield F,† A, C, or G	Lys-Ala$_{1-3}$	Rha, Gluc, Gal, GalNAc
S. sanguis	Lancefield H,‡ –	Lys-Ala$_{1-3}$	Rha, Gluc,
S. gordonii	Lancefield H,‡ –	Lys-Ala$_{1-3}$	Rha, Glyc
S. parasanguis	– (or Lancefield F, G, C or B)	ND	ND
S. crista	ND	ND	ND
S. oralis	–	Lys-direct	Gluc, Gal, GalNAc, (Rha), Rtl
S. mitis	– (Lancefield K or O)	Lys-direct	(Rha), Rtl
S. pneumoniae	C-polysaccharide capsular antigens	Lys-Ala$_2$ (Ser)	Gluc, (Gal), GalNAc, (Rha), Rtl
S. adjacens	–	ND	ND
S. defectivus	– (or Lancefield H)	ND	ND

Table 4.1 *continued*

Species groups	Serological markers	Murein type	Characteristic cell wall polysaccharide components*
Pyogenic streptococci			
S. pyogenes	Lancefield group A§	Lys-Ala_{1-3}	Rha
S. canis	Lancefield group G	Lys-Thr-Gly	ND
S. agalactiae	Lancefield group B	Lys-Ala_{1-3} (Ser)	Rha, Gal, Glucitol
S. dysgalactiae	Lancefield group C, G, L	Lys-Ala_{1-3}	Rha, GalNAc
S. parauberis	– or Lancefield group E, P	ND	ND
S. uberis	– (Lancefield group E, P, G)	Lys-Ala_{1-3}	Rha, Gluc
S. porcinus	Lancefield groups E, P, U, V	Lys-Ala_{2-4}	ND
S. iniae	–	Lys-Ala_{1-3}	Rha, Gluc, Gal,
S. equi subsp. *equi*	Lancefield group C	Lys-Ala_{1-3}	Rha, GalNAc
S. equi subsp. *zooepidemicus*	Lancefield group C	Lys-Ala_{2-3}	ND
S. hyointestinalis	–	Lys-Ala (Ser)	ND
Other streptococci			
S. alactolyticus	Lancefield group D	ND	ND
S. bovis	Lancefield group D	Lys-Thr-Ala	Rha, Gluc, Gal
S. equinus	Lancefield group D	Lys-Thr-Ala	ND
S. suis	Lancefield group R, S, RS, T	Lys-direct	Rha, Gluc, (Gal), (GalNAc)
S. acidominimus	–	Lys-Ser-Gly	Rha, Gal
S. intestinalis	– (or Lancefield group G)	ND	ND
S. caprinus	ND	ND	ND

*ND, not determined; Gal, galactose; GalNAc, *N*-acetyl galactosamine; Gluc, glucose; Glyc, glycerol; Rha, rhamnose; Rtl, ribitol; and ( ), trace amounts.

†Further subdivision of Lancefield Group F strains has been described on the basis of type-specific carbohydrate antigens (Ottens and Winkler, 1962).
‡Reactions with Group H antiserum vary according to the immunizing strain used.
§Further subdivision of Lancefield Group A strains on the basis of M, T and R antigens.

4.7 Genetics

It is beyond the scope of this chapter to review in any detail the large body of work on streptococcal genetics that has been published since the early studies on transformation in pneumococci by Avery *et al.* (1944). Several recent books and proceedings of conferences have been devoted to this topic and others are known to be in preparation (e.g. Ferretti and Curtiss, 1987; Dunny *et al.*, 1991; Orefici, 1992). A number of streptococcal genes have been cloned and sequenced, including those coding for various surface components and virulence determinants in *S. pyogenes* and other pathogenic species (Ferretti, 1992; Fischetti *et al.*, 1992), as well as transport systems and metabolic activities in *S. mutans* (Russell *et al.*, 1991, 1992). The molecular genetics of *S. pyogenes*, *S. agalactiae*, *S. pneumoniae* and several species amongst the oral streptococci has provided the focus for many of the reported studies in recent years (Ferretti and Curtis, 1987; Fischetti, 1989; Kehoe, 1991; Boulnois, 1992; Shiroza and Kuramitsu, 1993; Russell, 1994). Such studies have helped to cast new light on the molecular mechanisms behind some of the pathogenic and metabolic activities of the streptococci and will lead, hopefully, to improved methods for prevention and treatment of streptococcal infections and their sequelae.

4.8 Phylogeny

As mentioned previously, the results of DNA–DNA and DNA–rRNA hybridization studies, together with other chemotaxonomic data, led to the separation of the former 'enterococcal (or faecal)' and 'lactic' groups of streptococci into separate genera (Schleifer and Kilpper-Bälz, 1984, 1987; Schleifer *et al.*, 1985). These proposals were supported by subsequent comparison of 16S rRNA sequences from the redefined taxa (Collins *et al.*, 1989; Williams *et al.*, 1989). Apart from *Enterococcus* and *Lactococcus*, several other genera of gram-positive cocci have been described, some quite recently, which are phylogenetically distinct from *Streptococcus*. These include *Aerococcus*, *Alloiococcus*, *Atopobium*, *Dolosigranulum*, *Gemella*, *Helcococcus*, *Leuconostoc*, *Melissococcus*, *Pediococcus*, *Tetragenococcus* and *Vagococcus*. The phylogenetic relationship of some of these genera to *Streptococcus*, as revealed by 16S rRNA sequence analysis, is illustrated in Figure 4.1.

The intrageneric relationships between species within the genus *Streptococcus*, again determined from 16S rRNA sequence data, have been reported by Bentley *et al.* (1991). From this study, in which 31 of the 39 currently known species were included, it is evident that several clusters

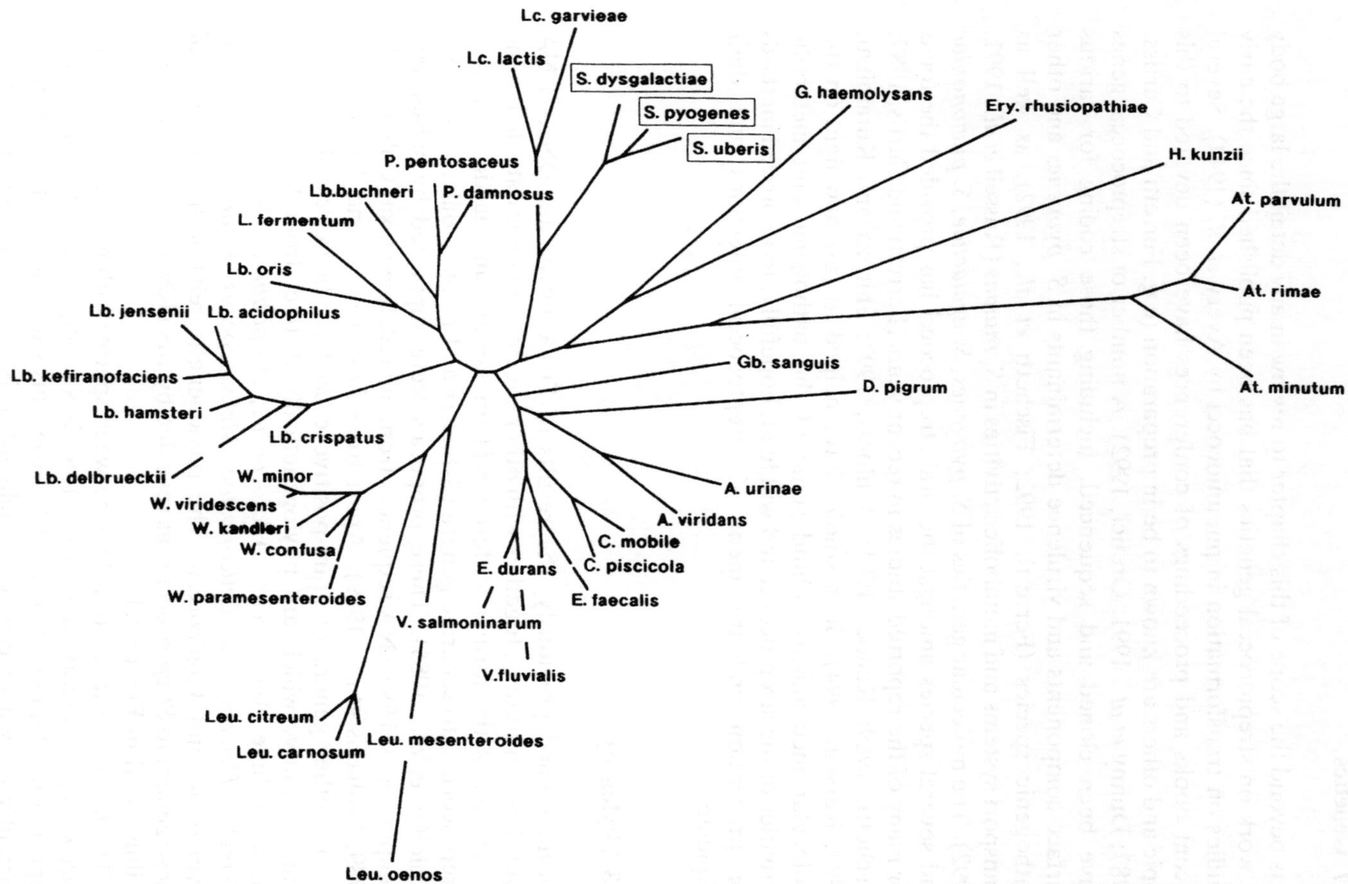

Figure 4.1 Phylogenetic network showing the relationships between *Streptococcus*, selected lactic acid bacteria and other low G+C Gram-positive taxa. *A*, *Aerococcus*; *At*, *Atopobium*; *C*, *Carnobacterium*; *D*, *Dolosigranulum*; *E*, *Enterococcus*; *Ery*, *Erysipelothrix*; *G*, *Gemella*; *Gb*, *Globicatella*; *H*, *Helococcus*; *Lb*, *Lactobacillus*; *Lc*, *Lactococcus*; *Leu*, *Leuconostoc*; *P*, *Pedicoccus*; *S*, *Streptococcus*; *V*, *Vagococcus*; *W*, *Weisella*. (Adapted, with permission, from Aguirre *et al*., 1993; Figure 1.)

can be discerned with the genus. In the main these correspond quite closely to species groupings revealed by other techniques, although a few exceptions were reported. The pyogenic group was found to include *S. agalactiae*, *S. parauberis*, *S. porcinus* and *S. uberis*, in addition to *S. pyogenes*, *S. equi*, *S. canis*, *S. dysgalactiae*, and *S. iniae*, but the position of *S. hyointestinalis* remained uncertain. A distinct cluster was formed by *S. bovis*, *S. equinus* and *S. alactolyticus* and the close relationship previously demonstrated by other methods between *S. bovis* and *S. equinus* was confirmed. Four groups were found amongst the oral streptococci, centred around *S. mutans*, *S. salivarius*, *S. anginosus* (often referred to as the '*S. milleri* group') and *S. oralis*, the last mentioned species being closely related to *S. pneumoniae* (which has misleadingly been included in the pyogenic group in previously published descriptions of the genus (Sneath *et al.*, 1986)). The species *S. acidominimus* and *S. suis* did not fall into any of the discernible clusters.

The unrooted tree showing the phylogenetic relationships between these species of streptococci is reproduced from the paper by Bentley *et al.* (1991) in Figure 4.2. The phylogenetic groupings and order of species shown have been utilized for the construction of tables presented later in this chapter. It would be useful to add to the database in the future by inclusion of 16S rRNA sequence data from the species not hitherto included in the published reports. To take one example, it would be particularly interesting to determine the phylogenetic relationships of the nutritionally variant streptococci, *S. adjacens* and *S. defectivus*, both to each other and to other species.

4.9 Importance of the genus

4.9.1 Normal commensal flora

Streptococci comprise a significant component of the commensal flora of man and animals, colonizing mucous membranes of the mouth, respiratory tract, alimentary tract and genitourinary tract. Some species are also found on the skin, and others may be isolated from milk and dairy products (in addition to lactococci and enterococci) (Skinner and Quesnel, 1978).

Data on the distribution of streptococcal species between different animal hosts, and of their specificities for particular body sites, are incomplete, although some comparative ecological studies have been reported (Devriese, 1991). There is a need for further investigations along the lines of those recently reported on the streptococcal flora of the tonsils in cattle (Cruz Colque *et al.*, 1993), dogs and cats (Devriese *et al.*, 1992a, b), pigs (Devriese *et al.*, 1994) and the intestinal flora of poultry (Devriese *et al.*, 1991), in order to obtain a fuller picture across the animal kingdom. Several of the streptococci associated with the oral flora in humans have

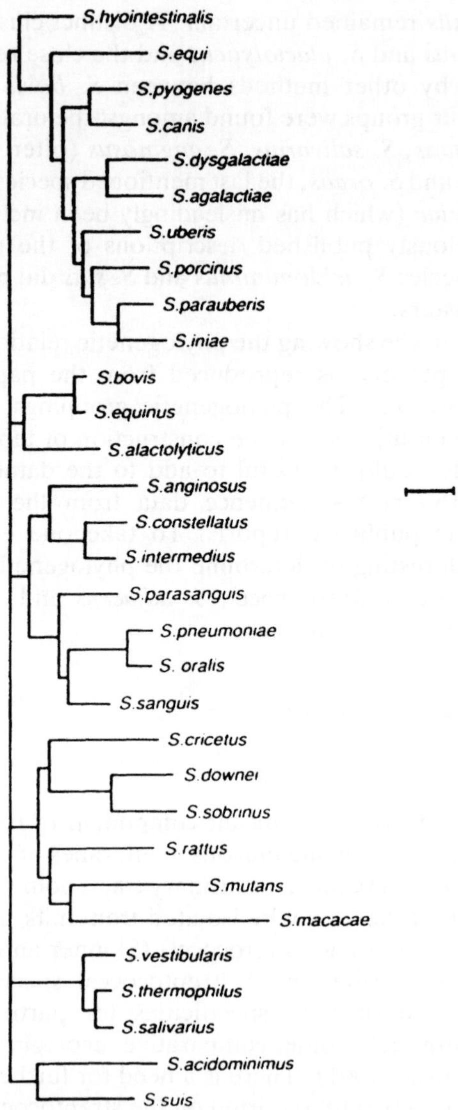

Figure 4.2 Unrooted tree showing the phylogenetic interrelationships of streptococci. The tree was based on a comparison of *c.* 1340 nucleotides (ranging from position 107 [G] to position 1431 [A] of the *E. coli* numbering system [1]). The evolutionary distance between any two species is the sum of the horizontal lines between them. Bar – K_{nuc}, 10^2. (From Bentley *et al.* (1991), reproduced by permission.)

also been isolated from the mouths of several animal species resident in a zoo (Dent *et al.*, 1978), but there is little available information about the streptococcal flora of wild animals living in their natural habitats. One recent study on feral goats in Australia led to the isolation of a new species from the rumen, *S. caprinus*, which is capable of degrading tannic acid–protein complexes (Brooker *et al.*, 1994). This unusual metabolic activity is significant because the goats browse on tannin-rich *Acacia* species. Tannin-degrading streptococci, identified as *S. bovis* biotype 1, have also been isolated from the caecum of koalas (Osawa and Mitsuoka, 1990).

A summary of the distribution of streptococcal species, albeit incomplete, is given in Table 4.2, together with their main disease association where known.

Studies on the streptococcal flora of the human oral cavity have shown that different species have a predilection for colonizing particular surfaces, such as the epithelia of the tongue, cheek and palate, or the hard tissues of the teeth (Hardie and Marsh, 1978a; Marsh and Martin, 1992). Several factors may be responsible for these variations, including surface structures (Handley, 1990), different adherence mechanisms (Gibbons and Van Houte, 1975), bacterial coaggregation (Kolenbrander and Andersen, 1986; Kolenbrander and London, 1993), and evasion of host mechanisms, such as by the production of IgA1 proteases (Kilian *et al.*, 1989b). The colonization of clean tooth surfaces to form dental plaque follows a recognized sequence of events (microbial succession), with species such as *S. mitis*, *S. oralis* and *S. sanguis* being prominent amongst the earliest bacteria to become established at these sites (Nyvad and Kilian, 1987). The ecological distribution of currently recognized streptococcal species in the mouth has been reported recently by Frandsen *et al.* (1991).

Less information is available about the streptococcal flora of the intestine in man and other animals, although several of the species normally associated with the oral cavity are known to be present, usually in low numbers (Mead, 1978). Similarly, there is limited information about the streptococci associated with the genitourinary tract.

4.9.2 Human diseases

The genus *Streptococcus* includes several species that are important pathogens in humans (Parker, 1978; Colman, 1990). In addition to the more highly virulent species, such as *S. pyogenes*, *S. pneumoniae* and *S. agalactiae*, many of the oral streptococci are capable of acting as opportunistic pathogens under appropriate circumstances. Because so many of the recognized species are found, at least on occasions, to be associated with disease, it is not possible to regard any of them as totally non-pathogenic. However, there are clearly different levels of disease-causing potential, and the more overt pathogens are known to possess a

Table 4.2 Ecological distribution and disease associations of streptococci

Species*	Main host	Main location†	Known disease associations
Oral streptococci			
S. mutans (–, E)	Man	Mouth, faeces	Caries, endocarditis
S. sobrinus (–)	Man	Mouth	Caries
S. cricetus (–)	Hamster, rats, man	Mouth	Caries
S. rattus (–)	Rats, man	Mouth	Caries
S. macacae (–)	Monkeys	Mouth	Caries (?)
S. downei (–)	Monkeys	Mouth	Caries
S. ferus (–)	Rats	Mouth	Caries
S. salivarius (–, K)	Man, animals	Mouth	Occasionally endocarditis
S. vestibularis (–)	Man	Mouth	–
S. thermophilus (–)	Milk, dairy products	?	–
S. intermedius (–, F, G)	Man	Mouth, URT	Abscesses
S. constellatus (–, A, C)	Man	Mouth, URT	Abscesses
S. anginosus (–, F, A, C, G)	Man	Mouth, URT, vagina	Abscesses
S. sanguis (–, H‡)	Man	Mouth, URT	Endocarditis
S. gordonii (–, H‡)	Man	Mouth, URT	Endocarditis
	Man	Mouth, URT	Endocarditis
S. parasanguis (–)	Man	Mouth, URT	?
S. crista (–)	Man	Mouth, URT	Endocarditis, infections in immunological compromised patients
S. oralis (–)			
S. mitis (–, K, O)	Man	Mouth, URT	Endocarditis, infections in immunological compromised patients
S. pneumoniae (–)	Man (domestic animals)	Mouth, URT	Pneumonia, meningitis, sinusitis, RTI, conjunctivitis, otitis media (occasional infectiosn in animals)
S. adjacens (–)	Man	Throat, urine	Endocarditis
S. defectivus (–)	Man	Throat, UGT, intestine	Endocarditis

Pyogenic streptococci			
S. pyogenes (A)	Man	Throat	Pharyngitis, tonsillitis, scarlet fever, pyoderma, invasive infections (rheumatic fever and acute glomerulonephritis are late complications)
S. canis (G)	Dogs, cats, cows and other animals	Skin, upper respiratory tract, udder	Mastitis
S. agalactiae (B)	Man	Genital tract, URT, faeces	Neonatal meningitis, septicaemia
S. agalactiae (B)	Cattle	Udder, milk	Mastitis
S. dysgalactiae (C)	Cattle	Udder, tonsils	Mastitis
S. dysgalactiae (C)	Man	URT. vagina, skin	?
S. dysgalactiae (C)	Pigs	Tonsils	
S. dysgalactiae (L)	Pigs		
S. dysgalactiae (G)	Man		
S. parauberis (–, E)	Cattle, milk	Lips, udder, skin	Mastitis
S. uberis (–, E)	Cattle	Udder, tonsils, lips, milk	Mastitis
S. porcinus (E, P, U, V)	Pigs	?	Cevical lymph node abscesses, pneumonia, septicaemia
S. iniae (–)	Freshwater dolphins	?	Subcutaneous abscesses
S. equi (C)	Horses, donkeys	?	Equine strangles, submaxillary gland abscess
S. hyointestinalis (–)	Pigs	Intestine	?
Other streptococci			
S. alactolyticus (D)	Pigs, chickens	Alimentary tract	?
S. bovis (D)	Cows, sheep, pigs, man, milk, dogs, pigeons	Alimentary tract, faeces, tonsils	Endocarditis in man (possibly colon cancer)
S. equinus (D)	Horses, other animals	Alimentary tract	?
S. suis (D, R, S, T)	Pigs, cattle, cats, dogs	Alimentary tract, tonsils	Bacteraemia, meningitis, respiratory disease
S. acidominimus (–)	Cattle	Vagina, skin, raw milk	?
S. intestinals (–, G)	Pigs	Alimentary tract	?
S. caprinus (–)	Feral goats	Rumen	?

*Lancefield Group antigens found.
†URT, upper respiratory tract; UGT, urogenital tract; RTI, respiratory tract infections; and ?, location or disease associations uncertain.
‡Group H varies according to immunizing strain employed.

number of important virulence determinants. As described by Parker (1978), streptococcal infections can present clinically in a number of ways. In some cases the infection is localized to a particular anatomical site, usually causing acute inflammation in the local tissues, but systemic spread, sometimes leading to the development of septicaemia, may also occur. Pus formation, as abscesses in various organs or within body cavities, is another characteristic of many streptococcal infections, hence the term pyogenic streptococci. An important feature of *S. pyogenes* infections, as described below, is the possible development of non-septic complications, such as scarlet fever, rheumatic fever and glomerulonephritis, following the initial acute condition.

4.9.2.1 Streptococcus pyogenes. The Lancefield Group A streptococcus, as *S. pyogenes* is often designated, is the most common cause of streptococcal infections in humans. It is highly communicable and can give rise to outbreaks or epidemics in susceptible populations. Patients most frequently present with either sore throat (pharyngitis, tonsillitis) or skin lesions (impetigo, pyoderma) as the primary infection, which may become invasive and lead to bacteraemia or septicaemia.

S. pyogenes produces a number of toxins and virulence determinants, including haemolysins (streptolysin O and streptolysin S), erythrogenic toxins (pyrogenic exotoxins), streptokinase (fibrinolysin), nucleases, hyaluronidases, proteinase, serum-opacity factor, nicotinamide adenine dinucleotidase and neuraminidase (Maxted, 1978; Colman, 1990). Group A streptococci also possess various surface proteins, such as the M, T and R proteins, which are utilized in serological typing schemes and may also be important virulence factors. M proteins are recognized as particularly significant because of their antiphagocytic activity and have been studied in considerable detail, several of their *emm* genes having been cloned and sequenced (Fischetti, 1989; Kehoe, 1991; Ferretti, 1992). There are over 90 known M antigens, and identification of these, together with the T antigens, forms the basis of current typing methods for investigating the epidemiology of Group A streptococcal infections (Colman, 1990; Colman *et al.*, 1993).

Some of the newer aspects of work on the pathogenicity have been discussed recently by Kehoe (1991). In this helpful review of an increasingly complex topic, the significance of some previously unrecognised factors, such as C5a peptidase is highlighted, in addition to current theories about the role of M, T and R proteins, adhesins, immunoglobulin-binding proteins, streptolysins, hyaluronidase, streptokinase and the pyrogenic exotoxins. At least one of the toxins (SPE A) is now considered to be a 'superantigen', grouped together with the staphylococcal enterotoxins, TSST-1 and exfoliative toxin (Marrack and Kappler, 1990).

One of the damaging and perplexing features of diseases caused by *S. pyogenes* is the possibility of developing later (post-streptococcal) complications following a primary infection of the upper respiratory tract or skin. Such conditions include scarlet fever (in which a skin rash is induced by the production of erythrogenic toxin in some strains), erysipelas (an erythematous skin lesion, usually on the face), rheumatic fever, acute glomerulonephritis, and toxic shock-like syndrome (Parker, 1978; Stevens *et al.*, 1989). The pathogenesis of these conditions which may follow *S. pyogenes* infections is complex and involves damaging immunological responses to the streptococci, either as a result of antigenic cross-reactivity between bacterial and host antigens (in the heart in rheumatic fever) or because of the deposition of antigen–antibody complexes in the kidney (in glomerulonephritis) (Maxted, 1978).

The prevalence of different serotypes of *S. pyogenes* in the UK over the period 1980–1990 has recently been reported, and shows that some M types (such as M1 and M49) are more often associated with epidemics. Serotypes M1 and M3 were found most commonly in invasive and fatal infections (Colman *et al.*, 1993), whilst M80 and M81 were most often isolated from patients with pyoderma. It has long been known that there is a connection between the type of *S. pyogenes* causing infection and the subsequent development of glomerulonephritis or rheumatic fever (Maxted, 1978). The occurrence of these serious, life-threatening conditions had declined in developed countries, but epidemiological studies have indicated a resurgence in recent years in several different geographical areas (Kaplan, 1992). Thus it is important to maintain surveillance on the pattern of streptococcal infections, both by *S. pyogenes* and other species, in order to detect significant shifts in their distribution. Such changes may develop as a result of alterations in the organisms themselves, or in the susceptibility of their human hosts (Barnham, 1989).

The UK is recovering from a feverish outburst of press and media interest in a small series of reports on serious cases of necrotizing fasciitis caused by *S. pyogenes*, emotively reported as 'The Killer Bug' and the 'Flesh-Eating bug (or virus!)'. Although some of these cases were from the same geographical area, no epidemiological connection between the strains isolated has so far been reported (Anon, 1994a, b). However, such episodes do serve to remind people of the potential seriousness of infections with Group A streptococci.

4.9.2.2 Streptococcus agalactiae. The streptococci of Lancefield Group B are associated with septicaemia and other infectious meningitis in man, and are a particularly important cause of infection in neonates (Jelinková, 1977; Ross, 1978; Henrichsen, 1985). Both early onset (24–36 h up to 5 days) and late onset (10 days or more after birth) forms of neonatal disease are recognized, the former having a considerably higher mortality rate due to rapid, fulminating progression of the infection. The most likely source of Group B streptococci in neonatal infections is the genital tract of the mother, although these bacteria are also carried in the intestine and upper respiratory tract (Ross, 1978).

Four serotypes of Group B streptocococci (Ia, Ib, II, III) were originally described by Lancefield (1934, 1938), but additional types (IV, V) have subsequently been identified, based on capsular polysaccharide antigens. Further serological subdivision is possible, using the protein antigens C, R and X (Rotta, 1986; Motlova *et al.*, 1986). The capsular antigens are thought to be virulence factors although several other potential virulence determinants have been investigated (Wibawan and Lämmler, 1991; Orefici, 1992). For further differentiation between strains within serotypes, a phage typing scheme has been described (Stringer, 1980), and, more recently, molecular typing methods based on pulsed-field electrophoresis and restriction enzyme analysis of chromosomal DNA have been reported (Gordillo *et al.*, 1993).

Strains of *S. agalactiae* are usually sensitive to penicillin, although some tolerant strains have been isolated. Resistance to tetracyclines and macrolides is not uncommon and may be determined by the presence of plasmids (Colman, 1990).

4.9.2.3 Streptococcus pneumoniae. Although phylogenetically and taxonomically close to some of the oral streptococci, it is necessary to distinguish the pneumococci because of their important role as a human pathogen. In the diagnostic laboratory, the pneumococci are usually differentiated from other α-haemolytic streptococci by observing a zone of inhibition around a paper disc containing 5 μg of optochin (ethylhydrocupreine). These cocci are normally arranged in pairs and surrounded by a polysaccharide capsule. Over 80 distinct types of capsular antigens have been described which form the basis of a serological typing scheme. The cell walls of pneumococci possess a choline-containing ribitol teichoic acid, referred to as the C-substance or C-polysaccharide, which has also been detected serologically in strains of the closely related species *S. oralis* and *S. mitis* (Gillespie *et al.*, 1993). Although serological typing has usually been employed for epidemiological investigations, DNA fingerprinting by means of pulsed-field gel electrophoresis can be considered as an alternative approach (Lefevre *et al.*, 1993). DNA probes and PCR-based methods for diagnosing *S. pneumoniae* infections have also been described (Denys and Carey, 1992; Rudolph *et al.*, 1993). *Streptococcus pneumoniae* infections are particularly important in the very young and the very old, and in patients who are debilitated in some way. It is the major cause of community-acquired pneumonia, especially lobar pneumonia, and is commonly involved in meningitis, sinusitis, and otitis media. Less frequently, pneumococci are found as aetiological agents in peritonitis, infective endocarditis and suppurative arthritis (Roberts, 1985; Colman, 1990).

Notwithstanding the role of the pneumococcus as a major human pathogen, it is also found as part of the normal commensal flora of the nasopharynx. How the organism reaches other sites, such as the lung, in order to initiate disease is not clearly understood at present (Brusse, 1991; Johnston, 1991).

A number of potential virulence factors from pneumococci have been described and the extensive published literature on them has been reviewed recently by Boulois (1992). The importance of the capsule has been recognized for many years and it is known to protect the pneumococci from phagocytosis by host cells. Vaccines based on selected capsular antigens have been effective in some population groups, although these polysaccharides are often only weakly immunogenic and may elicit poor antibody responses. Other virulence factors include the pneumococcal surface protein A, neuraminidase, the toxin pneumolysin, and autolysin. The latter cell-wall-associated enzyme may be particularly important because it facilitates the release of some of the other factors, such as pneumolysin and neuraminidase, which are located in the cytoplasm (Boulnois, 1992).

Since 1967 there has been an increasing number of pneumococci that are resistant to penicillin and other antibiotics (Hansman and Bullen, 1967; Spika *et al.*, 1991). Because of difficulties with detection of penicillin resistance by conventional disc-sensitivity tests, particular care must be taken in selecting the appropriate methodology (Marshall *et al.*, 1993).

4.9.2.4 The oral streptococci. Although the various species that have been grouped together within the oral streptococci are generally found as part of the commensal flora of the mouth and upper respiratory tract, almost all of them have been implicated as opportunistic pathogens. The types of infections associated with these streptococci include local conditions, such as dental caries and a variety of inflammatory conditions in the mouth, as well as more distant effects exemplified by infective endocarditis and abscesses in various organs (Hardie and Whiley, 1992). Some species, including *S. mitis*, *S. sanguis* and *S. oralis*, are also increasingly being recognized as aetiological agents in infections of immunologically compromised patients (Hardie and Whiley, 1994). Unlike the situation with some pathogens, it is generally not possible to link each individual species in this group with one specific disease for which it is the sole aetiological agent. In the following section, some examples of infections associated with oral streptococci are briefly discussed.

4.9.2.5 Infective endocarditis. Infective endocarditis, a serious and life-threatening infection of the endocardium of the heart, may be caused by a large variety of microoganisms. The condition can occur in an acute form in patients with previously undamaged heart valves, in which case it is usually associated with highly virulent pathogens such as *Staphylococcus aureus* or *Streptococcus pyogenes*, but it is more commonly found as a subacute disease in patients with pre-existing valvular abnormalities, which may be either congenital or acquired. In the latter situation, streptococci are the most frequently isolated aetiological agents, although many other genera and species have been reported on occasions, including enterococci, staphylococci and several Gram-negative genera.

It has been suggested that the proportion of cases of infective endocarditis due to oral streptococci may have fallen over the last 30 years or so (Bouvet and Acar, 1984), although analysis of several published retrospective surveys still suggests that they account for 60% or more of the total (Hardie and Whiley, 1992). Organisms which appear to have increased in frequency over this period include staphylococci and Lancefield group D cocci (encompassing both *S. bovis* and *Enterococcus* species). Such alterations in proportions of taxa isolated from infective endocarditis may be a reflection of changes in the age distribution of the disease, with increasing numbers of older individuals, and the introduction of new methods of treatment, such as valve replacements and other kinds of cardiac surgery.

It is generally assumed that when oral streptococci are implicated in endocarditis, their most likely portal of entry into the bloodstream is the mouth. It is certainly true that many forms of dental treatment, notably tooth extraction, periodontal surgery and deep scaling, will induce a transient bacteraemia which would put at risk patients with previously damaged and susceptible heart valves. However, even toothbrushing or chewing may carry some risk in people with unhealthy and inflamed gums. Unfortunately, however, it is extremely difficult in most cases to establish a definite cause-and-effect link between any particular treatment episode and the subsequent development of endocarditis in individual patients, and because of the inevitable time delay between these events the evidence is, at best, circumstantial. Nevertheless, the potential danger is recognized and all 'at risk' patients should be protected by appropriate antibiotic prophylaxis when undergoing dental treatment.

Other potential portals of entry for endocarditis-inducing bacteria include the skin, gastrointestinal tract and genitourinary tract, and these are more likely to be the source of infection for organisms such as staphylococci, enterococci, *S. bovis*, and coliforms.

Although there have been many reports on the species of streptococci associated with endocarditis, recent changes in the taxonomy and nomenclature of the oral group make it necessary to re-evaluate such data. From many of these studies it would appear that *S. sanguis* (which may include *S. gordonii*), *S. oralis* (under various names) and *S. mutans* were prominent amongst the streptococci identified, although other species were recorded on occasions. In one recent study of 47 endocarditis isolates, using current terminology, 31.9% were identified as *S. sanguis*, 29.8% as *S. oralis*, and 12.7% as *S. gordonii*, together with smaller numbers of *S. bovis* (6.4%), *S. parasanguis* (4.2%), *S. mutans* (4.2%), *S. mitis* (4.2%) and *S. salivarius* (4.2%) (Douglas *et al.*, 1993).

The nutritionally variant streptococci (NVS), now known as *S. defectivus* and *S. adjacens*, are estimated to account for 5% or more of cases of streptococcal endocarditis (Ruoff, 1991). Because of their requirement for growth media supplemented with pyridoxal, it is quite likely that these species have been underestimated or missed altogether in some studies.

Considerable interest has been shown in the possible virulence determinants of streptococci associated with infective endocarditis (Hardie and Whiley, 1994). Several features have been considered, including the production of extracellular polysaccharides, aggregation of platelets, and attachment to cellular components such as fibronectin and laminin (Herzberg *et al.*, 1990; Tart and Van de Rijn, 1991; Sommer *et al.*, 1992; Douglas *et al.*, 1993; Manning *et al.*, 1994). Further studies on such mechanisms, coupled with more extensive epidemiological surveys using the currently accepted classification schemes, should help in the search for more effective ways of preventing and treating this devastating disease.

Diagnosis of infective endocarditis depends to a large extent on isolation and identification of the causative agent from repeated blood cultures. However, the clinician needs a highly developed 'index of suspicion' in order to recognize the often insidious onset of the condition. Once the diagnosis has been made, appropriate antibiotic therapy should be instituted immediately. Early surgical intervention to replace severely damaged heart valves is increasingly recommended as part of the clinical management in many hospitals.

4.9.2.6 'Streptococcus milleri-*group*'. The species *S. anginosus*, *S. constellatus* and *S. intermedius*, which comprise the '*S. milleri*-group' (SMG), are found in the mouth, gastrointestinal and genitourinary tracts as part of the commensal flora, but have increasingly been recognized as significant pathogens (Gossling, 1988; Hardie and Whiley, 1992; Piscitelli *et al.*, 1992). They are mainly associated with purulent conditions, such as abscesses, from which they may be isolated as pure cultures or as part of a polymicrobial infection. The SMG have been reported from various oral infections, including dental abscesses, pericoronitis and Ludwig's angina; brain abscesses; ear, nose and throat infections; thoracic infections; abdominal infections (including liver abscesses); obstetric and neonatal infections; skin and subcutaneous infections; osteomyelitis and septic arthritis; and infections involving muscle (Gossling, 1988).

Because of confusion in the past about the nomenclature of these streptococci, it is not always possible to discern from published reports which of the three species has been isolated from the various clinical conditions. However, more recent studies have indicated that there is some specificity in their distribution (Whiley *et al.*, 1990b, 1992). The high level of association of *S. intermedius* with brain abscesses (23 out of 27 cases, 82%), and the relative frequency of occurrence of *S. anginosus* amongst genitourinary and intestinal isolates were particularly worthy of note. *Streptococcus anginosus* was also the species found most frequently in oral samples when examined on a nalidixic acid-sulphamethazine-containing selective medium (Whiley *et al.*, 1993).

A number of potential tissue-destroying enzymes, like hyaluronidase and chondroitin sulphatase, have been found in streptococci from this group, as well as surface-binding properties which may be of relevance to their pathogenicity (Beighton *et al.*, 1990; Homer *et al.*, 1993; Willcox *et al.*, 1993). There is clearly a need for further studies on these interesting and widespread opportunistic pathogens.

4.9.2.7 Infections in immunocompromised patients. Opportunistic infections are common in immunocompromised patients, whatever the underlying cause of their predisposition, and streptococci feature among the long list of microorganisms which may be involved. In some reports the exact identity of the streptococcal species is unclear, although it is apparent that they often belong to the 'oral' or 'viridans' group. However, from some of the more recent studies, it appears that members of the *S. oralis*-group, especially *S. oralis*, *S. mitis* and *S. sanguis*, are significant isolates from neutropenic patients (Classen *et al.*, 1990; McWhinney *et al.*, 1993; Hardie and Whiley, 1994). It will be interesting to determine from future studies whether there is any specificity in the streptococci associated with septicaemia and the adult respiratory distress syndrome (ARDS) in subjects who are immunocompromised.

4.9.2.8 Dental caries and the Streptococcus mutans *group.* It has been recognized since the last century that dental caries occurs as a result of the fermentation of carbohydrates by oral lactic acid bacteria, the acids they produce causing demineralization of the tooth surface (Miller, 1890). It is also well known that the mixed microbial community which colonizes both surfaces, in the form of dental plaque, is highly complex and contains a wide variety of bacteria with different metabolic activities (Hardie and Bowden, 1974; Marsh and Martin, 1992). A more detailed account of the lactic microflora of the oral cavity, and of their role in dental caries, is given in Volume 1 of this series (Hogg, 1992).

Many of the bacteria present in dental plaque, including several species of streptococci and lactobacilli, are capable of producing sufficient acid to decalcify dental enamel, and a number of these are also able to induce dental caries in experimental animals (Hardie, 1992). The search for a specific aetiological agent in human dental caries has largely been concentrated on the *S. mutans*-group, especially *S. mutans* and *S. sobrinus*, which are the species most commonly found in man. There is a considerable body of published work implicating these species in the disease, and many aspects of their biochemistry, physiology, antigenic structure, epidemiology and pathogenicity have been investigated (Hamada and Slade, 1980; Loesche, 1986; de Soet *et al.*, 1992). Increasingly such studies have used molecular approaches and several putative virulence determinants have been examined by genetic methods (Russell, 1994). Molecular typing methods have been applied successfully to the study of transmission of mutans streptococci, confirming earlier observations that these organisms are usually acquired by infants from their mothers (Caufield *et al.*, 1993).

Prevention of dental caries can be achieved in most cases by measures such as restriction of dietary sugar intake, use of systemic and topically applied fluorides, toothbrushing with fluoride toothpastes, and application of fissure sealants to susceptible teeth. The search for a caries vaccine, based on *S. mutans* antigens, has produced a large amount of valuable and interesting experimental data which have helped to develop understanding of the caries process and of immune responses in the mouth (Krasse *et al.*, 1987; Russell and Johnson, 1987; Klein and Scholler, 1988). However, despite successful results in experimental animals, no human trials of active immunization with such vaccines have so far been reported. Concerns about potential safety problems, and the availability of other effective caries preventive measures, have probably contributed to this lack of progress. However, the possibility of using preformed antibodies for the purposes of passive immunization against cariogenic streptococci is still under investigation in some laboratories.

Estimation of salivary levels of mutans streptococci and lactobacilli has been used as a method of assessing caries risk in patients, although this approach has not been universally recommended or adopted (Krasse, 1988; Johnson, 1991; Hardie, 1992). Commercial kits for this purpose have been developed and can be used by the dental practitioner, requiring only facilities for incubation of cultures (Davenport *et al.*, 1992). Levels of mutans streptococci of 2.5×10^5 cfu/ml of saliva, or greater, are often taken as being indicative of 'high risk' of caries, and may well be correlated to a high frequency of sucrose intake (Hardie, 1992).

4.9.3 Animal diseases

As indicated in Table 4.2, many species of streptococci are found in animals and some are responsible for important diseases. A more detailed account of streptococcal infections in animals can be found elsewhere (Buxton and Fraser, 1977; Wilson and Salt, 1978).

4.9.3.1 Cattle. Bovine mastitis is the most common and economically important condition in cattle which is frequently caused by streptococci. Species most often implicated in mastitis are *S. agalactiae*, *S. uberis*, *S. parauberis*, *S. dysgalactiae*, *S. canis*, although others are occasionally reported. Other types of streptococcal infections in cattle are less common, but may give rise to endocarditis, abortion, genitourinary infections or arthritis (Wilson and Salt, 1978).

4.9.3.2 Pigs. Streptococcal infections in pigs can take various forms, such as meningoencephalitis, arthritis, cervical lymphadenitis, endocarditis, abscesses, pneumonia and septicaemia. Several of the streptococci that have been reported in older studies on pig isolates as belonging to various Lancefield groups (e.g. E, P, U, V, R, S, T) have more recently been classified as either *S. porcinus* or *S. suis* (Devriese, 1991).

4.9.3.3 Horses. Equine strangles and other manifestations of streptococcal infection in horses are caused mainly by the Lancefield group C streptococci which belong to the species *S. equi*. The disease is characterized by the production of pharyngeal and submaxillary abscesses (Wilson and Salt, 1978).

4.9.3.4 Sheep. Suppurative arthritis, sometimes followed by bacteraemia and the development of endocarditis due to *S. dysgalactiae* (Lancefield group C), has been reported in lambs, but does not appear to be a common problem (Wilson and Salt, 1978). Ewes may develop mastitis, which can be caused by *S. agalactiae*.

4.9.3.5 Poultry and birds. Streptococci are not normally regarded as a major problem in poultry, although infections with *S. equi* subsp. *zooepidemicus* have been recorded. Recently, it has been shown that *S. bovis* is an important cause of septicaemia in pigeons (Devriese *et al.*, 1990).

4.10 Identification

The genus *Streptococcus* has rapidly undergone major taxonomic revision within recent years and currently consists of 39 recognized species. Undoubtedly the application of nucleic acid analyses to these studies has greatly advanced our understanding of the genetic relationships of these species at both intra- and intergeneric levels, as discussed earlier in this chapter. However, the rapid improvements in the classification of the genus have not been equalled in pace by the development of comprehensive identification schemes using phenotypic tests nor by the construction and application of species specific DNA probes. That this situation will undoubtedly improve has already been indicated by recent advances on both phenotypic and genotypic fronts; the shift in emphasis away from traditional tests such as Lancefield grouping reactions, production of haemolysis on blood agar and physiological tests for differentiating streptococci and the use, for example, of fluorogenic and chromogenic substrates for the rapid detection of preformed enzyme activities, has characterized several identification schemes aimed at particular species groups (Kilian *et al.*, 1989a; Whiley *et al.*, 1990b; Beighton *et al.*, 1991) as well as the complete genus (Freney *et al.*, 1992). In the latter investigation, a 32 test commercial, rapid identification kit was used, consisting of chromogenic substrates for detecting glycosidases together with carbohydrate fermentation tests, arylamidase reactions, alkaline phosphatase, arginine hydrolase, acetoin production, hippurate hydrolysis and urease production. This approach has the advantage of combining a sufficiently large number of tests to enable the identification of most of the recognized species of streptococci, together with a standardized test formulation to give increased confidence when comparing results between laboratories. The performance of the test kit was commendable with 413/433 (95.4%) of strains correctly identified, including 109 stains which required further tests for complete identification (16 strains remaining unidentified and four strains misidentified). Alternative approaches that have been applied to the problem of streptococcal identification include whole cell derived polypeptide patterns by SDS-PAGE (Whiley *et al.*, 1982), pyrolysis–mass spectrometry (Winstanley *et al.*, 1992) and monoclonal antibodies (de Soet *et al.*, 1990), although, as yet, none of these has been evaluated for the complete genus.

Some progress has also been made using DNA based approaches: restriction fragment polymorphisms of whole chromosomal digests stained with ethidium bromide (Rudney *et al.*, 1992) as well as RFLPs of rRNA genes (ribotyping) (Rudney and Larson, 1993), have been attempted but the ease of use of these techniques, especially in studies involving large numbers of strains, remains untested. The extensive application of ribosomal RNA sequencing in phylogenetic investigations of the streptococci and related genera in recent years has carried with it the implicit promise of allowing construction of species specific probes to facilitate identification. The study by Bentley *et al.* (1991) represents the most thorough application of this technology to date with most, but not all, streptococcal species sequenced. The extension of this strategy to the provision of 'working' rDNA probes has begun with the recent description of oligonucleotide probes for the differentiation of *Streptococcus uberis* and *S. parauberis* using PCR and dot blot formats (Bentley *et al.*, 1993; Harland *et al.*, 1993). However, the full impact of nucleic acid based techniques on routine laboratory identification of presumptive streptococci is still awaited.

4.11 Isolation and enumeration

Growth on non-enriched nutritient agar is usually poor, most species growing best on media supplemented with blood, serum, or with carbohydrates such as glucose or sucrose (Colman, 1990). A number of selective agents have been used in some isolation media, including crystal violet, thallous acetate, sodium azide (Hardie, 1986). More specific media, for isolation of particular groups or species of streptococci, have also been described (e.g. Barnes *et al.*, 1978; Whiley *et al.*, 1993), incorporating a variety of different antimicrobial agents as selective agents. For routine purposes, a good quality, non-selective blood agar will support the growth of most, if not all, species and also allows the recognition of haemolysis around the colonies. For selective isolation of extracellular polysaccharide producing streptococci, such as *S. bovis* and several of the oral streptococci, a sucrose-containing agar is useful (Hardie and Marsh, 1978b). Todd–Hewitt broth is commonly used as a liquid growth medium for streptococci.

4.12 Maintenance and preservation

Strains can usually be maintained by regular subculture on appropriate media. Many strains will survive storage for several days or even weeks on plates, either at room temperature or at 4°C. Streptococci can also be kept in agar stabs or in litmus milk + 1% chalk + 0.3% yeast extract + 1% glucose (Garvie *et al.*, 1981).

Long-term preservation can be achieved by freezing at −70°C or in liquid nitrogen, conveniently on beads, or by freeze-drying using standard methods (Hardie, 1986).

4.13 Species of the genus *Streptococcus*

All species of the genus *Streptococcus* are Gram-positive cocci, which may be spherical or ovoid in shape and are usually arranged in chains or pairs. They are non-motile and do not form endospores. Most are facultatively anaerobic, but some strains require CO_2 for growth, particularly on initial isolation. They are chemo-organotrophs, ferment carbohydrates with the production of lactic and other acids, and have complex nutritional requirements. They are catalase negative. The mol% G+C of the DNA is in the range 34–46%, and the type species is *Streptococcus pyogenes* (Rosenbach, 1884). Most streptococcal species occur as commensals or parasites on man and other animals, and several are highly pathogenic. Further descriptive details may be found in other reference works, such as *Bergey's Manual of Systematic Bacteriology* (Sneath *et al.*, 1986) and *The Prokaryotes* (Balows *et al.*, 1992).

Each of the 39 currently recognized species is listed alphabetically and described briefly in the following sections. Detailed phenotypic characteristics of these species are given for the oral, pyogenic, and 'other' groups of streptococci in Tables 4.3, 4.4 and 4.5, respectively.

4.13.1 Streptococcus acidominimus

First described by Ayers and Mudge (1922) from bovine udders, the taxonomic position of *S. acidominimus* has remained uncertain. Jones (1978) included this species within the 'other streptococci', a term used for a small group of mainly α-haemolytic streptococci not included within the pyogenic, oral, faecal, lactic or anaerobic streptococcal groups recognized at that time. Wilson and Miles (1975) considered *S. acidominimus* to be a variant of *S. uberis* but by 16S rRNA analysis this has been shown not be the case. Cells are cocci occurring in short chains. α-Haemolysis is produced on blood agar. Strains are weakly fermentative with most failing to decrease the pH of the growth medium below 6.0. The biochemical reactions of this species are shown in Table 4.5. DNA G+C content is 40 mol%. No group specific antigen has been demonstrated. 16S rRNA sequence analysis (Bentley *et al.*, 1991) has shown that *S. acidominimus* does not group with any other species with the possible exception of *S. suis*. Source/habitat: bovine vagina, skin of calves and raw milk. Type strain NCDO 2025.

4.13.2 Streptococcus adjacens

Streptococcus adjacens (Bouvet *et al.*, 1989) is one of two currently recognized species of nutritionally variant (pyridoxal dependant) streptococci (NVS). These clinically important streptococci were originally assumed to be variants of some already recognized α-haemolytic streptococci but taxonomic studies have shown them to be distinct species in their own right. Cells are 0.4–0.6 μm in diameter, small ovoid cocci occurring in chains of variable length, in pairs or singly in CDMT semi-synthetic medium. Stationary phase cells may tend to be rod shaped. However, strains may produce cocci, coccobacilli and rods within chains during growth on pyridoxal or cysteine-supplemented broth. Strains are

Table 4.3 Biochemical characteristics of the oral streptococci and closely related species*

	Streptococcus mutans group†						
	S. mutans	*S. sobrinus*	*S. cricetus*	*S. rattus*	*S. macacae*	*S. downei*	*S. ferus*
Acid from							
N-Acetyl-glucosamine	+	−	+	+	+	NT	+
Aesculin	+	−	+	+	NT	NT	+
Amygdalin	+(−)‡	−	+	+	+	NT	+
Arbutin	+	−	+	+	NT	NT	+
Cellobiose	+	+(−)‡	+	+	+	−	+
Erythritol	−	−	−	−	NT	NT	−
Fructose	NT	NT	NT	NT	+	+	NT
Galactose	+	−(+)	+	+	+	+	+
Glycerol	−	−	−	−	−	−	−
Glycogen	−	−	−	−	NT	−	−
Inulin	+	+(−)	+	+	−	+	−
Lactose	+	+(−)	+(−)	+(−)	NT	+	+
Maltose	+	+	+	+	+	+	+
Mannitol	+	+(−)	+	+	+	+	+
Melibiose	+(−)	−	+	+	−	+	−
Methyl-D-glucoside	+(−)‡	−	−	−	NT	NT	−
Pullulan	−	−	NT	NT	NT	−	NT
Raffinose	+	−	+	+	+	−	−
Ribose	−	−	−	−	−	NT	−
Salicin	+	−	+	+	NT	+	+
Sorbitol	+	−(+)‡	+	+	+	−	+
Starch	−	−	−	−	−	−	+
Tagatose	+(−)	+	NT	NT	NT	NT	NT
Trehalose	+	+	+(−)	+	+	+	+
Hydrolysis of							
Aesculin	+	+(−)	+(−)	+	+	−	+
Starch	+	NT	−	−	+§	−	NT
Arginine	−	−	−	+	−	−	−

Production of							
α-Arabinosidase	–	–	NT	NT	NT	NT	NT
Acid phosphatase	–(+)§	NT	NT	NT	NT	NT	NT
Alkaline phosphatase	–	–	NT	NT	NT	–	NT
α-Fucosidase	–	–	NT	NT	NT	NT	NT
β-Fucosidase	–	–	NT	NT	NT	NT	NT
α-Galactosidase¶	+	–	NT	NT	NT	NT	NT
β-Galactosidase¶	–	–	NT	NT	NT	NT	NT
Glycyl-tryptophan arylamidase	–	–	NT	NT	NT	NT	NT
α-Glucosidase¶	+	+	NT	NT	NT	NT	NT
β-Glucosidase¶	+	–	NT	NT	NT	NT	NT
β-Glucuronidase	–	–	NT	NT	NT	NT	NT
Neuraminidase	–	–	NT	NT	NT	NT	NT
N-Acetyl-*β*-glucosaminidase	–	–	NT	NT	NT	NT	NT
N-Acetyl-*β*-galactosaminidase	–	–	NT	NT	NT	NT	NT
Leucine arylamidase	NT	NT	NT	NT	NT	NT	NT
Pyrolidonyl arylamidase	–	–	NT	NT	NT	NT	NT
Valine arylamidase	NT	NT	NT	NT	NT	NT	NT
Urease	–	–	–	–	NT	NT	–
Hydrogen peroxide	–	+	–	–	–	–	–
Extracellular polysaccharide	+	+	+	+	+	–	+
IgA protease	–	NT	NT	NT	NT	NT	NT
Hyaluronidase	–	–	NT	NT	NT	NT	NT
Acetoin (VP)	+	+	+	+	+	+	+(–)
Amylase binding	–	–	NT	NT	NT	NT	NT

*Species are ordered in the table according to data from Bentley *et al.* (1991). No species reported produces acid from adonitol, arabinose, arabitol, cyclodextrin, dulcitol, gluconate, rhamnose, sorbose or xylose. All species produce acid from glucose, mannose and sucrose. No species hydrolyses hippurate.

Data taken from Anon (1991); Beighton *et al.* (1984, 1991); Bouvet *et al.* (1989); Colman and Ball (1984); Colman and Williams (1972); Coykendall (1983); Handley *et al.* (1991); Hardie (1986); Hardie and Whiley (1992); Kilian *et al.* (1989a); Kilpper-Bälz *et al.* (1985); Kral and Daneo-Moore (1981); Whiley (1987); Whiley, R.A., unpublished data; Whiley and Beighton (1991); Whiley and Hardie (1988, 1989); Whiley *et al.* (1990a, b).

†+, >90% of strains give a positive result; +(–), 50–89% of strains give a positive result; –(+), 11–49% of strains give a positive result; –, <10% of strains give a positive result; V, reported as 'variable'; and NT, not tested.

‡Proportion of strains reported as giving a positive result for this test differs between studies.

§Weak or slow reaction given by some strains.

¶Variation in results obtained depending on method of testing.

Table 4.3 *continued*

	S. salivarius group†			*S. milleri* group†		
	S. salivarius	*S. vestibularis*	*S. thermophilus*	*S. intermedius*	*S. constellatus*	*S. anginosus*
Acid from						
N-Acetyl-glucosamine	+	+(−)	−	+	−(+)	−(+)
Aesculin	+	NT	+	NT	NT	NT
Amygdalin	+	+(−)	−	+(−)	−(+)	+
Arbutin	+	−(+)	V	+	+	+
Cellobiose	+	V	−	+(−)	−(+)	+(−)
Erythritol	NT	NT	−	NT	NT	NT
Fructose	+	+	+	NT	NT	NT
Galactose	−(+)	+	V	NT	NT	NT
Glycerol	−	−	−	−	−	−
Glycogen	−	−	−	−	−	−
Inulin	+(−)	−	−	−	−	−
Lactose	+(−)	+(−)	+	+	+(−)	+
Maltose	+	+	−	+	+	+
Mannitol	−	−	−	−	−	−(+)
Melibiose	−	−	V	−	−	−
Methyl-D-glucoside	+(−)	−(+)	−	−(+)	+(−)	+
Pullulan	+	−	−	+	−	+(−)
Raffinose	−	−	V	−	−	−(+)
Ribose	−	−	V	−	−	−
Salicin	+	+	−	+(−)	+(−)	+§
Sorbitol	−	−	−	−	−	−
Starch	−	−	NT	NT	NT	NT
Tagatose	−	−	−	−	−	−
Trehalose	−(+)	V	−	+	+(−)	+
Hydrolysis of						
Aesculin	+	+(−)	−	+	+§	+
Starch	+	NT	V	NT	NT	NT
Arginine	−	−	−	+	+	+

Production of						
α-Arabinosidase	+	+	NT	−	−	−
Acid phosphatase	+	NT	−	NT	NT	NT
Alkaline phosphatase	+	−	−	+	+	+
α-Fucosidase	−	−	−	+	−	−
β-Fucosidase	+(−)	−	NT	+	−	−
α-Galactosidase¶	−	−	−	−	−	−(+)
β-Galactosidase¶	+(−)	+	+	+	−	−(+)
Glycyl-tryptophan arylamidase	NT	NT	NT	+(−)	−	−
α-Glucosidase¶	+	+(−)	−	+	+	−(+)
β-Glucosidase¶	+	−	−	−(+)	−	+
β-Glucuronidase	−	−	−	−	−	−
Neuraminidase	−	−	NT	+	−	−
N-Acetyl-*β*-glucosaminidase	−	−	−	+	−	−
N-Acetyl-*β*-galactosaminidase	−	−	NT	+	−	−
Leucine arylamidase	+	NT	+	NT	NT	NT
Pyrolidonyl arylamidase	−	−	−	−	−	−
Valine arylamidase	+	NT	−	NT	NT	NT
Urease	−(+)	+	NT	−	−	−
Hydrogen peroxide	−	+	NT	−	−	−(+)
Extracellular polysaccharide	+	−	NT	−	−	−
IgA protease	−	NT	NT	NT	NT	NT
Hyaluronidase	−	−	NT	+	+(−)	−
Acetoin (VP)	+(−)	−	+(−)	+	+	+
Amylase binding	V	−	NT	−	−	−

*Species are ordered in the table according to data from Bentley *et al.* (1991). No species reported produces acid from adonitol, arabinose, arabitol, cyclodextrin, dulcitol, gluconate, rhamnose, sorbose or xylose. All species produce acid from glucose, mannose and sucrose. No species hydrolyses hippurate.

Data taken from Anon (1991); Beighton *et al.* (1984, 1991); Bouvet *et al.* (1989); Colman and Ball (1984); Colman and Williams (1972); Coykendall (1983); Handley *et al.* (1991); Hardie (1986); Hardie and Whiley (1992); Kilian *et al.* (1989a); Kilpper-Bälz *et al.* (1985); Kral and Daneo-Moore (1981); Whiley (1987); Whiley, R.A., unpublished data; Whiley and Beighton (1991); Whiley and Hardie (1988, 1989); Whiley *et al.* (1990a, b).

†+, >90% of strains give a positive result; +(−), 50–89% of strains give a positive result; −(+), 11–49% of strains give a positive result; −, <10% of strains give a positive result; V, reported as 'variable'; and NT, not tested.

‡Proportion of strains reported as giving a positive result for this test differs between studies.

§Weak or slow reaction given by some strains.

¶Variation in results obtained depending on method of testing.

Table 4.3 *continued*

	S. sanguis/oralis group†							Nutritionally variant streptococci†	
	S. sanguis	*S. gordonii*	*S. parasanguis*	*S. crista*	*S. oralis*	*S. mitis*	*S. pneumoniae*	*S. adjacens*	*S. defectivus*
Acid from									
N-Acetyl-glucosamine	+	+	+	+	+	+	+(−)	NT	NT
Aesculin	+(−)	+	NT	NT	−	−(+)	NT	NT	NT
Amygdalin	−	+	−(+)	−	−	−	−	NT	NT
Arbutin	+	+	−(+)	+	−	−	−	NT	NT
Cellobiose	+(−)	NT	V	NT	−	+(−)	NT	NT	NT
Erythritol	NT	NT	NT	NT	−(+)	NT	−	NT	NT
Fructose	+	+	+	NT	+	+	+	NT	NT
Galactose	+	+	+	NT	+	+	+	NT	NT
Glycerol	−	−	−	NT	−‡	NT	+§	NT	NT
Glycogen	NT	NT	−	NT	V	−	+‡	−	−
Inulin	+	+(−)	−	−	−	−(+)‡	+	V	−
Lactose	+	+	+	+(−)	+	+	+	−	V
Maltose	+	+	+	+	+	+	+	+	+
Mannitol	−	−	−	−	−	−	−‡	−	−
Melibiose	+(−)	−(+)	+(−)	−	+‡	+(−)‡	−‡	−	−
Methyl-D-glucoside	−(+)	+(−)	NT	NT	−	−	−(+)	−	−
Pullulan	+(−)	−	NT	V	+	+(−)	NT	−	+
Raffinose	+(−)	−(+)	+(−)	−	+(−)	+(−)‡	+	−	V
Ribose	−	−	NT	−	+(−)‡	−(+)	−	−	−
Salicin	+	+	V	NT	−‡	−(+)	+(−)	NT	NT
Sorbitol	−(+)	−	−	−	−	−	−	−	−
Starch	−(+)	−	−	NT	−‡	−	−(+)	−	+
Tagatose	NT	NT	NT	−	−(+)	−	−	+(−)	−
Trehalose	+	+	V	+	−(+)	−	+	−	+
Hydrolysis of									
Aesculin	+(−)	+	−(+)	−	−	−	−(+)‡	NT	NT
Starch	+(−)	+(−)	+	NT	+	+(−)	+(−)	NT	NT
Arginine	+	+	+	+(−)	−	+(−)	−‡	−	−

Production of									
α-Arabinosidase	−	−	−(+)	−	−	−	NT	NT	NT
Acid phosphatase	+(−)	+	NT	NT	+	+(−)	NT	NT	NT
Alkaline phosphatase	−	+	+(−)	−	+	V	−	−	−
α-Fucosidase	−	+	−(+)	+	−	−	−(+)	NT	NT
β-Fucosidase	−(+)‡	−	−(+)	−	+§	+	−(+)	NT	NT
α-Galactosidase¶	+(−)	−(+)	+	−	NT	+(−)	+	−	+
β-Galactosidase¶	−(+)	+(−)‡,§	NT	−(+)	+‡	+(−)‡,§	+	−	+‡
Glycyl-tryptophan arylamidase	+(−)	+	NT	NT	+	+(−)	+	NT	NT
α-Glucosidase¶	−	+(−)‡,§	+	−	+	+‡	+	NT	NT
β-Glucosidase¶	−(+)‡,§	+	−(+)	−	−	−	−(+)	−‡	−
β-Glucuronidase	−	−	−	NT	−	−	−	−(+)	−
Neuraminidase	−	−	−	−	+	−(+)	+	NT	NT
N-Acetyl-*β*-glucosaminidase	+(−)‡,§	+	+	+	+	+	+(−)	−(+)	−
N-Acetyl-*β*-galactosaminidase	−	+(−)	+	+	+	−	+	NT	NT
Leucine arylamidase	+	+	+	NT	+	+	−(+)	+	+
Pyrolidonyl arylamidase	−	−	−	NT	−	−	−(+)	+(−)‡	+(−)‡
Valine arylamidase	NT	NT	NT	NT	+	+	NT	NT	NT
Urease	−	−	−	−	−	−	−	−	−
Hydrogen peroxide	+	+	+	+	+	+	+	NT	NT
Extracellular polysaccharide	+	+	−	NT	+(−)	−	−	−	−
IgA protease	+	−	NT	NT	+	−	+	NT	NT
Hyaluronidase	−	−	−	−	−	−	+	NT	NT
Acetoin (VP)	−	−	−	−	−	−	−	V	V
Amylase binding	−	+	+(−)	+	−	+	NT	NT	NT

*Species are ordered in the table according to data from Bentley *et al.* (1991). No species reported produces acid from adonitol, arabinose, arabitol, cyclodextrin, dulcitol, gluconate, rhamnose, sorbose or xylose. All species produce acid from glucose, mannose and sucrose. No species hydrolyses hippurate.

Data taken from Anon (1991); Beighton *et al.* (1984, 1991); Bouvet *et al.* (1989); Colman and Ball (1984); Colman and Williams (1972); Coykendall (1983); Handley *et al.* (1991); Hardie (1986); Hardie and Whiley (1992); Kilian *et al.* (1989a); Kilpper-Bälz *et al.* (1985); Kral and Daneo-Moore (1981); Whiley (1987); Whiley, R.A., unpublished data; Whiley and Beighton (1991); Whiley and Hardie (1988, 1989); Whiley *et al.* (1990a, b).

†+, >90% of strains give a positive result; +(−), 50–89% of strains give a positive result; −(+), 11–49% of strains give a positive result; −, <10% of strains give a positive result; V, reported as 'variable'; and NT, not tested.

‡Proportion of strains reported as giving a positive result for this test differs between studies.

§Weak or slow reaction given by some strains.

¶Variation in results obtained depending on method of testing.

Table 4.4 Biochemical characteristics of the pyogenic streptococci*

	S. pyogenes†	*S. canis*†	*S. agalactiae*†	*S. dysgalactiae*†	*S. parauberis*†
Acid from					
Amygdalin	−	NT	−(+)	NT	+
Arbutin	−	NT	−(+)	NT	+
Cyclodextrin	−(+)	−	−	NT	NT
Dulcitol	−	NT	−	NT	V
Cellobiose	+(−)	NT	−(+)	NT	+
Glycerol	−(+)	NT	−(+)	+(−)	−
Glycogen	−(+)	−	−	V	−
Inulin	−	NT	−	−	V
Lactose	+	+(−)	−(+)	+(−)	+
Maltose	+	+	+	+	+
Mannitol	−	−	−	−	+
Mannose	NT	NT	NT	NT	+
Melezitose	−	−	−	NT	V
Methyl-D-glucoside	+	+	+	NT	−
Methyl-D-xyloside	−	NT	−(+)	NT	−
Pullulan	+(−)	+	+	NT	NT
Raffinose	−	−	−	−	V
Rhamnose	NT	NT	NT	NT	−
Ribose	−	+	+(−)	+	+
Salicin	+	NT	V	V	+
Sorbitol	−	−	−	−(+)	+
Starch	NT	NT	NT	+	NT
Sucrose	+	+	+	+	+
Tagatose	−	−	−(+)	NT	V
Trehalose	+	−(+)	+	+	+

Hydrolysis of					
Aesculin	V	NT	–	–(+)	+
Gelatin	NT	NT	NT	NT	NT
Hippurate	–	–	+	–(+)	V
Starch	NT	NT	NT	NT	NT
Arginine	+	+	+	+	+
Production of					
Alkaline phosphatase	+	+	+	+	+
α-Galactosidase	–	+(–)	V	–	V
β-Galactosidase	–	–	–	–	NT
Glycyl-tryptophan arylamidase	–	–	–	NT	NT
β-Glucosidase	–	–	–	NT	NT
β-Glucuronidase	–(+)	–(+)	V	+	–
N-Acetyl-β-glucosaminidase	–	–	–	NT	NT
Leucine arylamidase	NT	NT	NT	+	+
Pyrolidonly arylamidase	+	–	–	–	+
Urease	–	–	–	NT	NT
Acetoin (VP)	–	–	+	–	+

*Species are ordered in the table according to data from Bentley *et al* (1991). No species reported produces acid from adonitol, arabinose, arabitol, erythritol, gluconate, melibiose, methyl-D-mannoside, sorbose or xylose. All species produce acid from *N*-acetyl-glucosamine, fructose, glucose and galactose.

Data taken from Anon (1991), Collins *et al.* (1984); Devriese *et al.* (1986, 1988); Farrow and Collins (1984b); Pier and Madin (1976).

†+, >90% of strains give a positive result; +(–), 50–89% of strains give a poisitive result; –(+), 11–49% of strains give a positive result; –, <10% of strains give a positive result; V, reported as 'variable'; and NT, not tested.

‡*S. equi* subsp. *zooepidemicus* strains give positive results in these tests.

§Reported results may vary between studies.

Table 4.4 *continued*

	S. uberis†	*S. porcinus*†	*S. iniae*†	*S. equi*†	*S. hyointestinalis*†
Acid from					
Amygadalin	+	+(−)	NT	NT	V
Arbutin	+	+(−)	NT	NT	+
Cyclodextrin	NT	−	NT	+	NT
Dulcitol	−	−	−	NT	−
Cellobiose	+	+(−)	NT	NT	V
Glycerol	−	+(−)	−	−	−
Glycogen	−(+)	−	NT	+	−
Inulin	+	−	−	−	NT
Lactose	+	+(−)	−	−‡	+
Maltose	+	+(−)	NT	+	+
Mannitol	+	+(−)	+	−	−
Mannose	+	+	+	NT	+
Melezitose	−	−	NT	−	NT
Methyl-D-glucoside	−(+)	+(−)	NT	+	−
Methyl-D-xyloside	−	−	NT	NT	NT
Pullulan	NT	+(−)	NT	+	−
Raffinose	−	−	−	−	V
Rhamnose	−	−	−	NT	−
Ribose	+	+(−)	NT	−‡	−
Salicin	+	+(−)	+	+	+
Sorbitol	+	+	−	−‡	−
Starch	NT	V	NT	+	+
Sucrose	+	+(−)	+	+	+
Tagatose	+(−)	−	NT	−	−
Trehalose	+	+	+	−	+

Hydrolysis of					
Aesculin	+	+(−)	+	+(−)	+
Gelatin	NT	V	−	−	−
Hippurate	+	−	−	−(+)§	−
Starch	NT	NT	+	NT	+
Arginine	+	+	NT	+	−
Production of					
Alkaline phosphatase	−(+)	+	NT	+	+
α-Galactosidase	−	+(−)	NT	−	V
β-Galactosidase	−	−	NT	−§	−
Glycyl-tryptophan arylamidase	NT	−	NT	NT	NT
β-Glucosidase	NT	+(−)	NT	−	NT
β-Glucuronidase	+	+	NT	+	−
N-Acetyl-β-glucosaminidase	NT	−	NT	−	NT
Leucine arylamidase	+	+	NT	+	+
Pyrolidonly arylamidase	+	−	NT	−	−
Urease	NT	NT	NT	−	NT
Acetoin (VP)	+	+(−)	NT		+

*Species are ordered in the table according to data from Bentley *et al* (1991). No species reported produces acid from adonitol, arabinose, arabitol, erythritol, gluconate, melibiose, methyl-D-mannoside, sorbose or xylose. All species produce acid from *N*-acetyl-glucosamine, fructose, glucose and galactose.

Data taken from Anon (1991); Collins *et al.* (1984); Debriese *et al.* (1986, 1988); Farrow and Collins (1984b); Pier and Madin (1976).

†+, >90% of strains give a positive result; +(−), 50–89% of strains give a positive result; −(+), 11–49% of strains give a positive result; −, <10% of strains give a positive result; V, reported as 'variable'; and NT, not tested.

‡*S. equi* subsp. *zooepidemicus* strains give positive results in these tests.

§Reported results may vary between studies.

Table 4.5 Biochemical characteristics of species referred to as 'other streptococci'*

	S. alactolyticus†	*S. bovis*†,‡		*S. equinus*†		*S. suis*†	*S. acidominimus*§	*S. intestinalis*†	*S. caprinus*†
Acid from									
N-Acetyl-glucosamine									
Aesculin	+	+	+(−)	+	+	NT	NT	NT	NT
Amygdalin	NT	NT	NT	NT	NT	NT	NT	NT	NT
Arbutin	+(−)	+(−)	−(+)	+(−)	+	NT	NT	NT	NT
Cellobiose	+(−)	+(−)	−(+)	+	+	NT	NT	NT	NT
Gluconate	+	+	−(+)	+(−)	+	NT	NT	+	+
Glycerol	−	−	−	−	−	NT	NT	NT	NT
Glycogen	−	−	−	−	−	−	−(+)	−	−
Inulin	−	+	+	+(−)	+(−)	+	−	NT	NT
Lactose	−	+(−)	−(+)	−(+)	+(−)	+	NT	NT	+
Maltose	−¶	+	+	+(−)	+(−)	+	+(−)	−	+
Mannitol	+	+	+	+	+	+	−(+)	+	+
Melibiose	+(−)	+(−)	−(+)	−	−	−	−(+)	−	+
Melezitose	+(−)¶	+(−)	−	+(−)	+(−)	V	−	NT	NT
Methyl-D-glucoside	−(+)	−(+)	−	−	−	−	−	NT	NT
Methyl-D-mannoside	−(+)	+(−)	−	−	NT	NT	−	NT	−
Pullulan	−	−(+)	−	−	−	NT	NT	NT	NT
Raffinose	−	NT	NT	NT	NT	+	−	NT	NT
Rhamnose	+	+(−)	+(−)	+	+(−)	V	−	−	+
Salicin	−	−(+)	−	−	−	NT	NT	NT	−
Sucrose	(+)−	+(−)	−(+)	+	+	+	NT	+	NT
Tagatose	+¶	+	+	+	+	+	+(−)	+	+
Trehalose	−	+(−)	−(+)	−	−	−	−	NT	NT
	+(−)¶	+(−)	+	−	−(+)	+	+(−)	−	+

Hydrolysis of									
Aesculin	+¶	+	+(−)	+	+	+	NT	+	NT
Hippurate	−	−	+(−)	−	−	−	NT	−	NT
Starch	NT	NT	NT	NT	NT	+	NT	−	NT
Arginine	−	−	−	−	−	+	−	−	NT
Production of									
α-Galactosidase	+	+(−)	−	+(−)	+(−)	+	−	NT	NT
β-Galactosidase	−	−(+)	−	−	−	V	−	NT	+
β-Glucosidase	+(−)	NT	NT	NT	NT	+(−)	−	NT	NT
β-Glucuronidase	−	−	−(+)	−	−	+	−(+)	NT	NT
N-Acetyl-β-glucosaminidase	−	NT	NT	NT	NT	+	−	NT	+
Pyrolidonyl arylamidase	−	−	−(+)	−	−	−(+)	−	NT	NT
Urease	+(−)	NT	NT	NT	NT	−	−	+	NT
Acetoin (VP)	+	+	−	+	+	−	−	NT	NT

*Species are ordered in the table according to data from Bentley *et al.* (1991). No species reported produces acid from adonitol, arabinose, arabitol, cyclodextrin, dulcitol, erythritol, sorbitol, starch or xylose. All species produce acid from fructose, galactose, glucose and mannose. No species produces alkaline phosphatase.

Data taken from Anon (1991); Brooker *et al.* (1994); Devriese *et al.* (1994); Farrow *et al.* (1984); Kilpper-Bälz and Schleifer (1987).

†+, >90% of strains give a positive result; +(−), 50–89% of strains give a positive result; −(+), 11–49% of strains give a positive result; −, <10% of strains give a positive result; V, reported as 'variable'; and NT, not tested.

‡Biochemical characteristics of the three DNA homology groups reported by Farrow *et al.* (1984).

§High final pH often results in difficulty when reading tests.

¶Number of strains giving a positive result varies between reports.

α-haemolytic on sheep blood agar forming tiny (0.2 mm diameter) colonies. No extracellular polysaccharide is produced on sucrose-containing medium. A red chromophore is produced, visualized by boiling the bacteria at pH 2.0 for 5 min. The biochemical reactions of this species are shown in Table 4.3. This species has complex growth requirements including the addition of one of the active forms of vitamin B_6 such as pyridoxal hydrochloride or pyridoxamine dihydrochloride. Also, satellitism can be observed around colonies of *S. epidermidis* on horseblood agar. Strains are ungroupable with Lancefield antisera. Cell walls are characterized by the absence of rhamnose and presence of ribitol teichoic acid. DNA G+C content is 36–37 mol%. DNA–DNA hybridization studies demonstrated *S. adjacens* to be a separate species and not to be variant strains of *S. mitis* or *S. sanguis II* as previously suggested. Unfortunately this species was not included in the 16S rRNA sequence study by Bentley *et al.* (1991). Source/habitat: human throat, urine and blood of patients with endocarditis. Type strain ATCC 49175.

4.13.3 Streptococcus agalactiae

Streptococcus agalactiae (Lehmann and Neumann, 1896) is synonymous with Lancefield Group B streptococcus. *Streptococcus agalactiae* is an important cause of mastitis in cattle and in the past few decades has also become recognized as an important pathogen of man, causing neonatal meningitis and septicaemia. Cells are 0.6–1.2 μm diameter, spherical or ovoid occurring frequently in very long chains. On blood agar most strains produce *β*-haemolysis although some strains are α- or non-haemolytic. Addition of starch to the medium, or anaerobic incubation, may enhance the production of yellow, orange or red pigments. Most strains grow in the presence of 40% bile and all strains hydrolyse hippurate. The other biochemical reactions of this species are shown in Table 4.4. Almost all strains give a positive CAMP (named after the initials of the authors who first described the test) reaction (Christie *et al.*, 1944). Cell wall peptidoglycan type is Lys-Ala_2(Ser). DNA G+C content is 34 mol%. Strains possess the Lancefield Group B specific carbohydrate antigen in the cell walls. Further serological division is possible on the basis of both capsular polysaccharide antigens and protein antigens of which the former are virulence factors. 16S rRNA sequence analysis places this species within the pyogenic group of streptococci and DNA hybridization has also demonstrated that Lancefield Group M streptococci are included within this species. Source/habitat: vaginal mucosa, upper respiratory tract, urine, faeces of man and in the milk and udder tissues of animals. Type strain NCTC 8181.

4.13.4 Streptococcus alactolyticus

This species was first described by Farrow *et al.* (1984) in a study of strains of *S. bovis* and *S. equinus*. DNA–DNA hybridization studies by these authors resulted in the recognition of six DNA homology groups of which one, comprising strains of *S. equinus* from pigs and chickens, was given the name *S. alactolyticus*. Cells are coccoid and form short chains or pairs. Colonies on blood agar are α- or non-haemolytic, circular, smooth and entire. Growth occurs at 45°C but not 50°C or in the presence of 6.5% NaCl. The biochemical characteristics are shown in Table 4.5. DNA G+C content is 40–41 mol%. Strains contain the Lancefield Group D antigen. 16S rRNA comparative sequencing has shown a relatively close phylogenetic relationship between *S. alactolyticus*, *S. bovis* and *S. equinus*. Strains of *S. alactolyticus* have previously been designated as *S. equinus*. Source/habitat: intestines of pigs and chickens. Type strain NCDO 1091.

4.13.5 Streptococcus anginosus

Although originally described at the beginning of the century by Andewes and Horder (1906), the taxonomic position of streptococci named *S. anginosus*, together with similar 'species', remained confused for many years (Jones, 1978). DNA–DNA hybridization studies (Kilpper-Bälz *et al.*, 1984; Whiley and Hardie, 1989) finally clarified the situation with the recognition of three closely related species, including *S. anginosus* present amongst these biochemically and serologically heterogeneous streptococci (Whiley and Beighton, 1991). Cells are 0.5–1 μm in diameter, forming short chains. On blood agar most strains produce α-haemolysis or are non-haemolytic, with some strains producing *β*-haemolysis. No extracellular polysaccharide is produced on sucrose containing media. The biochemical reactions of this species are shown in Table 4.3. Growth is enhanced in the presence of CO_2, reduced under aerobic conditions and some strains require anaerobic conditions. Most strains are serologically ungroupable with the majority of groupable isolates belonging to Lancefield Group F. Reactions with Lancefield Groups A, C and G antiserum are also occasionally found. The cell wall peptidoglycan type is Lys-Ala$_{1-3}$. DNA G+C content is 38–40 mol%. DNA homology studies (Whiley and Hardie, 1989) and 16S rRNA sequence analysis (Bentley *et al.*, 1991) in particular have shown that *S. anginosus* together with *S. constellatus* and *S. intermedius* form a group of closely related species sometimes referred to as the '*Streptococcus milleri*-group'. Strains resembling *S. anginosus* have previously been referred to as *Streptococcus milleri*, *Streptococcus* MG, haemolytic and non-haemolytic streptococci of Lancefield Group F, the minute colony-forming streptococci of Lancefield Groups F and G, *Streptococcus* MG-*intermedius* and *Streptococcus anginosus-constellatus*. Source/habitat: human oral cavity, upper respiratory tract and vagina. Frequently isolated from purulent infections of man. Type strain NCTC 10713 (ATCC 33397).

4.13.6 Streptococcus bovis

Originally described as a bovine bacterium that fermented arabinose, raffinose and starch but not mannitol (Orla-Jensen, 1919), *S. bovis* was later discovered to be of clinical importance to man as an aetiological agent in some cases of endocarditis and possible association with colon cancer (Facklam, 1972; Klein *et al.*, 1977). The biochemical heterogeneity presented by strains for a long time hindered any resolution of the taxonomy of these streptococci. The application of genetic approaches however confirmed that *S. bovis* was indeed made up of several distinct 'species' (Farrow *et al.*, 1984; Coykendall and Gustafson, 1985) although full descriptions of some of these are still awaited. Cells are spherical or ovoid and are 0.8–1.0 μm in diameter, occurring in moderate or long chains and also in pairs. Most strains give α-haemolysis on blood agar and produce large amounts of polysaccharide on sucrose containing media. This species comprises strains with heterogeneous properties that include anaerobic strains capable of growth in broth containing 6.5% NaCl and at pH 9.6, the production of urease by some strains, failure to grow at 45°C and sharing high DNA homology with *S. mutans*, fermentation of arabinose, xylose, mannitol, sorbitol, trehalose and inulin. Phenotypic similarity between *S. bovis* and *S. salivarius* has been noted by some authors. Strains from human sources have previously been designated as biotype I and II, the former being characterized by their ability to ferment mannitol and inulin and produce extracellular glucan on sucrose agar in contrast to the biotype II strains which are negative in these tests. The biochemical properties of this species are shown in Table 4.5. Peptidoglycan types Lys-Thr-Ala, Lys-Thr-Gly and Lys-Thr-Ala (Ser) occur in strains of *S. bovis*. Strains possess the Lancefield Group D antigen. DNA–DNA hybridization studies have revealed extensive genetic heterogeneity within strains classified as *S. bovis*. In the study of *S. bovis* and *S. equinus* strains by Farrow *et al.* (1984) six DNA homology groups were demonstrated: one group contained both the type strains of *S. bovis* and *S. equinus* leading these authors to propose that due to the priority of the name *S. equinus* the name '*S. bovis*' be reduced to synonymity. Another DNA homology group consisted of strains from cases of bovine mastitis capable of fermenting mannitol. Strains designated *S. bovis* were also grouped in another three DNA homology groups, one of which contained bovine strains and was proposed as a new species named *S. saccharolyticus*. *Streptococcus bovis* strains from human sources were included in unnamed DNA homology Group 4 of Farrow *et al.* (1984). 16S rRNA sequence data together with information from DNA–DNA hybridization have demonstrated the close relationship between *S. bovis*, *S. equinus* and *S. alactolyticus*. Source/habitat: alimentary tract of cow, sheep and other ruminants, faeces of pigs. Occasionally isolated from human faeces in large numbers, from raw and pasteurized milk and cheese and from some cases of endocarditis in humans. Type strain NCDO 597.

4.13.7 Streptococcus canis

Streptococci of Lancefield Group G include the so-called large colony, β-haemoloytic strains isolated from animals which differ from human Group G isolates within *S. dysgalactiae* on the basis of α- and β-galactosidase activities, lack of fibrinolysin, hyaluronidase, or β-glucuronidase and an inability to ferment trehalose. On the basis of DNA homology and phenotypic characterization Devriese *et al.* (1986) named these streptococci *S. canis*. Cells form chains or occur in pairs. β-Haemolysis is produced on blood agar. The strain is CAMP factor negative. Strains are facultatively anaerobic and good growth occurs at 37°C. No growth occurs in the presence of 6.5% w/v NaCl or 40% w/v bile. The biochemical reactions of this species are shown in Table 4.4. The cell wall peptidoglycan type is Lys-Thr-Gly. DNA G+C content is 39–40 mol%. Strains belong to Lancefield Group G. DNA–DNA hybridization studies and comparative 16S rRNA sequence analysis have shown *S. canis* to be within the pyogenic group of streptococci. Source/habitat: dogs (skin, upper respiratory tract and genitals) cows (udders) and probably cats. Isolated from dogs (neonates with septicaemia and from a wound exudate) and from cows suffering from mastitis. Not isolated from humans. Type strain DSM 20715.

4.13.8 Streptococcus caprinus

From studies on the bacteria inhabiting the digestive tracts of animals with tannin-rich diets has emerged the species description of *S. caprinus* (Brooker *et al.*, 1994) isolated from wild goats grazing tannin-rich shrubs in Australia. Cells occur mainly in short chains. Grows on nutrient agar plates containing 0.5% w/v tannic acid, forming large mucoid colonies surrounded by clear zones in the tannic acid agar. Able to grow in complex growth medium with at least 2.5% w/v condensed tannins from the acacia tree (*Acacia aneura*). The biochemical reactions of this species are shown in Table 4.5. The DNA G+C content and presence of a Lancefield group antigen have not been reported from this recently described species. Source/habitat: rumen of feral goats. Type strain ACM 2969.

4.13.9 Streptococcus constellatus

Streptococcus constellatus was the name given by Holdeman and Moore (1974) to strains isolated from clinical specimens and vaginal swabs, that closely resembled a species first described by Prevot (1924) as *Diplococcus constellatus*. These streptococci produced major amounts of lactic acid, fermented glucose, maltose, and sucrose but not lactose and hydrolysed aesculin. Subsequently a close resemblance was reported between these streptococci, and several other 'species' already described that included '*Streptococcus* MG', *Streptococcus intermedius*, and *S. anginosus* (Facklam, 1977). These were divided into two species on the basis of lactose fermentation: *S. anginosus-constellatus* (lac−) and *S.* MG-*intermedius* (lac+). Further taxonomic studies have revealed three distinct species within these biochemically and serologically heterogeneous streptococci which includes *S. constellatus*, *S. anginosus* and *S. intermedius* (Whiley and Beighton, 1991). Cells are 0.5–1 μm in diameter forming short chains. On blood agar strains can produce α-, *β*- or no (γ-)haemolysis. Extracellular polysaccharide is not produced on sucrose-containing medium. The biochemical reactions of this species are shown in Table 4.3. Growth is enhanced in the presence of CO_2, reduced under aerobic conditions and some strains require anaerobic conditions. Some strains react with Lancefield Groups A and C antisera with the majority of strains remaining ungroupable in this system. The cell wall peptidoglycan type is Lys-Ala_{1-3}. DNA G+C content is 37–38 mol%. DNA homology studies and 16S rRNA sequence analysis in particular have shown that *S. constellatus* together with *S. anginosus* and *S. intermedius* form a group of closely related species sometimes referred to as the '*Streptococcus milleri*-group'. Strains resembling *S. constellatus* have also been previously referred to as *Streptococcus milleri*, *Streptococcus* MG, haemolytic and non-haemolytic streptococci of Lancefield Group F, the minute colony-forming streptococci of Lancefield Groups F and G, *Streptococcus* MG-*intermedius* and *Streptococcus anginosus-constellatus*. Source/habitat: human oral cavity and upper respiratory tract. Frequently isolated from purulent infections in man. Type strain ATCC 27823 (NCDO 2226).

4.13.10 Streptococcus cricetus

Originally described as *S. mutans* serotype a from hamster and human dental plaque (Bratthall, 1970) these streptococci were shown to be genetically distinct from other mutans-like strains. Proposed initially as *S. mutans* subsp. *cricetus* (Coykendall, 1974) these streptococci were subsequently elevated to species status as *S. cricetus* (Coykendall, 1977). Cells are approximately 0.5 μm in diameter forming chains or occurring in pairs. Colonies formed on sucrose-containing agar are rough and heaped, with liquid glucan sometimes present. On blood agar most strains are non-haemolytic, whilst some are α-haemolytic. The biochemical characteristics of this species are shown in Table 4.3. Optimum growth is obtained with added CO_2 or under reduced O_2. Most strains possess the serotype a polysaccharide antigen (Bratthall, 1970). Cell wall peptidoglycan type is Lys-Thr-Ala. DNA G+C content is 42–44 mol%. *Streptococcus cricetus* is a species belonging to the mutans group of streptococci as demonstrated by DNA–DNA hybridization and rRNA studies. Strains of this species were previously designated as *Streptococcus mutans* serotype a. Source/habitat: oral cavities of wild rats, hamsters and man (occasionally). Type strain ATCC 19642.

4.13.11 Streptococcus crista

These streptococci isolated from the human oral cavity and throat were initially regarded as unusual strains of *S. sanguis* before DNA homology studies demonstrated that they constituted a new species named *S. crista* (Handley *et al.*, 1991). Cells are approximately 1 μm in diameter, spherical and form chains. By electron microscopy cells have fibrils arranged equatorially in lateral tufts. α-haemolysis is produced on blood agar and some strains produce glucan on sucrose-containing medium. The biochemical reactions of *S. crista* are given in Table 4.3. The peptidoglycan type of this species has not been determined. DNA G+C content is 42.6–43 mol%. Strains of this species have previously been referred to as the 'tufted fibril group', the 'CR group' and '*S. sanguis* I'. The phylogenetic position of *S. crista* has not been determined. Source/habitat: human throats and mouths. Type strain NCTC 12479.

4.13.12 Streptococcus defectivus

Streptococcus defectivus (Bouvet *et al.*, 1989) together with *S. adjacens* comprised the nutritionally variant (pyridoxal dependent) streptococci (NVS). Originally considered to be variant of already established α-haemolytic streptococci, both *S. defectivus* and *S. adjacens* have been shown to be distinct species in their own right. Cells are 0.4–0.55 μm in diameter, small ovoid cocci occurring in chains of varying length, in pairs or even singly in CDMT semi-synthetic medium. Stationary phase cells may tend to be rod shaped. However, strains may produce cocci, coccobacilli and rods within chains during growth on pyridoxal or cysteine-supplemented broth. Strains are haemolytic on sheet blood agar, forming tiny (0.2 mm diameter) colonies. No extracellular polysaccharide is produced on sucrose containing medium. A red chromophore is produced, visualized by boiling the bacteria at pH 2 for 5 min. The biochemical reactions of this species are shown in Table 4.3. Complex growth requirements including the addition of one of the active forms of vitamin B_6 such as pyridoxal hydrochloride or pyridoxamine dihydrochloride. Also, satellitism can be observed around colonies of *Staphylococcus epidermidis* on horse blood agar. Strains are serologically ungroupable against Lancefield antisera, with an occasional weak reaction against Group H antiserum. Cell walls are characterized by the absence of rhamnose and presence of ribitol teichoic acid. DNA G+C content is 46.0–46.6 mol%. DNA–DNA hybridization studies demonstrated *S. defectivus* to be a separate species and not to be variant strains of *S. mitis* or *S. sanguis* II as had been previously thought. Unfortunately this species was not included in the 16S rRNA sequence study by Bentley *et al.* (1991). Source/habitat: human throat, urogenital tract and intestine. Most frequently isolated from the blood of patients with bacteraemia or endocarditis. Type strain ATCC 49176.

4.13.13 Streptococcus downei

Following the initial division of streptococci resembling *S. mutans* into several distinct species (*S. mutans*, *S. sobrinus*, *S. cricetus* and *S. rattus*) (Coykendall, 1977) further studies revealed the existence of additional species within the 'mutans-group' of streptococci. *Streptococcus downei* (Whiley *et al.*, 1988) were isolated from monkey dental plaque and were characterized as mutans streptococci carrying a serologically distinct carbohydrate antigen designated h (Beighton *et al.*, 1981) before being recognized as a distinct species. On sucrose-containing agar colonies are large (2–3 mm diameter), conical and are usually surrounded by a white halo within the agar. Cells adhere to glass surfaces when grown in sucrose broth, indicating the production of extracellular polysaccharide, although no cell-free, ethanol precipitable polysaccharide has been demonstrated. The biochemical reactions of this species are shown in Table 4.3. No growth occurs at pH 9.6, at 45°C or in the presence of 6.5% w/v NaCl. Variable growth occurs on 10 and 40% w/v bile agar. The cellular long-chain fatty acid composition consists of major amounts of hexadecanoic (16:0 palmitic), octadecanoic (18:0 stearic), octadecenoic (18:1 vaccenic) and eicosenoic (20:1) acids. Minor amounts (<10% of total fatty acids present) of tetradecanoic (14:0 myristic), hexadecenoic (16:1 palmitoleic), octadecenoic (18:1 oleic), eicosenoic (20:0 arachidic) and cyclopropane (*cis*-9, 10-methyleneoctadecanoic acid). Cell wall peptidoglycan type is Lys-Thr-Ala. DNA G+C content is 41–42 mol%. A distinct polysaccharide antigen, designated h, is present in strains of this species. Monoinfected 'germ-free' rats develop dental caries. DNA–DNA hybridization and 16S rRNA comparative sequence analysis have demonstrated that *S. downei* is most closely related to *S. sobrinus* within the mutans group of streptococci. Previously strains were designated as *Streptococcus mutans* serotype h.

Source/habitat: dental plaque of monkey (*Macaca fascicularis*). Type strain NCTC 11391.

4.13.14 Streptococcus dysgalactiae *(including* 'S. equisimilis'*)*

Streptococcus dysgalactiae (Diernhofer, 1932) is a well-known cause of bovine mastitis but for reasons that remain unclear was not included on the Approved Lists of Bacterial Names (Skerman *et al.*, 1980) and was later revived by Garvie *et al.* (1983) following demonstration of its species status by DNA–DNA hybridization and studies of its lactate dehydrogenase. Cells are ovoid or coccal occurring in pairs or chains. A wide zone of β-haemolysis is produced on blood agar by some strains with others giving α-haemolysis. The biochemical reactions of this species are shown in Table 4.4. Strains will grow optimally at 37°C but not at 10°C or 45°C, at pH 9.6 or in the presence of either 6.5% NaCl or 0.1% methylene blue milk. This species does not survive heating at 60°C for 30 min. Cellular long-chain fatty acids composition consists of major amounts of hexadecanoic (C16:0) and octadecenoic (C18:1) acids. Cyclopropane-ring fatty acids and menaquinones are absent. Cell wall peptidoglycan type is Lys-Ala_{1-3}. DNA G+C content is 38.1–40.2 mol%. Strains may react with Lancefield Groups C, G or L antisera. DNA–DNA hybridization studies have demonstrated that *S. dysgalactiae* also includes streptococci designated as '*S. equisimilis*' as well as streptococci of Lancefield Groups C, G (large colony type) and L (Farrow and Collins, 1984b). 16S rRNA sequence analysis has confirmed *S. dysgalactiae* as a member of the pyogenic group of streptococci. Source/habitat: human respiratory tract, vagina and skin, throats and genital tracts of domestic animals. Isolated from mastitic bovine udders. Type strain NCDO 2023.

4.13.15 Streptococcus equi

This species is recognized as an important equine respiratory pathogen. The demonstration of the close relationship between the type strain of *S. equi* and '*S. zooepidemicus*' (Farrow and Collins, 1984b) but at the same time recognition of their respective phenotypes has resulted in the creation of the two subspecies *zooepidemicus* and *equi*.

4.13.15.1 S. equi *subsp.* equi. Cells are 0.6–1.0 μm in diameter, ovoid or spherical, sometimes resembling streptobacilli. Capsules can be demonstrated in young cultures. Wide zones of β-haemolysis are formed on blood agar. Growth is poor in media without serum. The biochemical properties of *S. equi* subsp. *equi* are shown in Table 4.4. Cell wall peptidoglycan type is Lys-Ala_{2-3}. Stains possess the Lancefield Group C antigen. DNA G+C content = 40–41 mol%. Streptococci previously assigned to '*S. zooepidemicus*' have been shown, by DNA homology studies, to belong to *S.*

equi but, because of their phenotypic differences (fermentation of lactose, ribose and sorbitol by subsp. *zooepidemicus*), were not reduced to synonymity. Source/habitat: isolated from equine strangles, abscesses in the submaxillary glands and in mucopurulent discharges of the lower respiratory system of horses and from their immediate environment. Type strain NCTC 9682.

4.13.15.2 S. equi *subsp.* zooepidemicus. The following description is based on that of Farrow and Collins (1984b). Cells are spherical or ovoid and can occur in chains or in pairs. Capsules may be present. Wide zone of β-haemolysis is produced on blood agar. Growth is optimal at 37°C and does not occur at 10°C or 45°C, after heating at 60°C for 30 min, in the presence of 6.5% NaCl, 10% bile or 0.1% methylene blue milk or pH 9.6. The biochemical properties of subsp. *zooepidemicus* are shown in Table 4.4. Major long-chain fatty acids are hexadecanoic (C18:0) and octadecenoic (C18:1) acids. Strains react with Lancefield Group C antiserum. Cell wall peptidoglycan type is Lys-Ala_{2-3}. DNA G+C content = 41.3–42.7 mol%. Source/habitat: isolated from the blood stream, inflammatory exudates and lesions of diseased animals. Type strain: NCDO 1358.

4.13.16 Streptococcus equinus

Streptococcus equinus was originally described by Andrewes and Horder (1906) as a saprophytic streptococcus chiefly from air, dust and horse dung. Although considered by Sherman (1937) to be a distinct species there was no general agreement between bacteriologists as to the species status of both *S. equinus* and *S. bovis* which shared many common phenotypic characteristics (Jones, 1978). This confusion has been clarified somewhat by the recent application of genetic approaches to the classification of *S. equinus* and related species (Farrow *et al.*, 1984). Cells occur in medium length chains especially in broth cultures. Weak α-haemolysis is produced on blood agar. No growth occurs in the presence of 4% w/v NaCl or 0.04% w/v potassium tellurite. Does not survive 60°C for 30 min. Strains contain the Lancefield Group D antigen and possess peptidoglycan type Lys-Thr-Ala. This species has been redefined on the basis of results from extensive DNA–DNA hybridization experiments and phenotypic characterization (Farrow *et al.*, 1984). These data demonstrated that the type strains of *S. bovis* and *S. equinus* belong in the same DNA homology group and, due to the priority that the same *S. equinus* has over '*S. bovis*', that the latter should be reduced to synonymity. Consequently *S. equinus* is defined according to Farrow *et al.* (1984) as follows: cells are spherical or ovoid in moderately long chains and producing α-haemolysis of varying intensity on blood agar. Growth occurs at 45°C but not at 50°C or at 10°C. Some strains survive heating at 60°C for 30 min. Growth occurs in 40% w/v bile but not at pH 9.6–0.1% methylene blue milk or in the presence of 6.5% w/v NaCl. The biochemical reactions of *S. equinus* as redefined are also shown in Table 4.5. Cells contain the Lancefield Group D antigen. The DNA G+C content is 36.2–38.6 mol%. Occurs in the alimentary tract of cows, horses, sheep and other ruminants, isolated occasionaly in large numbers from human faeces, and occasionally isolated from cases of human endocarditis. 16S rRNA comparative sequence analysis and DNA–DNA hybridization have shown a close relationship between strains of *S. equinus* and *S. bovis* which together with *S. alactolyticus* formed a distinct cluster within the 16S rRNA derived phylogenetic tree. Source/habitat: alimentary tract of horses. Type strain ATCC 9812 (NCDO 1037).

4.13.17 Streptococcus ferus

Following the recognition of four subspecies (*mutans*, *rattus*, *sobrinus* and *cricetus*) within *S. mutans* by Coykendall (1974) a new mutans-like *Streptococcus* was isolated from wild sucrose-eating rats living in sugar-cane fields (Coykendall *et al.*, 1974). These strains contained the serotype c antigen first described by Bratthall (1970), were found to have a relatively high DNA G+C content (43–45%) and to be genetically distinct by DNA–DNA hybridization. They were initially given the subspecific epithet *ferus* (Coykendall *et al.*, 1976) and later proposed as a separate species *S. ferus* (Coykendall, 1983). Cells are approximately 0.5 μm in diameter and occur in pairs or in chains. On sucrose-containing agar colonies are adherent and raised but without the presence of liquid glucan. Both extra- and intercellular polysaccharides are produced from sucrose. The biochemical reactions of this species are shown in Table 4.3. Strains do not grow at 45°C or in 6.5% NaCl. Serological studies have shown that strains react with *S. mutans* serotype c antiserum. Cell wall peptidoglycan type is Lys-Ala_{2-3}. DNA G+C content is 43–45 mol%. DNA–DNA and DNA–rRNA hybridization studies indicate that *S. ferus* is a species within the mutans group of streptococci although data from multilocus enzyme electrophoresis place it closer to *Streptococcus sanguis*. Source/habitat: oral cavity of wild rats. Type strain ATCC 33477.

4.13.18 Streptococcus gordonii

The classification of streptococci resembling *S. sanguis* has remained confused and unresolved until the recent application of genotypic analyses and extensive phenotypic characterization. This has resulted in the recognition of several new species and amended descriptions of these streptococci that include *S. gordonii* (Kilian *et al.*, 1989a). Cells are observed to form short chains in serum broth, to give α-haemolysis on horse blood agar plates and pronounced greening on chocolate agar. Most strains produce extracellular polysaccharide on sucrose containing medium. The biochemical reactions of this species and its biotypes are shown in Table 4.3. Cell wall peptidoglycan type is Lys-Ala_{1-3} and the cell wall contains rhamnose and glcyerol teichoic acid. DNA G+C content is 38–43 mol%. Strains react with Lancefield Group H antiserum raised against strain Blackburn or F90A. Strains of *S. gordonii* have previously been designated as *S. sanguis*, *S. sanguis* I, *Streptococcus* sbe, Group H *Streptococcus* and *S. mitis* (strain NCTC 3165). 16S rRNA comparative sequence analysis has not been reported for this species although DNA–DNA hybridization studies previously carried out using strains now known to belong to *S. gordonii* have demonstrated that this species is grouped within the '*S. oralis* group' of Schleifer and Kilpper-Bälz, (1987) that includes *S. sanguis*, *S. oralis*, *S. parasanguis*, *S. pneumoniae*, as well as *S. intermedius*, *S. constellatus* and *S. anginosus*. Strain NCTC 3165, previously designated as the type strain of *Streptococcus mitis* has also been shown to be a phenotypically atypical strain of *S. gordonii*. Source/habitat: human oral cavity and pharynx. Type strain NCTC 7865 (ATCC 10558).

4.13.19 Streptococcus hyointestinalis

Strains of this species were originally described as *S. salivarius* before recognition of their separate species status (Devriese *et al.*, 1988). *Streptococcus hyointestinalis* strains are also phenotypically distinct from another recently described *Streptococcus* isolated from pig intestines, *S. intestinalis*, the latter being *β*-haemolytic, having a lower G+C content and a different fermentation pattern. Cells form chains and produce sediment with clear supernatant when grown in broth. α-Haemolytic on blood agar. No growth occurs in the presence of 6.5% NaCl or 40% bile. Optimum growth at 37°C and under anaerobic conditions. The biochemical reactions of this species are shown in Table 4.4. No reaction with Lancefield grouping sera (A–G). Cell wall peptidoglycan type is Lys-Ala(Ser). DNA G+C content is 42–43 mol%. Source/habitat: pig intestines. Type strain DSM 20770.

4.13.20 Streptococcus iniae

This member of the pyogenic streptococci was isolated from abscesses on the thorax and abdomen of Amazon river-living freshwater dolphins (Pier and Madin, 1976). It has not however been shown to be pathogenic for other animals. Cells are 0.6–1 μm diameter spherical of ovoid forming medium to long chains. On blood agar colonies are 1 mm diameter with opaque centres and translucent borders and are *β*-haemolytic or α-haemolytic. The biochemical characteristics of this species are shown in

Table 4.4. No growth occurs at 45°C or in bile–esculin media. Good growth is obtained in Todd–Hewitt broth with overnight incubation at 37°C. DNA G+C content is 33 mol%. Contains a specific antigen extractable by HCl or formamide that does not react that Lancefield grouping sera A–V. 16S rRNA sequence analysis has demonstrated this species to belong to the pyogenic group of streptococci. Source/habitat: freshwater dolphin (*Inia geoffrensis*); isolates from subcultaneous abscesses on thorax and abdomen. Type strain ATCC 29178.

4.13.21 Streptococcus intermedius

The taxonomy and nomenclature of this species is, as with the other members of the '*S. milleri*-group', somewhat confused. *Streptococcus intermedius* (Holdeman and Moore, 1974) was reported as an amended description of the original published description (Prévot, 1925). The source of the original Prévot strain remains unknown and *S. intermedius* was described by Holdeman and Moore as being isolated from human clinical specimens and faeces. As described previously (see description of *S. constellatus*) the close resemblance between *S. intermedius* and several other biochemically and serologically heterogeneous 'species', resulted in the division of all such strains on the basis of lactose fermentation into *S.* MG-*intermedius* (lac+) and *S. anginosus-constellatus* (lac−) (Facklam, 1977). More recent taxonomic studies have shown that three distinct species exist within this group of streptococci and these have retained the names *S. anginosus*, *S. constellatus* and *S. intermedius* (Whiley and Beighton, 1991). An association between *S. intermedius* and abscesses of the brain has also been noted (Whiley *et al.*, 1992). Cells are 0.5–1 μm in diameter, forming short chains. Most strains are α- or non-haemolytic on blood agar. No extracellular polysaccharide is formed on sucrose-containing medium. The biochemical reactions of this species are shown in Table 4.3. Growth is enhanced in the presence of CO_2, reduced under aerobic conditions and some strains require an anaerobic environment for growth. Almost all strains are serologically ungroupable using Lancefield grouping antisera. The cell wall peptidoglycan type is Lys-Ala_{1-3}. DNA G+C content is 37–38 mol%. DNA homology studies and 16S rRNA sequence analysis in particular have shown that *S. intermedius* together with *S. anginosus* and *S. constellatus* form a group of closely related species sometimes referred to as the '*Streptococcus milleri*-group'. Strains resembling *S. intermedius* have also been previously referred to as *Streptococcus milleri*, *Streptococcus* MG, haemolytic and non-haemolytic streptococci of Lancefield Group F, the minute colony-forming streptococci of Lancefield Groups F and G, *Streptococcus* MG-*intermedius* and *Streptococcus anginosus-constellatus*. Source/habitat: human oral cavity and upper respiratory tact. Reported to be present in human faeces.

Isolated from purulent infections in man. Type strain ATCC 27335 (NCDO 2227).

4.13.22 Streptococcus intestinalis

This relatively recently described species of *Streptococcus* (Robinson *et al.*, 1988) comprises approximately 50% or more of the bacteria present in the colon of pigs. Of particular interest is the ability of strains to hydrolyse urea, an important aspect of nitrogen metabolism in animals. Cells form long chains, are often elongated and can occur in pairs of unequal cell size. On blood agar colonies are tiny (1 mm in diameter or less), white, flat to convex, circular, entire and β-haemolytic. The biochemical reactions of this species are shown in Table 4.5. Strains are characterized by the ability to hydrolyse urea. Growth occurs optimally at 37°C, can occur at 45°C and strains can survive 60°C for 30 min. However, no growth occurs at pH 9.6 or in the presence of 6.5% NaCl or 40% bile. DNA G+C content is 39–40%. Some strains react with Lancefield Group G antiserum. Source/habitat: intestines and faeces of pigs. Type strain ATCC 43492.

4.13.23 Streptococcus macacae

This species, first described by Beighton *et al.* (1984) from the dental plaque of monkeys, is one of the more recent additions to the 'mutans-streptococci' species group. However, the true taxonomic position of *S. macacae* within the genus *Streptococcus* remains to be determined. A chain-forming coccus that produces greening on horse blood agar when grown anaerobically or in candle jars. Dextran is produced from sucrose; on sucrose-containing agar 1–2 mm diameter colonies are formed that are easily removed but remain intact. Vivid white, crumbly colony variants can also arise. The biochemical reactions of *S. macacae* are shown in Table 4.3. Strains grow poorly in air and CO_2 stimulates growth. Growth does not occur in the presence of 6.5% w/v NaCl, at 45°C or at pH 9.6 but occurs in media containing 10% and 40% bile. This species is serologically ungroupable against Lancefield antisera. DNA G+C content is 35–36 mol%. Source/habitat: dental plaque of monkeys (*Macaca fascicularis*). Type strain NCTC 11558.

4.13.24 Streptococcus mitis

The name *S. mitis* was first used by Andrewes and Horder (1906) to describe a saprophytic *Streptococcus* present mainly in human saliva and faeces that was short chained, grew well at 20°C on gelatin, did not clot milk, often reduced neutral red and nearly always fermented lactose and saccharose but not the glucosides salicin and coniferin. Subsequent descriptions of *S. mitis* tended to be poor and ill-defined, with strains characterized mainly on negative criteria. Despite the lack of a clear description of this species, the name *S. mitis* was included in the Approved List of Bacterial Names (Skerman *et al.*, 1980) and has persisted in the literature. More recently Kilian *et al.* (1989a) published on amended description of *S. mitis* giving the name to a group of streptococci whose integrity as a species is better supported by phenotypic and genotypic data. Cells form short or long chains in serum broth, and give α-haemolysis on horse blood agar and pronounced greening on chocolate agar. Extracellular polysaccharide is not produced on sucrose-containing medium. The biochemical reactions of *S. mitis* and its biovars are shown in Table 4.3. Cell wall peptidoglycan type is Lys-direct and cell walls contain ribitol teichoic acid but lack rhamnose in significant amounts. DNA G+C content is 40–41 mol%. Strains may be serologically ungroupable using Lancefield grouping antisera, or may react with Group K or O antisera. *Streptococcus mitis* has not been compared with other streptococcal species using 16S rRNA sequence analyses but, from previous DNA–DNA hybridization studies that included strains now designated as *S. mitis*, it appears that this species belongs within the '*S. oralis* group' of Schleifer and Kilpper-Bälz (1987). Strains of *S. mitis* have been previously designated as '*Streptococcus viridans*', *Streptococcus* Groups O and K and '*Streptococcus mitior*'. Source/habitat: human oral cavity and pharynx. Type strain NCTC 12261.

4.13.25 Streptococcus mutans

Originally described by Clarke (1924) from carious teeth, *S. mutans* was reported as a significant factor in the aetiology of dental caries. Nevertheless, this species was virtually ignored until interest picked up again in the 1960s when experiments into the induction and transmission of dental caries in animals were initiated. Since that time there has been an enormous body of literature focused on *S. mutans*, and subsequent taxonomic studies have revealed that mutans-like cariogenic streptococci comprise a group that currently includes seven species (Coykendall, 1977). Cells are 0.5–0.76 μm in diameter cocci forming short or medium length chains and sometimes forming short rods on some solid media or under acid conditions in broths. Colonies on blood agar are sometimes hard with a tendency to adhere to the agar and are usually α- or non-haemolytic. Some strains produce *β*-haemolysis. On sucrose-containing agar strains produce extracellular polysaccharides to give colonies that are rough, heaped, and detachable, 1 mm in diameter, frequently with droplets of water-soluble polysaccharide. On TYC agar may yield yellow or white colonies. Strains produce both water-soluble and water-insoluble glucans as well as fructans when on sucrose-containing agar. Intracellular, glycogen-like glucan is also produced. The biochemical characteristics of this species are shown in Table 4.3. Optimum growth occurs under anaerobic conditions, at 37°C with some strains able to grow at 45°C but no strain growing at 10°C. There are three demonstrable polysaccharide antigens, designated c, e and f. Cell wall peptidoglycan type is Lys-Ala_{2-3}. DNA G+C content is 36–38 mol%. This species gives its name to a group of seven closely related species collectively referred to as the mutans streptococci. Many strains are thought to be cariogenic in man and also induce caries in experimental animals. This species is also isolated from blood cultures in some cases of infective endocarditis. Source/habitat: surfaces of teeth in man and can also be isolated from faeces. Type strain NCTC 10449 (ATCC 25175).

4.13.26 Streptococcus oralis

Bridge and Sneath (1982) originally gave the name *S. oralis* to a cluster of oral streptococci included in a numerical taxonomy study (Bridge and Sneath, 1983) some of which resembled *S. mitis*. However, the phenotypic heterogeneity apparent in the species description was confirmed by genetic analysis which revealed several centres of variation at the species level (Kilpper-Bälz *et al.*, 1985). Kilian *et al.* (1989a) amended the species description further by phenotypic and serological approaches in a taxonomic study of 151 viridans streptococci, many of which had been included in previous taxonomic studies, so that currently the name *S. oralis* is given to a well-defined species. Cells form long chains in serum broth, give α-haemolysis on horse blood agar and pronounced greening on chocolate agar. Extracellular polysaccharide production on sucrose containing medium is a variable characteristic of this species. The biochemical characteristics of this species are shown in Table 4.3. Cell wall peptidoglycan type is Lys-direct and cell walls contain ribitol teichoic acid but little or no rhamnose. DNA G+C content is 38–42 mol%. Streptococci corresponding to *S. oralis* have previously been referred to as '*S. mitior*', *S. mitis*, '*Streptococcus* sbe', '*S. sanguis* I' or '*S. sanguis* II'. 16S rRNA comparative sequencing and DNA–DNA hybridization studies have demonstrated that this species belongs to the so-called '*S. oralis*-group' of species that also include *S. sanguis*, *S. parasanguis*, *S. mitis*, *S. pneumoniae*, *S. intermedius*, *S. constellatus* and *S. anginosus*. Source/habitat: human oral cavity. Type strain NCTC 11427.

4.13.27 Streptococcus parasanguis

The application of DNA–DNA hybridization to atypical viridans streptococci revealed the existence of *S. parasanguis* within the species group that also includes *S. sanguis* and *S. oralis* (Whiley *et al.*, 1990a). Many of the strains that fell into this species had been included in unnamed DNA homology groups by previous authors. Cells are approximately 0.8–1 μm in diameter, coccoid and chain forming. α-Haemolysis is produced on blood agar. Extracellular polysaccharide is not produced on sucrose-containing medium. The biochemical reactions of this species are shown in Table 4.3. No growth is obtained in the presence of 4% w/v NaCl although most strains grow in the presence of 40% w/v bile and at 45°C. The cell wall peptidoglycan type of *S. parasanguis* has not been determined. DNA G+C content is 40.6–42.7 mol%. *Streptococcus parasanguis* has been shown to be most closely related to *S. sanguis* by DNA homology and 16S rRNA comparative sequence analysis. Source/habitat: human throat and clinical specimens (blood and urine). Type strain ATCC 15912.

4.13.28 Streptococcus parauberis

Streptococcus parauberis (Williams and Collins, 1990) was proposed following the demonstration by DNA–DNA hybridization and 16S rRNA sequencing that the important pathogenic species commonly responsible for bovine mastitis, *S. uberis* consisted of two phylogenetically distinct lines of descent. These had previously been designated *S. uberis* types I and II (Garvie and Bramley, 1979) and the latter were renamed *S. parauberis*. Cells are coccoid and form moderate length chains or occur in pairs. On blood agar strains are weakly α-haemoloytic or non-haemolytic. Growth occurs in the presence of 4% NaCl but not 6.5% NaCl or at pH 9.6. Some strains survive heating at 60°C for 30 min. The optimum temperature for growth is 35–37°C. The biochemical reactions of this species are shown in Table 4.4. DNA G+C content is 35–37 mol%. Some strains of *S. parauberis* have been shown to react against Lancefield E and P antisera (Garvie and Bramley, 1979). 16S rRNA comparative sequence analysis has demonstrated that *S. parauberis* falls into the pyogenic group of streptococci. Strains of this species have previously been designated *S. uberis* type II. Source/habitat: lips, skin and udder tissue of cattle, and in raw milk. Type strain NCDO 2020.

4.13.29 Streptococcus pneumoniae

This extremely important pathogenic species of *Streptococcus* causes pneumonia, meningitis, otitis media as well as being isolated from abscesses, pericarditis, conjunctivitis and other clinical conditions. Currently, this species is the focus of attention due to the emergence of penicillin resistant strains with increasing frequency worldwide (Klugman, 1990). Cells are spherical or ovoid, 0.5–1.25 μm in diameter and are usually seen in pairs or occasionally either as single cells or as short chains. Cells in pairs may be elongated of the distal ends. Strong α-haemolysis is produced on blood agar. Colonies can be mucoid due to production of a polysaccharide capsule particularly with fresh isolates, smooth due to decreased capsule production or occasionally rough. The temperature range for growth is 25–42°C and incubation under increased CO_2 tension prevents autolysis. In defined media this species requires choline for growth. Bile soluble. The biochemical reactions of this species are shown in Table 4.3. Cell wall peptidoglycan type is Lys-Ala_2(Ser). Variation of the stem peptide has been reported within penicillin resistant strains which carry branched-stem peptides with Ala-Ser or Ala-Ala on the epsilon-amino groups of the stem peptide lysine residue. DNA G+C content is 36–37 mol%. Capsular polysaccharide is an important virulence factor of this species and forms the basis of the antigenic division of strains into types and subtypes, antibody to a particular capsule conferring type specific immunity. 16S rRNA comparative sequencing shows *S. pneumoniae* to be closely related to *S. oralis*. Source/habitat: upper respiratory tract of normal humans and domestic animals and from the upper respiratory tract, inflammatory exudates and various body fluids of diseased humans. Type strain NCTC 7465 (ATCC 33400).

4.13.30 Streptococcus porcinus

β-Haemolytic streptococci of Lancefield Group E are important pathogens of pigs and have many biochemical characteristics in common with Lancefield Groups P, U and V strains. The somewhat controversial interrelationships of these streptococci were resolved with the demonstration that they should be included within a single species named *S. porcinus* (Collins *et al.*, 1984). Cells are ovoid and form small to medium length chains. On blood agar isolates produce *β*-haemolysis. No growth occurs at 10°C and at 45°C or after heating at 60°C for 30 min. The biochemical characteristics of *S. porcinus* are shown in Table 4.4. Cell wall peptidoglycan type is Lys-Ala_{2-4}. Major long-chain fatty acids are hexadecanoic (C16:0) and octadecenoic (*cis*-vaccenic). Menaquinones are absent. DNA G+C content is 37–38 mol%. Strains may react against Lancefield Group E, P, U or V antisera. 16s rRNA comparative sequencing has shown *S. porcinus* to belong within the pyogenic group of streptococci. Strains of *S. porcinus* have previously been referred to as Lancefield Group E, P, U or V streptococci, '*Streptococcus infrequens*', '*S. lentus*' or '*S. subacidus*'. Source/habitat: associated with diseases of pigs (abscesses of the cervical lymph nodes, pneumonia and septicaemia) and from milk. Type strain NCTC 10999.

4.13.31 Streptococcus pyogenes

Streptococcus pyogenes (Rosenbach, 1884), the type species of the genus *Streptococcus* is one of the most important human pathogens within the genus, giving rise to a number of pyogenic and septicaemic infections and is the only species of *Streptococcus* regularly causing epidemics in man (Maxted, 1978). Cells are 0.5–1 μm in diameter, spherical or ovoid, occurring in short to medium length chains or frequently as pairs in clinical samples. Broth cultures yield long chains. *β*-Haemolysis is produced on blood agar with three colonial types occurring: glossy, mucoid or matt (dehydrated mucoid). Growth is enhanced by the addition of blood or serum to broths and is optimum at 37°C. Strains do not grow at 10°C, 45°C or in the presence of 6.5% NaCl, 40% bile or at pH 9.6. The biochemical characteristics of this species are shown in Table 4.4. Cell wall peptidoglycan type is Lys-Ala_{2-3}. DNA G+C content is 35–39 (T_m). *Streptococcus pyogenes* possess the Lancefield Group A carbohydrate antigen and strains are also divided on the basis of M, T and R surface protein antigens. Extracellular products that are important biologically and diagnostically include streptolysins O (oxygen labile) and S (oxygen stable and responsible for the zone of *β*-haemolysis seen around colonies growing on blood agar), erythrogenic toxin (elicits the rash in scarlet fever), streptokinase, DNase, NADase, hyaluronidase and proteinase. 16S rRNA comparative sequencing shows that this species is grouped within the pyogenic streptococci. Source/habitat: upper respiratory tract in man, inflammatory exudates, skin lesions, blood and contaminated environmental dust. Type strain ATCC 12344.

4.13.32 Streptococcus rattus

This species within the mutans-like streptococci was first proposed by Coykendall (1977) for unusual strains of '*S. mutans*' that possessed the antigen b of Bratthall (1970) and produced ammonia from arginine. It should be noted that studies involving this species have invariably been limited to the same few strains. Cells are approximately 0.5 μm in diameter, occurring in chains or pairs. On sucrose-containing agar some strains form rubbery colonies or rough and heaped colonies, with liquid glucan present in beads or puddles. The biochemical characteristics of this species are shown in Table 4.3. Growth is improved under conditions of reduced O_2 or by the addition of CO_2. Strains of this species contain a polysaccharide antigen designated type b. Cell wall peptidoglycan type is Lys-Ala_{2-3}. DNA G+C content is 41–43 mol%. DNA–DNA hybridization studies and 16S rRNA comparative sequence analysis have shown *S. rattus* to be most closely related to *S. mutans* within the mutans streptococci. Strains of this species were previously designated *Streptococcus mutans* serotype b. Source/habitat: oral cavities of rat and man (occasionally). Type strain ATCC 19645.

4.13.33 Streptococcus salivarius

Streptococcus salivarius was first described by Andrewes and Horder (1906) from human saliva. Although not considered an important pathogenic species, *S. salivarius* has occasionally been isolated from infective endocarditis and some strains have been shown to be cariogenic in gnotobiotic animals. Cells are approximately 0.8–1 μm in diameter, spherical or ovoid and form chains of varying length. On blood agar strains are usually non-haemolytic with a few giving α- or *β*-haemolysis. On sucrose-containing media large mucoid colonies are formed due to extracellular polysaccharide production (soluble fructan:levan). In addition isolates can occasionally produce an insoluble glucan (dextran) with the relative proportions of these extracellular polysaccharides determining the resulting degree of roughness or smoothness of the colonial texture. The biochemical characteristics of this species are shown in Table 4.3. Growth can occur on complex media at 45°C but not at 10°C and ammonia and urea can serve as a source of nitrogen in media that include biotin, cysteine, glucose, nicotinic acid, riboflavin, thiamin, panthothenic acid and inorganic salts. Long-chain fatty acid analysis by capillary gas-liquid chromatography has demonstrated the presence of eicosenoic (C20:1) acids. Cell wall peptidoglycan type is Lys-Thr-Gly. DNA G+C content is 39–42 mol%. Some strains react with Lancefield Group K antiserum. A close relationship has been demonstrated between this species and *S. vestibularis* and *S. thermophilus* by DNA–DNA hybridization. These three species form a distinct cluster (species group) by comparative sequence analysis of 16S rRNA (Bentley *et al.*, 1991). Source/habitat: the oral cavities of man and animals, in particular the tongue and saliva. Type strain NCTC 8618 (ATCC 7073).

4.13.34 Streptococcus sanguis

Originally described by White and Niven (1946) from the blood of patients with endocarditis *S. sanguis* was shown to be biochemically, serologically and, more significantly, genetically heterogeneous before being redefined according to Kilian *et al.* (1989a). Cells usually grow as short chains in serum broth and produce alpha-haemolysis on blood agar and greening on chocolate agar. Extracellular polysaccharide (dextran) is produced on sucrose-containing agar, giving smooth, entire, hard and adherent colonies. The biochemical characteristics of this species and its biotypes are shown in Table 4.3. Cell wall peptidoglycan type is Lys-Ala_{1-3} and rhamnose and glycerol teichoic acid are present in the cell wall. DNA G+C content is 46 mol%. The majority of the strains react against Lancefield Group H antiserum raised against strain Blackburn but not with Group H antiserum raised against strain F90A. Strains of *S. sanguis* have also previously been designated *S. sanguis* I, '*S. sbe*' and group H *streptococcus*. 16S rRNA sequence analysis has revealed a relatively close phylogenetic relationship between *S. sanguis*, *S. oralis*, *S. pneumoniae*, *S. parasanguis* and the '*S. milleri*-group' (*S. anginosus*, *S. intermedius* and *S. constellatus*). Source/habitat: human oral cavity. Type strain NCTC 7863 (ATCC 10556).

4.13.35 Streptococcus sobrinus

This species was first described by Coykendall (1983) for mutans-like streptococci possessing the groups d or g antigens (Perch *et al.*, 1974). It is thought to be an aetiological agent of dental caries in man, together with *S. mutans*. Cells are 0.5 μm in diameter and form long chains or occur in pairs. Strains are mostly non-haemolytic on blood agar with some strains producing α-haemolysis. On sucrose-containing agar colonies are rough, heaped, approximately 1 mm in diameter and are surrounded by liquid containing glucan. The biochemical characteristics of this species are shown in Table 4.3. Strains of *S. sobrinus* belong to serotypes d or g on the basis of polysaccharide antigens. However, the type strain does not react with either type d or g antisera. Cell wall peptidoglycan type is Lys-Thr-Gly. DNA G+C content is 44–46 mol%. Within the mutans streptococci *S. sobrinus* is most closely related to *S. downei* as shown by DNA–DNA hybridization and 16S rRNA sequence comparisons. This species is associated with dental caries in man and is cariogenic in experimental animals. Previously *S. sobrinus* strains were designated *S. mutans* serotypes d or g. Source/habitat: tooth surface in human oral cavity. Type strain ATCC 33478.

4.13.36 Streptococcus suis

The group of streptococci brought together by Kilpper-Bälz and Schleifer (1987) into a single species named *Streptococcus suis* resolved the taxonomic position of a serologically heterogeneous collection of strains that constitute an important pathogen of pigs. Cells are ovoid, less than 2 μm in diameter, occurring mainly singly or in pairs and occasionally in short chains. Cells can sometimes also tend to form rods. *β*-haemolysis is produced on horse blood agar, whereas α-haemolysis is produced on sheep blood agar. The biochemical characteristics of this species are shown in Table 4.5. Some strains are resistant to 40% bile although no growth occurs in the presence of 6.5% NaCl, 0.04% tellurite or at 10°C or 45°C. Cell wall peptidoglycan type is usually lysine-direct with occasional strains possessing Lys-Ala_{1-2}. Glucose, galactose, glucosamine and rhamnose are present in the cell wall. Strains contain a lipid-bound teichoic acid cell wall antigen that is closely related to the Lancefield Group D antigen and results in a reaction with Lancefield Group D antiserum. Strains are groupable into Lancefield Groups R, RS, S and T or are non-groupable. Cross-reaction with Groups E and N antisera and between Group B and Group R antisera occur. In addition, strains can also be subdivided into one to eight capsular polysaccharide serotypes (serovars). Strains of *Streptococcus suis* have also been designated as streptococci of serological groups R, S or T. 16S rRNA sequence analysis has shown *S. suis* to be genetically distinct from other streptococcal species and species groups with the possible exception of *S. acidominimus*. Source/habitat: isolated from pigs with bacteraemia, meningitis or respiratory disease. Type strain NCTC 10234.

4.13.37 Streptococcus thermophilus

The taxonomic status of *S. thermophilus* (Orla-Jensen, 1919) has fluctuated in recent years due to the close relationship demonstrated between these streptococci and the species *S. salivarius*. This discovery resulted in the temporary inclusion of both in a single species *S. salivarius* as subsp. *salivarius* and subsp. *thermophilus* (Farrow and Collins, 1984a) until separate species status was reproposed by Schleifer *et al.* (1991) on the basis of both genetic and phenetic criteria. Cells are 0.7–1 μm in diameter, spherical or ovoid, forming chains or occurring in pairs. Growth at 45°C can give rise to irregular cells and segments. Strains are either α-haemolytic or non-haemolytic on blood agar. The biochemical characteristics of *S. thermophilus* are given in Table 4.3. No growth occurs at 15°C but all strains grow at 45°C and most are able to grow at 50°C. Survives heating for 30 min at 60°C. No growth occurs in 0.1% w/v methylene blue or at pH 9.6. Requires B-vitamins and some amino acids. A group antigen has not been demonstrated. Cell wall peptidoglycan type is Lys-Ala_{2-3}. DNA G+C content is 37–40 mol%. This species is closely related to *S. salivarius* and *S. vestibularis* and, as mentioned above, previously has been proposed as a subspecies of *S. salivarius* (*S. salivarius* subsp. *thermophilus*). 16S rRNA sequence data have demonstrated that *S. thermophilus* is one of a three-member species group that also includes *S. salivarius* and *S. vestibularis*. Source/habitat: milk (heated and pasteurized) – natural habitat unknown. Type strain ATCC 19258 (NCDO 573).

4.13.38 Streptococcus uberis

An important species occurring in bovine mastitis, *Streptococcus uberis* (Diernhofer, 1932) was later shown to include two distinct genetic groups called *S. uberis* type I and II. *Streptococcus uberis* II strains have now been proposed as a distinct species called *S. parauberis* (Williams and Collins, 1990). Cells form moderate length chains or pairs. Weak α-haemolysis or non-haemolysis is produced on blood agar. No growth occurs at 10°C or

45°C. May or may not survive heating at 60°C for 30 mins. Growth occurs in the presence of 4% NaCl but not 6.5% NaCl. The biochemical characteristics of *S. uberis* are shown in Table 4.4. Cell wall peptidoglycan type is Lys-Ala_{1-3}. DNA G+C content is 36–37.5 mol%. Some strains may react with Lancefield Groups E, P or G antisera (Garvie and Bramley, 1979). 16S rRNA comparative sequence analysis has demonstrated that *S. uberis* is a species within the pyogenic group of streptococci. Source/habitat: lips, skin and udder tissue of cows and raw milk. Type strain NCTC 3858 (ATCC 19436).

4.13.39 Streptococcus vestibularis

Streptococcus vestibularis (Whiley and Hardie, 1988) is a relatively new species of oral streptococcus, closely related to *S. salivarius* and *S. thermophilus*. Clinical significance of the species, if any, remains to be established. Cells are approximately 1 μm in diameter, chain-forming cocci. α-Haemolysis is produced on blood agar. Strains do not produce extra- or intracellular polysaccharide. The biochemical characteristics of *S. vestibularis* are shown in Table 4.3. Growth does not occur at 10°C or at 45°C, in the presence of 4% w/v NaCl, 0.0004% w/v crystal violet or in 40% w/v bile but most strains grow in the presence of 10% w/v bile. Long-chain fatty acid analyses have demonstrated major amounts of hexadecanoic (C16:0; palmitic) and octadecenoic (C18:1w7; *cis*-vaccenic) acids in addition to tetradecanoic (C14:0; myristic), hexadecenoic (C16:1), octadecanoic (C18:0; stearic), octadecenoic (C18:1W9; oleic) and eicosenoic (C20:1) acids. Cell wall peptidoglycan type is Lys-Ala_{1-3}. DNA G+C content is 38–40 mol%. Whole cell derived polypeptide patterns by SDS-PAGE, DNA–DNA hybridization and 16S rRNA studies have shown the close relationship between *S. vestibularis*, *S. salivarius* and *S. thermophilus*. Source/habitat: the human oral cavity, especially the vestibular mucosa. Type strain NCTC 12166.

References

Anon. (1991) *Rapid ID 32 Strep Analytical Profile Index*, 1st edn. BioMérieux s.a. Marcy-l'Etoile, France.

Anon. (1994a) Invasive group A streptococcal infections in Gloucestershire. *CDR Weekly Communicable Disease Report*, **4** (21), 97.

Anon. (1994b) Invasive group A streptococcal infections – update. *CDR Weekly Communicable Disease Report*, **4** (27), 123.

Aguirre, M., Morrison, D., Cookson, B.D., Gay, F.W. and Collins, M.D. (1993) Phenotypic and phylogenetic characterization of some *Gemella*-like organisms from human infections: description of *Dolosigranulum pigrum* gen. nov., sp. nov. *Journal of Applied Bacteriology*, **75**, 608–612.

Andrewes, F.W. and Horder, J. (1906) A study of the streptococci pathogenic for man. *Lancet*, **2**, 708–713.

Avery, O.T., Macleod, C.C. and McCarty, M. (1944) Studies on the chemical nature of the substance inducing transformation of pneumococcal types. Induction of transformation by a desoxyribonucleic acid fraction isolated from pneumococcus type III. *Journal of Experimental Medicine*, **79**, 137–158.

Ayers, S.H. and Mudge, C.S. (1922) The streptococci of the bovine udder. *Journal of Infectious Disease*, **31**, 40–50.

Balows, A., Trüper, H.G., Dworkin, M., Harder, W. and Schleifer, K.H. (1992) *The Prokaryotes*, Vol. II, 2nd edn. Springer-Verlag, New York, USA.

Barnes, E.H., Ross, P.W., Wilson, C.D., Hardie, J.M., Marsh, P.D., Mead, G.C., Sharpe, M.E., Mossel, D.A.A., Burman, N.P., Evans, A.W. and Ingram, M. (1978) Isolation media for streptococci: Proceedings of a discussion meeting. In *Streptococci* (Society for Applied Bacteriology Symposium Series No. 7) (eds Skinner, F.A. and Quenel, L.B.) Academic Press, London, UK, pp. 371–395.

Barnham, M. (1989) Invasive streptococcal infections in the era before the aquired immune deficiency syndrome: a 10 years' compilation of patients with streptococcal bacteraemia in North Yorkshire. *Journal of Infection*, **18**, 231–248.

Beighton, D., Russell, R.R.B. and Hayday, H. (1981) The isolation and characterization of *Streptococcus mutans* serotype h from dental plaque of monkeys (*Macaca fascicularis*). *Journal of General Microbiology*, **124**, 271–279.

Beighton, D., Hayday, H., Russell, R.R.B. and Whiley, R.A. (1984) *Streptococcus macacae* sp. nov. from dental plaque of monkeys (*Macaca fascicularis*). *International Journal of Systematic Bacteriology*, **34**, 332–335.

Beighton, D., Whiley, R.A. and Homer, K.A. (1990) Transferrin binding by *Streptococcus oralis* and other oral streptococci. *Microbial Ecology in Health and Disease*, **3**, 145–150.

Beighton, D., Hardie, J.M. and Whiley, R.M. (1991) A scheme for the identification of viridans streptococci. *Journal of Medical Microbiology*, **35**, 367–372.

Bentley, R.W., Leigh, J.A. and Collins, M.D. (1991) Intrageneric structure of *Streptococcus* based on comparative analysis of small-subunit rRNA sequences. *International Journal of Systematic Bacteriology*, **41**, 487–494.

Bentley, R.W., Leigh, J.A. and Collins, M.D. (1993) Development and use of species-specific oligonucleotide probes for differentiation of *Streptococcus uberis* and *Streptococcus parauberis*. *Journal of Clinical Microbiology*, **31**, 57–60.

Billroth, A.W. (1874) *Untersuchungen über die Vegetationsformen von Coccobacteria Septica*. Georg Reimer, Berlin, Germany.

Boulnois, G.J. (1992) Pneumococcal proteins and the pathogenesis of disease caused by *Streptococcus pneumoniae*. *Journal of General Microbiology*, **138**, 249–259.

Bouvet, A. and Acar, J.F. (1984) New bacteriological aspects of infective endocarditis. *European Heart Journal*, **5** (Suppl. C), 45–48.

Bouvet, A.F., Van de Rijn, I. and McCarty, M. (1981) Nutritionally variant streptococci from patients with endocarditis: growth parameters in a semisynthetic medium and demonstration of a chromophore. *Journal of Bacteriology*, **146**, 1075–1082.

Bouvet, A.F., Grimont, F. and Grimont, P.A.D. (1989) *Streptococcus defectivus* sp. nov. and *Streptococcus adjacens* sp. nov. nutritionally variant streptococci from human clinical specimens. *International Journal of Systematic Bacteriology*, **39**, 290–294.

Bratthall, D. (1970) Demonstration of five serological groups of streptococcal strains resembling *Streptococcus mutans*. *Odontologisk Revy*, **21**, 143–152.

Bridge, P.D. and Sneath, P.H.A. (1982) *Streptococcus gallinarum* sp. nov. and *Streptococcus oralis* sp. nov. *International Journal of Systematic Bacteriology*, **32**, 410–415.

Bridge, P.G. and Sneath, P.H.A. (1983) Numerical taxonomy of *Streptococcus*. *Journal of General Microbiology*, **129**, 565–597.

Brooker, J.D., O'Donovan, L.A., Skene, I., Clarke, K., Blackall, L. and Muslera, P. (1994) *Streptococcus caprinus* sp. nov. a tannin-resistant ruminal bacterium from feral goats. *Letters in Applied Microbiology*, **18**, 313–316.

Brown, J.H. (1919) *The Use of Blood Agar for the Study of Streptococci*. (Monograph No. 9). The Rockefeller Institute for Medical Research, New York.

Busse, W. (1991) Pathogenesis and sequelae of respiratory infections. *Reviews of Infectious Diseases*, **13**, S477–S485.

Buxton, A. and Fraser, G. (1977) *Animal Microbiology*, Vol. 1. Blackwell Scientific Publications, Oxford, UK, pp. 165–176.

Caufield, P.W., Cutter, G.R. and Dasanayake, A.P. (1993) Initial acquisition of mutans streptococci by infants; evidence for a discrete window of infectivity. *Journal of Dental Research*, **72**, 37–45.

Christie, R., Atkins, N.E. and Munch-Peterson, E. (1944) A note on the lytic phenomenon shown by Group B streptococci. *Australian Journal of Experimental Biology and Medical Science*, **22**, 197–200.

Clarke, J.K. (1924) On the bacterial factor in the aetiology of dental caries. *British Journal of Experimental Pathology*, **5**, 141–147.

Classen, D.C., Burke, J.P., Ford, C.D., Evershed, S., Aloia, M.R., Wilfahrt, J.K. and Elliott, J.A. (1990) *Streptococcus mitis* sepsis in bone marrow transplant patients receiving oral antimicrobial prophylaxis. *American Journal of Medicine*, **89**, 441–446.

Collins, M.D. and Wallbanks, S. (1992) Comparative sequence analyses of the 16s rRNA genes of *Lactobacillus minutus*, *Lactobacillus rimae* and *Streptococcus parvulus*: proposal for the creation of a new genus *Atopobium*. *FEMS Microbiology Letters*, **95**, 235–240.

Collins, M.D., Farrow, J.A.E., Katic, V. and Kandler, O. (1984) Taxonomic studies on streptococci of serological groups E, P, U and V: description of *Streptococcus porcinus* sp. nov. *Systematic and Applied Microbiology*, **5**, 402–413.

Collins, M.D., Ash, C., Farrow, J.A.E., Wallbanks, S. and Williams, A.M. (1989) 16S ribosomal ribonucleic acid sequence analysis of lactococci and related taxa. Description of *Vagococcus fluvialis* gen. nov., sp. nov. *Journal of Applied Bacteriology*, **67**, 453–460.

Colman, G. (1968) The application of computers to the classification of streptococci. *Journal of General Microbiology*, **50**, 149–158.

Colman, G. (1990) *Streptococcus* and *Lactobacillus*. In *Topley and Wilson's Principles of Bacteriology, Virology and Immunity*, Vol. 2, 8th edn. (eds Parker, M.T. and Collier, L.H.). Edward Arnold, London, UK, pp. 119–159.

Colman, G. and Ball, L.C. (1984) Identification of streptococci in a medical laboratory. *Journal of Applied Bacteriology*, **57**, 1–14.

Colman, G. and Williams, R.E.O. (1965) The cell walls of streptococci. *Journal of General Microbiology*, **41**, 375–387.

Colman, G. and Williams, R.E.O. (1972) Taxonomy of some human viridans streptococci. In *Streptococci and Streptococcal Disease* (eds Wannamaker, L.W. and Matsen, J.M.). Academic Press, London, UK, pp. 281–299.

Colman, G., Tanna, A., Efstratiou, A. and Gaworzewska, E.T. (1993) The serotypes of *Streptococcus pyogenes* present in Britain during 1980–1990 and their association with disease. *Journal of Medical Microbiology*, **39**, 165–178.

Coykendall, A.L. (1974) Four types of *Streptococcus mutans* based on their genetic, antigenic and biochemical characteristics. *Journal of General Microbiology*, **83**, 327–338.

Coykendall, A.L. (1977) Proposal to elevate the subspecies of *Streptococcus mutans* to species status, based on their molecular composition. *International Journal of Systematic Bacteriology*, **27**, 26–30.

Coykendall, A.L. (1983) *Streptococcus sobrinus* nom. rev. and *Streptococcus ferus* nom. rev.: habitat of these and other mutans streptococci. *International Journal of Systematic Bacteriology*, **33**, 883–885.

Coykendall, A.L. (1989) Classification and identification of the viridans streptococci. *Clinical Microbiology Reviews*, **2**, 315–328.

Coykendall, A.L. and Gustafson, K.B. (1985) Deoxyribonucleic acid hybridisations among strains of *Streptococcus salivarius* and *Streptococcus bovis*. *International Journal of Systematic Bacteriology*, **35**, 274–280.

Coykendall, A.L., Specht, P.A. and Samol, H.H. (1974) *Streptococcus mutans* in a wild, sucrose-eating rat population. *Infection and Immunity*, **10**, 216–219.

Coykendall, A.L., Bratthall, D., O'Connor, K. and Dvarskas, R.A. (1976) Serological and genetic examination of some non-typical *Streptococcus mutans* strains. *Infection and Immunity*, **14**, 667–670.

Cruz Colque, J.I., Devriese, L.A. and Haesebrouck, F. (1993). Streptococci and enterococci associated with tonsils of cattle. *Letters in Applied Microbiology*, **16**, 72–74.

Davenport, E.S., Day, S., Hardie, J.M. and Smith, J.M. (1992) A comparison between commercial kits and conventional methods for enumeration of salivary mutans streptococci and lactobacilli. *Community Dental Health*, **9**, 261–271.

Dent, V.E., Hardie, J.M. and Bowden, G.H. (1978) Steptococci isolated from dental plaque of animals. *Journal of Applied Bacteriology*, **44**, 249–258.

Denys, G.A. and Carey, R.B. (1992) Identification of *Streptococcus pneumoniae* with a DNA probe. *Journal of Clinical Microbiology*, **30**, 2725–2727.

de Soet, J.J., Van Dalen, P.J., Russell, R.R.B. and De Graaff, J. (1990) Identification of mutans streptococci with monoclonal antibodies. *Antonie van Leeuwenhoek*, **58**, 219–225.

de Soet, J.J., Van Steenbergen, T.J.M. and De Graaff, J. (1992) *Streptococcus sobrinus*: taxonomy, virulence and pathogenicity. *Alpe Adria Microbiology Journal*, **3**,127–145.

Devriese, L.A. (1991) Streptococcal ecovars associated with different animal species: epidemiological significance of serogroups and biotypes. *Journal of Applied Bacteriology*, **71**, 478–483.

Devriese, L.A., Hommez, J., Kilpper-Bälz, R. and Schleifer, K.H. (1986) *Streptococcus canis* sp. nov.: a species of group G streptococci from animals. *International Journal of Systematic Bacteriology*, **36**, 422–425.

Devriese, L.A., Kilpper-Bälz, R. and Schleifer, K.H. (1988) *Streptococcus hyointestinalis* sp. nov. from the gut of swine. *International Journal of Systematic Bacteriology*, **38**, 440–441.

Devriese, L.A., Uyttebroek, E., Gevaert, D., Vandekerckhove, P. and Ceyssens, K. (1990) *Streptococcus bovis* infections in pigeons. *Avian Pathology*, **19**, 429–434.

Devriese, L.A., Hommez, J., Wijfels, R. and Haesebrouck, F. (1991) Composition of the enterococcal and streptococcal intestinal flora of poultry. *Journal of Applied Bacteriology*, **71**, 46–50.

Devriese, L.A., Cruz Colque, J.I., De Herdt, P. and Haesebrouck, F. (1992a) Identification and composition of the tonsillar and anal enterococcal and streptococcal flora of dogs and cats. *Journal of Applied Bacteriology*, **73**, 421–425.

Devriese, L.A., Laurier, L., De Herdt, P. and Haesebrouck, F. (1992b) Enterococcal and streptococcal species isolated from faeces of calves, young cattle and dairy cows. *Journal of Applied Bacteriology*, **72**, 29–31.

Devriese, L.A., Hommez, J., Pot, B. and Haesebrouck, F. (1994) Identification and composition of the streptococcal and enterococcal flora of tonsils, intestines and faeces of pigs. *Journal of Applied Bacteriology*, **77**, 31–36.

Diernhofer, K. (1932) Aesculinbouillon als holfsmittel für die differenzierung von euter- und milchstreptokokken bei massenuntersuchungen. *Milchwirtsch Forsch.*, **13**, 368–374.

Douglas, C.W.I., Heath, J., Hampton, K.J. and Preston, F.E. (1993) Identity of viridans streptococci isolated from cases of infected endocarditis. *Journal of Medical Microbiology*, **39**, 179–182.

Dunny, G., McKay, L. and Cleary, P.P. (eds) (1991) *Streptococcal Genetics*. American Society for Microbiology, Washington, DC, USA.

Ellwood, D.C. (1976) Chemostat studies of oral bacteria. In *Microbial Aspects of Dental Caries* (Special Supplement to Microbiology Abstracts 3) (eds Stiles, H.M., Loesche, W.J. and O'Brien, T.C.). IRL Information Retrieval Inc., Washington DC, pp. 785–798

Facklam, R.R. (1972) Recognition of group D streptococcal species of human origin by biochemical and physiological tests. *Applied Microbiology*, **23**, 1131–1139.

Facklam, R.R. (1977) Physiological differentiation of viridans streptococci. *Journal of Clinical Microbiology*, **5**, 184–201.

Farrow, J.A.E. and Collins, M.D. (1984a) DNA base composition, DNA-DNA homology and long-chain fatty acid studies on *Streptococcus thermophilus* and *Streptococcus salivarius*. *Journal of General Microbiology*, **130**, 357–362.

Farrow, J.A.E. and Collins, M.D. (1984b) Taxonomic studies on streptococci of serological groups C, G and L and possibly related taxa. *Systematic and Applied Microbiology*, **5**, 483–493.

Farrow, J.A.E., Kruze, J., Phillips, B.A., Bramley, A.J. and Collins, M.D. (1984) Taxonomic studies on *Streptococcus bovis* and *Streptococcus equinus*: description of *Streptococcus alactolyticus* sp. nov. and *Streptococcus saccharolyticus* sp. nov. *Systematic and Applied Microbiology*, **5**, 467–482.

Ferretti, J.J. (1992) Molecular basis of virulence and antibiotic resistance in group A streptococci. In *New Perspectives on Streptococci and Streptococcal Infections* (Proceedings of the XI Lancefield Intenational Symposium on Streptococci and Streptococcal Diseases, September 1990, Siena) (ed. Orefici, G.). Gustav Fischer Verlag, Stuttgart, Germany, pp. 329–335.

Ferretti, J.J. and Curtiss, R. (eds) (1987) *Streptococcal Genetics*. American Society for Microbiology, Washington, DC, USA.

Fischetti, V.A. (1989) Streptococcal M protein: molecular design and biological behaviour. *Clinical Microbiology Reviews*, **2**, 285–314.

Fischetti, V.A., Danchol, V. and Schneedwind, O. (1992) Surface proteins from Gram-positive cocci share unique structural features. In *New Perspectives on Streptococci and Streptococcal Infections* (Proceedings of the XI Lancefield International Symposium on Streptococci and Streptococcal Diseases, September 1990, Siena) (ed. Orefici, G. Gustav). Fischer Verlag, Stuttgart, Germany, pp. 165–168.

Frandsen, E.V., Pedrazzoli, V. and Kilian, M. (1991) Ecology of viridans streptococci in the oral cavity and pharynx. *Oral Microbiology and Immunology*, **6**, 129–133.

Freney, J., Bland, S., Etienne, J., Desmonceaux, M., Boeufgras, J.M. and Fleurette, J. (1992) Description and evaluation of the semiautomated 4-hour Rapid ID STREP method for identification of streptococci and members of related genera. *Journal of Clinical Microbiology*, **30**, 2657–2661.

Garvie, E.I. and Bramley, A.J. (1979) *Streptococcus uberis*: an approach to its classification. *Journal of Applied Bacteriology*, **46**, 295–304.

Garvie, E.I., Farrow, J.A.E. and Phillips, B.A. (1981) A taxonomic study of some strains of streptococci which grow at 10°C but not at 45°C, including *Streptococcus lactis* and *Streptococcus cremoris*. *Zentralblatt für Bakteriologie Hygiene 1. Abteilung Originale*, **C2**, 151–165.

Garvie, E.I., Farrow, J.A.E. and Bramley, A.J. (1983) *Streptococcus dysgalactiae* (Diernhofer) nom. rev. *International Journal of Systematic Bacteriology*, **33**, 404–405.

Gibbons, R.J. and Van Houte, J. (1975) Bacterial adherence in oral microbial ecology. *Annual Reviews of Microbiology*, **29**, 19–44.

Gillespie, S.H., McWhinney, P.H.M., Patel, S., Raynes, J.G., McAdam, K.P.W.J., Whiley, R.A. and Hardie, J.M. (1993) Species of alpha-haemolytic streptococci possessing a C-polysaccharide phosphorylcholine-containing antigen. *Infection and Immunity*, **61**, 3076–3077.

Gordillo, M.E., Singh, K.V., Baker, C.J. and Murray, B.E. (1993) Typing of group B streptococci: comparison of pulsed-field gel electrophoresis and conventional electrophoresis. *Journal of Clinical Microbiology*, **31**, 1430–1434.

Gossling, J. (1988) Occurrence and pathogenicity of the *Streptococcus milleri* group. *Reviews of Infectious Diseases*, **10**, 257–285.

Hamada, S. and Slade, H.D. (1980) Biology, immunology and cariogenicity of *Streptococcus mutans*. *Microbiology Reviews*, **44**, 331–384.

Handley, P.S. (1990) Structure, composition and functions of surface structures of oral bacteria. *Biofouling*, **2**, 239–264.

Handley, P.S., Coykendall, A., Beighton, D., Hardie, J.M. and Whiley, R.A. (1991) *Streptococcus crista* sp. nov., a viridans streptococcus with tufted fibrils, isolated from the human oral cavity and throat. *International Journal of Systematic Bacteriology*, **41**, 543–547.

Hansman, D. and Bullen, M.M. (1967) A resistant pneumococcus. *Lancet*, **ii**, 264–265.

Hardie, J.M. (1986) Genus *Streptococcus*. In *Bergey's Manual of Determinative Bacteriology*, Vol. 2 (eds Sneath, P.H.A., Mair, N.S. and Sharpe, M.E.). Williams and Wilkins, Baltimore, pp. 1043–1071.

Hardie, J.M. (1992) Oral microbiology: current concepts in the microbiology of dental caries and periodontal disease. *British Dental Journal*, **172**, 271–278.

Hardie, J.M. and Bowden, G.H. (1974) The normal microbial flora of the mouth. In *The Normal Flora of Man* (eds Skinner, F.A. and Carr, J.G.). Academic Press, London, UK, pp. 47–83.

Hardie, J.M. and Marsh, P.D. (1978a) Streptococci and the human oral flora. In *Streptococci* (eds Skinner, F.A. and Quesnel, L.B.). Academic Press, London, UK, pp. 157–206.

Hardie, J.M. and Marsh, P.D. (1978b) Isolation media for streptococci. In *Oral Streptococci* (eds Skinner, F.A. and Quesnel, L.B.). Academic Press, London, UK, pp. 380–383.

Hardie, J.M. and Whiley, R.A. (1992) The genus *Streptococcus*-oral. In *The Prokaryotes*, Vol. II, 2nd edn (eds Balows, A., Trüper, H.G., Dworkin, M., Harder, W. and Schleifer, K.H.). Spring-Verlag, New York, USA, pp. 1421–1449.

Hardie, J.M. and Whiley, R.A. (1994) Recent developments in streptococcal taxonomy: their relation to infections. *Reviews in Medical Microbiology*, **5** (3), 151–162.

Harland, N.M., Leight, J.A. and Collins, M.D. (1993) Development of gene probes for the specific identification of *Streptococcus uberis* and *Streptococcus parauberis* based upon large subunit rRNA gene sequences. *Journal of Applied Bacteriology*, **74**, 526–531.

Henrichsen, J. (1985) The bacteriology of GBS. In *Neonatal Group B Streptococcal Infections* (eds Christensen, K.K., Christensen, P. and Ferrieri, P.). S. Karger Ag, Basel, Switzerland, pp. 53–56.

Herzberg, M.C., Gong, K.E., Macfarlane, G.D., Erickson, P.R., Soberay, A.H., Krebsbach, P.H., Manjula, G., Schilling, K. and Bowen, W.H. (1990) Phenotypic characterization of *Streptococcus sanguis* virulence factors associated with bacterial endocarditis. *Infection Immunity*, **58**, 515–522.

Hogg, S.D. (1992) The lactic microflora of the oral cavity. In *The Lactic Acid Bacteria in Health and Disease*, Vol. I (ed. Wood, B.J.B.). Elsevier Applied Science, London, UK, pp. 115–148.

Holdemann, L.V. and Moore, W.E.C. (1974) New genus, *Coprococcus*, twelve new species, and emended descriptions of four previously described species of bacteria from human feces. *International Journal of Systematic Bacteriology*, **24**, 260–277.

Homer, K.A., Denbow, L., Whiley, R.A. and Beighton, D. (1993) Chondroitin sulfate depolymerase and hyaluronidase activities of viridans streptococci determined by a sensitive spectorphotometric assay. *Journal of Clinical Microbiology*, **31**, 1648–1651.

Jelinková, J. (1977) Group B streptococci in the human population. *Current Topics in Microbiology and Immunology*, **76**, 127–165.

Johnson, R.B. (1991) Pathogenesis of pneumococcal pneumonia. *Reviews of Infectious Diseases*, **13**, S509–S517.

Johnston, N.W. (1991) *Risk Markers for Oral Diseases*, Vol. 3. Cambridge University Press, Cambridge, UK.

Jones, D. (1978) Composition and differentiation of the genus *Streptococcus*. In *Streptococci* (Society for Applied Bacteriology Symposium Series No. 7) (eds Skinner, F.A. and Quesnel, L.B.). Academic Press, London, UK, pp. 1–49.

Kaplan, E.L. (1992) Change in streptococcal infections in the late 20th century: the whats and whys. In *New Perspectives on Streptococci and Streptococcal Infections* (Proceedings of the XI Lancefield International Symposium on Streptococci and Streptococcal Diseases, September 1990, Siena) (ed. Orefici, G.). Gustav Fischer Verlag, Stuttgart, Germany, pp. 5–7.

Kehoe, M.A. (1991) New aspects of *Streptococcus pyogenes* pathogenicity. *Reviews in Medical Microbiology*, **2**, 147–152.

Kilian, M., Mikkelsen, L. and Henrichsen, J. (1989a) Taxonomic study of viridans streptococci: description of *Streptococcus gordonii* sp. nov. and emended descriptions of *Streptococcus sanguis* (White and Niven, 1946), *Streptococcus oralis* (Bridge and Sneath, 1982) and *Streptococcus mitis* (Andrewes and Horder 1906). *International Journal of Systematic Bacteriology*, **39**, 471–484.

Kilian, M., Reinholdt, J., Nyvad, B., Frandsen, E.V. and Mikkelsen, L. (1989b) IgA1 proteases of oral streptococci: ecological aspects. *Immunological Investigations*, **18**, 161–170.

Kilpper-Bälz, R. and Schleifer, K.H. (1987) *Streptococcus suis* sp. nov, nom. rev. *International Journal of Systematic Bacteriology*, **37**, 160–162.

Kilpper-Bälz, R. and Schleifer, K.H. (1988). Transfer of *Streptococcus morbillorum* to the genus *Gemella* as *Gemella morbillorum* comb. nov. *International Journal of Systematic Bacteriology*, **38**, 442–443.

Kilpper-Bälz, R., Williams, B.L., Lütticken, R. and Schleifer, K.H. (1984) Relatedness of '*Streptococcus milleri*' with *Streptococcus anginosus* and *Streptococcus constellatus*. *Systematic and Applied Microbiology*, **5**, 494–500.

Kilpper-Bälz, R., Wenzig, P. and Schleifer, K.H. (1985) Molecular relationships and classification of some viridans streptococci as *Streptococcus oralis* and emended description of *Streptococcus oralis* (Bridge and Sneath, 1982). *International Journal of Systematic Bacteriology*, **35**, 482–488.

Klein, J.P. and Scholler, M. (1988) Recent advances in the development of a *Streptococcus mutans* vaccine. *European Journal of Epidemiology*, **4**, 419–425.

Klein, R.S., Recco, R.A., Catalano, M.T., Edberg, S.C., Casey, J.J. and Steigbigel, N.H. (1977) Association of *Streptococcus bovis* with carcinoma of the colon. *New England Journal of Medicine*, **297**, 800–802.

Klugman, K.P. (1990) Pneumococcal resistance to antibodies. *Clinical Microbiology Reviews*, **3** 171–196.

Kolenbrander, P.E. and Andersen, R.N. (1986) Multigeneric aggregations among oral bacteria: a network of independant cell-to-cell interactions. *Journal of Bacteriology*, **168**, 851–859.

Kolenbrander, P.E. and London, J. (1993) Adhere today, here tomorrow: oral bacterial adherence. *Journal of Bacteriology*, **175**, 3247–3252.

Kral, T.A. and Daneo-Moore, L. (1981) Biochemical differentiation of certain oral streptococci. *Journal of Dental Research*, **60**, 1713–1718.

Krasse, B. (1988) Biological factors as indicators of future caries. *International Dental Journal*, **38**, 219–225.

Krasse, B., Emilson, C-G. and Gahnberg, L. (1987) An anticaries vaccine: report on the status of research. *Caries Research*, **21**, 255–276.

Krause, R.M. (1972) The antigens of Group D streptococci. In *Streptococci and Streptococcal Diseases* (eds Wannamaker, L.W. and Matsen, J.M.). Academic Press, New York, USA, pp. 67–74.

Krause, R.M. and McCarty, M. (1962) Studies on the chemical structure of the streptococcal cell wall II. The composition of Group C cell walls and chemical basis for serological specificity of the carbohydrate moeity. *Journal of Experimental Medicine*, **115**, 49–62.

Lancefield, R.C. (1933) A serological differentiation of human and other groups of haemolytic streptococci. *Journal of Experimental Medicine*, **57**, 571–595.

Lancefield, R.C. (1934) A serological differentiation of specific types of bovine hemolytic streptococci (group B). *Journal of Experimental Medicine*, **59**, 441–458.

Lancefield, R.C. (1938) Two serological types of group B hemolytic streptococci with related but not identical, type-specific substances. *Journal of Experimental Medicine*, **67**, 25–39.

Lefevre, J.C., Faucon, G., Sicard, A.M. and Gasc, A.M. (1993) DNA fingerprinting of *Streptococcus pneumoniae* strains by pulse-field gel electrophoresis. *Journal of Clinical Microbiology*, **31**, 2724–2728.

Lehmann, K.B. and Neumann, R.O. (1896) *Atlas und Grundriss der Bakteriologie und Lehrbuch der Speciellen Bakteriologischen Diagnostik*, 1st edn. J.F. Lehmann, Munich, Germany, pp. 1–448.

Loesche, W.J. (1986) Role of *Streptococcus mutans* in human dental decay. *Microbiology Reviews*, **50**, 353–380.

Manning, J.E., Hume, E.B.H., Hunter, N. and Knox, K.W. (1994) An appraisal of the virulence factors associated with streptococcal endocarditis. *Journal of Medical Microbiology*, **40**, 110–114.

Marrack, P. and Kappler, J. (1990) The staphylococcal enterotoxins and their relatives. *Science*, **248**, 1066.

Marsh, P.D. and Martin, M. (1992) *Oral Microbiology*, 3rd edn. Chapman and Hall, London, UK.

Marshall, K.J., Musher, D.M., Watson, D., Mason, Jr, E.O. (1993) Testing of *Streptococcus pneumoniae* for resistance to penicillin. *Journal of Clinical Microbiology*, **31**, 1246–1250.

Maxted, W.R. (1978) Group A Streptococci: pathogenesis and immunity. In *Streptococci* (Society for Applied Bacteriology Symposium Series No. 7) (eds Skinner, F.A and Quesnel, L.B.). Academic Press, London, UK, pp. 107–125.

McWhinney, P.H.M., Patel, S., Whiley, R.A., Hardie, J.M., Gillespie, S.H. and Kibbler, C.C . (1993) Activities of potential therapeutic and prophylactic antibiotics against blood culture isolates of viridans group streptococci from neutropenic patients receiving ciprofloxacin. *Antimicrobial Agents and Chemotherapy*, **37**, 2493–2495.

Mead, G.C. (1978) Streptococci in the intestinal flora of man and other non-ruminant animals. In *Streptococci* (Society for Applied Bacteriology Symposium Series No. 7) (eds Skinner, F.A. and Quesnel, L.B.). Academic Press, London, UK, pp. 245–261.

Miller, W.D. (1890) *The Micro-organisms of the Human Mouth.* The S.S. White Dental Manufacturing Company (Reprinted in 1973 by S. Karger, Basel, Switzerland).

Motlová, J., Wagner, M. and Jelinková, J. (1986) A search for new group-B streptococcal serotypes. *Journal of Medical Microbiology*, **22**, 101–105.

Nyvad, B. and Kilian, M. (1987) Microbiology of the early colonization of human enamel and root surfaces *in vivo*. *Scandinavian Journal of Dental Research*, **95**, 369–380.

Orefici, G. (ed.) (1992) *New Perspectives on Streptococci and Streptococcal Infections* (Proceedings of the XI Lancefield International Symposium on Streptococci and Streptococcal Diseases, September 1990, Siena), Gustav Fischer Verlag, Stuttgart, Germany.

Orla-Jensen, S. (1919) *The Lactic Acid Bacteria.* Host and Son, Copenhagen, Denmark.

Osawa, R. and Mitsuoka, T. (1990) Selective medium for the enumeration of tannin protein complex-degrading *Streptococcus* sp. in feces of koalas. *Applied and Environmental Microbiology*, **56**, 3609–3611.

Ottens, H. and Winkler, K.C. (1962) Indifferent and haemolytic streptococci possessing group-antigen F. *Journal of General Microbiology*, **28**, 181–191.

Parker, M.T. (1978) The pattern of streptococcal disease in man. In *Streptococci* (Society for Applied Bacteriology Symposium Series No. 7) (eds Skinner, F.A. and Quesnel, L.B.). Academic Press, London, UK, pp. 71–106.

Perch, B., Kjems, E. and Ravn, T. (1974) Biochemical and serological properties of *Streptococcus mutans* from various human and animal sources. *Acta Pathologica et Microbiologia Scandinavica (B)*, **82**, 357–370.

Pier, G.B. and Madin, S.H. (1976) *Streptococcus iniae* sp. nov., a beta hemolytic *Streptococcus* isolated from an Amazon freshwater dolphin, *Inia geoffrensis*. *International Journal of Systematic Bacteriology*, **26**, 545–553.

Piscitelli, S.C., Schwed, J., Schreckenberger, P. and Danziger, L.H. (1992) *Streptococcus milleri* group: renewed interest in an elusive pathogen. *European Journal of Clinical Microbiology*, **11**, 491–498.

Prévot, A.R. (1924) *Diplococcus constellatus* (n. sp.) *Compte-Rendu de la Société de Biologie* (Paris), **91**, 426–428.

Prévot, A.R. (1925) Les streptocoques anaérobies. *Annals de l'Institute Pasteur* (Paris), **39**, 415–447.

Roberts, R.B. (1985) *Streptococcus pneumoniae*. In *Principles and Practices of Infectious Diseases*, (eds Mandell, G.L., Douglas, R.G. and Bennett, J.E.). Churchill Livingstone, New York, USA, pp. 1142–1152.

Robinson, I.M., Stromley, J.M., Varel, V.H. and Cato, E.P. (1988) *Streptococcus intestinalis*, a new species from the colons and faeces of pigs. *International Journal of Systematic Bacteriology*, **38**, 245–248.

Rosenbach, F.J. (1884) *Mikro-organismen bei den Wund-Infections-Krankheiten des Menschen*. J.F. Bergmann, Wiesbaden, Germany.

Ross, P.W. (1978) Ecology of Group B streptococci. In *Streptococci* (Society for Applied Bacteriology Symposium Series No. 7) (eds Skinner, F.A. and Quesnel, L.B.). Academic Press, London, UK, pp. 127–142.

Rotta, J. (1986). Pyogenic hemolytic streptococci. In *Bergey's Manual of Systematic Bacteriology*, Vol. 2 (eds Sneath, P.H.A., Mair, N.S., Sharpe, M.E. and Holt, J.G.). Williams and Wilkins, Baltimore, pp. 1047–1054.

Rudney, J.D. and Larson, C.J. (1993) Species identification of oral viridans streptococci by restriction fragment polymorphism analysis of rRNA genes. *Journal of Clinical Microbiology*, **31**, 2467–2473.

Rudney, J.D., Neuvar, E.K. and Soberay, A.H. (1992) Restriction endonuclease-fragment polymorphisms of oral viridans streptococci, compared by conventional and field-inversion gel electrophoresis. *Journal of Dental Research*, **71**, 1182–1188.

Rudolph, K.H., Parkinson, A.J., Black, C.M. and Mayer, L.W. (1993) Evaluation of polymerase chain reaction of diagnosis of pneumococcal pneumonia. *Journal of Clinical Microbiology*, **31**, 2661–2666.

Ruoff, K.L. (1991) Nutritionally variant streptococci. *Clinical Microbiology Reviews*, **4**, 184–190.

Ruoff, K.L. (1992) The genus *Streptococcus*-medical. In *The Prokaryotes*, Vol. II, 2nd edn (eds Balow, A., *et al.*). Springer-Verlag, New York, USA, pp. 1450–1464.

Russell, R.R.B. (1994) The application of molecular genetics to the microbiology of dental caries. *Caries Research*, **28**, 69–82.

Russell, R.R.B. and Johnson, N.W. (1987) The prospects for vaccination against dental caries. *British Dental Journal*, **162**, 29–34.

Russell, R.R.B., Aduse-Opoku, J., Tao, L. and Ferretti, J.J. (1991) A binding protein-dependent transport system in *Streptococcus mutans*. In *Genetics and Molecular Biology of Streptococci, Lactococci and Enterococci* (eds Dunny, G., Cleary, P. and McKay, L.). American Society for Microbiology, Washington, DC, USA, pp. 244–247.

Russell, R.R.B., Aduse-Opoku, J., Sutcliffe, I.C., Tao, L. and Ferretti, J.J. (1992) A binding protein-dependent transport system in *Streptococcus mutans* responsible for multiple sugar metabolism. *The Journal of Biological Chemistry*, **267**, 4631–4637.

Schleifer, K.H. and Kandler, O. (1972) Peptidoglycan types of bacterial cell walls and their taxonomic implications. *Bacteriology Reviews*, **36**, 407–477.

Schleifer, K.H. and Kilpper-Bälz, R. (1984) Transfer of *Streptococcus faecalis* and *Streptococcus faecium* to the genus *Enterococcus* nom. rev. as *Enterococcus faecalis* comb. nov. and *Enterococcus faecium* comb. nov. *International Journal of Systematic Bacteriology*, **34**, 31–34.

Schleifer, K.H. and Kilpper-Bälz, R. (1987) Molecular and chemotaxonomic approaches to the classification of streptococci, enterococci and lactococci: a review. *Systematic Applied Microbiology*, **10**, 1–19.

Schleifer, K.H. and Seidl, P.H. (1985) Chemical composition and structure of murein. In *Chemical Methods in Bacterial Systematics* (eds Goodfellow, M. and Minnikin, D.E.). Academic Press, London, UK, pp. 201–219.

Schleifer, K.H., Kraus, J., Dvorak, C., Kilpper-Bälz, R., Collins, M.D. and Fischer, W. (1985) Transfer of *Streptococcus lactis* and related streptococci to the genus *Lactococcus* gen. nov. *Systematic and Applied Microbiology*, **6**, 183–195.

Schleifer, K.H., Ehrmann, M., Krusch, U. and Neve, H. (1991) Revival of the species *Streptococcus thermophilus* (*ex* Orla-Jensen, 1919) nom. rev. *Systematic and Applied Microbiology*, **14**, 386–388.

Schottmüller, H. (1903) Die Artunterscheidung der für den Menschen pathogenen Streptokokken durch Blutagar. *Münchener Medizinische Wochenschrift*, **50**, 849–853, 909–912.

Sherman, J.M. (1937) The streptococci. *Bacteriology Reviews*, **1**, 3–97.

Shiroza, T. and Kuramitsu, H.K. (1993) Construction of a model secretion system for oral streptococci. *Infection and Immunity*, **61**, 3745–3755.

Skerman, V.B.D., McGowan, V. and Sneath, P.H.A. (1980) Approved lists of bacterial names. *International Journal of Systematic Bacteriology*, **30**, 225–420.

Skinner, F.A. and Quesnel, L.B. (eds) (1978) *Streptococci*. Academic Press, London, UK.

Sneath, P.H.A., Mair, N.S. and Sharpe, M.E. (eds) (1986) *Bergey's Manual of Systematic Bacteriology*, Vol. 2. Williams and Wilkins, Baltimore, MD, USA.

Sommer, P., Gleyzal, C., Guerret, S., Etienne, J. and Grimaud, J.-A. (1992) Induction of a putative laminin-binding protein of *Streptococcus gordonii* in human infective endocarditis. *Infection and Immunity*, **60**, 360–365.

Spika, J.S., Facklam, R.R., Plikaytis, B.D., Oxtoby, M.J. and the Pneumococcal Surveillance Working Group (1991) Antimicrobial resistant *Streptococcus pneumoniae* in the United States 1979–1987. *Journal of Infectious Diseases*, **163**, 1273–1278.

Stevens, D.L., Tanner, M.H., Winslip, J., Swarts, R., Reis, K.M., Schlievert, P.M. and KLaplan, E. (1989) Severe group A streptococcal infections associated with toxic shock-like syndrome and scarlet fever toxin. *Annals of the New England Journal of Medicine*, **321**, 1–7.

Stringer, J. (1980) The development of a phage typing system for group B streptococci. *Journal of Medical Microbiology*, **13**, 133–143.

Tart, R.C. and Van de Rijn, I. (1991) Analysis of adherence of *Streptococcus defectivus* and endocarditis-associated streptococci to extracellular matrix. *Infection and Immunity*, **59**, 857–862.

Whiley, R.A. (1987). A taxonomic study of oral streptococci. PhD thesis, University of London, London, UK.

Whiley, R.A. and Beighton, D. (1991) Emended descriptions and recognition of *Streptococcus constellatus*, *Streptococcus intermedius*, and *Streptococcus anginosus* as distinct species. *International Journal of Systematic Bacteriology*, **41**, 1–5.

Whiley, R.A. and Hardie, J.M. (1988) *Streptococcus vestibularis* sp. nov. from the human oral cavity. *International Journal of Systematic Bacteriology*, **38**, 335–339.

Whiley, R.A. and Hardie, J.M. (1989) DNA–DNA hybridisation studies and phenotypic characteristics of strains within the '*Streptococcus milleri* group'. *Journal of General Microbiology*, **135**, 2623–2633.

Whiley, R.A., Hardie, J.M. and Jackman, P.J.H. (1982) SDS–polyacrylamide gel electrophoresis of oral streptococci. In *Proceedings of the VIIIth Intenational Symposium on Streptococci and Streptococcal Diseases* (eds Holm, S.E. and Christensen, P.). Reedbooks, Chertsey, Surrey, UK, pp. 61–62.

Whiley, R.A., Russell, R.R.B., Hardie, J.M. and Beighton, D. (1988). *Streptococcus downei* sp. nov. for strains previously described as *Streptococcus mutans* serotype h. *International Journal of Systematic Bacteriology*, **38**, 25–29.

Whiley, R.A., Fraser, H.Y., Douglas, C.W.I., Hardie, J.M., Williams, A.M. and Collins, M.D. (1990a) *Streptococcus parasanguis* sp. nov. An atypical viridans streptococcus from human clinical specimens. *FEMS Microbiology Letters*, **68**, 115–122.

Whiley, R.A., Fraser, H.Y., Hardie, J.M. and Beighton, D. (1990b) Phenotypic differentiation of *Streptococcus intermedius*, *Streptococcus constellatus*, and *Streptococcus anginosus* strains within the '*Streptococcus milleri* group'. *Journal of Clinical Microbiology*, **28**, 1497–1501.

Whiley, R.A., Beighton, D., Winstanley, T.G., Fraser, H.Y. and Hardie, J.M. (1992) *Streptococcus intermedius*, *Streptococcus constellatus*, and *Streptococcus anginosus* (the *Streptococcus milleri* group): association with different body sites and clinical infections. *Journal of Clinical Microbiology*, **30**, 243–244.

Whiley, R.A., Freemantle, L., Beighton, D., Radford, J.R., Hardie, J.M. and Tillotsen, G. (1993) Isolation, identification and prevalence of *Streptococcus anginosus*, *S. intermedius* and *S. constellatus* from the human mouth. *Microbial Ecology in Health and Disease*, **6**, 285–291.

White, J.C. and Niven, Jr, C.F. (1946) *Streptococcus* S.B.E.: a steptococcus associated with subacute bacterial endocarditis. *Journal of Bacteriology*, **51**, 717–722.

Wibawan, I.W.T. and Lämmler, C. (1991) Influence of capsular neuraminic acid on properties of streptococci of serological group B. *Journal of General Microbiology*, **137**, 2721–2725.

Willcox, M.D.P., Patrikakis, M., Loo, C.Y. and Knox, K.W. (1993) Albumin-binding proteins on the surface of the *Streptococcus milleri* group and characterization of the albumin receptor of *Streptococcus intermedius* C5. *Journal of General Microbiology*, **139**, 2451–2458.

Williams, A.M. and Collins, M.D. (1990) Molecular taxonomic studies on *Streptococcus uberis* types I and II. Description of *Streptococcus parauberis* sp. nov. *Journal of Applied Bacteriology*, **68**, 485–490.

Williams, A.M., Farrow, J.A.E. and Collins, M.D. (1989) Reverse transcriptase sequencing of 16S ribosomal RNA from *Streptococcus cecorum. Letters in Applied Microbiology*, **8**, 185–190.

Wilson, G.S. and Miles, A.A. (eds) (1975) *Topley & Wilson's Principles of Bacteriology and Immunity*, Vol. 1, 6th edn. Arnold, London, UK.

Wilson, C.D. and Salt, G.F.H. (1978) Streptococci in animal disease. In *Streptococci* (Society for Applied Bacteriology Symposium Series No. 7) (eds Skinner, F.A. and Quesnel, L.B.). Academic Press, London, UK, pp. 143–156.

Winstanley, T.G., Magee, J.T., Limb, D.I., Hindmarch, J.M., Spencer, R.C., Whiley, R.A., Beighton, D. and Hardie, J.M. (1992) A numerical taxonomic study of the '*Streptococcus milleri* group' based upon conventional phenotypic tests and pyrolysis mass spectrometry. *Journal of Medical Microbiology*, **36**, 149–155.

5 The genus *Pediococcus*, with notes on the genera *Tetratogenococcus* and *Aerococcus*

W.J. SIMPSON and H. TAGUCHI

5.1 Introduction

Pediococci are the only lactic acid bacteria that divide alternately in two perpendicular directions to form tetrads (Figure 5.1). They are invariably spherical and produce lactic acid, but no gas, from glucose. The genus is heterogeneous and includes organisms able to grow in beer and those active during soya sauce manufacture. A number of papers address various aspects of the genus (Pederson, 1949; Pederson *et al.*, 1954; Nakagawa and Kitahara, 1959; Sakaguchi and Mori, 1969; Garvie, 1974, 1986a; Eschenbecher and Back, 1976; Back, 1978a; Rainbow, 1981; Bergan *et al.*, 1984; Priest, 1987; Raccach, 1987; Weiss, 1991; Teuber, 1993) which currently contains eight species. Information on the genus *Aerococcus* can be found in the review by Weiss (1991). Table 5.1 lists the species of pediococci, together with common synonyms and a brief description of

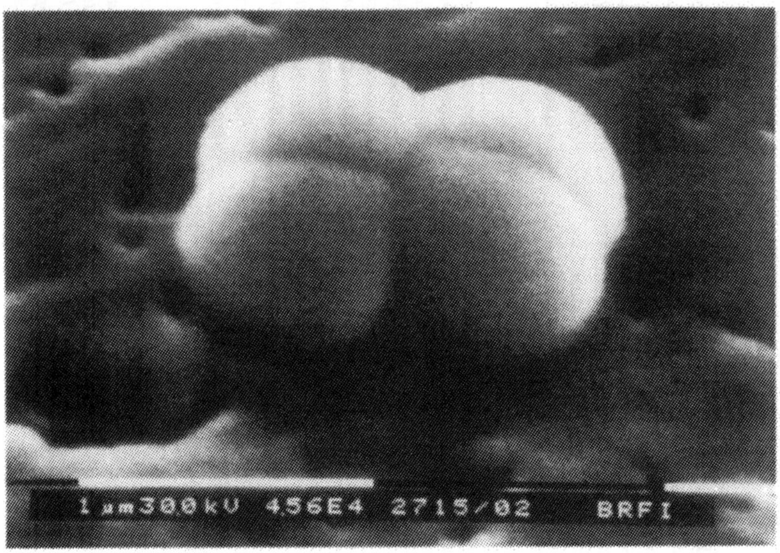

Figure 5.1 Tetrad formation of pediococci. The scanning electron micrograph shows cells of *Pediococcus pentosaceus* BSO 347 (BSO, beer-spoilage organism collection, BRF International, UK). Scale bar represents 1 μm.

Table 5.1 Names, synonyms and descriptions of *Pediococcus* species

Species and priority	Type strain*	Synonyms	Description
Pediococcus acidilactici (Lindner, 1887)	DSM 20284 (proposed by Garvie (1986b))	'*Pediococcus linderi*' '*Pediococcus cerevisiae*' '*Streptococcus lindneri*'	Grow at 50°C. Ferment ribose, arabinose and/or xylose. Unable to utilize maltose; DL-lactate produced from glucose. Hydrolyse arginine. Mol% G+C 38–44. Associated mainly with plant materials.
Pediococcus damnosus (Claussen, 1903)	NCDO 1832† (ATCC 29358; DSM 20331)	'*Pediococcus cerevisiae*' '*Pediococcus cerevisiae* subsp. *mevalovorus*' '*Pediococcus viscosus*' '*Pediococcus perniciousus*' '*Pediococcus sarcinaeformis*' '*Pediococcus odoris mellisimilis*' '*Pediococcus mevalovorus*' '*Streptococcus damnosus*' '*Streptococcus damnosus* var. *limosus*'	Unable to ferment ribose or hydrolyse arginine. No acid from starch, no acid or gas from gluconate, no growth at pH 8.0 or at 35°C. Most strains hop-tolerant and able to grow in beer; DL-lactate produced from glucose. Mol% G+C 37–42. Associated mainly with beer and breweries.
Pediococcus dextrinicus (Coster and White, 1964; Back, 1978b)	DSM 20335 (NCDO 1561; ATCC 33087)	'*Pediococcus cerevisiae* subsp. *dextrinicus*' '*Streptococcus damnosus* var. *diastaticus*'	Unable to ferment ribose or hydrolyse arginine. Acid from starch, acid and gas from gluconate, growth at pH 8.0. L(+)-Lactate produced from glucose. Mol% G+C 40–41. Associated with fermenting plant materials.
Pediococcus halophilus (Mees, 1934)	NCDO 1635 (ATCC 33315; DSM 20339)	'*Pediococcus soyae*' '*Pediococcus acidilactici* var. *soyae*' '*Tetracoccus no. 1*' '*Tetracoccus halophilus*' '*Sarcina hamaguchiae*' '*Tetratogenococcus halophilus*'	Grow in presence of 15% NaCl and at pH 9.0. L(+)-Lactate produced from glucose. Mol% G+C 34–36.5. Associated with salty environments.

Pediococcus inopinatus (Back, 1978a)	DSM 20285	'*Pediococcus cerevisiae*'	Unable to ferment pentoses and lactose. Does not hydrolyse arginine. DL-Lactate produced from glucose. Mol% G+C 39–40. Associated with beer and alcoholic beverages.
Pediococcus parvulus (Günther and White, 1961a)	NCDO 1634 (ATCC 19371; DSM 20332)	None	Grows at pH 4.5. Unable to utilize pentoses, lactose or starch. Does not hydrolyse arginine. Forms DL-lactate from glucose. Mol% G+C 40.5–41.6. Associated with fermented plant materials, cider and wine.
Pediococcus pentosaceus (Mees, 1934)	NCDO 990 (ATCC 33161; DSM 20336)	'*Pediococcus hennebergi*' '*Pediococcus citrovorum*' '*Pediococcus cerevisiae*' '*Pediococcus acidilacti*' '*Streptococcus acidi-lactici*'	Ferment pentoses (except strains belonging to *P. pentosaceus* subsp. *intermedius*). Ferment maltose. Do not ferment starch or melizitose. Hydrolyse arginine. Maximum temperatures for growth 39–45°C. DL-Lactate produced from glucose. Mol% G+C 35–39. Associated with plant materials.
Pediococcus urinae-equi (ex Mees) nom. rev.	NCDO 1636 (ATCC 29723; DSM 20341)	'*Pediococcus cerevisiae* var. *urinae-equi*' '*Pediococcus urinae-equi*' '*Aerococcus viridans*'	Grow at pH 9.0. Produces L(+)-lactate from glucose. Grow in the absence of fermentable carbohydrate. Mol% G+C 39.6–39.7. Associated with horse urine and animal faeces.

*DSM; Deutsche Sammlung von Mikrorganismen, Munich, Germany. NCDO; National Collection of Dairy Organisms, Reading, UK. ATCC; American Type Culture Collection, Rockville, Maryland, USA.

†Type strain of the genus.

each. One of these (*P. urinae-equi*) clearly belongs to the genus *Aerococcus* and proposals have been made that *P. halophilus* should also be placed in a new genus, '*Tetratogenococcus*' (Collins *et al.*, 1990). If the suggestions to exclude these acid-sensitive species from the genus are adopted, then the general description might be altered to read 'pediococci are the only acidophilic, homofermentative, lactic acid bacteria that divide alternately in two perpendicular directions to form tetrads'.

5.2 Morphology

In a single culture, the cells of pediococci are spherical and of uniform size, 0.36–1.43 μm in diameter (Günther and White, 1961a). They are never elongated. In contrast, cells of *Leuconostoc* spp. are often elongated and arranged in chains.

The mode of division of pediococci has frequently been the subject of dispute. Balcke (1884) stated that the cells divided in *one plane* and, in recognition of this, derived the name *Pediococcus* from two Greek nouns: *pedium*, meaning a plane surface, and *coccus*, meaning a berry. Later descriptions of tetrad-forming cocci described them as dividing in two planes, rather than in one, as Balcke had suggested (Günther, 1959; Herrmann, 1965). This may have been to differentiate their mode of division from that of chain-forming streptococci. Others, notably Shimwell (1948a, 1949), disputed the fact that tetrad formation in pediococci was the result of an unusual method of cell division, proposing that the cells divided normally to form chains, but then re-arranged to form tetrads. Shimwell (1949) pointed out that tetrads were more noticeable than chains of cells, and this may have explained the preoccupation of bacteriologists with such morphological features. Günther (1959) used time-lapse photography to show that tetrad formation did not result from rearrangement of the cells. However, she interpreted the mode of division as being in two planes. Pediococci are spherical, so it is not possible for them to divide in more than one plane while undergoing only two cell divisions (Simpson, 1994) since a plane is defined *as a surface containing all straight lines passing through a fixed point and also intersecting a straight line in space*. Thus, at any point in each of the two divisions needed to form a tetrad, the centres of all cells must lie within one plane (Figure 5.2). Of course, the cells must divide in two *directions*, each approximately at right angles to the other.

Pediococci are nonmotile, do not form spores and are not capsulated. When grown on a rich medium, such as de Man, Rogosa, Sharpe (MRS) agar (de Man *et al.*, 1960), the colonies are typically 1–3 mm in diameter, generally smooth-edged and invariably not pigmented (Back, 1978a). In stab culture, the cells grow along the line of the stab with little surface growth.

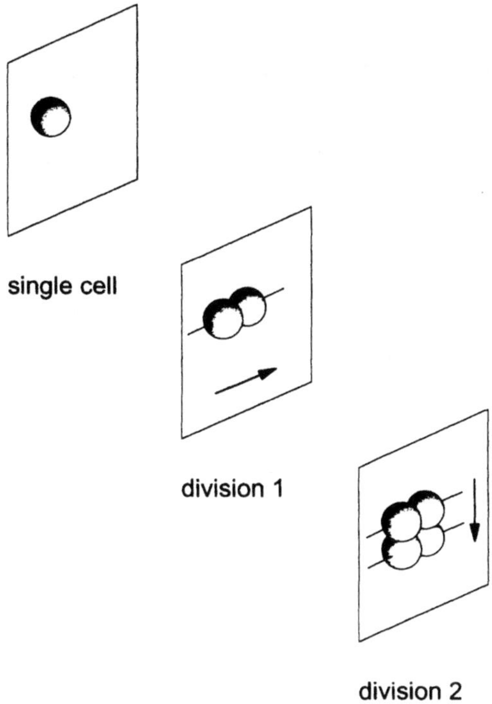

Figure 5.2 Tetrad formation takes place by division of the cells in two directions in a single plane.

In broth culture, growth is uniform throughout the medium (Nakagawa and Kitahara, 1959).

5.3 Physiology

5.3.1 Carbohydrate metabolism

Under anaerobic conditions, pediococci ferment glucose to give optically inactive (DL) or dextrorotary (L[+]) lactate. A wide range of carbohydrates can be used by various species, ranging from pentoses such as arabinose, ribose and xylose; hexoses, such as fructose and mannose; disaccharides, such as maltose; trisaccharides, such as maltotriose; and polymers, such as starch (Table 5.2). All species, except *P. urinae-equi*, are unable to grow in the absence of carbohydrate (Deibel and Niven, 1960). *Pediococcus pentosaceus* transports glucose using the phosphoenolpyruvate:phosphotransferase system and metabolizes it via the Embden–Meyerhof–Parnas pathway (Romano *et al.*, 1979). The metabolism of glucose by other pediococci has not been reported. Under certain conditions, metabolic

Table 5.2 Physiological characteristics of the pediococci*

Character†	*P. acidilactici*	*P. damnosus*	*P. dextrinicus*	*P. halophilus* (syn. '*Tetragenococcus halophilus*')	*P. inopinatus*	*P. parvulus*	*P. pentosaceus*	'*P. pentosaceus* subsp. *intermedius*'	*P. urinae-equi*
Growth at 35°C	+	–	+	+	+	+	+	+	+
40°C	+	–	+	+/– (weak)	+/– (weak)	–	+/– (weak)	+/– (weak)	+
45°C	+	–	+/– (weak)	–	–	–	+/– (weak)	+/– (weak)	+/– (weak)
50°C	+	–	–	–	–	–	–	–	–
Maximum NaCl concentration for growth	10%	5%	6%	>18%	8%	8%	10%	10%	10%
Growth at pH 4.5	+	+	+/–	–	+	+	+	+	–
pH 5.0	+	+	+	–	+	+	+	+	–
pH 7.5	+	–	+	+	+/–	+/–	+	+	+
pH 8.0	+	–	–	+	–	–	+	+	+
pH 8.5	+/–	–	–	+	–	–	+/–	+/–	+
Catalase activity	–	–	–	–	–	–	+/–	+/–	+/–
Gas from gluconate	–	–	+	–	–	–	–	–	–
Arginine hydrolysis	+	–	–	–	–	–	+	+	–
Hippurate hydrolysis	–	–	–	–	–	–	–	–	+
Production of acetoin	+/–	+/–	+/–	–	+/–	–	+/–	+/–	–
Lactate configuration	DL	DL	L(+)	L(+) (3% D(–))	DL	DL	DL	DL	L(+)
Litmus milk reaction									
Acid	+/–	–	+/–	–	+/–	–	+	+	ND
Reduction	+/–	–	+/–	–	+/–	–	+	+	ND
Clotting	+/–	–	–	–	–	–	+/–	+	ND

Acid produced from or splitting of									
Arabinose	+/−	−	−	+	−	−	+	−	+/−
Ribose	+	−	−	+	−	−	+	+	ND
Xylose	+	−	−	−	−	−	+/−	−	+/−
Fructose	+	+	+	+	+	+	+	+	+
Rhamnose	+/−	−	−	−	−	−	+/−	−	ND
Glucose	+	+	+	+	+	+	+	+	+
Mannose	+	+	+	+	+	+	+	+	+
Galactose	+	+/−	+	+/−	+	+/−	+	+	+
Maltose	−	+/−	+	+	+	+/−	+	+	+
Trehalose	+/−	+/−	+/−	+	+	+	+	+/−	+/−
Cellobiose	+	+	+	+	+	+	+	+	ND
Sucrose	+/−	+/−	+/−	+	−	−	−	+/−	+
Lactose	+/−	−	+/−	−	+	−	+	+	+/−
Melibiose	−	−	−	−	−	−	+/−	+/−	ND
Melezitose	−	+/−	−	+	−	−	−	−	ND
Raffinose	+/−	−	−	+/−	−	−	+/−	+/−	+
Maltotriose	−	+/−	+	+	+/−	+/−	+/−	+/−	ND
Dextrin	+/−	−	+	+/−	+/−	+/−	−	−	+/−
Starch	−	−	+	−	−	−	−	−	−
Inulin	−	−	+/−	−	−	−	+/−	+/−	−
Glycerol	+/−	−	−	+/−	−	−	+/−	+/−	−
Mannitol	+/−	−	−	−	−	−	−	−	+/−
Sorbitol	−	−	−	−	−	−	−	−	+/−
α-Methyl glucoside	−	+/−	+/−	+	+/−	−	−	−	ND
Salicin	+/−	+/−	+	+	+	+	+	+/−	+
Amygdalin	+/−	+/−	+	+	+	+/−	+	+/−	ND

+, >90% strains positive; +/−, 10–90% strains positive; −, <10% strains positive; ND, not determined.
†Characters in bold are useful for discrimination of species.

products other than lactate are formed. For example, *P. pentosaceus* produces an equimolar mixture of acetate and lactate from pentose sugars (Fukui *et al.*, 1957). Back (1978a) showed that the differences in lactate configuration result from the activities of different lactate dehydrogenases. Indeed, the electrophoretic mobility of such enzymes provides an aid to differentiation of *Pediococcus* species (see below). *Pediococcus pentosaceus* forms D(–) and L(+)-lactate from glucose but converts malic acid to L(+)-lactate (Radler *et al.*, 1970). The ability of *P. halophilus* to metabolize organic acids, such as citrate and malate, has been investigated by Kanbe and Uchida (1982, 1987b). The metabolism of citrate by this organism differs from that in lactic streptococci. Acetate and formate are the main products of citrate metabolism: no acetoin or diacetyl are produced (Kanbe and Uchida, 1987b). Many strains of *P. damnosus* form diacetyl (Shimwell and Kirkpatrick, 1939).

All strains of *P. dextrinicus* grow on starch. It is not known whether starch breakdown occurs as a result of α-amylase, glucoamylase or other enzymic activities, or whether the extracellular enzymes involved are secreted into the growth medium, or located at the cell surface.

Pediococci oxidize some substrates. For example, *P. pentosaceus* uses glycerol when O_2 is available, producing lactic acid, acetic acid, acetoin and CO_2 (Dobrogosz and Stone, 1962a). Similarly, this organism can oxidize lactate to acetate and CO_2 (Thomas *et al.*, 1985).

5.3.2 Nitrogen metabolism

Pediococci grow best in rich media. Most strains need a range of amino acids including alanine, aspartic acid, glutamic acid, arginine, histidine, isoleucine, phenylalanine, proline, threonine, tyrosine, valine, tryptophan, cystine, glycine and leucine. Some strains need lysine, methionine and serine (Jensen and Seeley, 1954; Sakaguchi, 1960). Many strains grow poorly, or not at all, without a complex source of nitrogen such as peptides (Nakagawa and Kitahara, 1959). Aminopeptidases are produced by some strains. For example, Tzanetakis and Litopolou-Tzanetaki (1989), using the API ZYM test kit, showed that *P. pentosaceus* produces leucine aminopeptidase and valine aminopeptidase.

Bhowmik and Marth (1990b) examined the intracellular protease, endopeptidase, dipeptidase, dipeptidyl aminopeptidase and carboxypeptidase activities of six strains of *P. pentosaceus* and two strains of *P. acidilactici*. All the strains produced proteases, dipeptidases, dipeptidyl aminopeptidases and aminopeptidases, but did not produce carboxypeptidases or endopeptidases. Crude cell-free extracts of most strains partially hydrolysed α_{s1}-casein. *Pediococcus pentosaceus* ATCC 996 completely hydrolysed this protein. *β*-Casein was completely hydrolysed by some strains, but only partially by others.

Little information is available about the proteolytic activities of other pediococci, with the exception of a report by Davis *et al.* (1988) indicating that *P. parvulus* does not possess proteolytic activity, and one by Uhl and Kühbeck (1969) indicating that growth of *P. damnosus* in beer is associated with endo-enzymic hydrolysis of short peptides and exo-enzymic hydrolysis of polypeptides.

5.3.3 Vitamin and organic base requirements

All species need nicotinic acid, pantothenic acid and biotin for growth. Thiamine, *p*-aminobenzoic acid and cobalamin are not essential. Some strains need riboflavin, pyridoxine and folinic acid (Sakaguchi and Mori, 1969). Pyridoxin stimulates growth of most strains of *P. damnosus* and is essential for growth of some (Solberg and Clausen, 1973b). Most do not need preformed organic bases. Adenine, guanine, uracil and xanthine do not stimulate growth of pediococci in a defined medium (Sakaguchi and Mori, 1969).

An interesting coda relating to the vitamin requirements of pediococci concerns the case of '*Leuconostoc citrovorum*' strain 8081. Dunn *et al.* (1947) first studied the nutrition of this organism and found that, in addition to a requirement for 16 amino acids, a 'concentrate' of folic acid was required for optimal growth. Sauberlich and Baumann (1948) showed that the strain needed a growth factor, found in liver extract, which they named 'citrovorum factor'. High concentrations of folic acid could replace this unknown factor, thus explaining why Dunn *et al.* (1947) had not been aware of the requirement. The substance was later identified and referred to both as folinic acid-SF and leucovorin. It is now known as 5-formyl-tetrahydrofolic acid (5-formyl-THF). ('Citrovorum factor' is sometimes incorrectly referred to as folic acid (e.g. Raccach, 1987).) Confusion temporarily arose when it was discovered that the requirement for 5-formyl-THF was not generally found among other leuconostocs and that all pediococci which had been studied up to that time needed the factor for growth. Further studies showed that the requirement was restricted only to some strains of pediococci (Günther and White, 1961a). The paradox was resolved when it was shown that '*Leuconostoc citrovorum*' strain 8081 was not a leuconostoc at all, but belonged to the pediococci (Felton and Niven, 1953). Initially, it was named '*P. cerevisiae*' but would now be classified as *P. pentosaceus*.

Tetrahydrofolate derivatives usually play a metabolic role in transfer of single-carbon units. The significance of 5-formyl-THF in bacterial metabolism is not presently clear. However, the compound has been used in the treatment of malignant tumours in mammals, including man (Metzler, 1977).

5.3.4 Mineral requirements

All pediococci studied so far (*P. acidilactici*, *P. pentosaceus*) need large quantities of manganese for growth, and in this respect differ from *Enterococcus* spp. (Efthymiou and Joseph, 1972). In common with *Lactobacillus* spp., they have no requirement for iron (Archibald, 1986). Whether this is true of *P. halophilus* and *P. urinae-equi* is not known. Raccach (1981) studied the ability of different metal ions to stimulate the fermentative activity of *P. pentosaceus*. Stimulation followed the sequence $Mn^{2+} > Ca^{2+} > Fe^{2+} > Zn^{2+} = Fe^{3+} > Mg^{2+}$.

5.3.5 Reaction to oxygen

Pediococci are aero-tolerant anaerobes. Growth of some strains is improved by anaerobic incubation, particularly on primary isolation. Most pediococci are unable to control the redox potential (rH) of the growth medium (Nakagawa and Kitahara, 1959). *Pediococcus halophilus*, however, is exceptional in this respect. Kanbe and Uchida (1987a) correlated the ability of different strains of *P. halophilus* to control rH with possession of an NADH dehydrogenase. This species also produces pyruvate oxidase, which catalyses a direct reaction between pyruvate and molecular oxygen (Kanbe and Uchida, 1985).

In general, pediococci are catalase-negative, but some strains of *P. pentosaceus* produce a 'pseudo-catalase' which gives false positive reactions when the cells are tested with H_2O_2 (Felton *et al.*, 1953; Whittenbury, 1964). Forty-nine out of 75 strains of *P. pentosaceus*, isolated by Tzanetakis and Litopolou-Tzanetaki (1989) from raw goat's milk and Feta and Kaseri cheeses, gave weak catalase reactions. 'Pseudo-catalase' differs from true catalase in that it is insensitive to inhibition by azide and does not contain a haem group. Cells grown on media with a low glucose content are most likely to produce pseudo-catalase. *Pediococcus acidilactici* forms catalase when provided with haemin (Whittenbury, 1964). In spite of reports suggesting that cytochromes are produced by pediococci (Jensen and Seeley, 1954; Whittenbury, 1964), it is now accepted that this is not the case (Garvie, 1986a). Dobrogosz and Stone (1962a,b) suggest that a flavoprotein enzyme system donates electrons to oxygen, resulting in H_2O_2 formation. For example, an α-glycerophosphate oxidase has been identified in cells of *P. pentosaceus* (Dobrogosz and Stone, 1962a).

Pediococci do not possess a superoxide dismutase. Instead they protect themselves against damage by oxygen radicals using high concentrations of Mn(II) (Archibald, 1986).

5.3.6 *Cell wall chemistry*

The cell walls of pediococci have interpeptide bridges of the L-Lys-L-Asp type between the alanine and lysine residues. They have D-Asp linkages between positions three and four of the two peptide bridges (Kandler, 1970). However, *P. urinae-equi*, like *Aerococcus viridans*, has only one type of peptidoglycan polypeptide, with no interpeptide bridge. In these organisms the D-Ala carboxyl residue binds to the amino group of the adjacent L-Lys (Bergan *et al.*, 1984). Pediococci do not have techoic acids in their cell walls (Garvie, 1986a).

5.3.7 *Miscellaneous metabolic features*

Pediococci neither reduce nitrate nor produce indole from tryptophan (Nakagawa and Kitahara, 1959). In general, they do not hydrolyse hippurate, although some strains belonging to *P. urinae-equi* can (Tanasupawat and Daengsubha, 1983). Two species (*P. acidilactici*, *P. pentosaceus*) produce ammonia from arginine.

Lipase activity is generally weak, or absent, in pediococci (Davis *et al.*, 1988; Tzanetakis and Litopolou-Tzanetaki, 1989). Some strains of *P. damnosus* need mevalonic acid for growth (Kitahara and Nakagawa, 1958), while others have a requirement for CO_2 (Nakagawa and Kitahara, 1959).

Pediococci differ in their tolerance to NaCl. Strains of the salt-tolerant species *P. halophilus* grow in the presence of >18% NaCl. Other species are less tolerant; their sensitivity varies with the composition of the growth medium and conditions of incubation (Nakagawa and Kitahara, 1959; Coster and White, 1964).

Some species (e.g. *P. damnosus*, *P. inopinatus*, *P. parvulus*) are tolerant to ethanol as evidenced by their ability to grow in alcoholic beverages, such as beer, wine and cider. For example, all 23 strains of *P. parvulus* isolated from wine by Davis *et al.* (1988) could grow in the presence of 12.5% (w/v) ethanol, while five of the strains could grow in the presence of 15% ethanol.

Some strains of *P. damnosus* are eight- to 20-fold more resistant than sensitive strains to the antibacterial action of hop bitter acids and are thus better equipped to grow in hopped beer (Simpson and Fernandez, 1992).

5.4 Genetic features

The genus *Pediococcus* is genetically heterogeneous. Mol% G+C values of *Pediococcus* spp., determined by various methods, lie in the range 34–44 (see species descriptions for individual values – Sakaguchi and Mori, 1969; Kocur *et al.*, 1971; Solberg and Clausen, 1973a; Back, 1978a). Even if *P. urinae-equi* and *P. halophilus* are excluded from the genus, as suggested by some (Bergan *et al.*, 1984; Collins *et al.*, 1990); the range is similarly broad (35–44%).

Some strains harbour plasmids which range in size from 4.5–40 MDa (Graham and McKay, 1985; Torriani *et al.*, 1987). Some code for production of bacteriocins (see below), others for fermentation of carbohydrates. In *P. pentosaceus*, the ability to ferment raffinose, melibiose and sucrose is associated with three different plasmids. Sucrose hydrolase and α-galactosidase activities are associated with plasmid-encoded raffinose utilization (Gonzalez and Kunka, 1986). Lactose fermentation in some strains of *P. pentosaceus*, and sucrose fermentation in some strains of *P. acidilactici*, may be plasmid-linked (Hoover *et al.*, 1988). Kayahara *et al.* (1989) found that 92 of 160 strains of *P. halophilus*, isolated from miso and soya sauce factories, harboured plasmids. In *P. acidilactici*, erythromycin resistance is coded for by a 40 MDa plasmid (Torriani *et al.*, 1987).

Pediococci can be transformed by electroporation or conjugation (Kim *et al.*, 1992). Plasmids can be transferred from genera such as *Enterococcus*, *Streptococcus* and *Lactococcus* to *Pediococcus* spp. and vice versa (Gonzalez and Kunka, 1983).

Bacteriophage attack of *P. halophilus* has been observed (Uchida and Kanbe, 1993) but phages that attack other pediococci are not known. Those attacking *P. halophilus* have a narrow host spectrum and several phage types of this species can be discriminated. This is consistent with the observation that *P. halophilus* is a heterogeneous species, consisting of many biovars which can be discriminated on the basis of their carbohydrate utilization patterns (Uchida, 1982).

5.5 Immunochemistry

Common precipitins are associated with *P. damnosus*, *P. parvulus* and '*P. cerevisiae*' (*P. pentosaceus* and *P. acidilactici*), but not with *P. halophilus* (Günther and White, 1961b). Antisera prepared against pediococci do not react with extracts prepared from closely related genera such as *Streptococcus* spp. and *Leuconostoc* spp. Coster and White (1964) found that antisera prepared against *P. parvulus* and *P. damnosus* reacted against '*P. cerevisiae*' (*P. pentosaceus*). Extracts of *P. halophilus* strains did not react with antisera prepared against other pediococci except for that prepared from one strain of '*P. cerevisiae*' (Coster and White, 1964). Antisera prepared against Coster and White's Group III strains (now classified as *P. dextrinicus*) showed cross-reactions with '*P. cerevisiae*', *P. parvulus* and some *P. damnosus* strains. Group III extracts gave no cross-reactions with antisera prepared against other pediococci.

London and co-workers used antibodies raised to aldolase enzymes to elucidate phylogenetic relationships among lactic acid bacteria, including *Pediococcus* spp. (London *et al.*, 1975; London and Chace, 1976, 1983).

Bhunia and Johnson (1992b) prepared monoclonal antibodies to several bacteriocin-producing strains of *P. acidilactici*. These antibodies did not react to other lactic acid bacteria or to other Gram-positive or Gram-negative organisms. A protein of M_r 116 000, located on the surface of *P. acidilactici* cells, was the antigenically reactive site (Bhunia and Johnson, 1992b). No reactions were obtained with proteins of identical M_r situated on the surfaces of *P. pentosaceus* cells, indicating that a specific epitope on the protein was responsible for antigenicity.

5.6 Historical aspects

Historical aspects of the genus *Pediococcus* have been thoroughly dealt with by a number of workers (Shimwell and Kirkpatrick, 1939; Shimwell, 1949; Pederson *et al.*, 1954; Garvie, 1974; Eschenbecher and Back, 1976). Salient features are summarized in Table 5.3.

Certain aspects of the literature relating to pediococci can be confusing, since the use of species names has lacked consistency. In particular, the name '*P. cerevisiae*', first used by Balcke (1884) for beer-spoilage strains (probably *P. damnosus*) that had only been observed microscopically and not isolated in pure culture, was used for plant pediococci (*P. pentosaceus*, *P. acidilactici*) by Pederson (1949). (Pederson mistakenly believed that plant pediococci and beer pediococci were one and the same.) Nakagawa and Kitahara (1959) used the name '*P. cerevisiae*' for beer-spoilage pediococci. In the literature spanning the 1960s, and to some extent even to the present day, this name has been applied to both groups of organisms. Günther *et al.* (1962) had proposed that ATCC 8081 be designated as the type strain of '*P. cerevisiae*' Balcke (this organism was originally known as '*Leuc. citrovorum*', see above). However, the description of this plant pediococcus was inconsistent with that described by Balcke (1884). As a result of a request from Garvie (1974), the Judicial Committee of the International Committee on Systematic Bacteriology issued an opinion in 1976 to the effect that the type species of the genus should be *P. damnosus* (Claussen, 1903) and the neotype strain Be.1 (NCDO 1832) (Judicial Commission, 1976). This strain was, in fact, isolated from lager beer by D.H. Williamson in our laboratories (then the Brewing Industry Research Foundation) more than 50 years ago. The name '*P. cerevisiae*' is no longer used.

The relationship between acidophilic beer-spoilage tetrad-forming cocci

Table 5.3 Historical development of the genus *Pediococcus**

Organism name	Reference	Comment
'Beer sarcinae'	Hansen (1879)	Produce 'sarcina sickness' in beer.
'*Pediococcus cerevisae*'	Balcke (1884)	Cell division in one plane, successive cell divisions at 90° to each other. Acid in sugar-containing media. Grow in beer. Optimum growth temperature 20–25°C.
Pediococcus acidi-lactici	Lindner (1888)	Optimum growth temperature 41°C. Produce large amounts of lactic acid in sugar-containing media.
'*Pediococcus sarcinaeformis*'	Reichard (1894)	Tetrad-forming coccus isolated from beer. Optimal growth temperature 20–25°C. Cells form clusters, or packets, under acidic conditions.
Pediococcus damnosus	Claussen (1903)	Cells grow in wort and pasteurized beer. Resistant to fluoride.
'*Pediococcus perniciousus*'	Claussen (1903)	Similar to *P. damnosus* but cells smaller. More vigorous growth in beer than *P. damnosus*.
'*Pediococcus hennebergi*'	Sollied (1903)	Optimal growth temperature 40°C. Maltose, galactose, glucose, arabinose and xylose fermented to give optically-inactive lactic acid. Differs from *P. acidilactici* in ability to ferment sucrose and arabinose.
'*Sarcina hamaguchiae*'	Saito (1907)	Salt-tolerant lactic acid-producing tetrad-forming coccus isolated from Japanese soya sauce mash.
'*Pediococcus damnosus* var. *perniciousus*'	Mees (1934)	Tetrad-forming cocci that produce DL-lactate from glucose.
'*Pediococcus damnosus* var. *salicinaceus*'	Mees (1934)	Similar to '*P. damnosus* var. *perniciousus*' but ferments salicin.
Pediococcus pentosaceus	Mees (1934)	Ferments arabinose. Grows at 45°C.
Pediococcus halophilus	Mees (1934)	Salt-tolerant tetrad-forming cocci.
Pediococcus urinae-equi	Mees (1934)	Produces less lactic acid than other pediococci and grows at alkaline pH values.

'*Streptococcus damnosus*'	Shimwell and Kirkpatrick (1939)	Synonym of *P. damnosus*.
'*Streptococcus tetragenus*'	Walters (1940)	Pentose-fermenting tetrad-forming cocci isolated from beer.
'*Streptococcus damnosus* var. *diastaticus*'	Andrews and Gilliland (1952)	Dextrin-degrading tetrad-forming cocci isolated from beer. Impart a bitter flavour and strong bitter after-flavour to beer.
'*Pediococcus mevalovorus*'	Kitahara and Nakagawa (1958)	Tetrad-forming cocci resembling *P. damnosus* but which require mevalonic acid for growth.
Pediococcus parvulus	Günther and White (1961a)	Tetrad-forming coccus which formed very small colonies on media used for isolation. (Later shown that growth could be improved by addition of Tween 80.) Serologically distinct from other pediococci.
Pediococcus damnosus var. '*damnosus*' var. '*diastaticus*' var. '*limosus*'	Coster and White (1964)	Proposed subspecies of *P. damnosus*.
Pediococcus inopinatus	Back (1978a)	New species of beer-spoiling pediococci. Identified, in addition to phenotypic behaviour, on the basis of DNA–DNA homology tests and electrophoretic behaviour of LDHs.
'*Pediococcus pentosaceus* var. *intermedius*'	Back (1978a)	Subspecies of *P. pentosaceus* identified on the basis of inability to use certain pentose sugars, DNA/DNA homology tests and electrophoretic mobility of LDHs.
Pediococcus dextrinicus	Back (1978b)	Valid publication of description of starch-degrading pediococci.
'*Tetratogenococcus halophilus*'	Collins *et al.* (1990)	Proposed new genus and species to accommodate strains described as *P. halophilus*.

*Note: This table is not comprehensive. For further details on the history of the development of the genus *Pediococcus* see Shimwell and Kirkpatrick (1939); Shimwell (1949); Garvie (1974); Eschenbecher and Back (1976).

and their acid-sensitive counterparts (*P. urinae-equi*) has long been the subject of debate and confusion. Indeed, many references can be found in the early brewing science literature to the relationship between organisms able to grow in horse urine and those able to grow in beer. Until the 1930s, most beer was transported by horse. Consequently, stables were maintained within the confines of the brewery. In finding that many beers suffered spoilage by tetrad-forming cocci, many brewery bacteriologists searched for the source of such contaminants. They found brewery stables to be heavily contaminated with such bacteria, and the urine of horses to be a particularly good source. However, it remained unknown whether the cocci that could be isolated from this source could spoil beer. A large number of studies were carried out in breweries to establish a link.

Typical of such studies was that of Stockhausen and Stege (1925) who showed that 'sarcinae' isolated from horse urine could grow in pasteurized beer and produce the 'usual symptoms' of sarcina sickness after 3–4 weeks at 25°C. Conversely, they showed that 'sarcinae' isolated from beer developed in sterile horse urine, of neutral pH value, forming a cloudiness after 6–9 days incubation. As a result of these studies they concluded that 'urinary sarcinae, which originally find their most favourable habitat in an alkaline medium, are capable, without gradual adaptation, of producing sarcina sickness in beer, a point of practical importance in regard to the location of stables in breweries'. In light of current knowledge of pediococci, this finding seems most likely to have been caused by use of mixed cultures rather than by adaptation of a pure culture.

Back (1978a) made a substantial taxonomic study of the genus, isolating 840 pediococcus colonies from a range of sources. He identified them on the basis of biochemical characteristics and genetic attributes. In addition to confirming the known species of *P. pentosaceus*, *P. acidilactici*, *P. parvulus*, *P. damnosus* and *P. halophilus*, Back identified two further species for which he proposed the names *P. inopinatus* (Back, 1978a) and *P. dextrinicus* (Back, 1978b). In addition, he found that some strains, which had been isolated from plant materials and had been identified as *P. damnosus* on the basis of biochemical attributes (in particular, their inability to use pentoses) had a high level of genetic homology with *P. pentosaceus*. He proposed that these organisms were closely related to *P. pentosaceus*, but distinct from *P. damnosus*, and named them '*P. pentosaceus* subsp. *intermedius*' (Back, 1978a).

5.7 Phylogenetic relationships

DNA–DNA homology assays (Back and Stackebrandt, 1978; Dellaglio *et al.*, 1981; Dellaglio and Torriani, 1986) have been used to elucidate phylogenetic relationships among pediococci. Table 5.4 shows the homology values obtained for representatives of each species. These support the groupings made on the basis of physiological tests. For example, although *P. pentosaceus* and *P. acidilactici* can sometimes be difficult to separate using phenotypic tests, DNA–DNA homology assays reveal only 5–35% homology, thus justifying separation of the species (Back and Stackebrandt, 1978; Dellaglio and Torriani, 1986). *Pediococcus damnosus* shows a significant degree of homology with *P. inopinatus* and *P. parvulus* (41–54 and 34–36%, respectively), but little homology to other species. *Pediococcus dextrinicus* has little genetic homology with the other pediococci (0–8%). Likewise, *P. urinae-equi* had no detectable homology with other members of the genus. Although they differ with respect to their phenotypic properties, strains belonging to *P. pentosaceus* and '*P. pentosaceus* subsp. *intermedius*' are clearly related as evidenced by homology values of 88–97%. Dellaglio and Torriani (1986) isolated three strains of pediococci from maize silage that resembled *P. pentosaceus* phenotypically but did not show significant DNA homology with strains belonging to this species. These isolates remain unidentified.

Stackebrandt *et al.* (1983), on the basis of 16S rRNA oligonucleotide cataloguing, suggested that *Pediococcus* spp. and *Leuconostoc* spp. were phylogenetically related to lactobacilli but distinct from streptococci. In this respect, they confirmed earlier immunological studies reported by London and Chace (1976). They suggested that descriptions of the genus *Lactobacillus* should be extended to include cocci occurring in pairs and chains (Stackebrandt *et al.*, 1983). This suggestion has not been generally accepted. The genus *Pediococcus* is placed within the Gram-positive cocci in the ninth edition of *Bergey's Manual* published in 1986. Kandler and Weiss (1986) stated that more work was needed concerning the phylogenetic relationships between the genus *Lactobacillus* and other lactic acid bacteria before the suggestion of Stackebrandt *et al.* could be considered.

Phylogenetic relationships between members of the genus *Pediococcus* and other genera including *Lactobacillus, Enterococcus*, *Vagococcus*, *Carnobacterium* and *Aerococcus*, were explored by Collins *et al.* (1990) who analysed 16S rRNA sequences. Calculation of sequence homologies allowed the species to be compared (Table 5.5). These values showed that the genus *Pediococcus*, as presently constituted, was phylogenetically heterogeneous. *Pediococcus acidilactici*, *P. damnosus*, *P. parvulus* and *P. pentosaceus* formed a distinct group, but *P. halophilus* had little homology with other pediococci. Homology values comparing *P. halophilus* and other pediococci (with the exception of *P. urinae-equi*) did not exceed 89.7%. In fact, *P. halophilus* had a closer affinity with members of the genus *Enterococcus* and *Carnobacterium* than with members of the genus *Pediococcus* or with other lactic acid bacteria. Collins *et al.* (1990) suggested that this organism be reclassified in a new genus and named '*Tetratogenococcus halophilus*'. *Pediococcus urinae-equi* was found to be

Table 5.4 DNA–DNA homology among pediococci*

	P. acidilactici	*P. damnosus*	*P. dextrinicus*	*P. inopinatus*	*P. halophilus*	*P. parvulus*	*P. pentosaceus*	*P. pentosaceus* subsp. *intermedius*	*P. urinae-equi*
P. acidilactici	100								
P. damnosus	0–7	100							
P. dextrinicus	0–5	4–5	100						
P. inopinatus	0–7	41–54	7	100					
P. halophilus	0–2	0–2	6	3–5	100				
P. parvulus	0–7	34–36	8	30–40	4	100			
P. pentosaceus	5–35	0–18	6	7–8	4	7	100		
'*P. pentosaceus* subsp. *intermedius*'	17–19	0–7	5	6–7	3	6	88–97	100	
'*P. urinae-equi*'	0	0	0	0	0	0	0	0	100

*Data compiled from Back and Stackebrandt (1978), Dellaglio *et al.* (1981), and Dellaglio and Torriani (1986).

Table 5.5 Percentage homology for a 1340-nucleotide region of 16S rRNAs of *Pediococcus* spp. and *Aerococcus viridans**

	P. acidilactici NCDO 2767	*P. damnosus* NCDO 1832	*P. dextrinicus* NCDO 1561	*P. halophilus* NCIB 12011	*P. parvulus* NCDO 1634	*P. pentosaceus* NCDO 990	*P. urinae-equi* NCDO 1636	*A. viridans* NCDO 1225
P. acidilactici NCDO 2767	100							
P. damnosus NCDO 1832	96.6	100						
P. dextrinicus NCDO 1561	93.8	94.0	100					
P. halophilus NCIB 12011	89.7	88.7	88.6	100				
P. parvulus NCDO 1634	97.0	98.7	94.5	87.4	100			
P. pentosaceus NCDO 990	98.3	96.5	93.2	88.3	96.7	100		
P. urinae-equi NCDO 1636	90.3	89.3	90.5	90.4	89.8	89.6	100	
A. viridans NCIDO 1225	89.3	89.9	89.6	89.7	89.6	89.0	99.9	100

*Data from Collins *et al.* (1990).

very closely related to *A. viridans* (99.9% sequence homology) confirming the view of Sakaguchi and Mori (1969) that the organisms belonged to the same species. The relationships between pediococci and other lactic acid bacteria, highlighted by 16S rRNA cataloguing, are summarized in the unrooted phylogenetic tree shown in Figure 5.3.

Many of these findings are supported by other chemotaxonomic studies. Thus, pediococci (with the exception of *P. urinae-equi*) differ from members of the genus *Leuconostoc* in that the former contain aspartic acid residues in their cell walls while the latter do not (Kandler, 1970). The cellular fatty acid spectra of *Pediococcus* spp. can be used to discriminate such organisms from those belonging to other genera and to differentiate between members of the genus (Uchida and Mogi, 1972; Bergan *et al.*, 1984). Bergan *et al.* (1984) divided the pediococci into three groups on the basis of their fatty acid composition. Group I, containing *P. urinae-equi* (which they considered identical to *A. viridans*), was characterized by the presence of C:14,1 and C:12 fatty acids. An unidentified peak, representing a fatty acid containing 12 or 13 carbon atoms, was also detected. This peak was not present in any other group. Group II strains (*P. damnosus*, *P. dextrinicus*, *P. halophilus*, *P. inopinatus* and *P. parvulus*) contained significant quantities of cyclopropane fatty acids (C:17cy; C:19cy). These were absent from other pediococci. Strains belonging to Group III (*P. acidilactici*, *P. pentosaceus* and '*P. pentosaceus* subsp. *intermedius*') lacked all of the fatty acids characteristic of Groups I and II, viz. C:12, C:14, C:17cy and C:19cy. However, they contained a unique small peak, probably a C:15 fatty acid, and had a much lower content of C:14 fatty acids than strains belonging to other groups. *Pediococcus dextrinicus*

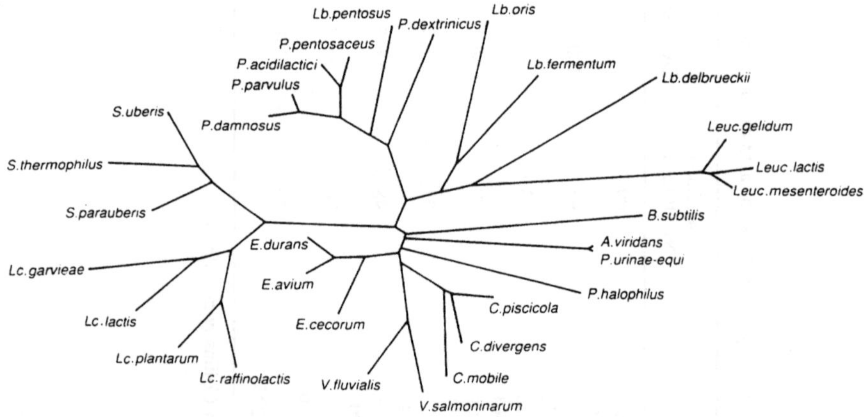

Figure 5.3 Unrooted phylogenetic tree showing the relationship between pediococci and other species. (From Collins *et al.* (1990), reproduced by permission of the Federation of European Microbiological Societies).

differed from other Group III strains in that it contained a higher concentration of C:18,1 and no C:19 fatty acids. A similar division of the pediococci based on fatty acid composition has also been described by Uchida and Mogi (1972).

5.8 Importance of the genus

5.8.1 Beer spoilage

Only two species, *P. damnosus* and *P. inopinatus*, grow in beer, but not all strains belonging to these species are able to do so. Others merely survive, or grow only in those of low alcohol content, high pH value, or low content of hop bitter acids (Back, 1978a; Lawrence and Priest, 1981).

The ability of an isolate to grow in beer has been given substantial taxonomic importance in the past. However, the beers used to determine the result of this test have varied considerably. They have included a French 'country beer' (Bière de garde) (Claussen, 1903), lager beer of the Pilsen type (Hansen, 1879; Back, 1978a) and stout (Andrews and Gilliland, 1952). Beers differ in their ability to support growth of pediococci (Dolezil and Kirsop, 1980). Fermentable carbohydrate, amino nitrogen (in the form of both amino acids and peptides), some vitamins, and minerals such as manganese, are limiting to growth (Uhl and Kühbeck, 1969; Rainbow, 1981). Beer has a relatively low pH value (typically 3.8–4.6) and contains ethanol and CO_2. Perhaps the most significant factor affecting the ability of pediococci to grow in beer is the presence of antibacterial compounds derived from hops (Figure 5.4). The antibacterial activity of these compounds is pH-dependent, being greater at low pH (Simpson and Smith, 1992). Thus, in tests designed to compare the sensitivity of different strains to such compounds, the pH value of the medium must be carefully controlled and reported alongside the results. At pH 5.2 in modified MRS, strains of *P. damnosus* which are resistant to hop bitter acids have minimum inhibitory concentrations (MIC) of about 100 μM, while sensitive strains, including those belonging to *P. pentosaceus*, have MIC of 20 μM or less (Fernandez and Simpson, 1993). Most beer spoilage lactic acid bacteria are unable to grow in beer unless they are grown under specialized conditions. For example, growth of the cells in the presence of a subinhibitory concentration of iso-α-acids, before inoculation into beer stimulates their ability to grow in beer (Simpson and Fernandez, 1992). The ability of different species or strains to grow in 'beer' is clearly an unreliable characteristic unless care is taken to describe the conditions of pre-culture and the chemical composition of the beer.

Pediococci are ubiquitous contaminants of breweries. *Pediococcus acidilactici* and *P. pentosaceus* are found on malt and can grow during the

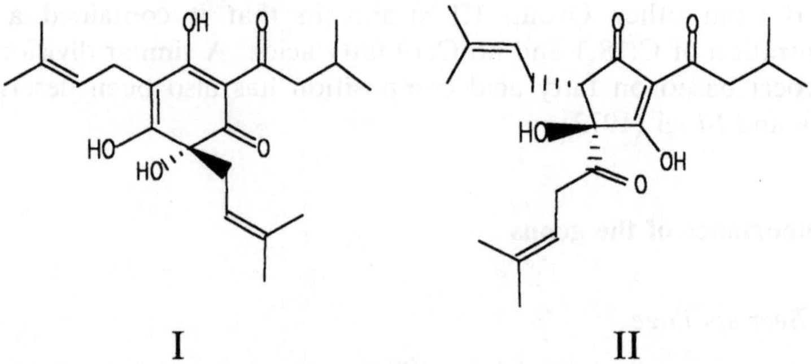

Figure 5.4 Hop bitter acids. α-Acids such as (−)-humulone (I) are found in hops; iso-α-acids, such as *trans*-isohumulone (II) are found in beer. Both are inhibitory to Gram-positive bacteria but many strains of *P. damnosus* are relatively resistant to both.

early stages of wort production when the temperature is of the order of 50°C and no hop compounds are present. Growth of these bacteria is not known to cause any defect in the beer produced from such worts. However, if the mash temperature is not controlled, growth of pediococci and thermophilic lactobacilli can result in acidification.

Strains belonging to *P. damnosus* and *P. inopinatus* can grow during fermentation and survive to be harvested with the yeast crop. For example, in this laboratory we have found *P. damnosus* at a level of up to 2×10^5 cfu/10^6 yeast cells in a contaminated commercial culture. The organisms were active during the fermentation process in the affected brewery, causing high levels of diacetyl to accumulate in the beer. More typically, such organisms are present at a level of 0–100 cfu/10^6 yeast cells.

McCaig and Weaver (1983) examined the physiological properties of a range of pediococci from a lager brewery. They isolated *P. damnosus* from yeast and throughout the fermentation process, in addition to the later stages of processing and in the final unpasteurized beer. Organisms which they referred to as *P. damnosus* 'var. 1' (probably *P. inopinatus*) could be isolated only from yeast and the early stages of fermentation. Occasionally, they were found in the later stages of fermentation, but never during conditioning or in unpasteurized beer. Those belonging to *P. pentosaceus* could be isolated from all stages and from final beer, but more commonly in the early stages of processing.

Rope (extracellular polymer) can be formed in beer by pediococci (Shimwell, 1948b; Kulka *et al.*, 1949). In some cases, rope production is so great that it is possible to draw the beer out into strings of almost 1 m in length without breaking them (Shimwell, 1948b). Nakagawa and Kitahara (1959) did not isolate any rope-forming strains of pediococci, but found that some strains produced rope when they were inoculated into media containing 0.05% agar.

Some strains now classified as *P. damnosus* (formerly '*P. mevalovorus*') require mevalonic acid for growth (Kitahara and Nakagawa, 1958). Mevalonic acid is not found in wort but is produced during the fermentation process. These strains can thus grow in beer, but are unable to grow in wort.

The source of *P. damnosus* contamination in breweries is frequently contaminated pitching yeast, beer, or equipment. Green and Gray (1949) isolated pediococci from the air of one brewery using an electrostatic sampler.

5.8.2 Wine and cider

Some pediococci produce off-flavours in wine, caused by diacetyl and acetoin (Pilone and Kunkee, 1965). Others spoil wine by producing rope, a D-glucan consisting of a trisaccharide repeating unit of D-glucose, that increases the viscosity of the wine to such an extent as to make it unpalatable (Llaubères *et al.*, 1990). Wines with higher than average pH values are more susceptible to growth of pediococci. Edwards and Jensen (1992) isolated nine strains of *P. parvulus* from wine and tentatively identified one strain as *P. inopinatus*.

Pediococci play a minor role in cider microbiology. Carr (1970) isolated several strains from spoiled ciders which he identified as '*P. cerevisiae*', some of which could form slime. These isolates can now be classified as *P. inopinatus* on the basis of their biochemical characteristics.

5.8.3 Soft drinks

Unidentified *Pediococcus* spp. were among the most prevalent of bacteria isolated from carbonated soft drinks in a Nigerian factory (Odunfa, 1987). Pediococci do not usually cause problems in soft drink production.

5.8.4 Soya sauce and miso

Citric acid is the major organic acid produced in the early stages of moromi fermentation, the second stage of the two-stage soya sauce process. Kanbe and Uchida (1987b) found that the ability of different strains of *P. halophilus* to utilize citrate correlated with production of the inducible enzyme, citrate lyase. The main products of citrate metabolism were acetate and formate; diacetyl was not formed.

Immobilized cells of *P. halophilus* can be used to prepare soya sauce. Osaki *et al.* (1985) described a two-stage process, in which *P. halophilus* cells were immobilized in one bioreactor and yeast cells immobilized in a second bioreactor. This protocol allowed the soya sauce production time to be reduced from 6 months to 2 weeks (Osaki *et al.*, 1985). Subsequently, *P.*

halophilus has been immobilized within porous alumina ceramic beads to produce a feedstock for yeast fermentation in soya sauce production (Iwasaki *et al.*, 1993).

Salt-tolerant pediococci also play a role in the fermentation of miso, a fermented food prepared from mould, rice, soyabeans and salt (Shibasaki and Hesseltine, 1962).

5.8.5 Cheese

The presence of pediococci in cheese was first reported by Dacre (1958a). Starter cultures of *Pediococcus* spp. have since been employed in cheese production (Dacre, 1958b; Bhowmik and Marth, 1990a). Pediococci represent only a small proportion of the total lactic acid bacteria in cheese (Litopolou-Tzanetaki *et al.*, 1989). Their precise influence on cheese quality is not yet fully understood (Bhowmik and Marth, 1990a; Fox *et al.*, 1990; Olson, 1990).

5.8.6 Meat and fish products

Starter cultures, consisting of a selected strain of *P. acidilactici*, are used in the preparation of semi-dry sausages. An ability to metabolize in conditions of reduced a_w favours pediococci in this matrix. Benefits of the use of such cultures include improvements in sausage uniformity and a reduction in process times (Everson *et al.*, 1970). Both lyophilized cultures (Deibel *et al.*, 1961) and frozen cell concentrates (Porubcan and Sellars, 1979) have been used. *Pediococcus pentosaceus* can be used instead of *P. acidilactici*. It is better suited to dry sausage fermentation as it has a lower optimal growth temperature and a lower minimum temperature for fermentation (Raccach, 1987).

Pediococci can be used to protect other types of sausage from pathogens, such as *Listeria monocytogenes*, since they produce bacteriocins (Berry *et al.*, 1991; Yousef *et al.*, 1991; Foegeding *et al.*, 1992; Luchansky *et al.*, 1992). Commercial starter cultures of *P. acidilactici* (e.g. 'Accel' from Muller in Germany) have been used to prevent meat spoilage (Gibbs, 1987).

Pediococci are present in various fermented foods including buroung dalag (a Philippino dish prepared from dalag fish and rice), marinated herrings (Blood, 1975), pla-som (Thai fermented fish), som-fak (Thai fish cake), nham (Thai fermented pork), fermented shrimps and a wide range of Thai fermented foods (Tanasupawat and Daengsubha, 1983).

5.8.7 Miscellaneous roles in fermentation processes

Costilow and Gerhardt (1983) used *P. pentosaceus* in a dialysis fermentation system to prepare fermented brined cucumbers and green beans. The inoculum and vegetables were separated by a semi-permeable dialysis membrane. The process was unsatisfactory with respect to the rate of acid production, pH reduction and utilization of carbohydrates (Costilow and Gerhardt, 1983). Use of a mixed inoculum containing *Lactobacillus plantarum*, *Streptococcus faecium* and an unidentified pediococcus improved control of silage fermentation, reducing losses and restricting growth of undesirable organisms (Weinberg *et al.*, 1988). Fitzsimons *et al.* (1992) obtained similar results using pure cultures of *P. acidilactici*. This organism and *P. pentosaceus* also effect some of the microbiological transformations that occur in uninoculated silage (Langston and Bouma, 1960; Lin *et al.*, 1992). *Pediococcus pentosaceus* has been used to ferment heat-treated soya milk (Raccach, 1987). Tou-pan-chiang is a traditional Chinese fermented food. In a study of the microorganisms found in Tou-pan-chiang mash, Hwang *et al.* (1988) isolated 88 strains of lactic acid bacteria, 33 of which were salt-tolerant strains of pediococci. Strains of *P. pentosaceus* are also associated with some vegetable fermentations. Examples include pickled cucumbers (Etchells *et al.*, 1975), olives and sauerkraut (Stammer, 1975).

5.8.8 Pediocin production

Pediocins, bacteriocins produced by pediococci, have been discussed in Volume 1 of this series (Earnshaw, 1992; Vandevoorde *et al.*, 1992). They include pediocin AcH (Bhunia *et al.*, 1988; Biswas *et al.*, 1991), bacteriocin PA-1 (Gonzalez and Kunka, 1987; Pucci *et al.*, 1988) and pediocin SJ-1 (Schved *et al.*, 1993) produced by strains of *P. acidilactici*, pediocin A produced by *P. pentosaceus* (Fleming *et al.*, 1975; Daeschel and Klaenhammer, 1985) and an unnamed bacteriocin (Hoover *et al.*, 1988) produced by an unidentified *Pediococcus*. Pediocins can be separated on the basis of their sensitivity to different proteolytic enzymes, chromatographic behaviour and molecular structure (Ray, 1992a). A broad spectrum antibacterial activity produced by three strains of *P. damnosus* isolated from beer was studied by Skyttä *et al.* (1993). The antibacterial compounds inhibited growth of both Gram-positive and Gram-negative bacteria, were thermotolerant, non-proteinaceous, and thus atypical bacteriocins. Their identification awaits further work. Pediocins of *P. acidilactici* have been extensively reviewed (Ray, 1992a) as have those of *P. pentosaceus* (Daeschel, 1992).

Pediocin production is often plasmid-linked. For example, the ability of *P. acidilactici* to produce pediocin AcH is correlated to possession of an 11.1 kb plasmid (pSMB74): the plasmid does not encode pediocin resistance (Kim *et al.*, 1992; Ray *et al.*, 1992). Pediocin AcH is a basic polypeptide (p*I* 9.6) that contains 44 amino acids and two disulphide bonds. Post-translational modification cleaves 18 amino acids from a 'prepediocin' of 62 amino acids to form the active pediocin (Ray, 1992b). A microbiological overlay technique used after separation by SDS-PAGE can be used to identify it (Bhunia and Johnson, 1992a).

Marugg *et al.* (1992) cloned the genes responsible for production of pediocin PA-1, a 44 amino acid polypeptide, and expressed them in *Escherichia coli*. Production of pediocin PA-1 depends on the presence of four clustered open reading frames (*ped*A, *ped*B, *ped*C and *ped*D). The *ped*A gene encodes a 62 amino acid precursor of pediocin PA-1; the *ped*B and *ped*C genes encode proteins of 112 and 174 amino acids, respectively; their function is not known. The *ped*D gene may be involved in translocation of pediocin PA-1 and, possibly, also of pediocin AcH (Marugg *et al.*, 1992).

Bhunia *et al.* (1991) investigated the mode of action of pediocin AcH against sensitive bacteria. Following exposure to the bacteriocin, sensitive cells released K^+ and UV-absorbing materials into the medium and became more permeable to larger molecules such as *o*-nitrophenol-*β*-D-galactopyranoside. Binding of the bacteriocin molecules to specific receptors on the surface of sensitive bacteria preceded its bactericidal effects. Non-specific and specific receptors for pediocin AcH are present on the surface of sensitive cells (Bhunia *et al.*, 1991). In the case of mutant strains of Gram-positive bacteria which are resistant to pediocin AcH, the specific receptors may either be absent, or not available for binding. In the case of Gram-negative bacteria, which are insensitive to pediocin AcH, both non-specific and specific receptor sites are absent (Bhunia *et al.*, 1991).

Kalchayanand *et al.* (1992) showed that pediocin AcH-resistant Gram-negative and Gram-positive bacteria could be sensitized to the pediocin if they were first exposed to a sublethal stress, such as heat, freeze–thawing or acid treatment.

No pediocin has yet been approved for food use by any regulatory authority. However, a US patent has been granted that relates to use of pediocin PA-1 for prevention of spoilage of salad and salad dressings (Gonzalez, 1989). Ray (1992b) sees no reason why pediocins should not be approved for food use.

5.8.9 Biological assays

Pediococcus acidilactici NCIMB 6990 is used to assay pantothenic acid. A greater degree of accuracy can be achieved using this organism than when *Lactobacillus plantarum* is used (Solberg *et al.*, 1975).

5.8.10 Public health considerations

Pediococci are non-pathogenic. A few strains decarboxylate histidine to histamine, but most have low, or undetectable, activities of histidine decarboxylase and tyrosine decarboxylase (Radler, 1975). Biogenic amines can cause illness and Raccach (1987) suggested that pediococci should be tested for their ability to produce such compounds before using them to prepare foods. Bravo Abad (1990) showed that higher levels of biogenic amines are found in beers which have been contaminated with *Lactobacillus* spp. and *Pediococcus* spp. than in uncontaminated beers.

5.9 Isolation and enumeration of pediococci

A variety of media can be used to isolate pediococci. Because the genus is heterogeneous, no single medium or incubation conditions can be used for all species. For general purposes, MRS medium (de Man *et al.*, 1960), YGP medium (Garvie, 1978) and TGE medium (Biswas *et al.*, 1991) suffice. *Pediococcus halophilus* can be grown in YGP medium supplemented with 5% NaCl (Collins *et al.*, 1990). Tanasupawat and Daengsubha (1983) used GYP-calcium carbonate medium (pH 6.8), with or without 5% NaCl, to isolate a range of pediococci, including *P. urinae-equi* and *P. halophilus*, from foodstuffs. They screened for acid-forming colonies by checking for zones of clearing in the calcium carbonate, after aerobic incubation at 30°C. Nakagawa and Kitahara (1959) used end-fermented beer to which the pH value had been raised to 5.0 with sodium acetate, and which contained D-mannose as carbon source, to isolate pediococci from a variety of environments.

Nakagawa later developed a semi-solid agar that contains unhopped beer as a base (Nakagawa, 1970). Eto and Nakagawa (1975) isolated a strain of '*P. cerevisiae*' from beer which grew slowly on this medium. Enrichment with tomato juice extract or unhopped wort allowed good growth. A single substance, 4′-*o*-(β-D-glucopyranosyl)-D-pantothenic acid, was responsible for growth stimulation. Eto and Nakagawa (1975) recommended that 2% tomato juice should be added to media used to grow pediococci, or that, alternatively, the media should be dissolved in unhopped beer to meet the requirement of some strains for this component. In addition to providing the pantothenyl derivative required by some strains, tomato juice also stimulates growth of pediococci by providing manganese, together with complex nitrogen compounds, such as adenine and adenine derivatives.

Nakagawa (1964) also described a medium, based on unhopped beer, designed to detect beer-spoilage pediococci, including those strains that need mevalonic acid. In addition to unhopped beer, the medium contained mannose or salicin, a high concentration (20 g/litre) of sodium acetate, ascorbic acid, cycloheximide and agar.

Back (1978a) found that MRS medium supported good growth of most pediococci. However, 41 of 519 strains of *P. damnosus* isolated from beer,

beer yeast, breweries, wine and cider grew poorly (A_{578} 0.3–0.6). If the MRS was mixed 1:1 with lager beer, all strains grew well. Back speculated that this was related to a requirement of the strains for mevalonic acid. MRS medium supplemented with 4% NaCl, and adjusted to pH 7.0, was suitable for isolation of *P. halophilus* (Back, 1978a). In both media, most pediococci could be isolated at an incubation temperature of 28°C, except for beer-spoilage strains for which 22°C was more suitable (Back, 1978a).

Back (1978a) also successfully enriched cultures for certain species of pediococci by using melezitose, ribose or dextrin as carbon source, incubating the cultures at 50°C, or by including 10% NaCl to the growth medium.

Other media for detection of pediococci include NBB (Nachweismedium für bierschädliche Bakterien) (Dachs, 1981), VLB-S7 (Emeis, 1969), Raka–Ray agar no. 3 (Saha *et al.*, 1974), Hsu's rapid medium (Hsu *et al.*, 1975b) and Hsu's *Lactobacillus–Pediococcus* medium (Hsu *et al.*, 1975a), Lee's multidifferential agar (LMDA) (Lee *et al.*, 1975), sucrose agar (Boatwright and Kirsop, 1983) and KOT medium (Taguchi *et al.*, 1990). For *P. halophilus*, PAT agar (Uchida, 1982) or YGP broth supplemented with 5% NaCl can be used. The American Society of Brewing Chemists (1992) recommends the use of LMDA, Raka–Ray medium, Barney–Miller brewery medium and MRS agar for detection of brewery pediococci. The Institute of Brewing (1991) recommends a modified form of MRS and Raka–Ray medium for this purpose.

Carr (1970) used apple juice–yeast extract medium to isolate and maintain strains of '*P. cerevisiae*' (probably *P. inopinatus*) from spoiled cider.

On primary isolation, some pediococci are intolerant of oxygen; in addition, some have a requirement for CO_2. Commonly, cultures are incubated under anaerobic conditions. Anaerobic jars or cabinets, filled with a mixture of CO_2 and N_2, are suitable.

Pediococci can be isolated in the presence of lactobacilli by using MRS in which glucose has been replaced by 1% mannose, cellobiose or salicin (Back, 1978a). In the case of pediococci associated with plants (*P. acidilactici*, *P. pentosaceus*, '*P. pentosaceus* subsp. *intermedius*') incubation of primary cultures in Rogosa's SL medium at 45°C resulted in rapid initial growth of pediococci, with the result that they outgrew lactobacilli and other organisms that were present (Mundt *et al.*, 1969).

Selective agents used to assist isolation of pediococci from primary culture include cycloheximide and crystal violet to inhibit growth of yeasts and 2-phenylethanol, sorbic acid and acetic acid, to inhibit growth of both yeasts and Gram-negative bacteria. Thallous acetate inhibits growth of most microorganisms other than lactic acid bacteria (Sharpe, 1955). Vancomycin inhibits growth of Gram-positive bacteria other than those belonging to the genera *Pediococcus* and *Leuconostoc*, and some members of the genus *Lactobacillus* (Simpson *et al.*, 1988). In the case of beer-spoilage strains of *P. damnosus*, hop bitter acids can be used as selective agents (Simpson and Hammond, 1991).

DNA probe techniques and immunochemical methods can be used to simultaneously identify and enumerate pediococci in test samples. For example, a monoclonal antibody colony immunoblot method, specific for bacteriocin-producing *P. acidilactici*, can detect such organisms in foods (Bhunia and Johnson, 1992b). DNA probe methods are available for detection and identification of glucan-forming wine-spoilage strains of *P. damnosus* (Lonvaud-Funel *et al.*, 1993), for non-glucan-forming strains of *P. damnosus* and *P. pentosaceus* in fermenting grape must and wine (Lonvaud-Funel *et al.*, 1991), and for strains of *P. pentosaceus* that colonize silage (Cocconcelli *et al.*, 1991).

Dolezil and Kirsop (1976) used a commercially available antiserum from Group D streptococci to detect *Pediococcus* spp. in brewers' yeast, wort and beer. Whiting *et al.* (1992) used an immunofluorescent antibody technique to detect diacetyl-producing pediococci in brewery pitching yeast.

Flow cytometry has been used to enumerate *P. damnosus* cells in brewery samples (Hutter, 1991). Interference from other microorganisms was minimized by using fluorescent dyes conjugated to antibodies (Hutter, 1992).

5.10 Maintenance and preservation of pediococci

Cultures of pediococci can be preserved in several ways. Most strains survive on agar slopes at 4°C provided they are sub-cultured regularly (usually every 3 months). The storage characteristics of the cultures are improved by addition of calcium carbonate (1%) to the growth medium to neutralize the acid produced by the organisms. Pediococci can also be preserved by lyophilization (Garvie, 1986a). Cells from a late logarithmic or early stationary phase should be suspended in horse serum, containing glucose (7.5%), prior to lyophilization. Alternatively, cells can be stored at −20°C in a mixture of growth medium and glycerol (1:1) (Weiss, 1991).

5.11 Identification of pediococci

Figure 5.5 shows a simple key to discrimination of *Pediococcus* spp. from other Gram-positive bacteria. This is based on production of catalase, anaerobic and aerobic growth, ability to produce gas from glucose, and cell morphology. Unlike other Gram-positive cocci (including *Streptococcus* spp., *Lactococcus* spp. and *Enterococcus* spp.), pediococci can grow in the

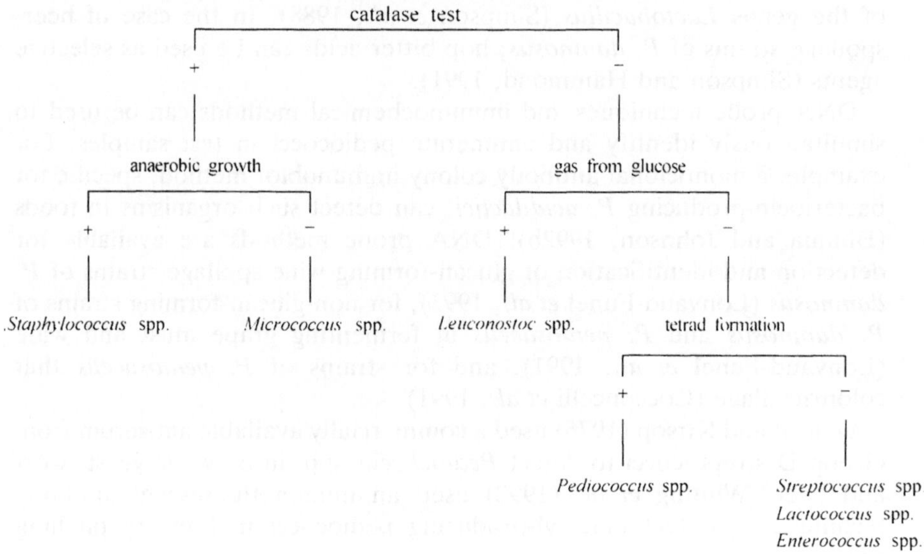

Figure 5.5 Key to differentiation of pedicocci from other Gram-positive bacteria.

presence of vancomycin or avoparcin (50 mg/litre) (Simpson *et al.*, 1988). The biochemical basis of such resistance is not known. The reaction of *P. halophilus* and *P. urinae-equi* to vancomycin is not clear. Fourteen strains of *P. halophilus* isolated from porcine faeces were sensitive to vancomycin, but these strains were atypical with respect to several other phenotypic characters (Molitoris *et al.*, 1986). Pediococci can also be discriminated from lactococci by the fact that the former do not react to Group N streptococcal antiserum, while the latter give a positive reaction (Gonzalez and Kunka, 1983).

Pilone *et al.* (1991) described a single broth culture test, based on detection of heterofermentative metabolism, production of mannitol from fructose and production of ammonia from arginine, to differentiate between *Lactobacillus* spp., *Leuconostoc* spp. and *Pediococcus* spp. isolated from wine.

Figure 5.6 shows the key to identification of each of the eight *Pediococcus* species. Satisfactory differentiation can be achieved in most cases on the basis of growth at pH 8.5 and 4.2, at 40°C or 50°C, and in the presence or absence of 5% or 10% NaCl. Confirmatory tests, especially useful in the case of brewery pediococci, include those for acid from ribose, maltose, lactose or starch, hydrolysis of arginine, acid and gas from gluconate, and ability to grow at 35°C. Fuller details of differentiation between species of pediococci are given in the descriptions below.

An alternative scheme, proposed by Back (1978a), is shown in Figure

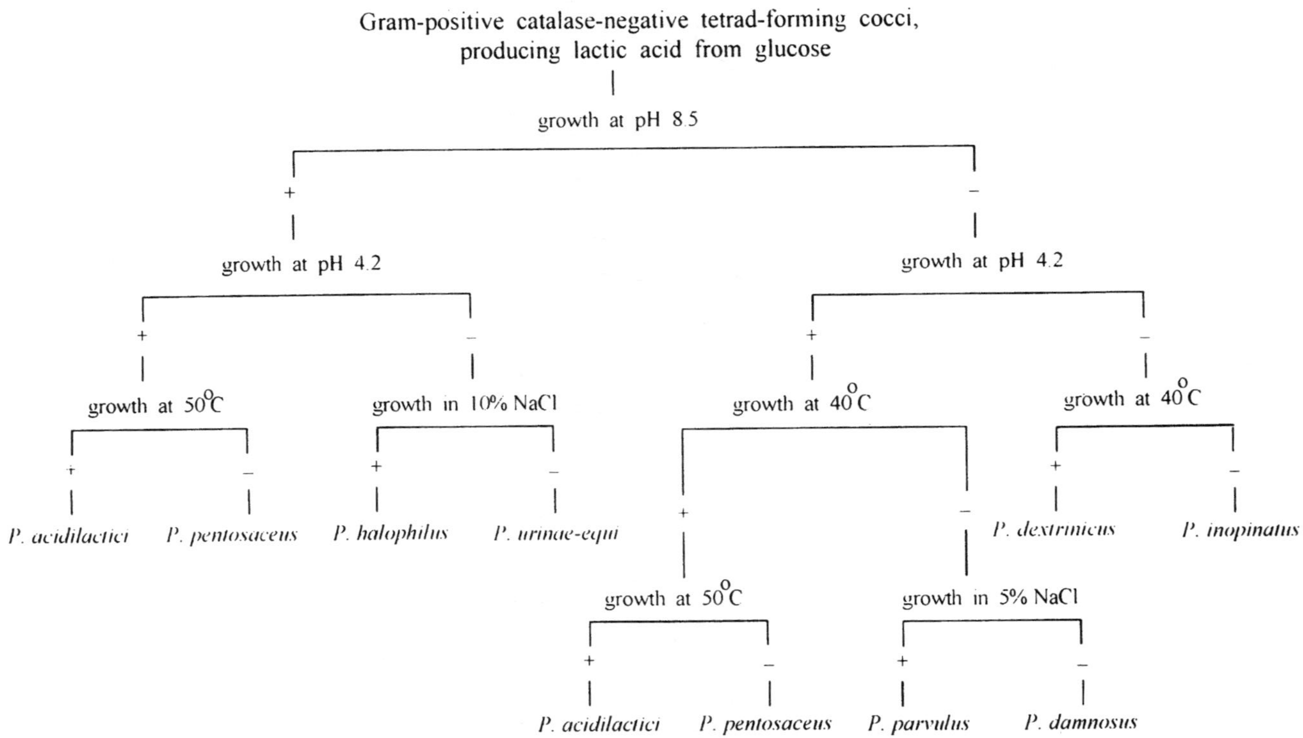

Figure 5.6 Key to the differentiation of the species of pediococci.

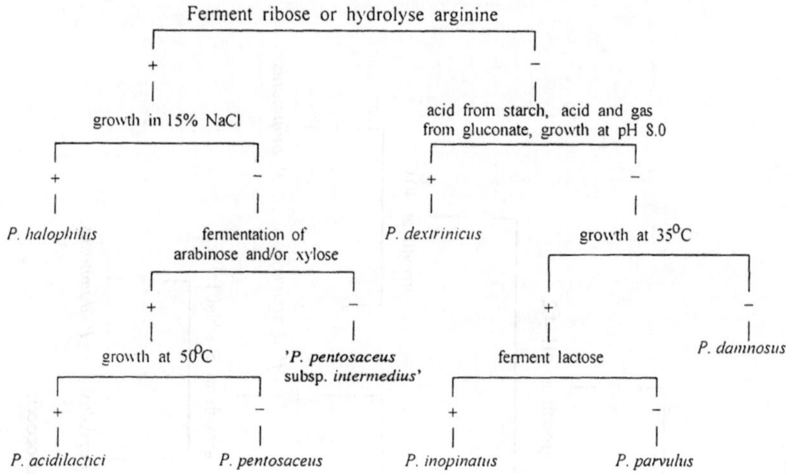

Figure 5.7 Alternative key to differentiation of pediococci. (Adapted from Back (1978a).)

5.7. Attributes in this scheme include the ability to hydrolyse arginine, fermentation of ribose, arabinose, xylose, lactose or starch, tolerance to 15% NaCl, ability to grow at 35°C or 50°C or at pH 8.0, and ability to produce acid and gas from gluconate. Confirmation of species identity can be obtained from sugar utilization profiles and from measurements of the electrophoretic mobility and characteristics of the cells' lactate dehydrogenase (LDH) (Back, 1978a). Pediococci produce three types of LDH: D-LDH and two types of L-LDH, one of which is activated by fructose-1,6-diphosphate. The electrophoretic mobilities of the different isozymes differ sufficiently to allow species differentiation (Table 5.6). DNA/DNA hybridization assays can also be used to confirm species identity (Back and Stackebrandt, 1978). Dolezil and Kirsop (1977) used the API lactobacillus system (now the API 50CH system) to discriminate between strains of pediococci. Lawrence and Priest (1981) confirmed the value of this kit for identification of brewery pediococci while Davis *et al.* (1988) used it to show the heterogeneity within the range of *P. parvulus* strains isolated from wine. Tzanetakis and Litopolou-Tzanetaki (1989) used the API ZYM test kit for biochemical characterization of 83 *P. pentosaceus* strains of dairy origin. Leucine aminopeptidase and valine aminopeptidase reactions were positive with all strains and of greatest intensity of the characters tested. *β*-Galactosidase and *N*-acetyl-*β*-glucosamidase were found in most strains. *β*-Glucosidase, esterase, esterase:lipase, lipase, phosphoamidase, cystine aminopeptidase and acid phosphatase activities were found in some strains. The information obtained from the API ZYM tests did not allow individual strains of *P.*

Table 5.6 Relative mobilities of the L-LDH and D-LDH enzymes of different pediococci*

Species	Relative mobility†	
	L-LDH	D-LDH
P. acidilactici	1.39	1.29
P. damnosus	0.92	1.16
P. dextrinicus	1.02‡	—§
P. halophilus	0.82	—
P. inopinatus	1.18	1.36
P. parvulus	0.97	1.42
P. pentosaceus	1.36	1.32
'*P. pentosaceus* subsp. *intermedius*'	1.38	1.23
P. urinae-equi	ND¶	ND

*Data from Back (1978a).
†Relative mobilities determined by acrylamide disc electrophoresis relative to rabbit heart L-LDH isoenzyme I.
‡Fructose 1,6-diphosphate-activated enzyme.
§—, activity not detected.
¶ND, not determined.

pentosaceus to be discriminated from one another, but the authors suggested that the kit may be useful for discriminating *P. pentosaceus* from other organisms.

Schisler *et al.* (1979) identified brewery bacteria by analysing metabolic end-products by gas chromatography. They could differentiate *Pediococcus* spp. from other brewery bacteria but not between pediococci. Similar attempts to identify brewery microorganisms, including *Pediococcus* spp., were made by Hug and Bosio (1991), who employed high performance liquid chromatography to measure metabolic end-products.

Luchansky *et al.* (1992) successfully used clamped homogeneous electric field (CHEF) electrophoresis to discriminate strains of *P. acidilactici* used in sausage fermentation after low frequency-cleavage of chromosomal DNA by endonuclease *Asc*I.

5.12 Description of species which comprise the genus *Pediococcus*

The following information on the individual species of pediococci has been compiled mainly from the data of Sakaguchi (1958), Nakagawa and Kitahara (1959), Günther *et al.* (1962), Günther and White (1961a), Sakaguchi and Mori (1969), Dellaglio *et al.* (1974), Garvie (1974), Back (1978a,b), Dellaglio and Torriani (1986) and Tjandraatmadja *et al.* (1990) unless otherwise indicated. For synonyms, type strain details and priority see Table 5.1. For full details of the characteristics of each species, refer to Table 5.2.

5.12.1 Pediococcus acidilactici

The species name is derived from the Latin nouns, *acidium lacticum*, meaning lactic acid. Thus, *acidilactici* means 'of lactic acid'. Cells of *P. acidilactici* are 0.6–1 μm in diameter occurring singly, in pairs, tetrads or irregular clusters.

Growth occurs at pH 4.2 and 8.0 and sometimes at pH 8.5. Maximum temperature for growth is 50–53°C; all strains grow at 50°C. Optimum growth temperature is 40°C. Cells grow in the presence of 9–10% NaCl. The sugar fermentation reactions of *P. acidilactici* resemble those of *P. pentosaceus*. The inability of *P. acidilactici* to ferment maltose and its ability to grow at 50°C differentiate it from *P. pentosaceus*. An inability to hydrolyse starch and produce gas from gluconate separates the species from *P. dextrinicus*. An ability to hydrolyse arginine separates it from all other pediococci except for *P. pentosaceus*. Cells of this species are heat resistant. At 70°C, 10 min is needed to kill all cells in a culture. D and Z values have not been reported.

It is not always possible to differentiate between strains of *P. acidilactici* and *P. pentosaceus* on the basis of morphological, cultural and physiological traits alone. That this is so is highlighted by the fact that the type strain of *P. acidilactici* proposed by Garvie as the type species of the genus *Pediococcus* was later shown to belong to *P. pentosaceus* using DNA hybridization studies (Back and Stackebrandt, 1978). DL-Lactate is produced from glucose. Final pH in MRS broth is between 3.5 and 3.8. Mol% G+C is 38–44 (T_m).

Strains of *P. acidilactici* are widely distributed in fermenting plant material, including silage, cereal mashes and pickles, potato mashes and sake mashes, barley, malt, dried leaves and hay. Some have been isolated from salami. *Pediococcus acidilactici* has some application in semi-dry sausage production, miscellaneous fermentation processes and vitamin assays. Some strains produce pediocins.

5.12.2 Pediococcus damnosus

The species name is derived from the Latin adjective *damnosus*, which means 'destructive'. Cells of *P. damnosus* are 0.7–1.0 μm in diameter occurring singly, in pairs, tetrads or irregular clusters. Growth is slow, even on rich media.

Cells grow at pH 4.2 but not at pH 8.5. The maximum pH for growth is typically 6.5–7.0, optimum pH being in the range 4.0–6.0. Growth occurs in the range 8–30°C. At 25°C, growth is slow: 7–10 days are needed in some cases for colonies to reach their maximum size on MRS agar. Anaerobic incubation is essential for growth of most strains on agar media. Even at

6°C, most grow well in MRS after 3–5 weeks incubation. No growth occurs at 35°C, or above. Some strains grow in the presence of 4% NaCl. Few grow in the presence of 5% NaCl, but when they do, growth is slow and weak. None grow in the presence of 5.5% NaCl. The inability of *P. damnosus* to utilize starch and produce gas from gluconate allows it to be differentiated from *P. dextrinicus*. Its inability to hydrolyse arginine separates it from *P. acidilactici* and *P. pentosaceus*. Cells of *P. damnosus* are hop-tolerant. Exposure of some strains to hop bitter acids lead to formation of 'giant' cells, 5–15 μm in diameter (Nakagawa and Kitahara, 1962). Final pH in MRS broth is 3.7–4.2. Some strains form slime. DL-Lactate is produced from glucose. When galactose, sucrose or maltotriose are used as carbon source, the final pH in MRS broth is 4.8–5.0. Mol% G+C is 37–42 (T_m).

Pediococcus damnosus is associated with breweries and brewery products. It can be isolated from brewers' yeast, fermenting wort and beer, wine and cider (Back, 1978a). Cells of *P. damnosus* produces antibacterial substances which have a broad spectrum of activity, are not affected by proteolytic enzymes or catalase and are very heat resistant (Skyttä *et al.*, 1993). This property has not yet been exploited.

5.12.3 Pediococcus dextrinicus

The species name is derived from the Latin noun *dextrinosum*, which means dextrin. Thus, *dextrinicus* means 'relating to dextrin'. Cells are *c.* 1 μm diameter, occurring singly, in pairs, tetrads or irregular clusters. Occasionally short chains are formed.

Growth does not occur at pH 4.2, but does at pH 8.0. Maximum temperature for growth is 43–45°C. Optimum growth temperature is 32°C. Cells grow in the presence of 6% NaCl. In contrast to all other pediococci, *P. dextrinicus* utilizes starch and produces acid and gas from gluconate. The inability of *P. dextrinicus* to ferment pentoses allows the species to be differentiated from *P. acidilactici*, *P. pentosaceus*, *P. halophilus* and *P. urinae-equi*. Their inability to hydrolyse arginine separates them from *P. acidilactici* and *P. pentosaceus*. Uniquely among pediococci, *P. dextrinicus* possesses a fructose 1,6-diphosphate-activated L-LDH. L(+)-lactate is produced from glucose. *Pediococcus dextrinicus* is less anaerobic than other pediococci. 16S rRNA cataloguing has shown that the species is not closely related to other pediococci. Colonies develop on agar under aerobic conditions, but growth is improved by anaerobic incubation. Final pH in MRS broth is about 4.4. Mol% G+C is 40–41 (T_m).

Pediococcus dextrinicus has been isolated from silage, brewers' spent grains, beer and empty beer bottles. There are no applications for this organism at present.

5.12.4 Pediococcus halophilus

Note: Phylogenetic studies (Collins *et al.*, 1990) indicate that *P. halophilus* is more closely related to enterococci and carnobacteria than to other pediococci. Collins *et al.* (1990) propose that members of this species be transferred to a new genus '*Tetratogenococcus*' and named '*T. halophilus*'.

The species name currently used is derived from the Greek noun *halos*, which means 'salt', and the adjective *philus*, which means loving. These combine to give *halophilus*, which means salt-loving. Colonies of *P. halophilus* develop slowly on agar media under both aerobic and anaerobic conditions. Growth in broth is slow and less confluent than that seen with other *Pediococcus* spp.; 4–5 days is required for the cells to reach stationary phase.

Pediococcus halophilus is a heterogeneous species consisting of numerous biovars. Growth occurs at pH 9.0. Maximum temperature for growth is 37–40°C. Optimum temperature is 25–30°C: growth does not occur at 10°C or 45°C. Cells grow in the presence of 18% NaCl: some strains can grow with 20–26% NaCl. Growth of *P. halophilus* in MRS broth, and in other acidic media, is poor. Arginine is not usually hydrolysed, although strains have been isolated from pig faeces which are positive in this respect (Molitoris *et al.*, 1986). The salt-tolerance of *P. halophilus* allows members of this species to be easily separated from other pediococci. L(+)-Lactate is the major end-product of glucose metabolism. About 3% D(–)-lactate is also formed. Final pH in MRS broth is about 5.0. Mol% G+C is 34–36.5 (T_m).

Pediococcus halophilus can be isolated from soya sauce mashes, pickling brines, pickled anchovies, pig faeces and Tou-pan-chiang mash. Selected strains are used to inoculate fermented soya products and salted fermented foodstuffs.

5.12.5 Pediococcus inopinatus

The species name is derived from the Latin adjective *inopinatus*, which means 'unexpected'. Cells are 0.5–0.8 μm in diameter, occurring singly, in pairs, tetrads or irregular clusters. Growth of the organisms is slow and colonies on agar media take 5 days or more to reach their maximum size on MRS agar.

No growth occurs at pH 4.2. The maximum pH value for growth is 7.5. Maximum temperature for growth is 37–40°C. The optimum growth temperature lies between 30 and 32°C. Cells grow in the presence of 6–8% NaCl. *Pediococcus inopinatus* can be differentiated from *P. pentosaceus* and *P. acidilactici* by its inability to utilize pentoses and hydrolyse arginine. It differs from *P. dextrinicus* in being unable to grow on starch or produce acid and gas from gluconate. Unlike some strains of *P. damnosus*, it cannot utilize melezitose. An ability to utilize lactose separates *P. inopinatus* from *P. parvulus*. DL-Lactate is produced from glucose. Some strains form slime. The final pH in MRS broth is about 4.0. Mol% G+C is 39–40 (T_m).

Pediococcus inopinatus has been isolated from sauerkraut, beer yeast, hops, wine, empty beer bottles, fermented beans. There are no uses for this organism at present.

5.12.6 Pediococcus parvulus

The species name is derived from the Latin adjective *parvulus*, which means 'very small'. This name was originally chosen by Günther and White (1961a) because the organisms formed very small colonies on agar media. However, the size of the colonies can be increased by use of anaerobic incubation and by inclusion of Tween 80 in the growth medium. Cells are 0.7–1.1 μm in diameter occurring singly, in pairs, tetrads or irregular clusters.

Growth occurs at pH 4.5. The upper pH limit for growth lies between 7.0 and 7.5. Optimum pH for growth is about 6.5. The optimum growth temperature is about 30°C. Maximum temperature for growth is 37–39°C. Growth occurs in the presence of 5.5–8% NaCl. The inability of *P. parvulus* to utilize pentoses separates it from *P. pentosaceus* and *P. acidilactici*. Unlike *P. dextrinicus*, this species is unable to utilize starch or hydrolyse arginine. An inability to utilize lactose separates it from *P. inopinatus*. An inability to tolerate high NaCl concentrations separates the species from *P. halophilus*. DL-Lactate is formed from glucose. Some strains form slime. Final pH in MRS broth is 3.9–5.5 (the wide range is probably a reflection of the fact that some strains grow poorly). Mol% G+C is 40.5–41.6 (T_m).

Strains of *P. parvulus* have been isolated from sauerkraut, fermented vegetables, fermented beans, beer, cider and wine. Attempts have been made to use the organism to effect a malolactic fermentation of wine, with limited success.

5.12.7 Pediococcus pentosaceus

The species name is derived from the Latin noun *pentosum*, which means 'pentose'. Thus, *pentosaceus* means 'relating to a pentose'. Cells are 0.6–1.0 μm in diameter, occurring singly, in pairs, tetrads, or irregular clusters.

Growth occurs at pH 8.0 and pH 4.5. Maximum temperature for growth is 39–45°C. Cells of *P. pentosaceus* are less heat-resistant than those of *P. acidilactici*. Optimum growth temperature is 28–32°C. Cells grow in the presence of 9–10% NaCl. The sugar fermentation reactions of *P. pentosaceus* resemble those of *P. acidilactici*. The ability of *P. pentosaceus* to ferment maltose and its lower growth temperature differentiate it from *P. acidilactici* (although not invariably). Its inability to hydrolyse starch and produce gas from gluconate separates it from *P. dextrinicus*. Lack of sucrose and melizitose fermentation separates this species from most other pediococci. Its ability to hydrolyse arginine separates it from all other pediococci with the exception of *P. acidilactici*.

Garvie (1986a) points out that, in some circumstances, strains of *P. pentosaceus* could be confused with micrococci since they form small colonies on sugar-free agar media, grow at pH 9.0 and, on media with low glucose content, may be weakly (pseudo-) catalase-positive. The rapid growth of *P. pentosaceus*, together with the low final pH which the organisms produce in broth media and their lack of cytochromes, serve to discriminate them from micrococci.

They produce DL-lactate from glucose. Final pH in MRS broth is between 3.5 and 3.8. Mol% G+C is 35–39 (T_m).

Pediococcus pentosaceus can be isolated from various plant materials including barley, malt, hops, dried leaves, hay, citrus fruits, apples and strawberries. The type strain (NCDO 990) was isolated from beer yeast. Selected strains of *P. pentosaceus* have been used to inoculate various fermentation processes including semi-dry sausage fermentations, cucumber and green bean fermentations, soya milk fermentations and silage. Some strains produce pediocins.

5.12.8 'Pediococcus pentosaceus *subsp.* intermedius'

Cells are 0.6–1 μm in diameter, occurring singly, in pairs, tetrads, or irregular clusters. Growth occurs at pH 4.5 and 8.0. Maximum temperature for growth is about 39°C. Most strains grow in the temperature range 8–45°C.

The sugar utilization reactions of '*P. pentosaceus* subsp. *intermedius*' are very similar to those of *P. pentosaceus*, except that arabinose, xylose and rhamnose are not fermented. Most strains give positive catalase reactions. DL-Lactate is formed from glucose. Final pH in MRS broth is between 3.5 and 3.8.

'*Pediococcus pentosaceus* subsp. *intermedius*' has been isolated from barley, malt, beer yeast, empty beer bottles, dried leaves, citrus fruits, strawberries, hay and other plant materials. There are no known uses for this subspecies.

5.12.9 Pediococcus urinae-equi

Note: The description of *P. urinae-equi* is given only for historical completeness. Strains fitting this description clearly belong to the genus *Aerococcus*.

The species name is derived from the Latin nouns *urina*, which means urine, and *equs*, which means horse. Thus, *urinae-equi* refers to horse urine. Cells are 0.8–1.0 μm in diameter, occurring singly, in pairs, tetrads or irregular clusters. On agar media, they form circular colonies, 1–2 mm in diameter, which are greyish-white in colour and raised. On agar stab cultures, growth occurs along the stab with limited surface growth.

The optimum pH value for growth is between 8.5 and 9.0. Growth will commence in media of initial pH 6.5–7.0. Final pH in broth is about 5.0. Temperature optimum is 25–30°C (max. 42°C). Cells produce L(+)-lactate from glucose. The electrophoretic behaviour of the LDH of this species has not been studied. Unlike other pediococci, growth of *P. urinae-equi* can take place in the absence of fermentable carbohydrate. Mol% G+C is 39.6–39.7 (T_m).

Reported isolations of *P. urinae-equi* are rare. Strains have been isolated from horse urine, rabbit dung and phak-gard-dong (Thai pickled vegetables) (Tanasupawat and Daengsubha, 1983). There are no known uses for this species.

5.13 Concluding remarks

On the basis of the evidence presently available, it seems appropriate to include the following species in the genus *Pediococcus*: *P. acidilactici*, *P. damnosus*, *P. dextrinicus*, *P. inopinatus*, *P. parvulus* and *P. pentosaceus*. The proposal of Back (1978a) that a subspecies of *P. pentosaceus* ('*P. pentosaceus* subsp. *intermedius*') be established, to accommodate strains of this species which are unable to utilize certain pentoses, also seems warranted.

Collins *et al.* (1990) have proposed that strains currently named *P. halophilus* be reclassified as '*Tetratogenococcus halophilus*' on the basis of 16S rRNA cataloguing of strain NCIMB 12011. It may be advisable to compare data from other strains of *P. halophilus* before adopting the suggestion, since *P. halophilus* is a heterogeneous species (Uchida, 1982).

The taxonomic position of strains belonging to *P. urinae-equi* is not in doubt. Both phenotypic and genotypic evidence points to their close association with the aerococci. These strains clearly do not belong within the genus *Pediococcus* and a review of their nomenclature would be appropriate.

Acknowledgements

The authors thank George Brown and George Henderson for assistance in obtaining some of the original publications cited in this chapter; Thomas

Bühler and Evelyne Canterranne for German and French translations, respectively; Nigel Davies, Sachin Chandra, Gareth Davies and Valerie Simpson for helping to prepare the figures and tables; M.D. Collins for allowing us to reprint Figure 5.3; Charlie Bamforth and Jacqui Fernandez for critically reviewing the manuscript; and the Directors of BRF International and Kirin Brewery Company Ltd for permission to publish.

References

American Society of Brewing Chemists (1992) Differential culture media. In *Methods of Analysis of the American Society of Brewing Chemists*, 8th edn. ASBC, MN, USA.

Andrews, J. and Gilliland, R.B. (1952) Super-attenuation of beer: a study of three organisms capable of causing abnormal attenuations. *Journal of the Institute of Brewing*, **58**, 189–196.

Archibald, F. (1986) Manganese: its acquisition by and function in the lactic acid bacteria. *CRC Critical Reviews in Microbiology*, **13**, 63–109.

Back, W. (1978a) Zur Taxonomie der Gattung *Pediococcus*. Phänotypische und genotypische Abgrenzung der bisher bekannten Arten sowie Beschreibung einer neuen bierschädlichen Art: *Pediococcus inopinatus*. *Brauwissenschaft*, **31**, 237–250, 312–320, 336–343.

Back, W. (1978b) Elevation of *Pediococcus cerevisiae* subsp. *dextrinicus* Coster and White to species status [*Pediococcus dextrinicus* (Coster and White) comb. nov.]. *International Journal of Systematic Bacteriology*, **28**, 523–527.

Back, W. and Stackebrandt, E. (1978) DNS–DNS-homologiestudien innerhalb der Gattung *Pediococcus*. *Archives of Microbiology*, **118**, 79–85.

Balcke, J. (1884) Über häufig vorkommende Fehler in der Bierbereitung. *Wochenschrift für Brauerei*, **1**, 181–184.

Bergan, T., Solberg, R. and Solberg, O. (1984) Fatty acid and carbohydrate cell composition in pediococci and aerococci, and identification of related species. In *Methods in Microbiology*, Vol. 16 (ed. Bergan, T.). Academic Press, London, UK, pp. 179–211.

Berry, E.D., Hutkins, R.W. and Mandigo, R.W. (1991) The use of bacteriocin-producing *Pediococcus acidilactici* to control postprocessing *Listeria monocytogenes* contamination of frankfurters. Journal of Food Protection, **54**, 681–686.

Bhowmik, T. and Marth, E.H. (1990a) Role of *Micrococcus* and *Pediococcus* species in cheese ripening: a review. *Journal of Dairy Science*, **73**, 859–866.

Bhowmik, T. and Marth, E.H. (1990b) Peptide-hydrolysing enzymes of *Pediococcus* species. *Microbios*, **62**, 197–211.

Bhunia, A.K. and Johnson, M.G. (1992a) A modified method to directly detect in SDS-PAGE the bacteriocin of *Pediococcus acidilactici*. *Letters in Applied Microbiology*, **15**, 5–7.

Bhunia, A.K. and Johnson, M.G. (1992b) Monoclonal antibody-colony immunoblot method specific for isolation of *Pediococcus acidilactici* from foods and correlation with pediocin (bacteriocin) production. *Applied and Environmental Microbiology*, **58**, 2315–2320.

Bhunia, A.K., Johnson, M.C. and Ray, B. (1988) Purification, characterization and antimicrobial spectrum of a bacteriocin produced by *Pediococcus acidilactici*. *Journal of Applied Bacteriology*, **65**, 261–268.

Bhunia, A.K., Johnson, M.C., Ray, B. and Kalchayanand, N. (1991) Mode of action of pediocin AcH from *Pediococcus acidilactici* H on sensitive bacterial strains. *Journal of Applied Bacteriology*, **70**, 25–33.

Biswas, S.R., Ray, P., Johnson, M.C. and Ray, B. (1991) Influence of growth conditions on the production of a bacteriocin, pediocin AcH, by *Pediococcus acidilactici* H. *Applied and Environmental Microbiology*, **57**, 1265–1267.

Blood, R.M. (1975) Lactic acid bacteria in marinated herring. III. 1. In *Lactic Acid Bacteria in Beverages and Foods* (eds Carr, J.G., Cutting, C.V. and Whiting, G.C.). Academic Press, London, UK, pp. 195–208.

Boatwright, J. and Kirsop, B.H. (1983) Sucrose agar – a growth medium for spoilage organisms. *Journal of the Institute of Brewing*, **82**, 343–346.

Bravo Abad, F. (1990) Biogenic amines in beer. *Cerveza Malta*, **27**, 18–23.

Carr, J.G. (1970) Tetrad-forming cocci in ciders. *Journal of Applied Bacteriology*, **33**, 371–379.

Claussen, N.H. (1903) Étude sur les bactéries dites sarcines et sur les maladies qu'elles provoquent dans la bière. *Comptes-rendus des Travaux du Laboratoire Carlsberg*, **6**, 64–83.

Cocconcelli, P.S., Triban, E., Basso, M. and Bottazzi, V. (1991) Use of DNA probes in the study of silage colonization by *Lactobacillus* and *Pediococcus* strains. *Journal of Applied Bacteriology*, **71**, 296–301.

Collins, M.D., Williams, A.M. and Wallbanks, S. (1990) The phylogeny of *Aerococcus* and *Pediococcus* as determined by 16S rRNA sequence analysis: description of *Tetratogenococcus* gen nov. *FEMS Microbiology Letters*, **70**, 255–262.

Coster, E. and White, H.R. (1964) Further studies of the genus *Pediococcus*. *Journal of General Microbiology*, **26**, 185–197.

Costilow, R.N. and Gerhardt, P. (1983) Dialysis pure-culture process for lactic acid fermentation of vegetables. *Journal of Food Science*, **48**, 1632–1636.

Dachs, E. (1981) NBB – Nachweismedium für bierschädliche Bakterien. *Brauwelt*, **121**, 1778–1782, 1784.

Dacre, J.C. (1958a) Characteristics of a presumptive *Pediococcus* occurring in New Zealand Cheddar cheese. *Journal of Dairy Research*, **25**, 409.

Dacre, J.C. (1958b) A note on the pediococci in New Zealand Cheddar cheese. *Journal of Dairy Research*, **25**, 409–413.

Daeschel, M.A. (1992) Bacteriocins of lactic acid bacteria. In *Food Biopreservatives of Microbial Origin* (ed. Ray, B. and Daeschel, M.). CRC Press, Boca Raton, FL. USA, pp. 323–345.

Daeschel, M.A. and Klaenhammer, T.R. (1985) Association of a 13.6-megadalton plasmid in *Pediococcus pentosaceus* with bacteriocin activity. *Applied and Environmental Microbiology*, **50**, 1528–1541.

Davis, C.R., Wibowo, D., Fleet, G.H. and Lee, T.H. (1988) Properties of wine lactic acid bacteria: their potential enological significance. *American Journal of Enology and Viticulture*, **39**, 137–142.

Deibel, R.H. and Niven Jr, C.F. (1960) Comparative study of *Gaffkya homari*, *Aerococcus viridans*, tetrad forming cocci from meat curing brines, and the genus *Pediococcus*. *Journal of Bacteriology*, **79**, 175–180.

Deibel, R.H., Wilson, G.D. and Niven Jr, C.F. (1961) Microbiology of meat curing. IV. A lyophilized *Pediococcus cerevisiae* starter culture for fermented sausage. *Applied Microbiology*, **9**, 239–243.

Dellaglio, F. and Torriani, S. (1986) DNA–DNA homology, physiological characteristics and distribution of lactic acid bacteria isolated from maize silage. *Journal of Applied Bacteriology*, **60**, 83–92.

Dellaglio, F., Bottazzi, V. and Battistotti, B. (1974) Carrateri e distribuzione della microflora pedioccoica in alcuni formaggi italiani. *Annati di Microbiologia ed Enzimologia*, **24**, 325–334.

Dellaglio, F., Trovatelli, L.G. and Sarra, P.G. (1981) DNA–DNA homology among representative strains of the genus *Pediococcus*. *Zentralblatt für Bakteriologie, Mikrobiologie und Hygiene. 1. Abteilung. Originale C, Allgemeine, angewante und ökologische Mikrobiologie*, **2**, 140–150.

Dobrogosz, W.J. and Stone, R.W. (1962a) Oxidation metabolism in *Pediococcus pentosaceus*. I. Role of oxygen and catalase. *Journal of Bacteriology*, **84**, 716–723.

Dobrogosz, W.J. and Stone, R.W. (1962b) Oxidation metabolism in *Pediococcus pentosaceus*. II. Factors controlling the formation of oxidative activities. *Journal of Bacteriology*, **84**, 724–729.

Dolezil, L. and Kirsop, B.H. (1976) The detection and identification of pediococcus and micrococcus in breweries, using a serological method. *Journal of the Institute of Brewing*, **82**, 93–95.

Dolezil, L. and Kirsop, B.H. (1977) The use of the A.P.I. Lactobacillus system for the characterization of pediococci. *Journal of Applied Bacteriology*, **42**, 213–217.

Dolezil, L. and Kirsop, B.H. (1980) Variations amongst beers and lactic acid bacteria relating to beer spoilage. *Journal of the Institute of Brewing*, **86**, 122–124.

Dunn, M.S., Shankman, S., Camien, M.N. and Block, H. (1947) The amino acid requirements of twenty-three lactic acid bacteria. *Journal of Biological Chemistry*, **168**, 1-22.

Earnshaw, R.G. (1992) The antimicrobial action of lactic acid bacteria: natural food preservation systems. In *Lactic Acid Bacteria*, Vol. 1 (ed. Wood, B.J.B.). Elsevier, London, UK, pp. 211–232.

Edwards, C.G. and Jensen, K.A. (1992) Occurrence and characterization of lactic acid bacteria from Washington State Wines: *Pediococcus* spp. *American Journal of Enology and Viticulture*, **43**, 233–238.

Efthymiou, C.J. and Joseph, S.W. (1972) Difference between manganese ion requirements of pediococci and enterococci. *Journal of Bacteriology*, **112**, 627–628.

Emeis, C.C. (1969) Methoden der brauereibiologischen Betriebskontrolle. III VLB-S7-Agar zum Nachweis bierschädlicher Pediokokken. *Monatsschrift für Brauerei*, **22**, 8–11.

Eschenbecher, F. and Back, W. (1976) Erforschung und Nomenklatur der bierschädlichen Kokken. *Brauwissenschaft*, **29**, 125–131.

Etchells, J.L., Fleming, H.P. and Bell, T.A. (1975) Factors influencing the growth of lactic acid bacteria during the fermentation of brined cucumbers. V.2. in *Lactic Acid Bacteria in Beverages and Food* (ed. Carr, J.G., Cutting, C.V. and Whiting, G.C.). Academic Press, London, UK, pp. 281–305.

Eto, M. and Nakagawa, A. (1975) Identification of a growth factor in tomato juice for a newly isolated strain of *Pediococcus cerevisiae*. *Journal of the Institute of Brewing*, **81**, 232–236.

Everson, C.W., Danner, W.E. and Hammes, P.A. (1970) Improved starter culture for semi-dry sausage. *Food Technology*, **24**, 42–44.

Felton, E.A. and Niven Jr, C.F. (1953) The identity of '*Leuconostoc citrovorum* strain 8081'. *Journal of Bacteriology*, **65**, 482–483.

Felton, E.A., Evans J.B. and Niven Jr, C.F. (1953) Production of catalase by pediococci. *Journal of Bacteriology*, **65**, 481–482.

Fernandez, J.L. and Simpson, W.J. (1993) Aspects of the resistance of lactic acid bacteria to hop bitter acids. *Journal of Applied Bacteriology*, **75**, 315–319.

Fitzsimons, A., Duffner, F., Brophy, G., O'Kiely, O. and O'Connell, M. (1992) Assessment of *Pediococcus acidilactici* as a potential silage inoculant. *Applied and Environmental Microbiology*, **58**, 3047–3052.

Fleming, H.P., Etchells, J.L. and Costilow, R.N. (1975) Microbial inhibition by an isolate of *Pediococcus* from cucumber brines. *Applied and Environmental Microbiology*, **30**, 1040–1042.

Foegeding, P.M., Thomas, A.B., Pilkington, D.H. and Klaenhammer, T.R. (1992) Enhanced control of *Listeria monocytogenes* by *in-situ*-produced pediocin during dry sausage production. *Applied and Environmental Microbiology*, **58**, 884–890.

Fox, P.F., Lucey, J.A. and Cogan, T.M. (1990) Glycolysis and related reactions during cheese manufacture and ripening. *Critical Reviews in Food Science and Nutrition*, **29**, 237–253.

Fukui, S., Obayashi, ÔI.A and Kitahara, K. (1957) Studies on the pentose metabolism by microorganisms. A new type – lactic acid fermentation of pentoses by lactic acid bacteria. *Journal of General and Applied Microbiology*, **3**, 258–268.

Garvie, E.I. (1974) Nomenclature problems of the pediococci. Request for an opinion. *International Journal of Systematic Bacteriology*, **24**, 301–306.

Garvie, E.I. (1978) *Streptococcus raffinolactis* Orla Jensen and Hansen, a group N *Streptococcus* found in raw milk. *International Journal of Systematic Bacteriology*, **28**, 190–193.

Garvie, E.I. (1986a) Genus *Pediococcus* Claussen 1903, 68^{AL}. In *Bergey's Manual of Systematic Bacteriology*, Vol. 2. Williams and Wilkins, Baltimore, MD, USA, pp. 1075–1079.

Garvie, E.I. (1986b) Request for an opinion. Conservation of the name *Pediococcus acidilactici* with DSM 20284 as the neotype strain and rejection of the previous neotype strain NCDO 1859 (= IFO 3884 = DSM 20333 = ATCC 33314). *International Journal of Systematic Bacteriology*, **36**, 579–580.

Gibbs, P.A. (1987) Novel uses for lactic acid fermentation in food preservation. *Journal of Applied Bacteriology Symposium Supplement*, **63**, 51S–58S.

Gonzalez, C.F. (1989) Methods for inhibiting bacterial spoilage and resulting composition. US Patent 4883673. Cited in Ray (1992b).
Gonzalez, C.F. and Kunka, B.S. (1983) Plasmid transfer in *Pediococcus* spp.: Intergeneric and intrageneric transfer of pIP501. *Applied and Environmental Microbiology*, **46**, 81–89.
Gonzalez, C.F. and Kunka, B.S. (1986) Evidence for plasmid linkage of raffinose utilization and associated galactosidase and sucrose hydrolase activity in *Pediococcus pentosaceus*. *Applied and Environmental Microbiology*, **51**, 105–109.
Gonzalez, C.F. and Kunka, B.S. (1987) Plasmid associated bacteriocin production and sucrose fermentation in *Pediococcus acidilactici*. *Applied and Environmental Microbiology*, **53**, 2534–2538.
Graham, D. and McKay, L.L. (1985) Plasmid DNA in strains of *Pediococcus cerevisiae* and *Pediococcus pentosaceus*. *Applied and Environmental Microbiology*, **50**, 532–534.
Green, S.R. and Gray, P.P. (1949) Tracing air-borne infection by a new technique. The detection of beer cocci. *Wallerstein Laboratories Communications*, **12**, 325–333.
Günther, H.L. (1959) Mode of division of Pediococci. *Nature*, **183**, 903–904.
Günther, H.L. and White, H.R. (1961a) The cultural and physiological characters of the pediococci. *Journal of General Microbiology*, **26**, 185–197.
Günther, H.L. and White, H.R. (1961b) Serological characters of the pediococci. *Journal of General Microbiology*, **26**, 199–205.
Günther, H.L., Coster, E. and White, H.R. (1962) Designation of the type strain of *Pediococcus parvulus* Günther and White. *International Bulletin of Bacteriological Nomenclature and Taxonomy*, **12**, 189–190.
Hansen, E.C. (1879) Bitrag til kundskab om hvilke Organismer, der kunne forekomme og leve i Øl og Ølurt. *Comptes-rendus des Travaux du Laboratoire Carlsberg*, **1**, 185–292.
Herrmann, C. (1965) Morphologie der Biersarcina. *European Brewery Convention* (Proceedings of the 10th Congress, Stockholm). Elsevier, Amsterdam, The Netherlands, pp. 454–459.
Hoover, D.G., Walsh, P.M., Kolaetis, K.M. and Daly, M.M. (1988) A bacteriocin produced by *Pediococcus* species associated with a 5.5-megadalton plasmid. *Journal of Food Protection*, **51**, 29–31.
Hsu, W.P., Taparowsky, J.A. and Brenner, M.W. (1975a) Two new media for culturing of brewery organisms. *Brewers' Digest*, **50**, 52–54, 56–57.
Hsu, W.P., Taparowsky, J.A. and Brenner, M.W. (1975b) Rapid culturing of brewery lactic acid bacteria. *Brauwissenschaft*, **28**, 157–160.
Hug, H. and Bosio, E. (1991) Hilfsmittel der mikrobiologischen Betriebskontrolle. *Brauerei Rundschau*, **102**, 225–229.
Hutter, K.-J. (1991) Simultane Identifizierung von *L. brevis* und *P. damnosus* im filtrierten Bier. *Brauwelt*, **131**, 1797–1798, 1800–1802.
Hutter, K.-J. (1992) Simultane mehrparametrige durchflußzytometrische Analyse verschiedener Mikroorganismenspezies. *Monatsschrift für Brauwissenschaft*, **45**, 280–284.
Hwang, G.-R., Chou, C.-C. and Wang, Y.-J. (1988) Isolation, identification and screening of lactic acid bacteria as well as yeast from Tou-pan-chiang mash. *Journal of the Chinese Agricultural Chemical Society*, **26**, 447–456.
Institute of Brewing (1991) *Recommended Methods of Analysis*. Institute of Brewing, London, UK.
Iwasaki, K., Nakajima, M. and Sasahara, H. (1993) Rapid continuous lactic acid fermentation by immobilised lactic acid bacteria for soy sauce production. *Process Biochemistry*, **28**, 39–45.
Jensen, E.M. and Seeley, H.W. (1954) The nutrition and physiology of the genus *Pediococcus*. *Journal of Bacteriology*, **67**, 484–488.
Judicial Commission (1976) Opinion 52. Conservation of the generic name *Pediococcus* Claussen with the type species *Pediococcus damnosus* Claussen. *International Journal of Systematic Bacteriology*, **26**, 292.
Kalchayanand, N., Hanlin, M.B. and Ray, B. (1992) Sublethal injury makes Gram-negative and resistant Gram-positive bacteria sensitive to the bacteriocins, pediocin AcH and nisin. *Letters in Applied Microbiology*, **15**, 239–243.
Kanbe, C. and Uchida, K. (1982) Diversity in the metabolism of organic acids by *Pediococcus halophilus*. *Agricultural and Biological Chemistry*, **46**, 2357–2359.

Kanbe, C. and Uchida, K. (1985) Oxygen consumption by *Pediococcus halophilus*. *Agricultural and Biological Chemistry*, **49**, 2931–2937.

Kanbe, C. and Uchida, K. (1987a) NADH dehydrogenase activity of *Pediococcus halophilus* as a factor determining its reducing force. *Agricultural and Biological Chemistry*, **51**, 507–514.

Kanbe, C. and Uchida, K. (1987b) Citrate metabolism by *Pediococcus halophilus*. *Applied and Environmental Microbiology*, **53**, 1257–1262.

Kandler, O. (1970) Amino acid sequence of the murein and taxonomy of the genera *Lactobacillus*, *Bifidobacterium*, *Leuconostoc* and *Pediococcus*. *International Journal of Systematic Bacteriology*, **20**, 491–507.

Kandler, O. and Weiss, N. (1986) Genus *Lactobacillus* Beijerinck 1901, 212[AL]. In *Bergey's Manual of Systematic Bacteriology*, Vol. 2. Williams and Wilkins, Baltimore, MD, USA, pp. 1209–1234.

Kayahara, H., Yasuhira, H. and Sekiguchi, J. (1989) Isolation and classification of *Pediococcus halophilus* plasmids. *Agricultural and Biological Chemistry*, **53**, 3039–3041.

Kim, W.J., Ray, B. and Johnson, M.C. (1992) Plasmid transfers by conjugation and electroporation in *Pediococcus acidilactici*. *Journal of Applied Bacteriology*, **72**,201–207.

Kitahara, K. and Nakagawa, A. (1958) *Pediococcus mevalovorus* nov. spec. isolated from beer. *Journal of General and Applied Microbiology*, **4**, 21–30.

Kocur, M., Bergan, T. and Mortensen, N. (1971) DNA base composition of Gram-positive cocci. *Journal of General Microbiology*, **69**, 167–183.

Kulka, D., Cosbie, A.J.C. and Walker, T.K. (1949) *Streptococcus mucilaginosus* Kulka, Cosbie and Walker (Spec. nov.). *Journal of the Institute of Brewing*, **55**, 315–320.

Langston, C.W. and Bouma, C. (1960) A study of the microorganisms from grass silage. I. The cocci. *Applied Microbiology*, **8**, 212–222.

Lawrence, D.R. and Priest, F.G. (1981) Identification of brewery cocci. In *European Brewery Convention* (Proceedings of the 18th Congress, Copenhagen). IRL Press, Oxford, UK, pp. 217–227.

Lee, S.Y., Jangaard, N.O., Coors, J.H., Hsu, W.P., Fuchs, C.M. and Brenner, M.W. (1975) Lee's multi-differential agar (LMDA): a culture medium for enumeration and identification of brewery bacteria. *Proceedings of the American Society of Brewing Chemists*, **33**, 18–25.

Lin, C.L., Bolsen, K.K. and Fung, D.Y.C. (1992) Epiphytic lactic acid bacteria succession during the pre-ensiling periods of alfalfa and maize. *Journal of Applied Bacteriology*, **73**, 375–387.

Lindner, P. (1887) Über ein neues in Malzmaischen vorkommendes, milchsäurebildenes Ferment. *Wochenschrift für Brauerei*, **4**, 437–440.

Lindner, P. (1888) Die Sarcina-organismen der Gärunggewerbe. *Zentralblatt für Bakteriologie und Parasitenkunde Infektionskrankheiten und Hygiene (II)*, **4**, 427–429.

Litopolou-Tzanetaki, E., Graham, D.C. and Beyatli, Y. (1989) Detection of pediococci and other nonstarter organisms in American Cheddar cheese. *Journal of Dairy Science*, **72**, 854–858.

Llaubères, R.M., Richard, B., Lonvaud, A. and Dubourdieu, D. (1990) Structure of an exocellular *β*-D-glucan from *Pediococcus* sp., a wine lactic acid bacterium. *Carbohydrate Research*, **203**, 103–107.

London, J. and Chace, N.M. (1976) Aldolases of lactic acid bacteria. Demonstration of immunological relationships among eight genera of gram positive bacteria using an anti-pediococcal aldolase serum. *Archives of Microbiology*, **110**, 121–128.

London, J. and Chace, N.M. (1983) Relationships among lactic acid bacteria demonstrated with glyceraldehyde-3-phosphate dehydrogenase as an evolutionary probe. *International Journal of Systematic Bacteriology*, **33**, 723–737.

London, J., Chace, N.M. and Kline, K. (1975) Aldolase of lactic acid bacteria: immunological relationships among aldolases of Streptococci and Gram-positive nonsporeforming anaerobes. *International Journal of Systematic Bacteriology*, **25**, 114–123.

Lonvaud-Funel, A., Joyeux, A. and Ledoux, O. (1991) Specific enumeration of lactic acid bacteria in fermenting grape must and wine colony hybridization with non-isotopic DNA probes. *Journal of Applied Bacteriology*, **71**, 501–508.

Lonvaud-Funel, A., Guilloux, Y. and Joyeux, A. (1993) Isolation of a DNA probe for

identification of glucan-producing *Pediococcus damnosus* in wines. *Journal of Applied Bacteriology*, **74**, 41–47.
Luchansky, J.B., Glass, K.A., Harsono, K.D., Degnan, A.J., Faith, N.G., Cauvin, B., Baccus-Taylor, G., Arihara, K., Bater, B., Maurer, A.J. and Cassens, R.G. (1992) Genomic analysis of *Pediococcus* starter cultures used to control *Listeria monocytogenes* in turkey summer sausage. *Applied and Environmental Microbiology*, **58**, 3035–3059.
Man, J.C. de, Rogosa, M. and Sharpe, M.E. (1960) A medium for the cultivation of lactobacilli. *Journal of Applied Bacteriology*, **23**, 130–135.
Marugg, J.D., Gonzalez, C.F., Kunka, B.S., Ledeboer, A.M., Pucci, M.J., Toonen, M.Y., Walker, S.A., Zoetmulder, L.C.M. and Vandenbergh, P.A. (1992) Cloning, expression, and nucleotide sequence of genes involved in production of pediocin PA-1, a bacteriocin from *Pediococcus acidilactici* PAC1.0. *Applied and Environmental Microbiology*, **58**, 2360–2367.
McCaig, R. and Weaver, R.L. (1983) Physiological studies on *Pediococcus*. *MBAA Technical Quarterly*, **20**, 31–38.
Mees, R.H. (1934) Onderzoekingen over de biersarcina. Dissertation, Technical University of Delft. Cited in Eschenbecher and Back (1976).
Metzler, D.E. (1977) Tetrahydrofolic acid and other pterin coenzymes. In *Biochemistry. The Chemical Reactions of Living Cells*. Academic Press, London, UK, pp. 493–515.
Molitoris, E., Krichesvsky, M.I., Fagerberg, D.J. and Quarles, C.L. (1986) Effects of dietary chlortetracycline on the antimicrobial resistance of porcine faecal streptococcaceae. *Journal of Applied Bacteriology*, **60**, 111–120.
Mundt, J.O., Beattie, W.G. and Wieland, F.R. (1969) Pediococci residing on plants. *Journal of Bacteriology*, **98**, 938–942.
Nakagawa, A. (1964) A simple method for the detection of beer-sarcinae. *Bulletin of Brewing Science*, **10**, 7–10.
Nakagawa, A. (1970) UK patent no. 1193975.
Nakagawa, A. and Kitahara, K. (1959) Taxonomic studies on the genus *Pediococcus*. *Journal of General and Applied Microbiology*, **5**, 95–126.
Nakagawa, A. and Kitahara, K. (1962) Pleomorphism in bacterial cells. 2. Giant cell formation in *Pediococcus cerevisiae* induced by hop resins. *Journal of General and Applied Microbiology*, **8**, 142–148.
Odunfa, S.A. (1987) Microbial contaminants of carbonated soft drinks produced in Nigeria. *Monatsschrift für Brauwissenschaft*, **40**, 220–222.
Olson, N.F. (1990) The impact of lactic acid bacteria on cheese flavor. *FEMS Microbiology Reviews*, **87**, 131–147.
Osaki, K., Okamato, Y., Akao, T., Nagata, S. and Takamatsu, H. (1985) Fermentation of soy sauce with immobilized whole cells. *Journal of Food Science*, **50**, 1289–1292.
Pederson, C.S. (1949) The genus *Pediococcus*. *Bacteriological Reviews*, **13**, 225–232.
Pederson, C.S., Albury, M.N. and Breed, R.S. (1954) *Pediococcus cerevisiae*, the beer sarcina. *Wallerstein Laboratories Communications*, **17**, 7–16.
Pilone, G.J. and Kunkee, R.E. (1965) Sensory characterization of wines fermented with several malo-lactic strains of bacteria. *American Journal of Enology and Viticulture*, **16**, 224–230.
Pilone, G.J., Clayton, M.G. and Van Duivenboden, R. (1991) Characterization of wine lactic acid bacteria: single broth culture for tests of heterofermentation, mannitol from fructose, and ammonia from arginine. *American Journal of Enology and Viticulture*, **42**, 153–137.
Porubcan, R.S. and Sellars, R.L. (1979) Lactic starter culture concentrates. In *Microbial Technology*, Vol. 1, 2nd edn (ed. Peppler, H.J. and Perlman, D.). Academic Press, New York, USA, pp. 59–68.
Priest, F.G. (1987) Gram-positive brewery bacteria. In *Brewing Microbiology* (eds Priest, F.G. and Campbell, I.). Elsevier Applied Science, London, UK, pp. 121–154.
Pucci, M.J., Vedamuthu, E.R., Kunka, B.S. and Vandenbergh, P.A. (1988) Inhibition of *Listeria monocytogenes* by using bacteriocin PA-1 produced by *Pediococcus acidilactici* PAC1.0. *Applied and Environmental Microbiology*, **54**, 2349–2353.
Raccach, M. (1981) Method for fermenting vegetables. US Patent 4, 342, 786. Cited in Raccach (1987).

Raccach, M. (1987) Pediococci and Biotechnology. *CRC Critical Reviews in Microbiology*, **14**, 291–309.
Radler, F. (1975) The metabolism of organic acids by lactic acid bacteria. In *Lactic Acid Bacteria in Beverages and Food* (eds Carr, J.G., Cutting, C.V. and Whiting, G.C.). Academic Press, London, UK, pp. 17–27.
Radler, F., Schütz, M. and Doell, H.W. (1970) Die beim Abbau von L-Äpfelsäure durch Milchsäurebakterien entstehenden Isomeren der Milchsäure. *Naturwissenschaften*, **57**, 672.
Rainbow, C. (1981) Beer spoilage organisms. In *Brewing Science*, Vol. 2 (ed. Pollock, J.R.A.). Academic Press, London, UK, pp. 491–550.
Ray, B. (1992a) Bacteriocins of starter culture bacteria as food preservatives. In *Food Biopreservatives of Microbial Origin* (eds Ray, B. and Daeschel, M.). CRC Press, Boca Raton, FL, USA, pp. 177–205.
Ray, B. (1992b) Pediocin(s) of *Pediococcus acidilactici* as a food biopreservative. In *Food Biopreservatives of Microbial Origin* (eds Ray, B. and Daeschel, M.). CRC Press, Boca Raton, FL, USA, pp. 265–322.
Ray, B., Motlagh, A., Johnson, M.C. and Bozoglu, F. (1992) Mapping of *pSMB74*, a plasmid-encoding bacteriocin, pediocin AcH, production (Pap$^+$) by *Pediococcus acidilactici* H. *Letters in Applied Microbiology*, **15**, 35–37.
Reichard, A. (1894) Studien über einen Sarcinaorganismus des Bieres. *Zeitschrift für das gesamte Brauwesen*, **17**, 257–259, 265–267, 275–276, 283–286, 291–293, 299–301.
Romano, A.H., Trifone, J.D. and Brustolon, M. (1979) Distribution of the phosphoenolpyruvate: glucose phosphotransferase system in fermentative bacteria. *Journal of Bacteriology*, **139**, 93–97.
Saha, R.B., Sondag, R.J. and Middlekauff, U.E. (1974) An improved medium for the selective culturing of lactic acid bacteria. *Journal of the American Society of Brewing Chemists*, **32**, 9–10.
Saito, K. (1907) *Centralblatt für Bakteriologie, Parasitenkinde und Infektionskrankheiten. 2. Abteiling*. Cited in Sakaguchi and Mori (1969).
Sakaguchi, K. (1958) Studies on the activities of bacteria in soy sauce brewing. Part 3. Taxonomic studies on *Pediococcus soyae* nov. sp., the soy sauce lactic acid bacteria. *Bulletin Agricultral Chemical Society of Japan*, **22**, 353–362.
Sakaguchi, K. (1960) Vitamin and amino acid requirements of *Pediococcus soyae* and *Pediococcus acidilactici* Kitahara's strain. *Bulletin of the Agricultural Chemical Society of Japan*, **24**, 638–643.
Sakaguchi, K. and Mori, H. (1969) Comparative study on *Pediococcus halophilus*, *P. soyae*, *P. homari*, *P. urinae-equi* and related species. *Journal of General and Applied Microbiology*, **15**, 159–167.
Sauberlich, H.E. and Baumann, C.A. (1948) A factor required for growth of *Leuconostoc citrovorum*. *Journal of Biological Chemistry*, **176**, 165–173.
Schisler, D.O., Mabee, M.S. and Hann, C.W. (1979) Rapid identification of important beer microorganisms using gas chromatography. *Journal of the American Society of Brewing Chemists*, **37**, 69–76.
Schved, F., Lalazar, A., Henis, Y. and Juven, B.J. (1993) Purification, partial characterization and plasmid-linkage of pediocin SJ-1, a bacteriocin produced by *Pediococcus acidilactici*. *Journal of Applied Bacteriology*, **74**, 67–77.
Sharpe, M.E. (1955) The selective action of thallous acetate for lactobacilli. *Journal of Applied Bacteriology*, **18**, 274–283.
Shibasaki, K. and Hesseltine, C.W. (1962) Miso fermentation. *Ecomonic Botany*, **16**, 180–195.
Shimwell, J.L. (1948a) A rational nomenclature for the brewing lactic acid bacteria. *Journal of the Institute of Brewing*, **54**, 100–104.
Shimwell, J.L. (1948b) A study of ropiness in beer. Part II. Ropiness due to tetrad-forming cocci. *Journal of the Institute of Brewing*, **54**, 237–244.
Shimwell, J.L. (1949) Brewing bacteriology. VI – The lactic acid bacteria (Family Lactobacteriaceae). *Wallerstein Laboratories Communications*, **12**, 71–88.
Shimwell, J.L. and Kirkpatrick, W.F. (1939) New light on the 'Sarcina' question. *Journal of the Institute of Brewing*, **45**, 137–145.

Simpson, W.J. (1994) Comments on the mode of division of *Pediococcus* spp. *Letters in Applied Microbiology*, **18**, 69–70.

Simpson, W.J. and Fernandez, J.L. (1992) Selection of beer-spoilage lactic acid bacteria and induction of their ability to grow in beer. *Letters in Applied Microbiology*, **14**, 13–16.

Simpson, W.J. and Hammond, J.R.M. (1991) Antibacterial action of hop resin materials. In *European Brewery Convention*, Proceedings of the 23rd Congress, Lisbon. pp. 185–192. London: IRL Press.

Simpson, W.J. and Smith, A.R.W. (1992) Factors affecting antibacterial activity of hop compounds and their derivatives. *Journal of Applied Bacteriology*, **72**, 327–334.

Simpson, W.J., Hammond, J.R.M. and Miller, R.B. (1988) Avoparcin and vancomycin: useful antibiotics for the isolation of brewery lactic acid bacteria. *Journal of Applied Bacteriology*, **64**, 299–309.

Skyttä, E., Haikara, A. and Mattila-Sandholm, T. (1993) Production and characterization of antibacterial compounds produced by *Pediococcus damnosus* and *Pediococcus pentosaceus*. *Journal of Applied Bacteriology*, **74**, 134–142.

Solberg, O. and Clausen, O.G. (1973a) Classification of certain pediococci isolated from brewery products. *Journal of the Intitute of Brewing*, **79**, 227–230.

Solberg, O. and Clausen, O.G. (1973b) Vitamin requirements of certain pediococci isolated from brewery products. Journal of the Institute of Brewing, **79**, 231–237.

Solberg, O., Hegna, I.K. and Clausen, O.G. (1975) *Pediococcus acidilactici* NCIB 6990, a new test organism for microbiological assay of pantothenic acid. *Journal of Applied Bacteriology*, **39**, 119–123.

Sollied, P.R. (1903) Studien über den Einfluss von Alkohol auf die an verschiedenen Brauerei- und Brennereimaterialien sich vorfindenden Organismen, sowie Beschreibung einer gegen Alkohol sehr widerstandsfähigen neuen Pediokokkus-Art (*Pediococcus hennebergi*, n.sp.). *Centralblatt für Bakteriologie, Parasitenkunde und Infektionskrankheiten. 2. Abteilung*, **11**, 708–712.

Stackebrandt, E., Fowler, V.J. and Woese, C.R. (1983) A phylogenetic analysis of lactobacilli, *Pediococcus pentosaceus* and *Leuconostoc mesenteroides*. *Systematic and Applied Microbiology*, **4**, 326–337.

Stammer, J.R. (1975) Recent developments in the fermentation of sauerkraut. In *Lactic Acid Bacteria in Beverages and Food* (eds Carr, J.G., Cutting, C.V. and Whiting, G.C.). Academic Press, London, UK, pp. 267–280.

Stockhausen, F. and Stege, E. (1925) Sarcina. *Wochenschrift für Brauerei*, **42**, 240–244, 253–257, 261–263, 268–272. (Abstracted *Journal of the Institute of Brewing*, **31**, 636–637.)

Taguchi, H., Ohkochi, M., Uehara, H., Kojima, K. and Mawatari, M. (1990) KOT medium. A new medium for the detection of beer spoilage lactic acid bacteria. *Journal of the American Society of Brewing Chemists*, **48**, 72–75.

Tanasupawat, S. and Daengsubha, W. (1983) Pediococcus species and related bacteria found in fermented foods and related materials in Thailand. *Journal of General and Applied Microbiology*, **29**, 487–506.

Teuber, M. (1993) Lactic acid bacteria. In *Biotechnology*, Vol. 1 (eds Rehm, H.-J. and Reed, G.). (With the cooperation of A. Pühler and P. Stadler). UCH, Weinheim, Germany, pp. 325–366.

Thomas, T.D., McKay, L.L. and Morris, H.A. (1985) Lactate metabolism by pediococci isolated from cheese. *Applied and Environmental Microbiology*, **49**, 908–913.

Tjandraatmadja, M., Norton, B.W. and Macrae, I.C. (1990) A numerical taxonomic study of lactic acid bacteria from tropical silages. *Journal of Applied Bacteriology*, **68**, 543–553.

Torriani, S., Vescovo, M. and Dellaglio, F. (1987) Tracing *Pediococcus acidilactici* in ensiled maize by plasmid-encoded erythromycin resistance. *Journal of Applied Bacteriology*, **63**, 305–309.

Tzanetakis, N. and Litopolou-Tzanetaki, E. (1989) Biochemical activities of *Pediococcus pentosaceus* isolates of dairy origin. *Journal of Dairy Science*, **72**, 859–863.

Uchida, K. (1982) Multiplicity in soy pediococci carbohydrate fermentation and its application to analysis of their flora. *Journal of General and Applied Microbiology*, **28**, 215–223.

Uchida, K. and Kanbe, C. (1993) Occurrence of bacteriophages lytic for *Pediococcus halophilus*, a halophilic lactic-acid bacterium in soy sauce fermentation. *Journal of General and Applied Microbiology*, **39**, 429–437.

Uchida, K. and Mogi, K. (1972) Cellular fatty acid spectra of pediococcus species in relation to their taxonomy. *Journal of General and Applied Microbiology*, **18**, 109–129.

Uhl, A. and Kühbeck, G. (1969) Conditions, especially nitrogen utilization, for *Pediococcus cerevisiae* growth in beer. *Brauwissenschaft*, **22**, 121–129, 199–208, 248–254. Abstracted in *Journal of the Institute of Brewing*, **75**, 487.

Vandevoorde, L., Woestyne, M.V., Brunyeel, B., Christianens, H. and Verstraete, W. (1992) Critical factors governing the competitive behaviour of lactic acid bacteria in mixed cultures. In *Lactic Acid Bacteria*, Vol. 1 (ed. Wood, B.J.B.). Elsevier, London, UK, pp. 447–475.

Walters, L.S. (1940) A note on the characters of a coccus isolated from South Australian stout. *Journal of the Institute of Brewing*, **46**, 11–14.

Weinberg, Z.G, Ashbell, G. and Azrieli, A. (1988) The effect of applying lactic acid bacteria at ensilage on the chemical and microbiological composition of vetch, wheat and alfalfa silages. *Journal of Applied Bacteriology*, **64**, 1–7.

Weiss, N. (1991) The Genera *Pediococcus* and *Aerococcus*. In *The Prokaryotes*, Vol. 2, 2nd edn (eds Balows, E., Trüper, H.G., Davorkin, M., Harder, W. and Schliefer, K.-H.). Springer-Verlag, New York, USA, pp. 1502–1507.

Whiting, M., Crichlow, M., Ingeldew, W.M. and Ziola, B. (1992) Detection of *Pediococcus* spp. in brewing yeast by a rapid immunoassay. *Applied and Environmental Microbiology*, **58**, 713–716.

Whittenbury, R. (1964) Hydrogen peroxide formation and catalase activity in the lactic acid bacteria. *Journal of General Microbiology*, **35**, 13–26.

Yousef, A.E., Luchansky, J.B., Degnan, A.J. and Doyle, M.P. (1991) Behaviour of *Listeria monocytogenes* in wiener exudates in the presence of *Pediococcus acidilactici* H or pediocin AcH during storage at 4 or 25°C. *Applied and Environmental Microbiology*, **57**, 1461–1467.

6 The genus *Lactococcus*

M. TEUBER

6.1 History

Lactococci are coccoid Gram-positive, anaerobic bacteria which produce L(+)-lactic acid from lactose in spontaneously fermented raw milk which is left at ambient temperatures around 20–30°C for 10–20 h. They are commonly called 'mesophilic lactic streptococci'. It is tempting to suggest that the first isolation, identification and description of the chemical entity lactic acid by Carl Wilhelm Scheele from sour milk in Sweden in the year 1780, was actually L(+)-lactic acid produced by lactococci. The microbial nature of lactic fermentation was recognized in 1857 by Louis Pasteur. The first bacterial pure culture on earth, obtained and scientifically described by Joseph Lister (1873) was *Lactococcus lactis*, at that time called: '*Bacterium lactis*'.

> Admitting then that we had here to deal with only one bacterium, it presents such peculiarities both morphologically and physiologically as to justify us, I think, in regarding it a definite and recognizable species for which I venture to suggest the name *Bacterium lactis*. This I do with diffidence, believing that up to this time no bacterium has been defined by reliable characters. Whether this is the only bacterium that can occasion the lactic acid fermentation, I am not prepared to say.

Around 1890, Storch in Copenhagen and Weigmann in Kiel isolated the mesophilic lactic streptococci responsible for the spontaneous fermentation of sour cream, sour milk and cheese and paved the way for their application as starter cultures for the dairy industry (for a detailed history, see von Milczewski, 1990). In 1909, Löhnis renamed '*Bacterium lactis*' as '*Streptococcus lactis*' mainly on the basis of the strains discovered in fermented dairy products. In the elegant taxonomy of lactic acid bacteria by Orla-Jensen (1919) the mesophilic lactic streptococci have a firm standing as '*Streptococcus lactis*' and '*Streptococcus cremoris*'. The serological differentiation scheme of the streptococci by Lancefield (1933) put the lactic streptococci into the group N which clearly separated them from the pathogenic streptococci (e.g. groups A, B, C) and enterococci (group D). Unfortunately, since 1993 the group N antiserum previously available from the Pasteur Institute (Paris) is no longer on the market.

The taxonomic confusion generated by the fact that quite unrelated bacteria were put and maintained in the genus *Streptococcus* on merely morphological criteria ended finally when modern methods of chemical taxonomy were successfully applied. Based on nucleic acid hybridization studies and immunological relationships of superoxide dismutase, Schleifer's laboratory (Schleifer *et al.*, 1985) separated the mesophilic lactic streptococci from the true streptococci (genus *Streptococcus*) and the enterococci (genus *Enterococcus*) and generated the new genus *Lactococcus*. This name is an ingenious description of the function and morphology of these bacteria which, of course, was obviously influenced by the already existing genus *Lactobacillus* which contains most of the rodshaped lactic acid bacteria (for review, see Teuber, 1993). A basic phylogenetic positioning of the lactococci as an own cluster within the *Clostridium* subvision of bacterial evolution (low GC Gram-positive bacteria) became evident from the analysis of oligonucleotide similarities of ribosomal RNAs (Stackebrandt and Teuber, 1988). The historical development of commercially available starter cultures for the dairy industry started with fresh liquid cultures as described by Weigmann (1905–1908), continued with the invention of viable freeze-dried cultures after World War II, and stands today at a peak with the offer of concentrated deep-frozen cultures in cans or the form of pellets which contain up to 10^{12} living bacteria per ml and are suitable for direct vat inoculation (Teuber, 1993).

Genetics and genetic engineering of lactococci started with the detection of plasmids in all investigated strains by McKay's Laboratory and others (McKay *et al.*, 1972; Pechmann and Teuber, 1980; Davies *et al.*, 1981).

Table 6.1 Lactococci as components of starter cultures for fermented dairy products*

Type of product	Composition of starter culture
1. Cheese types without eye formation (Cheddar, Camembert, Tilsit)	*Lc. lactis* subsp. *cremoris*, 95–98% *Lc. lactis* subsp. *lactis*, 2–5%
2. Cottage cheese, quarg, fermented milks, cheese types with few or small eyes (e.g. Edam)	*Lc.* lactis *subsp. cremoris*, 95%; *Leuconostoc mesenteroides* subsp. *cremoris*, 5%; or *Lc. lactis* subsp. *cremoris*, 85–90%; *Lc. lactis* subsp. *lactis*, 3%; *Leuc. mesenteroides* subsp. *cremoris*, 5%
3. Cultured butter, fermented milk, buttermilk, cheese types with round eyes (e.g. Gouda)	*Lc. lactis* subsp. *cremoris*, 70–75%; *Lc. lactis* subsp. '*diacetylactis*', 15–20%; *Leuc. mesenteroides* subsp. *cremoris*, 2–5%
4. Taette (Scandinavian ropey milk)	*Lc. lactis* subsp. *cremoris* (ropey strain)
5. Viili (Finnish ropey milk)	*Oidium lactis* (covers surface); *Lc. lactis* subsp. *cremoris* (ropey strain)
6. Casein	*Lc. lactis* subsp. *cremoris*
7. Kefir	Kefir grains containing lactose-fermenting yeasts (e.g. *Candida kefir*) *Lb. kefir*, *Lb. kefiranofacians*, *Lc. lactis* subsp. *lactis*

*Data taken from Teuber *et al* (1991). The quantitative composition has been taken from the culture catalogue of a major worldwide supplier.

Two facts have mainly contributed to a rapid progress: (1) the presence of technologically important functions on plasmids (e.g. lactose utilization, casein degradation, citrate uptake, bacteriocin production, bacteriophage resistance and slime formation); and (2) the occurrence of natural gene transfer mechanisms in the form of conjugation and transduction.

Today, the basic structure of the chromosome is known, gene transfer by conjugation and electroporation is routinely applied, cloning and expression vectors are available, gene structures and functions are elucidated at levels coming close to well-characterized species like *Escherichia coli* and *Bacillus subtilis* (including genetic engineering and expression of heterologous proteins) (Gasson and de Vos, 1994).

Since 1982, biochemical and genetic research on lactococci has a high and continuing priority within several research programs of the European Community/Union (BAP, BEP, BRIDGE, BIOTECH) (Aguilar, 1991, 1992). The lactococci used in starter cultures for the dairy industry are listed in Table 6.1

6.2 Morphology

The not very spectacular morphology of the Gram-positive lactococci is characterized by spheres of ovoid cells occurring singly, in pairs or in chains, and being often elongated in the direction of the chain. One of the early micrographs of Weigmann (1899) is shown in Figure 6.1. With the imagination of a chemist (that he was by education) he called his bacteria *Kiel I* and *Kiel II* (the place of his laboratory). The length of the chain is mainly strain dependent, sometimes also influenced by the growth

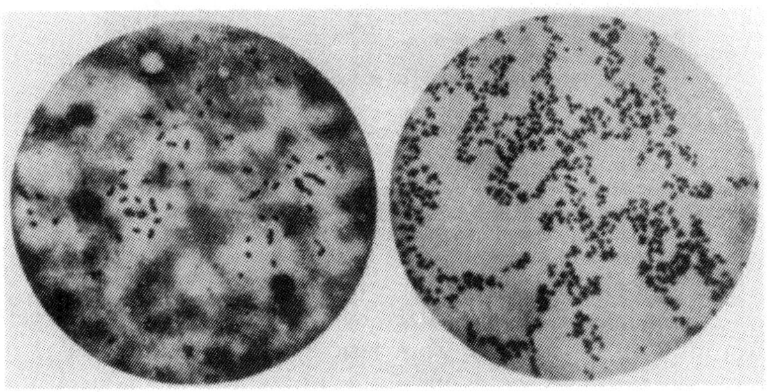

Figure 6.1 Early photomicrograph of mesophilic lactic streptococci isolated by Weigmann from sour milk and sour cream. (Courtesy of the Institute of Microbiology, Federal Dairy Research Centre, Kiel, Germany.)

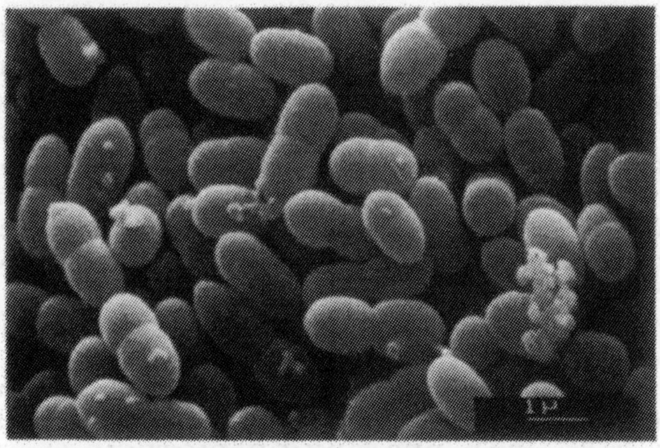

Figure 6.2 Scanning electron micrograph of *Lactococcus lactis* subsp. *lactis* Bu2–60. (Courtesy of Horst Neve, Kiel, Germany.)

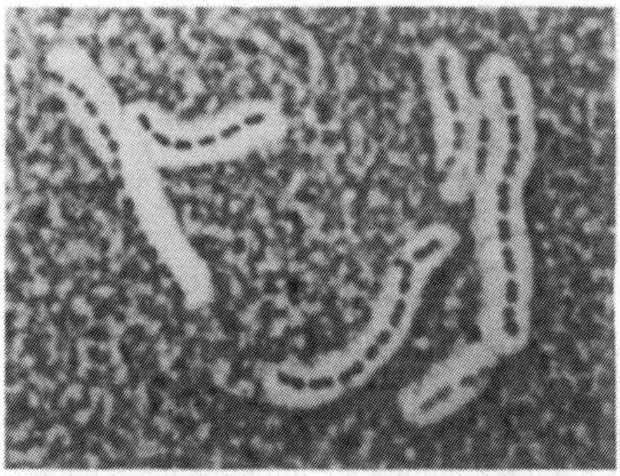

Figure 6.3 Micrograph of ropey strain of *Lactococcus lactis* isolated from Swedish sour milk (longfil). (Courtesy of Horst Neve, Kiel, Germany.)

medium. The ovoid shape is shown in the scanning electron micrograph of Figure 6.2. Some slime-forming strains excrete extracellular polysaccharides which are easily visualized by immersion of the cell suspension in Indian ink (see Figure 6.3). This basic morphology is typical, but not suitable to differentiate the genus *Lactococcus* from the genera *Streptococcus* or *Enterococcus*, or to differentiate between the five *Lactococcus* species (*Lc. lactis*, *Lc. garviae*, *Lc. raffinolactis*, *Lc. plantarum*, *Lc. piscium*). Lactococci do not form endospores nor do they have flagella.

Motile streptococci previously regarded as belonging to the genus *Lactococcus* due to the presence of a group N antigen have been shown to represent a genus on their own (*Vagococcus*; Collins *et al.*, 1989). This genus has a closer relationship with *Enterococcus*.

6.3 Biochemistry and physiology

The metabolism of milk ingredients (lactose, caseins, citrate, and others) by the two subspecies *lactis* and *cremoris* of *Lactococcus lactis* provides the basis for the spontaneous and industrial fermentations of milk into sour milk, sour cream and many different types of cheeses. Although lactococci do not possess a citric acid cycle and a respiratory chain and in that sense are true anaerobic bacteria, they are able to grow in the presence of oxygen partially due to certain oxygen metabolizing enzymes like superoxide dismutase or NADH oxidases. The resulting microaerophilic phenotype does not require a completely oxygen free substrate (e.g. milk) and facilitates the large scale of industrial application of lactococci.

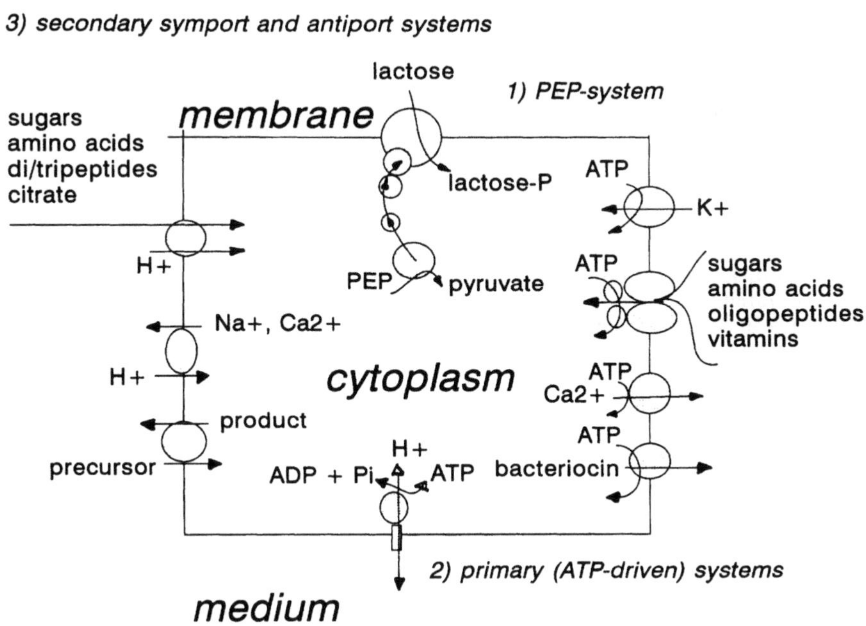

Figure 6.4 Solute transport mechanisms in lactococci. Primary (ATP-driven) transport systems are assembled at the right-hand side, secondary symport and antiport systems at the left, the PEP-dependent sugar:phosophotransferase system on top. (Modified from Poolman, 1993.)

6.3.1 *Solute transport and energy transduction (Poolman, 1993)*

A prerequisite of metabolism of low-molecular-weight substrates is their efficient transport from the medium through the cytoplasmic membrane into the bacterial cytoplasm, the location of the catabolic and anabolic enzymatic pathways. In addition, fermentation products like lactic acid and protons have to be transported from the cytoplasm into the medium. The solute transport mechanisms detected in *Lactococcus lactis* and its main subspecies are shown in Figure 6.4.

The main substrates transported by these different systems are listed in Table 6.2. The most prominent system in lactococci is group translocation of carbohydrates by phosphorylation of the sugar concomitant with transport by the phosphoenol pyruvate: sugar phosphotransferase system (PTS). Lactose is accumulated in lactococci as lactose 6-phosphate (for a review, see Hengstenberg *et al.*, 1993) which is the substrate for further

Table 6.2 Transport mechanisms in *Lactococcus lactis**

Substrate specificity	Proteins
H^+-Symport	
Leucine, isoleucine, valine	–
Alanine, glycine	AlaT
Serine, threonine	–
Lysine, ornithine	–
Phenylalanine, tyrosine, tryptophane	–
Di- and tripeptides	DtpT
Citrate	CitP
Galactose	–
TMG	–
H^+-Antiport	
Calcium	–
Substrate/product antiport (exchange)	
Arginine/ornithine	–
Sugar-phosphate/phosphate	–
Malate/lactic acid	–
ATP-driven uptake	
Glutatamate, glutamine	–
Asparagine	–
Proline, glycine, betaine	–
Phosphate	–
Oligopeptides	OppFDCBA
ATP-driven efflux	
Proton	F_0F_1
BCECF†	–
Calcium	–
(Lactococcin A)	LcnCD
(Nisin)	NisT

*Data taken from Poolman (1993).
†2′,7′-bis-(2-carboxyethyl)-5-carboxyfluorescein.

metabolism (see below). Initial uptake rates for glucose in *Lactococcus lactis* are about 400 μmol sugar/g dry weight cells per min compared to about 30 μmol in PTS-defective mutants. A sophisticated model of how lactose uptake is regulated has been developed by Thompson (1988): fructose diphosphate is a positive effector for pyruvate kinase and HPr-kinase, inorganic phosphate a negative effector for pyruvate kinase, but an activator of HPr(ser)P-phosphatase. This leads to a build-up of a PEP-potential in starved cells which allows immediate uptake of fermentable substrate if it should become available. Ion-linked sugar transport plays a minor role in lactococci.

An interesting system is the sugar 6-phosphate/phosphate antiport in *Lc. lactis*: (1) It catalyses homologous and heterologous exchange of phosphate and sugar 6-phosphates, (2) arsenate can replace phosphate, (3) affinity constants decrease from mannose 6-phosphate, glucose 6-phosphate, fructose 6-phosphate, glucosamine 6-phosphate to ribose 5-phosphate (4) homologous exchange is five times faster than heterologous exchange, (5) it favours monovalent phosphate, and (6) the exchange is electro-neutral. The system could be useful in habitats containing phosphorylated sugar intermediates (like the intestine).

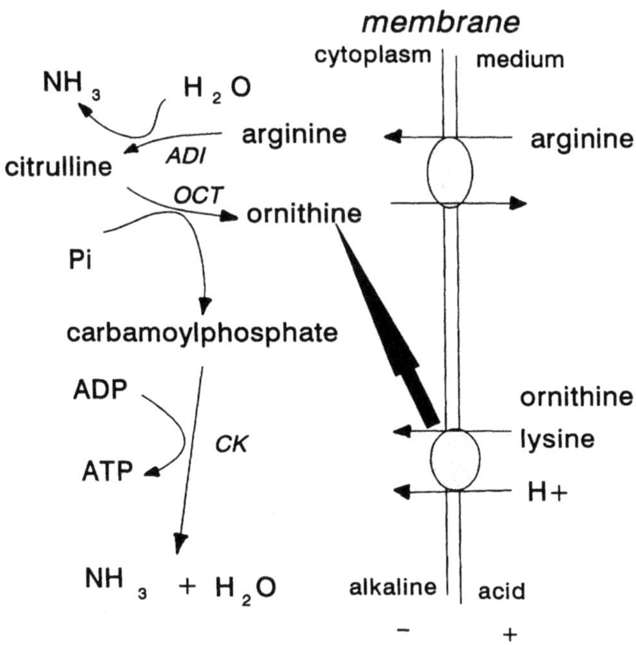

Figure 6.5 Arginine/ornithine antiport and the arginine deiminase pathway in *Lactococcus lactis* subsp. *lactis*. Accumulation of ornithine (lysine) via the Δp-driven lysine transport system is also shown. ADI, arginine deiminase; OCT, ornithine carbamoyl-transferase; CK, carbamate kinase. (Modified from Poolman, 1993.)

Since *Lc. lactis* subsp. *lactis* (in contrast to *Lc. lactis* subsp. *cremoris*) has the ability to metabolize arginine via the arginine deiminase pathway, the arginine/ornithine antiport should be briefly characterized. As shown in Figure 6.5, this pathway (when coupled with a product/precursor exchange) yields overall 1 mol of ornithine and carbon dioxide, 2 mol of ammonia, and 1 mol of ATP by substrate level phosphorylation per mol of arginine metabolized. When part of the transported arginine is used for biosynthetic purposes, *Lc. lactis* takes advantage of a Δp-driven lysine transport system that can accept ornithine. The antiporter can also catalyse heterologous exchange of arginine for lysine, in addition to exchange of arginine for ornithine. Accumulation of lysine via Δp-driven transport in combination with exchange of lysine for arginine via the antiporter results in cyclic transport of lysine and net accumulation of arginine at the expense of Δp.

The malolactic fermentation carried out by some *Lc. lactis* strains (see Figure 6.6) is another system conserving metabolic energy. In this pathway L-malate enters the cells and is decarboxylated by malolactic enzyme to yield L(+)-lactic acid and CO_2, after which the products leave the cells. The decarboxylation reaction does not yield energy directly. The free energy of the decarboxylation reaction, however, can be conserved via an indirect

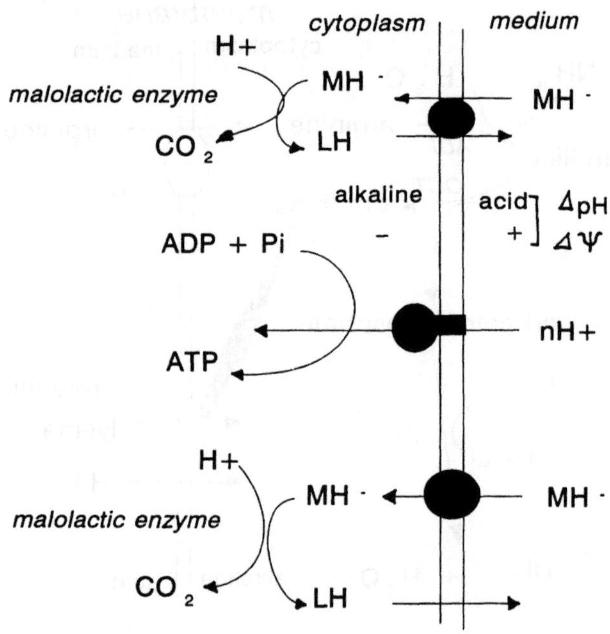

Figure 6.6 Metabolic energy conservation by malolactic fermentation in *Lactococcus*. MH^-, monoanionic malate; LH, lactic acid. Malate uptake is shown as MH^-/LH exchange and MH^- uniport. Exit of lactic acid can occur via the exchange reation or by passive and/or carrier-mediated diffusion. (Modified from Poolman, 1993.)

H^+ pump mechanism. In *Lc. lactis*, L-maltate utilization results in the formation of a Δp that is sufficiently high to drive ATP synthesis via the F_0F_1-ATPase.

Although the mechanism by which hydrolysis of ATP is coupled to solute transport is still not clear, the fact that in *Lc. lactis* transport of glutamate/glutamine, asparagine, phosphate and oligopeptides proceeds in the absence of a Δp is strong evidence for an ATP-driven undirectional nutrient uptake.

The opposite, an ATP-driven excretion system probably is used by *Lc. lactis* subsp. *lactis* biovar. *diacetylactis* to excrete the bacteriocin lactococcin A (see below).

The major lactococcal F_0F_1-ATPase can pump protons at the expense of ATP or also function as ATP synthetase. When Δp is high as in the case of product efflux or solute decarboxylation, the metabolic energy in form of an electrochemical H^+ gradient can be used for ATP synthesis. If ATP is formed by substrate level phosphorylation (e.g. glycolysis), the F_0F_1-ATPase will function as a hydrolase in order to generate a Δp.

The branched chain aliphatic amino acids (Leu, Ile, Val), the neutral amino acids (Ala, Gly), the aliphatic amino acids with a hydroxyl side chain (Ser, Thr), and the aromatic amino acids (Phe, Tyr, Trp) are transported by separate H^+-linked mechanisms. In addition, di- and tripeptides are taken up by a Δp-driven peptide transporter essential for casein utilization (see below) and for growth of lactococci in milk.

The plasmid-encoded citrate transport gene (citP, see below) of *Lc. lactis* subsp. '*diacetylactis*' is homologous to the citrate–H^+–Na^+ symporter (citS) of *Klebsiella pneumoniae*, however, CitP seems not to need Na^+ as an additional coupling ion. Intracellular Ca^{2+} and Na^+ concentrations are kept below those of the surrounding medium. *Lactococcus lactis* subsp. *lactis* uses a Ca^{2+}ATPase while subsp. *cremoris* excretes Ca^{2+} by Ca^{2+}/H^+ antiport. Na^+/H^+ antiport activities have not yet been demonstrated in lactococci.

6.3.2 Carbohydrate metabolism

The production of acid (formic, lactic or acetic acid, see below) from carbohydrates is an important and indispensible property used in the identification and differentiation of individual *Lactococcus* species. The details are given in section 6.11. Using the API 50 CH system, all lactococci produce acid from glucose, fructose, mannose, and *N*-acetyl glucosamine. Acid is not produced from D-arabinose, arabitol, adonitol, 2-keto-gluconate, 5-keto-gluconate, dulcitol, erythritol, D-fucose, L-fucose, glycerol, glycogen, inositol, D-lyxose, α-methyl-D-mannoside, *β*-methylxyloside, rhamnose, L-xylose and xylitol (Schleifer *et al.*, 1985; Williams *et al.*, 1990).

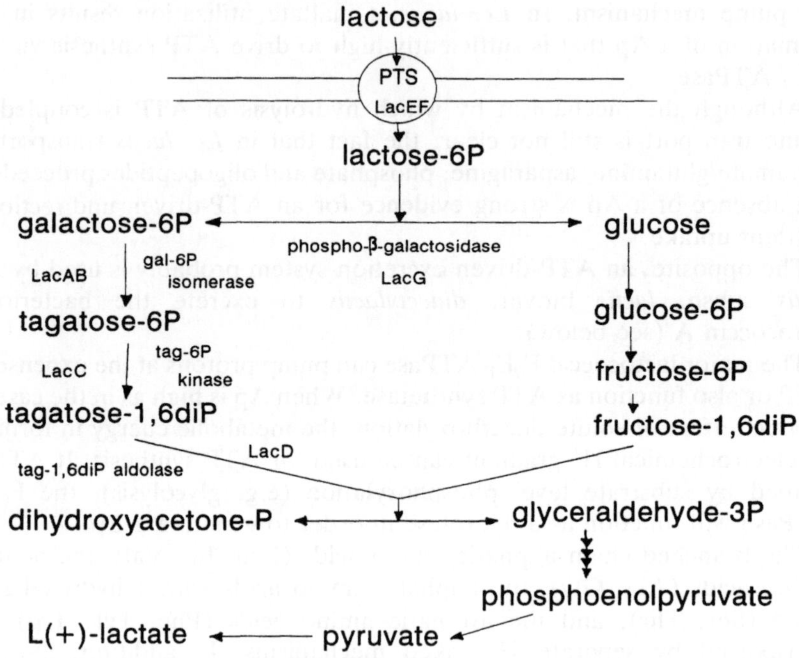

Figure 6.7 Pathway of lactose utilization by industrial *Lactococcus lactis* strains. The enzymes encoded by the *lac* operon are indicated (compare Figure 6.11). (Modified from de Vos and Simons, 1994.)

The biochemical pathway of lactose which is the basis of the industrial application of lactococci has been elucidated (see Figure 6.7). Lactose 6-phosphate, the product of group translocation by a specific PTS system (see above), is split by phospho-*β*-galactosidase into glucose and galactose 6-phosphate which is further metabolized via the tagatose 6-phosphate pathway into triose phosphate. Both homofermentative pathways lead to the formation of L(+)-lactic acid which is excreted into the medium. Konings and co-workers (Konings and Otto, 1983) have clearly shown that in the stationary phase of growth (at about 170–180 mg cell protein/litre complex medium in batch culture with an initial lactose concentration of 2 g/litre) the lactate gradient and the resulting Δp, $\Delta\Psi$ and ΔpH values approach zero (explaining the growth stop by lack of energy supply). In milk, dairy lactococci may produce about 0.8% of L(+)-lactic acid and a pH value of about 4.6 after 15–20 h of growth at room temperature. Viable cell numbers are around 5×10^9 cfu/ml. The viability of lactococci tends to decay rapidly under such acidic conditions.

A key compound in the intermediate metabolism in lactococci is pyruvate. Under normal anaerobic conditions of glycolysis, pyruvate is reduced to lactate by lactic dehydrogenase in order to regenerate NAD^+,

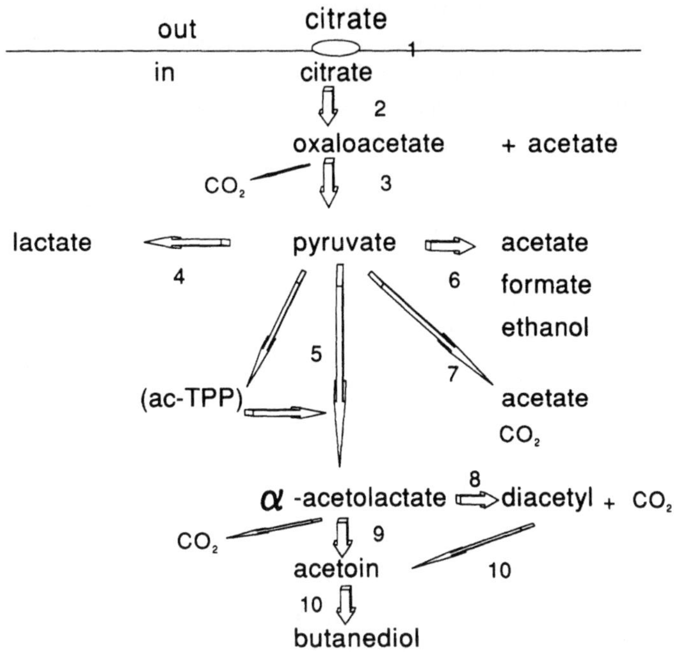

Figure 6.8 Citrate metabolism in *Lactococcus lactis* subsp. *lactis* biovar diacetylactis involving four different pathways. (A) Lactate production via lactate dehydrogenase (4). (B) Formate and acetate or ethanol production via pyruvate formate lyase (6). (C) Acetate and CO_2 production via pyruvate dehydrogenase (7). (D) α-Acetolactate production via α-acetolactate synthase (5) and subsequent acetoin and butanediol production by α-acetolactate decarboxylase (9) and acetoin reductase (10). Production of diacetyl is by spontaneous chemical disintegration of α-acetolactate. The citrate permease (1) is coded on plasmids, citrate lyase (2) and oxaloacetate decarboxylase (3) are coded on the chromosome. Resulting diacetyl concentration in sour milk, cream and butter are between 1.5 and 2.5 mg/kg. (Modified from Hugenholtz, 1993.)

i.e. it serves as an electron acceptor for the substrate level phosphorylation which is driven by the intramolecular oxidoreduction of glyceraldehyde 3-phosphate by glyceraldehyde-3-phosphate-dehydrogenase to 1,3-phosphoglycerate. However, lactococci possess three more pathways for pyruvate as shown in Figure 6.8 in addition to lactate dehydrogenase:

(1) lactate dehydrogenase:
pyruvate + NADH → L-lactate + NAD
(2) pyruvate formate lyase:
pyruvate → acetylP + formate
(3) pyruvate dehydrogenase:
pyruvate + NAD → acetylCoA + CO_2 + NADH
(4) α-acetolactate synthase:
2 pyruvate → α-acetolactate + CO_2

The last reaction is important for the formation of diacetyl (see Figure 6.8). This typical butter-flavour compound is produced by the metabolism of citrate in *Lactococcus lactis* subsp. *lactis* biovar. *diacetylactis*. This biovar is differing from normal *Lc. lactis* subsp. *lactis* in that it possesses a plasmid encoded citrate permease (see below). The permease has a narrow pH range between pH 5 and 6 corresponding with the citrate-metabolizing activities of *Lc. lactis* (and *Leuconostoc*). Below pH 5, citrate utilization is low in these organisms due to a low permease activity and low metabolic activity in general (see above remarks on the lactate Δp and $\Delta\Psi$ potential). In citrate-utilizing *Lc. lactis* citrate is converted initially into oxaloacetate and acetate by the enzyme citrate lyase. This enzyme seems to be unique for the citrate-utilizing *Lc. lactis* subsp. *lactis* biovar. *diacetylactis* since it is not found in non-citrate-utilizing *Lc. lactis*. Acetate is a good indicator in citrate-containing cultures of homofermentative lactic acid bacteria that citrate is metabolized.

In lactococci, oxaloacetate is decarboxylated by oxaloacetate decarboxylase into pyruvate and CO_2 the enzyme being similar in structure to the *Salmonella typhimurium* enzyme. Fermentation of citrate generally leads to a mixture of products including lactate, CO_2, acetate, formate, and C4-compounds (acetoin, diacetyl and butandiol).

α-Acetolactate is extremely unstable.

6.3.2.1 Regulation of citrate metabolism in Lactococcus lactis *subsp.* lactis biovar. diacetylactis (*for a review, see Hugenholtz, 1993*) (Figure 6.8). The citrate permease is constitutively expressed and citrate lyase is not regulated (in contrast to *Enterobacter aerogenes*). Lactate dehydrogenase (LDH) is regulated by two positive activators: fructose 1,6-diphosphate and NADH. Since both compounds are not present in cells supplied with citrate as sole carbon source, a low activity of this enzyme results, and products other than lactate are formed. Pyruvate formate lyase (PFL) is inactivated by already low oxygen levels, therefore formate and ethanol disappear completely upon switch of a *Lactococcus* culture from anaerobic to aerobic conditions. In addition, the enzyme has a narrow pH optimum at 7.5. Mixed acid fermentation proceeds accordingly only when the bacteria grow at pH values near neutrality. No formate is produced from lactose or citrate below pH 6.0.

The pyruvate dehydrogenase complex (PDC) of *Lc. lactis* has been purified and characterized. Of the three different enzymes constituting the complex (enzyme 1, decarboxylation of pyruvate; enzyme 2, regeneration of cofactors and thiamin C_0A pyrophosphate; enzyme 3, oxidation reaction) enzyme 3, which is strongly inhibited by NADH, has an unusually low activity compared to other bacteria. *In vivo*, activity of the complex was only observed under strong aeration.

α-Acetolactate synthase (ALS) from *Lc. lactis* was purified and characterized. The enzyme measures 172 kDa consisting of three identical 62 kDA monomers. It catalyses the thiamine pyrophosphate-dependent condensation reaction of two pyruvate molecules to the C5 compound α-acetolactate with the release of carbon dioxide: one pyruvate molecule is decarboxylated with thiamine pyrophosphate (TPP) acting as the co-enzyme, resulting in the formation of hydroxyethyl TPP ('active acetaldehyde') which reacts with the other pyruvate molecule to α-acetolactate. ALS has a very low affinity (K_m = 50 mM), but a strong positive cooperativity for pyruvate. α-Acetolactate will only be produced when pyruvate is accumulated inside the cells. When citrate is added to active cultures of *Lc. lactis*, rapid uptake and conversion takes place resulting in internal accumulation of pyruvate to concentrations of 50 mM and higher.

α-Acetolactate is further metabolized by α-acetolactate decarboxylase into acetoin and CO_2, and chemically decarboxylated into diacetyl. A previous claim for a diacetylsynthase condensing hydroxyethyl thiamine pyrophosphate and acetyl-CoA could not be substantiated.

Diacetyl and acetoin can be further reduced by one enzyme (an acetoin/diacetyl reductase) to butanediol with NAD as reducing factor. The reduction of diacetyl to acetoin is irreversible, the reduction of acetoin by butanediol is reversible. The K_m of the enzyme for acetoin is 0.2 mM, for diacetyl it is 9 mM. This difference helps the accumulation of diacetyl in fermented dairy products like butter and sour cream. A starter culture with a high diacetyl production potential contained a *Lc. lactis* strain which lacked α-acetolactate decarboxylase. When this strain was incubated in 10 mM citrate without, with low, and with high air in the gas atmosphere, α-acetolactate concentrations increased from about 0.3–4 mM (low air) to 6 mM (high air) α-acetolactate, respectively. Acetoin increased from 0.1 over 1 to 2 mM.

6.3.3 Nitrogen metabolism

The content of free amino acids in cow's milk is too low to provide growth of *Lc. lactis* to more than about 10^7 cfu/ml. In order to achieve a sufficient lactic fermentation (with about 2–5 times 10^9 cells/ml in the stationary growth phase) the dairy lactococci have developed a proteolytic system, which in most strains is specific for β-casein, consisting of

(1) a plasmid coded cell wall protease,
(2) efficient peptide and amino acid transport systems (see above), and
(3) a set of specific peptidases (for a review, see Pritchard and Coolbear, 1993).

Two different types of proteases located at the outer surface of the cells are known in *Lc. lactis*: type P I, preferentially cleaving β-casein, and type P

III cleaving α_{s1}-caseins in addition. Their specificity is summarized in Table 6.3. Both enzyme types can also cleave K-caseins. The cell envelope protease is encoded by a single gene carried generally on a plasmid. Its structure, function and properties are described in the genetic section below.

The peptides liberated by the wall protease are further cleaved by a set of peptidases as summarized in Table 6.4. Although most of these peptidases seem to be localized in the cytoplasm, a localization of small amounts of some peptidases in the cell wall can not be completely excluded, since it still seems unlikely that some of the larger peptides (more than seven amino acids) are transported by the oligopeptide transport system. A model of how the complex proteolytic system of dairy lactococci might function is shown in Figure 6.9.

Generally, lactococci are auxotrophic for the aminoacids isoleucine, valine, leucine, histidine, methionine, arginine and proline. A detailed list for different strains is given in Table 6.5 which clearly shows that additional amino acids may be necessary for growth. For exponential growth, *Lc. lactis* does need a complement of 19 amino acids (in addition to seven vitamins; Jensen and Hammer, 1993) (see Table 6.15).

Regarding nucleotide metabolism, *Lc. lactis* has the ability to utilize uracil, uridine, deoxyuridine, cytidine, and deoxycytidine, but not cytosine as sole pyrimidine source by pathways similar to the ones in *Escherichia*

Table 6.3 Peptide bonds in β-casein which are cleaved by the cell wall proteinases from six different strains of *Lactococcus lactis**

Cleaved bond	P_I-type proteinases				P_{III}-type proteinases	
	763	AC1	HP	H2	SK11	AM1
Tyr193-Gln194	+ +	+ +	+ +	+ +	+ +	+ +
Leu192-Tyr193	+	−	−	±	+ +	+ +
Gln182-Arg183	+ +	+ +	+ +	+ +	+	+
Gln175-Lys176	+ +	+ +	+ +	+ +	±	±
Ser168-Lys169	+	+	−	+	±	−
Ser166-Gln167	+ +	+ +	+ +	+ +	−	−
Leu165-Ser166	+	−	+	+ +	−	−
Leu163-Ser164	+	+	+	+	+	+ +
Gln160-Ser161	+	−	ND	+	−	−
Gln141-Ser142	+	−	ND	+	+	+
Asn132-Leu133	+	−	ND	+	+ +	+ +
Gln123-Ser124	−	−	ND	−	+ +	+ +
Met93-Gly94	−	+	ND	+	+	+
Asn68-Ser69	+	+	ND	+	+	−
Gln56-Ser57	+	+	ND	+	−	−
Phe52-Ala53	+	+	ND	+	+ +	+ +
Gln46-Asp47	−	−	ND	±	+ +	+ +
Asp43-Glu44	−	−	ND	−	+ +	+ +

*Data taken from Pritchard and Coolbear (1993).

Table 6.4 Peptidases purified and characterized from lactococci*

Peptidase	Abbreviation	Specificity	M_r (native) (kDa)	M_r (subunit) (kDa)
X-propyl dipeptidyl aminopeptidase	PepX	X-Pro ↓ Y- . . .	160–190	82–90
			117	ND
Aminopeptidase N	PepN	X ↓ Y-Z . . .	93–95	95
Aminopeptidase C	PepC	X ↓ Y-Z . . .	300	50
Aminopeptidase A	GAP	Asp(Glu) ↓ Y-Z . . .	130	43
			245	41
			440–520	approx. 40–43
Pyrrolidone carboxylyl peptidase	PCP	pGlu ↓ Y-Z . . .	80	40
Prolyl imino-peptidase	PIP	Pro ↓ Y-Z	110	50
Dipeptidase	DIP	X ↓ Y	49	49
			100	ND
Prolidase	PRD	X ↓ Pro	42–43	ND
Tripeptidase	TRP	X ↓ Y-Z	103–105	52–55
Endopeptidases	PepO	. . . W-X ↓ Y-Z . . .	70	70
	NOP		70	70
	LEP I		98	98
	LEP II		80	40
	MEP		180	70 (or 35)

*Data taken from Pritchard and Coolbear (1993).

Table 6.5 Amino acid requirements of some lactococci*

	Lc. lactis subsp. *lactis*†							*Lc. lactis* subsp. *cremoris*†		
Number of strains tested	4	8	1	1	36	60	18	31	18	1
Isoleucine	+	+	+	+	+		±	+	+	+
Valine	+	+	+	+	+		+	+	+	+
Leucine	+	+	+	+	+		±	+	+	+
Histidine	+	+	+	+	+	±		+	+	+
Methionine	+	+	+	+	−			+	+	+
Arginine	+	−	+	−	−			±	±	+
Proline	±	−	+	−	−			±	+	+
Phenylalanine	−	±	−	−	−			±	±	+
Glutamate	−	+	+		−			−	+	
Glutamine	±		−		−			±		
Glu or Gln	+			+				+		+
Threonine	−	−	−	−	−			±	−	−
Glycine	−	−	−	−	−			±	±	−
Lysine	−	−	−	−	−			±	±	+
Alanine	−	−	−	−	−			±	±	−
Asparagine	−		+	−	−			±		−
Aspartate	−	−	−	−	−			−	−	−
Tyrosine	−	−	−	−	−			±	±	+
Cysteine	−		−	−	−			−		−
Tryptophan	−	−	−	−	−			−	±	−
Serine	−	−	−	−	−			−	±	−

*Data taken from Chopin (1993).
†+, required for growth; −, not required for growth; ±, strain-dependent requirement.

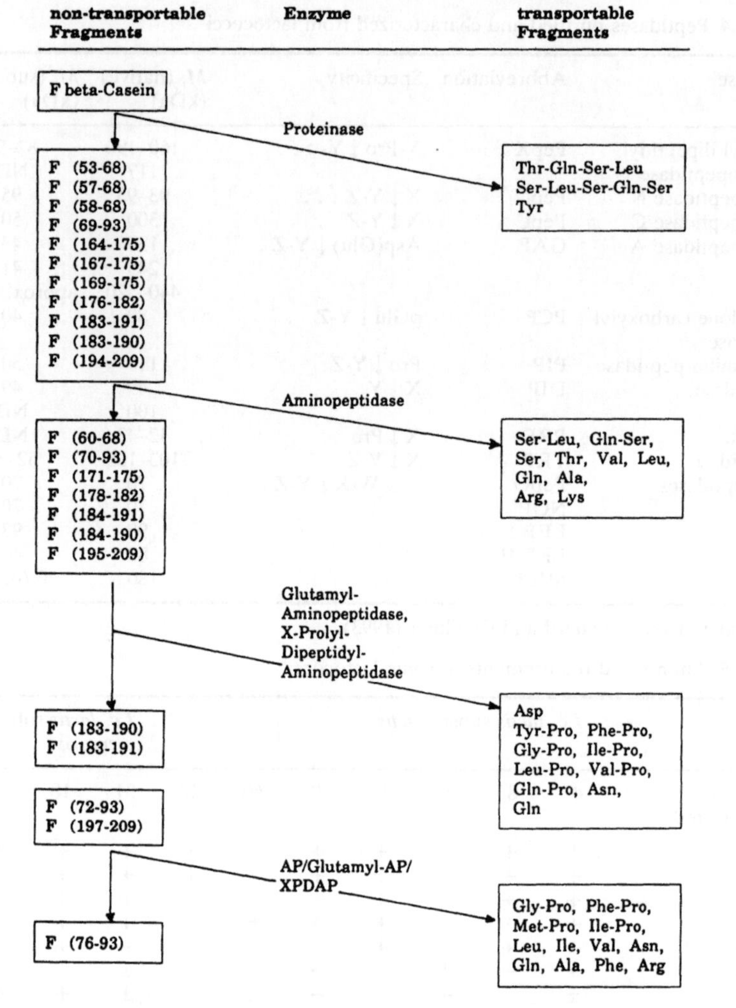

Figure 6.9 Scheme of the proteolytic breakdown of *β*-casein by *Lactococcus lactis* subsp. *cremoris*. The *β*-casein fragments (F) shown in the left column are generated by the cell wall proteinase. Transportable fragments can be produced by the indicated peptidases (see Table 6.4). It is still a matter of discussion whether at least parts of peptidase activities are located outside the cytoplasmic membrane in the cell wall. (Modified from Smid *et al.*, 1991.)

coli. Cytosine can not be used since *Lc. lactis* lacks the enzymes cytidine deaminase in addition to deoxycytidine kinase and dCMP deaminase (Martinussen *et al.*, 1994).

6.3.4 Vitamin and mineral requirements (Jensen and Hammer, 1993)

For optimum growth, *Lc. lactis* requires biotin, pyridoxal, folic acid, riboflavin, niacinamid, thiamine and pantothenate. Niacin, panthotenate and biotin are essential. Micronutrients should include molybdate, borate, cobalt, copper, manganese and zinc (see defined minimal growth medium (SA) in section 6.8, Table 6.15).

6.3.5 Reaction to oxygen (Zitzelsberger et al., *1984)*

Six oxygen metabolizing enzymes have been described in lactic acid bacteria: (1) NADH: H_2O_2 oxidase; (2) NADH: H_2O oxidase; (3) pyruvate oxidase; (4) α-glycerophosphate oxidase; (5) superoxide dismutase; and (6) NADH peroxidase (Condon, 1987). Whereas lactobacilli seem to have no true superoxide dismutase (SOD) (but inorganic manganese instead), the investigated lactococci (*Lc. lactis* and *Lc. raffinolactis*) demonstrated high levels of enzymatic superoxide dismutase activities in the order of 5.4–11.8 units/mg protein. The immunological cross-reactivities of this Mn-SOD of lactococci, streptococci and enterococci corresponded to their phylogenetic distance (Schleifer *et al.*, 1985). In addition, all investigated lactococci contained NADH oxidase activity. The hydrogen peroxide (about 10–20 mg/litre) generated by SOD and NADH oxidase in *Lc. lactis* accumulates in the medium since no catalase or NADH peroxidase is present in these bacteria. Inhibitory levels of H_2O_2 in starter culture production can be reduced by addition of catalase to the growth medium. The only other enzyme activity of oxygen metabolism reported in lactococci is NADH peroxidase in *Lc. raffinolactis*.

6.3.6 Cell wall chemistry and cellular fatty acids (Schleifer et al., *1985).*

Lactococcus lactis and *Lc. garviae* contain a peptidoglycan of the Lys-D-Asp type, *Lc. raffinolactis* Lys-Thr-Ala, and *Lc. plantarum* Lys-Ser-Ala, respectively. The murein type of *Lc. piscium* has not been reported.

Lactococcus lactis possesses a 1,3-linked poly(glycerophosphate) lipoteichoic acid substituted with α-D-galactopyranosyl residues and D-alanine ester. *Lactococcus garviae* has a poly[Gal(α1–6) Gal (α1,3)-Gro-1-P (2←1αGal)]-lipoteichoic acid. Poly(glycerophosphate) lipteichoic acids solely substituted with D-alanine esters are present in *Lc. raffinolactis* and *Lc. plantarum*.

With regard to polar lipid composition, all species contain phosphatidyl glycerol and cardiolipin. *Lactococcus lactis* and *Lc. garviae* lack aminophospholipids whereas *Lc. raffinolactis* and *Lc. plantarum* contain D-alanyl- and L-lysyl-phosphatidyl glycerol. Long-chain fatty acid composition has been reported. Hydroxylated fatty acids are not detected. *Lc. lactis*, *Lc. garviae* and *Lc. plantarum* have hexadecenoic, *cis*-11,12-octadecenoic, and *cis*-11,12-methylene octadecanoic acids as the main fatty acids. *Lactococcus raffinolactis* lacks cyclopropane-ring acids.

The neutral lipid fraction of *Lc. raffinolactis* and *Lc. plantarum* seemed to lack menaquinones. *Lactococcus lactis* and *Lc. garviae* contained low levels of menaquinones, unsaturated compounds with nine isoprene units predominating with additional significant levels of MK-7, MK-8, and MK-10, respectively.

6.4 Genetics and genetic engineering (for an extensive review, see Gasson and de Vos, 1994)

The species *Lc. lactis* is characterized by the presence of plasmids in every investigated strain isolated from starter cultures. Between one and 12 plasmids of different sizes can be found in different strains. Sizes vary from about 2 to more than 100 kb (see Figure 6.10). The most exciting discoveries about plasmids, however, were two observations. (1) Many of the functions encoded on plasmids turned out to be related to or necessary for growth of lactococci in milk. (2) Most *Lc. lactis* strains could exchange plasmids by natural conjugation or transduction at reasonable rates. These two basic properties of dairy lactococci initiated, facilitated and enabled their genetic exploration. Six laboratories, active in this field (one each in Germany, France, UK, Netherlands, New Zealand and the USA) in the year 1978, have sparked ongoing genetic activities in probably several hundred laboratories in every corner of the world.

6.4.1 The plasmids of Lactococcus lactis

The following functions have been identified as plasmids encoded in *Lc. lactis*:

- lactose transport and metabolism
- casein degradation by cell wall protease
- citrate and oligopeptide transport (permease)
- bacteriophage protection by restriction/modification and abortive infection
- formation of extracellular polysaccharides (slime; see Figure 6.3)
- bacteriocin production and immunity
- insertion (IS) element dependent recombination and cointegrate formation
- antibiotic resistance
- conjugal transfer and mobilization of plasmids
- plasmid replication

Plasmids can be lost by growth without lactose or casein, by growth at high sublethal temperatures (about 38–42°C), by ethidium bromide treatment, by freezing and thawing or by freeze-drying. The stabilities of plasmids under these different treatments is very much strain dependent. Under

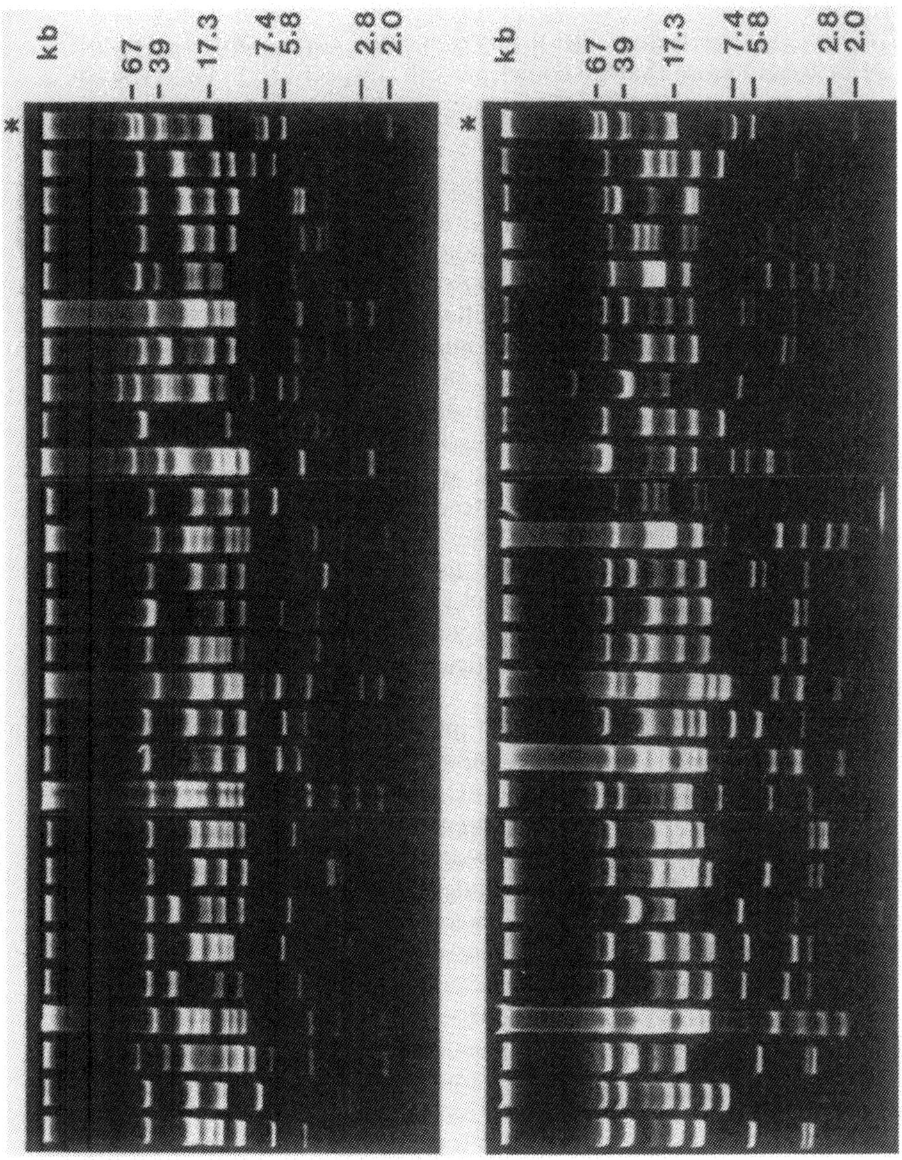

Figure 6.10 Plasmid profiles obtained from 54 *Lactococcus lactis* strains isolated from one commercial mixed strain starter culture. The asterisk denotes the reference plasmid profile of *Lc. lactis* subsp. *cremoris* AC 1. (Courtesy of Andresen *et al.*, 1984.)

controlled conditions, however, plasmid profiles are excellent tools for strains' characterization and identification and for the characterization of starter culture composition (see Figure 6.10).

6.4.1.1 Plasmid replication. Both low and high copy number plasmids exist (1 to 10 and more than 50 copies per cell, respectively). Two types of replication have been identified in lactococci: plasmids replicated by the σ (1) or θ (2) modes.

(1) The small (cryptic) latococcal plasmic pSH 71 (2060 nucleotides) contains a 1.8 kb minimal replicon with the rep-operon comprising two genes, the *repC* gene encoding a 6 kDa repressor involved in replication control, and the *repA* gene coding for a 27 kDa replication protein being rate limiting in replication. It is a rolling-circle-replication (RCR)-type plasmid and produces single-stranded DNA-intermediates in *E. coli*, *B. subtilis* and *Lc. lactis*. The pSH71 replicon is widely disseminated and present in one out of 10 industrial strains. It differs from plasmid pWVO1 (2178 bp) only by a few nucleotides and the absence of a direct repeat region. Both plasmids have been developed into cloning vectors of the pGK- and pNZ-series (see Table 6.6). The segregational stability of these vectors appears to be high in *Lc. lactis* (plasmid loss less than 10^{-5} per generation). During industrial food fermentations representing about 25 generations, such a segregational instability does not pose a problem and explains the high efficiency of plasmid coded functions in industrial fermentations.
(2) The 8.7 kb *Lc. lactis* plasmid pCI 305 represents the θ-type replication mode. The 1.6 kb minimal replicon contains the *repB* gene encoding a trans-acting 46 kDa replication protein, and a 0.3 kb *cis*-acting AT-rich region designated *repA*, containing a 22 bp sequence that was tandemly repeated three and a half times. No single-stranded DNA intermediates are detected. This type of replicon is present in many *Lc. lactis* strains, e.g. the universal 8 kb citrate plasmid, the 47 kb lactose plasmid pSK11L, the 40 kb lactose–proteinase plasmid pUCL22 and the cryptic 3 kb pWVO2 in strain Wg2. A remarkable property of replicons of this type is that more than one of these θ-type plasmids may reside in a single strain of *Lc. lactis*, indicating that they are compatible and explaining the many plasmids of different sizes found in some lactococcal strains. This compatibility is probably due to a sequence heterogeneity in the mentioned repeat sequences.

The following heterologous plasmids have been experimentally introduced into *Lc. lactis*. The *Enterococcus faecalis* 26.5 kb self-transmissible plasmid pAMβ1 codes for resistance to macrolides, lincosamides and streptogramin

Table 6.6 Properties of cloning vectors used in lactococci*

Vector	Replicon	Size markers† (kb)	Copy number
pGB301	pIP501	9.8 Cm^R Em^R	Low
pSA3	pIP501 pACYC184	10.2 Cm^R Em^R	Low
pHV1301	pAMβ1	13.0 Em^R	Low
pIL252	pAMβ1	4.7 Em^R	Low
pIL253	pAMβ1	5.0 Em^R	High
pVA838	pVA380-1 pACYC184	9.2 Cm^R Em^R	Low
pDL278	pVA380-1 pUC19	6.9 Sp^R	High
pGK12	pWV01	4.4 Cm^R Em^R	Low
pGL3	pWV01	5.0 Cm^R Km^R	Low
pGKV2	pWV01	5.0 Cm^R Em^R	Low
pNZ12	pSH71	4.3 Cm^R Km^R	High
pNZ121	pSH71	4.3 Cm^R Km^R	Low
pNZ123	pSH71	2.8 Cm^R	High
pCK1	pSH71	5.5 Cm^R Km^R	High
pFX1	pDI25	5.5 Cm^R	High
pFX3	pDI25	4.5 Cm^R	High
pCI374	pCI305	4.9 Cm^R	Low

*Data taken from de Vos and Simons (1994). The size and the copy numbers of the vectors in *Lc. lactis* are shown (low, approximately 5–10 copies per cell; high, more than approximately 50 copies per cell).
†Resistances to chloramphenicol (Cm^R), erythromycin (Em^R), kanamycin (Km^R), spectinomycin (Sp^R) are indicated.

B (MLS). Its θ-type replicon can be maintained (after transfer by, for example, conjugation) in many Gram-positive bacteria, e.g. *Bacillus*, *Clostridium*, *Lactococcus*, *Listeria*, *Streptococcus*, *Enterococcus*, *Staphylococcus*, *Lactobacillus*, *Leuconostoc* and *Pediococcus*. Conjugal transfer into *Lc. lactis* has been described early in the history of lactococcal genetics (Gasson and Davies, 1980). Other heterologous plasmid-replicons maintained in *Lc. lactis* are the small 4.2 kb cryptic plasmid pVA 380-1 from *Streptococcus ferus*; and the 30 kb pIP501 from *Streptococcus agalactiae* which is related to pAMβ1. There is some evidence that lactococci carrying antibiotic resistance genes on plasmids (or transposons) can be isolated from raw milk cheeses (Teuber and co-workers, unpublished).

6.4.2 Gene transfer in Lactococcus lactis

The following mechanisms have been identified or developed:

(1) conjugation – discussed under (1) below
(2) transduction with temperate or lytic bacteriophage – discussed under (2) below
(3) protoplast transformation – discussed under (3) below

(4) protoplast fusion – of historical interest only
(5) transformation by electroporation – discussed under (3) below
(6) transposition – discussed under (4) below

(1) *Conjugation*. Transfer of plasmids between different strains of *Lc. lactis* by conjugation is a common phenomenon (Neve *et al*., 1984, 1987). Highest efficiencies (up to 10^{-2}) are obtained by surface matings on agar or filter. Using plasmid-cured derivatives as receptors the functional properties of plasmids can be easily identified in transconjugants. The biochemical and ultrastructural basis of homologous and heterologous conjugation in lactococci and most Gram-positive bacteria is still unknown. However, one system which has been intensively studied, involves a clumping phenomenon (Gasson and Fitzgerald, 1994). It is dependent on co-integrate formation of the transferred plasmid with either another plasmid or a 'sex-factor' excised from the chromosome in some strains of the *Lc. lactis* subsp. *lactis* 712/ML3/C2 family. Cointegrate formation is due to the presence of insertion elements (IS*S1S* and IS*S1T*) in the transferred chromosome. The mentioned *Enterococcus* plasmid pAMβ1 is an example for efficient heterologous conjugation in *Lactococcus*. pAMβ1 has been used to mobilize several lactococcal plasmids carrying genes for bacteriophage resistance, proteinase production, lactose catabolism and polysaccharide production. In some instances, co-transfer involved co-integrate formation as described above for some homologous conjugation systems.
(2) *Transduction*. Transduction, i.e. transfer of genetic material by bacteriophages, has been achieved by the use of lytic and temperate bacteriophages (Anderson and Elliker, 1953) which are very common in dairy lactococci (see below). McKay's laboratory transduced lactose (Lac) metabolism and proteinase activity (Prt) from UV-induced cells of *Lc. lactis* subsp. *lactis* C2 with a lytic phage into Lac$^-$ derivatives (McKay *et al*., 1973). It was then shown that 20–22 MDa plasmids had been transferred into the Lac$^+$ Prt$^+$ transductants which could be converted into Lac$^-$Prt$^-$ forms by plasmid curing. Investigation of the original C2 donor plasmids demonstrated that a transductional shortening of an originally 30 MDa plasmid had occurred (McKay *et al*., 1976). Obviously, the prophage heads can only package plasmid DNA which in size is equivalent to the size of phage DNA (see below). Gasson observed a similar phenomenon in the closely related *Lc. lactis* subsp. *lactis* strain LP712. His laboratory also proved that specific portions of the donor plasmids had to be deleted in order to achieve transduction at high frequencies (HFT). The study of these deletion events could be used to localize the *lac* and *prt* genes in pLP712 (Gasson *et al*., 1987).

Transduction also led to incorporation of transferred *lac* and *prt* genes into the lactococcal chromosome. Similarly, the well-studied temperate phage BK5-T from *Lc. lactis* subsp. *cremoris* was able to transduce both plasmid and chromosomal markers.

(3) *Transformation.* Protoplast transformation has been achieved, however, it has been almost completely eliminated as a genetic tool for lactococci by the invention and development of electrotransformation ('electroporation') (Harlander, 1987). The application of high-voltage electric field pulses to lactococci (5–12.5 kV/cm) suspended in buffers containing sucrose (0.5 M), $MgCl_2$ or $CaCl_2$, and 10% glycerol may yield more than 10^7 transformants per μg of DNA. The capacitance used is generally 25 μF (the highest setting in most available instruments). The transformation efficiency is dependent on the size of the DNA used (usually decreasing with increasing size of DNA), the presence of restriction/modification systems and incompatible plasmids. Electroporation can also be applied for the isolation of plasmids from a donor cell. The harvested electroporation supernatant can then be applied in a second electroporation step to the recipient cells. By this technique, electrotransfer of a plasmid shuttle vector pFX3 between *E. coli* and *Lc. lactis* was realized in both directions (Ward and Jarvis, 1991). Clearly, electrotransformation is now the basic gene transfer technique in genetic and genetic engineering studies of lactococci. Transfection of protoplasts with bacteriophage DNA was discovered by Geis (1982).

(4) *Transposition.* A number of insertion sequences has been discovered in *Lc. lactis*. The first IS element identified was ISS*IS* associated with lactose plasmic cointegrate formation. Similar elements have been found adjacent to the *prtM* gene in proteinase plasmids (ISI/W and ISS*IN*) and a bacteriophage resistance plasmid (IS946). IS904 was found to be located near one end of the conjugative 70 kb transposon Tn5301 that encodes genes for nisin biosynthesis, sucrose fermentation, conjugation transposition, phage insensibility phenomenon and the enzyme N5-(carboxyethyl) ornithine synthase. Multiple copies of different IS elements are found on different plasmids and fewer copies on the chromosome. The characteristic features of the various *Lactococcus* IS elements are summarized in Table 6.7.

IS elements seem to contribute substantially to the long-known genetic flexibility of dairy lactococci. In addition, analysis of the distribution of IS elements by specific probes is valuable for strain identification. Heterologous transposons introduced into *Lc. lactis* include Tn916 (originally isolated from *E. faecalis* D16) and Tn919 from *Streptococcus sanguis* (Hill *et al.*, 1985). Insertial inactivation of genes for maltose metabolism, citrate utilization and malolactic

Table 6.7 IS elements in lactococci*

IS element	Size (bp)	Inverted repeats (bp)	Target duplication (bp)	Possible transposase (aa'O)	Homology	Known distribution	Known iso-elements
IS*S1S*	808	18	8	226	IS*26*	*Lactococcus*	IS*S1T*; IS*S1N* IS*S1W* IS*946*
IS*904*	1241	39	4	253	IS2/IS*3*	*Lactococcus*	IS*1076*
IS*981*	1222	40	5	280	IS2/IS*3*	*Lactococcus*	
IS*905*	1313	28	8	392	IS*256*	*Lactococcus*	

*Data taken from Gasson and Fitzgerald (1994).

fermentation was demonstrated. Tn917, a widely exploited genetic tool in Gram-positive bacteria, has been developed to serve as a 'forced' transposition system in *Lc. lactis* (Israelson and Hansen, 1993).

6.4.3 *Genetic engineering in* Lactococcus

Cloning and expression of homologous and heterologous genes as well as secretion of the gene products have been successfully achieved. Based on the briefly characterized homologous and heterologous plasmids described in section 6.4.1, cloning, expression and secretion vectors have been developed. First experiments started with vectors applied in the genera *Enterococcus* and *Streptococcus* (see Table 6.6). Later on, small cryptic plasmids of *Lc. lactis* itself were constructed. Most of them use antibiotic resistant traits as selection markers which, however, are not accepted in food. For that reason, food grade selection markers have been introduced like the *lacF* marker present in pNZ1125 used to express the *pep*N gene (a debittering peptidase with potential for accelerated cheese ripening). Table 6.8 lists some heterologous genes expressed in *Lc. lactis.*

6.4.4 *Organization of lactococcal genes (Van de Guchte* et al., *1992)*

Transcription initiation and termination, translation initiation and codon usage, control and regulation as well as protein excretion (sec dependent and independent) events have been studied and determined in great detail (see also de Vos and Simons, 1994). The mastering of genetic engineering in *Lc. lactis* was dependent on this knowledge. In the context of this review it must suffice to characterize only a few of the many different functions, especially those necessary for the application of lactococci in the dairy industry.

6.4.4.1 Lactose utilization (see Figure 6.7). The *lac* genes for the lactose-specific enzymes of the PTS (Lac EF), the phospho-β-galactosidase (Lac

Table 6.8 Heterologous genes expressed by *Lactococcus lactis**

Genes	Source (plasmid)	Vector(s)	Secreted
Cm^R; *cat-194*	*S. aureus* (pC194)	pGK12	–
Cm^R; *cat-86*	*B. pumilis*	pGKV210	–
Em^R; *ery-194*	*S. aureus* (pE194)	pGK12	–
Em^R; *ery-β1*	*E. faecalis* (pAMβ1)	pIL253	–
Km^R; *knt-110*	*S. aureus* (pUB110)	pNZ12	–
Km^R; *aphA3*	*S. faecalis* (pJH1)	pMG36	–
Tc^R; *tetM*	*E. faecalis* (Tn1545)	pVE6002	–
Tc^R; *tetL*	*E. faecalis* (pJH1)	pNZ280	–
β-Galactosidase; *lacZ*	*E. coli*	pNZ262	–
Egg-white lysozyme	Hen	pIL253	–
T4 lysozyme	*E. coli* phage T4	pMG36e	–
λ-Lysozyme	*E. coli* phage λ	pMG36e	–
β-Glucuronidase; *gusA*	*E. coli*	pNZ123	–
β-Galactosidase	*C. acetobutylicum*	pFX1	–
Luciferase; *luxAB*	*V. fischerii*	pCK17	–
Prochymosin	Bovine	pNZ18	–
T7 RNA polymerase	*E. coli*	pIL253	–
Prochymosin	Bovine	pKM1363	Yes
Prochymosin	Bovine	pNZ18	Yes
Streptodornase; *sdc*	*S. equisimilis*	pDG13	Yes
Neutral protease; *nprE*	*B. subtilis*	pMG36e	Yes
β-Lactamase; *bla*	*E. coli*	pVS2	Yes
α-Amylase; *amyS*	*B. stearothermophilus*	pNZ123	Yes
α-Amylase; *amyL*	*B. licheniformis*	pGA14	Yes
α-Galactosidase	*C. tetragonoloba*	pGK13	Yes
Surface protein	*S. mutans*	pSA3	Yes
Phage lysin	*L. monocytogenes*	pCK17	Yes
Tetanus toxin fragment C	*C. tetanus*	pLET1	–

*Date taken from de Vos and Simons (1994).

G) and the enzymes of the tagatose 6-phosphate pathway (Lc ABCD) are found on several indigenous plasmids. In the 23.7 kb plasmid pMG 820 of *Lc. lactis* strain ME 1820, they are organized in a 7.8 kb operon with the gene order *lac* ABCDFEGX followed by an iso-IS*SI* element (see Figure 6.11). The function of the distal *lac* X is still unknown. The *lac* genes are transcribed as two 6 and 8 kb polycistronic transcripts (*lac* ABCDFE, *lac* ABCDFEGX). An inverted repeat between *lac* E and *lac* G could function as an intercistronic terminator. Transcription of the *lac* operon is regulated by the divergently transcribed product of the 0.8 kb *lac* R gene. The presumed amino acid sequence of the 28 kDa Lac R protein is highly similar to several *E. coli* repressors of the DeoR family. Growth on lactose leads to a 10-fold induction of transcription of the *lac* operon. Tagatose 6-phosphate is the inducer of Lac R. Regulation of *lac* gene expression presumably functions in three stages:

Stage 1. Growth on glucose activates transcription of the lacR promotor due to binding of the LacR repressor at *lac* operator 1 (O1).

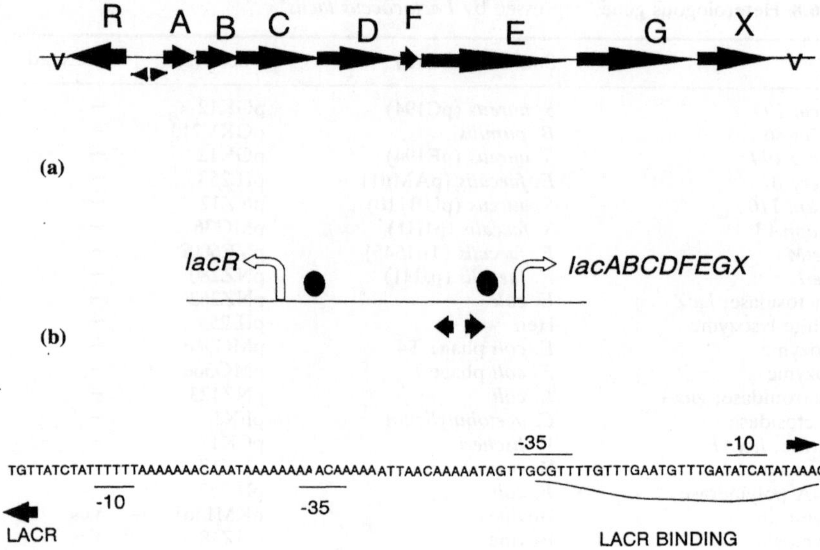

Figure 6.11 (a) Genetic and transcriptional organization of the *lac* operon of the lactose plasmid pMG820. The intercistronic promoter region is enlarged. The black arrowheads indicate the transcription initiation sites, the white triangles the transcription terminators, and the black dots the position of *lac*O2 (left) and *lac*O1 (right). (b) Nucleotide sequence of the *lac* promotor region. The canonical sequences and transcriptional initiation sites of the *lac* operon and *lacR* promotor are indicated as are the nucleotides protected from DNAase digestion by LacR. (Modified from de Vos and Simons, 1994.)

Stage 2. Binding of LacR to *lac* operator 2 (O2) at increasing LacR concentrations during growth on glucose results in repression of the *lac*R gene and *lac* operon expression. Since *lac* O2 has a lower affinity for LacR than *lac* O1, it will be bound at increasing LacR concentrations.

Stage 3. Binding of the LacR repressor to tagatose 6-phosphate during growth in lactose results in dissociation of LacR-operator complex concomitant with the induction of the *lac* operon expression. A second transcriptional control system that reacts to glucose is a tetradecanucleotide sequence that conforms to the consensus sequence involved in catabolite repression in Gram-positive bacteria (at position +12 to +25 in the *lac* promotor region).

6.4.4.2 Plasmid copy number control. The copy number control system of the broad host range plasmid pSH71 (see above) was the first repressor–operator system characterized in lactic acid bacteria. The 6 kDa repressor encoded by the *rep*C gene contains a helix–turn–helix motif, and interacts with an operator sequence around the transcription intitiation site of the *rep*CA promotor. Thereby, it controls its own transcription leading to a steady state level of RepC and RepA in the cell. A stable plasmid copy number is maintained since the replication protein RepA interacts with the plasmid origin of replication and limits the rate of replication (see Figure 6.12).

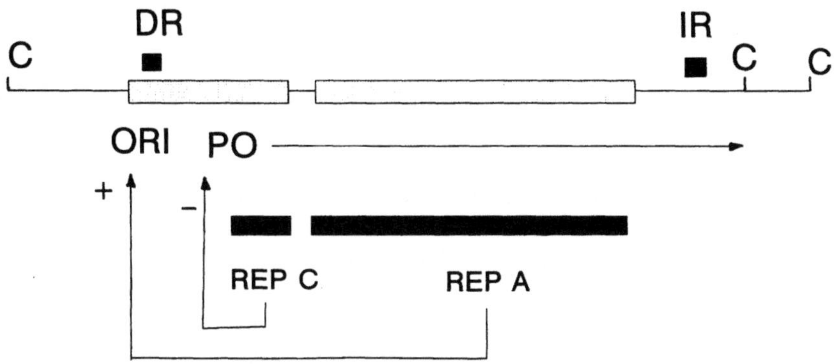

Figure 6.12 Copy number control of the replicon of the *Lactococcus lactis* plasmid pSH71 by the interaction of the repressor RepC and the promotor–operator of the *repCA* operon. (From de Vos *et al.*, 1989 and de Vos and Simons, 1994.)

(a)

Protein	Function	Leader Sequence
Ss1	unknown	M-K-K-I-L-I-I-G-L-G-L-I-G-S-S-I-A-L-G↑-I
Ss30	unknown	M-K-I-W-T-K-L-G-L-L-S-L-V-G-L-T-L-T-A-C-G↑-S
Ss38	unknown	M-K-K-I-L-I-T-T-T-L-A-L-A-L-L-S-L-G-A↑-C
Ss45	unknown	M-K-K-I-N-L-S-K-L-I-I-I-A-L-I-I-I-I-A-A-M-S-A-I-F-I-S-A↑-K
Ss80	unknown	M-K-K-K-I-F-I-A-L-M-A-S-V-S-L-F-T-L-A-A↑-C

(b)

Protein	Function	Leader Sequence
PrtP	proteinase	M-Q-R-K-K-K-G-L-S-I-L-L-A-G-T-V-A-L-G-A-L-A-V-L-P-V-G-E-I-Q-A-K-A↑-A
NisP	proteinase	M-K-K-I-L-G-F-I-V-C-S-L-G-L-S-A-T-V-H-G↑-E
Usp45	unknown	M-K-K-K-I-I-S-A-I-L-M-S-T-V-I-L-S-A-A-A-P-L-S-G-V-Y-A↑-D
PrtM	maturation	M-K-K-K-M-R-L-K-V-L-L-A-S-T-A-T-A-L-L-L-L-S-G↑-C
NisI	immunity	M-R-R-Y-L-I-L-I-V-A-L-I-G-I-T-G-L-S-G↑-C

(c)

Protein	Function	Leader Sequence
NisA/NisZ	lantibiotic	M-S-T-K-D-F-N-L-D-L-V-S-V-S-K-K-D-S-G-A-S-P-R↑-I
Lct	lantibiotic	M-K-N-K-L-F-N-L-L-Q-E-V-T-E-S-E-L-D-L-I-L-G-A↑-K
LcnA	bacteriocin	M-K-N-Q-L-N-F-N-I-V-S-D-E-E-L-S-E-A-N-G-G↑-K
LcnB	bacteriocin	M-K-N-Q-L-N-F-N-I-V-S-D-E-E-L-A-E-V-N-G-G↑-S
LcnMa	bacteriocin	M-K-N-Q-L-N-F-E-I-L-S-D-E-E-L-Q-G-I-N-G-G↑-I

Figure 6.13 Lactococcal leader sequences with defined cleavage sites. (a) sequences obtained by screening using signal sequence vectors; (b) sec-dependent signal sequences derived from defined genes; (c) sec-independent leader sequences derived from defined genes. (de Vos and Simons, 1994.)

6.4.4.3 Protein secretion. Sec-dependent and independent protein transport has been characterized. The leader sequences of defined genes of both types are shown in Figure 6.13. Sec-dependent systems are the well studied, plasmid encoded 200 kDa cell wall proteinase (see above), the transposon coded serine proteinase NisP, and a 45 kDa secreted protein (Usp45) of unknown functions. All sequences have a consensus leader peptidase I cleavage site. Two other proteins (PrtM and Nis I) found outside the lactococcal cell, contain the consensus sequence L-S-G-C characteristic of lipoproteins that is recognized by signal peptidase II. Sec-independent secretion is known for some bacteriocins (see below) similar to the ABC-type translocator of the haemolysin excretion system of Gram-negative bacteria.

6.4.4.4 The cell wall protease (Kok and de Vos, 1994). The genetic organization of the plasmid coded cell wall proteases of *Lc. lactis* subsp. *cremoris* and *lactis* has been elucidated in great detail. The proteinase region codes for a maturation protein (prtM) and the proteinase protein. The maturation protein is anchored with a lipid-modified cystein in the cytoplasmic membrane. The pre-pro-proteinase (corresponding to a calculated molecular weight of about 220 kDa) is transported through the membrane into the cell wall with loss of the leader sequence (see also Figure 6.13). It is anchored into the membrane with its carboxy terminal end. In the cell wall region, the maturation protein (prtM) splits off the *N*-terminal pro-region of the proteinase generating the active cell wall proteinase. The organization of the proteinase structural gene is shown in Figure 6.14. The specificity of the enzyme is determined by a specific region almost midway between the *N*- and carboxy terminals (see Table 6.3 for the consequences).

In Ca^{2+} free buffer, self digestion of the proteinase leads to the release of an enzymatically active fraction (MW 135 kDa) into the medium, having a specificity indistinguishable from that of the cell-bound enzyme.

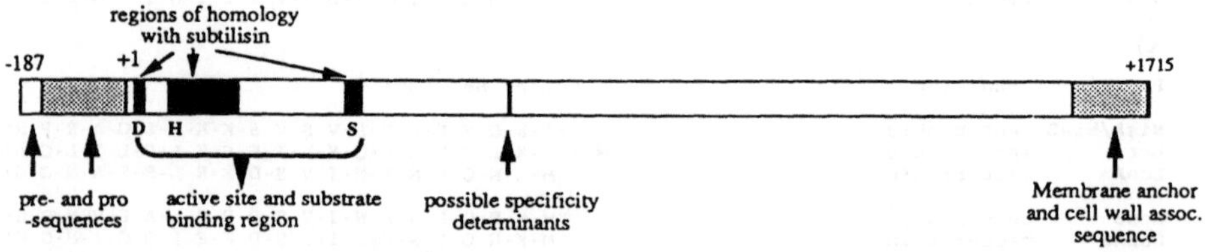

Figure 6.14 Functional regions of the cell wall proteinase of *Lactococcus lactis* deduced from the nucleotide sequence of the gene. D, H, S denote positions of aspartate, histidine and serine residues constituting the catalytic triad typical for the subtilisin class of proteases. (Modified from Pritchard and Coolbear, 1993.)

6.4.5 The Lactococcus lactis *chromosome (Le Bourgeois* et al., *1993)*

The development of pulsed-field-gel electrophoresis and other DNA-based techniques has enabled the investigation and elucidation of genome size and organization of the chromosome of *Lc. lactis*. Chromosome sizes of 35 strains of *Lc. lactis* subsp. *lactis* and *cremoris* varied between 2.0 and 2.7 Mb. Restriction endonucleases ApaI and SmaI due to cleavage specification for G+C-rich sequences appeared to be most suitable for mapping since the resulting pattern of cleavage fragments was not too complex. The deduced genome size of *Lc. lactis* is one of the smallest within eubacterial chromosomes.

The physical map of the chromosome of *Lc. lactis* supsp. *lactis* strain DL 11 was constructed on the basis of PFGE of NotI, SmaI and SarI fragments of chromosomal DNA (Tulloch *et al.*, 1991). In contrast, the chromosomal map of *Lc. lactis* subsp. *lactis* IL1403 was obtained after integration of plasmid pRL1 (carrying an erythromycin resistance and a functional insection sequence IS*SI*) into the host chromosome by replicative transposition. Then 35 derivatives carrying random pRL1 insertions were analysed. The result is shown in Figure 6.15, which also presents the position of 26 marker genes. Most notably, of the six ribosomal RNA loci five are clustered (*rrn* A, B, C, D, E), one is 90° apart and orientated in the opposite direction. Genes involved in amino acid biosynthesis and some peptidase genes (*pep*N, C, X) are not clustered at a single site. PFGE of chromosomal DNA of *Lc. lactis* digested with suitable restriction endonucleases (e.g. SmaI and ApaI) is an excellent tool to produce fingerprints of lactococcal chromosomes. The strength of this technique was exemplified by application to the already mentioned strain family of *Lc. lactis* subsp. *lactis* strain 712/ML3/C2. NotI and SmaI profiles of these strains handled for many years in laboratories around the world turned out to be very similar. The main difference was a 60 kb deletion in strains C2, LM2301 and NCDO 2031. It correlated with loss of the above described sex factor.

6.5 Phylogeny

The phylogenetic position of the genus *Lactococcus* has been unequivocally determined by several methods. The earliest definition was based on the comparison of the similarity values of oligonucleotide patterns from ribosomal RNA (Stackebrandt and Teuber, 1988). The genus *Lactococcus* is an offspring of the branch leading to the genus *Streptococcus*. *Lactococcus* is also separated from the *Lactobacillus–Pediococcus–Leuconostoc* branch, from the (aerobic) *Listeria–Enterococcus–Bacillus–Staphylococcus* and the (anaerobic) *Clostridium* branch (Figure 6.16).

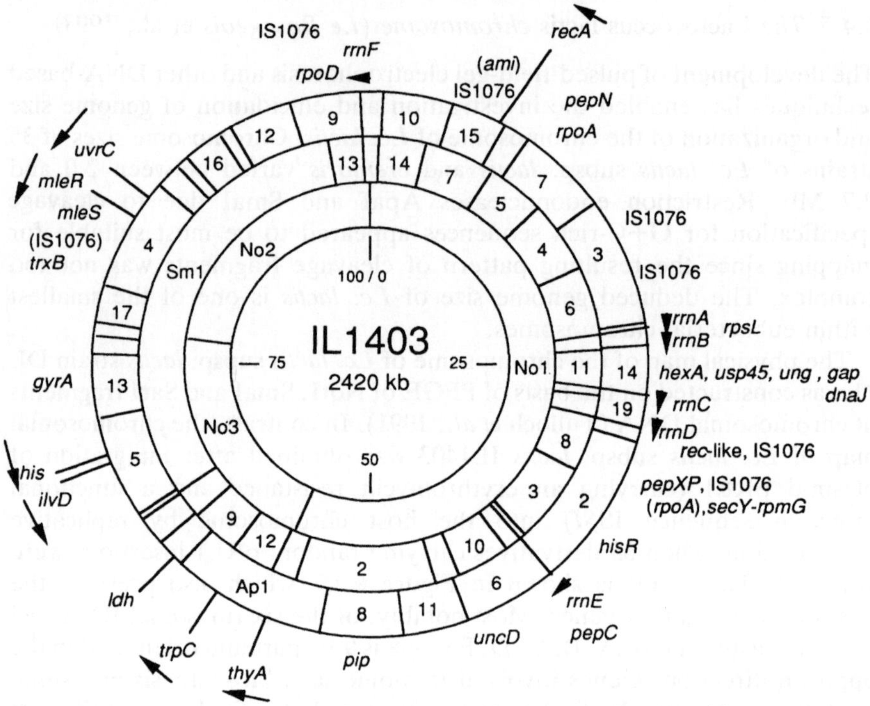

Figure 6.15 The physical and genetic map for *Lactococcus lactis* subsp. *lactis* chromosome determined by Le Bourgeois *et al.* (1993). Numbers represent the largest restriction fragments present in *Not*1, *Sma*1 and *Apa*1 digests of the chromosome. Bars on the pointers indicate the range over which genes are actually located, reflecting the current limits of mapping data. The mapped homologous genes are for biosynthesis of isoleucine and valine (*ilv*), histidine (*his*), tryptophan (*trp*), X-propyldipeptidyl aminopeptidase (*pepXP*), malolactic enzyme (*mleR*), the highly secreted protein Usp45 (*usp*45) as well *trxB* and *uvrC*. The heterologous 'housekeeping genes' are for 16s (*rrn*), ribosomal protein (*rpsL*), RNA polymerase (*rpoA*), ATPase (*uncD*), heat shock protein (*groE*), recombination (*rec*), DNA mismatch repair (*hexA* and *ung*), DNA topology (*gyrA*) and oligonucleotide transport (*ami*).

Comparison of the 16S rRNA sequences has confirmed their position (Williams *et al.*, 1990). 16S RNA sequences have also been used to determine the close relationships within the *Lactococcus* side branch. *Lactococcus garviae* and *Lc. lactis* cluster together and are separate from the *Lc. raffinolactis*, *Lc. plantarum* and *Lc. piscinum* cluster (see Figure 6.17). This evaluation clearly showed that all members of the genus *Lactococcus* have a common ancestor.

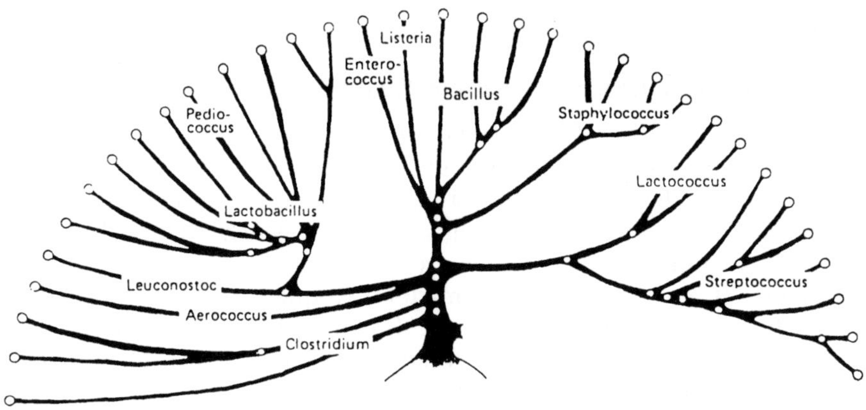

Figure 6.16 Phylogenetic tree of the *Clostridium* subdivision of Gram-positive bacteria as determined by similarity values (S_{AB}) of oligunucleotides of the ribosomal RNAs (Stackebrandt and Teuber, 1988). The lactococci branch of the streptococci at about 0.55 S_{AB}. The *Lactococcus–Streptococcus* branch diverges from the *Leuconostoc–Lactobacillus–Pediococcus* branch at about 0.42 S_{AB} at the same level as the *Enterococcus–Bacillus–Staphylococcus* branch. (From Teuber, 1987.)

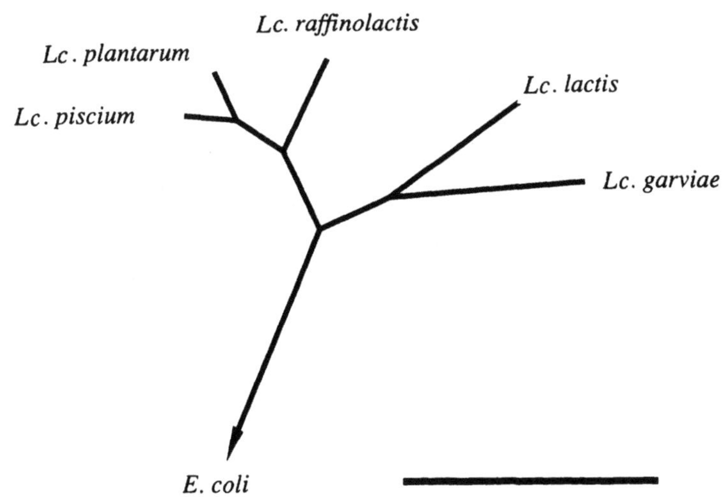

Figure 6.17 Phylogenetic tree reflecting the close relationships within the genus *Lactococcus*. The tree was constructed by a maximum likelihood analysis of 16S rRNA sequences. The bar indicates 10% estimated sequence differences. (Courtesy of Wolfgang Ludwig, Munich, Germany.)

6.6 Importance

6.6.1 The use of Lactococcus lactis *in the dairy industry*

The subspecies of *Lc. lactis* (*lactis*, *cremoris* and biovar. *diacetylactis*) contribute substantially to human nutrition and well-being. Economically and technically, they are of outmost importance as starter cultures for the production of fermented milk and its derivatives as indicated in Table 6.1: sour milk, sour cream, lactic butter, fresh cheeses (cottage cheese, quarg), soft cheeses (Camembert, Brie, Romadour, Roquefort), hard cheeses with and without eyes (Cheddar, Gouda, Edam, Tilsit types), casein, and specific products like Scandinavian ropey milk (Taette, Viili). They are also part of the complex microflora of the kefir grain.

About one third of the annual world output of milk of around 500 million tons is transformed into fermented products. Although exact numbers are not available regarding products made with thermophilic cultures (e.g. Swiss cheese, yogurt) as compared to the above mentioned ones made with mesophilic lactococci, the author estimates that two-thirds of milk fermentation is of the mesophilic type, i.e. 100 million tons of milk are annually inoculated with *Lc. lactis*. If the initial level of inoculation is put at 10^7 cells/ml and the final level after fermentation at 5×10^9 ml, 10^{21} cfu of lactococci are added to 100 million tons ($= 10^{14}$ ml) of milk which multiply about 500-fold (i.e. in nine generations) to 5×10^{23} cfu, representing a microbial bimass of about $5 \times 10^{23} \times 10^{-12}$ g, i.e. about 500 000 tons. Starting with one *Lactococcus* cell, nine generations will represent 519 single cell division events. Starting with 10^{21} cells will consequently yield in nine generations 5.19×10^{23} cell divisions, ample opportunities for genetic adaptation to environmental pressures and consequently evolutionary progress.

6.6.1.1 Functions of lactococci in milk fermentations. The main early function is a sufficiently fast and quantitatively defined conversion of lactose into lactic acid. Later functions include formation of diacetyl, CO_2, and other aroma compounds, some possibly generated by a controlled proteolysis during cheese ripening (on the basis of the described cell wall protease and the peptidases listed in Table 6.4). In the beginning, these processes are surely driven by living bacteria and their enzymes. During ripening and after salting, most of the lactococcal cells will die, but further biochemical changes could occur due to remaining bacterial enzyme activities. The number of live cells included in cheese curd during coagulation of milk has been estimated to be at least 10^8/ml (for an extensive review, see Lawrence *et al.*, 1976), enough to induce substantial biochemical changes in cheese even at low temperatures (about 10–12°C) but during long ripening periods of up to several months.

The bacterial enzyme systems are thereby severely influenced by the salt concentration, the pH value, the water activity and the ripening temperature as well as a contaminating secondary microflora. Dysfunctions – if not caused by improper biophysical conditions or handling – may have one or more of three microbiological reasons:

(1) Irreversible loss of technologically important functions from the culture, e.g. by plasmid curing during culture preparation.
(2) Inactivation of the culture by bacteriophages (see section 6.6.2)
(3) Inactivation of the culture by residual antibiotics and disinfectants in milk.

6.6.2 The Lactococcus-*specific bacteriophages (Hill, 1993; Teuber, 1993)*

Bacteriophages are a common and constant threat to milk fermentation by lactococci, since most fermentations can not be performed under sterile conditions, since heat sterilization damages the casein micelles preventing proper coagulation and syneresis of the cheese curd. Even pasteurized milk may still contain phages and some residual lactic acid bacteria.

Hundreds of such phages have been isolated and described during the last five decades. However, only after progress in molecular biology could these phages be differentiated on the basis of, for example, DNA–DNA hybridization and protein profiles. As a consequence, the International Subcommission on the Taxonomy of Bacteriophages of *Lactococcus* was able to agree on the establishment of defined bacteriophage species and type phages which are available from an international reference laboratory (see Table 6.9). The 11 defined species do not show detectable DNA-homology which is quite astonishing on the basis of the quite similar morphology especially of species 1 to 4, 6 and 7, commonly named 'small isometric head' phages (see Figure 6.18). Most of these species have been isolated as virulent or lytic phages, only BK5-T is a temperate phage, and P335 reveals strong DNA homology with some temperate phages. With regard to host strains, between 50 and 70% of the investigated bacterial strains could be induced to release between one and three morphologically distinct prophage particles upon induction with mitomycin C or UV irradiation. Isolation of prophage cured strains, however, is extremely difficult, due to immediate relysogenization of the cured derivatives in the induction assay. By electron microscopy, at least 19 morphologically distinguishable prophage particles have been detected. The taxonomic diversity of the lactococcal bacteriophages suggests, together with the observed diversity of lactococcal strains in mixed strain starter cultures, a long and rapid evolution.

6.6.2.1 Biology of host–virus interaction. Typically, milk to be fermented with *Lactococcus* is seeded with about 10^7 cfu/ml, which grow up to about

Table 6.9 Phage species, type phages and some characteristics of lactococcal bacteriophages*

Family	Morphological type	Phage species	Type phage	Morphology			Appendages, unusual features	Phage genome		G+C (%)
				Head diameter	Tail diameter	Tail length		Size (kb)	Cohesive or non-cohesive ends	
1. Siphoviridae	B1	936	P008	53	11	159	Collar, whiskers	29.7	C	37.5
2. Siphoviridae	B1	P335	P335	54	11	145	Collar, large base plate	36.4	NC	
3. Siphoviridae	B1	P107	P107	55	11	152	Long tail fibre	51.5		
4. Siphoviridae	B1	1483	1483	52	10	127		36	NC	
5. Siphoviridae	B1	P087	P087	65	16	200		54.5		
6. Siphoviridae	B1	1358	1358	53	10	109	Distinctive cross bars on tail	40	NC	
7. Siphoviridae	B1	BK-T	BK5-T	58	11	233		37.6	NC circular	
8. Siphoviridae	B1	949	949	79	14	527		52		38.8
9. Siphoviridae	B2	C2	c6A	60 × 41	10	87		21.9		36.7
10. Podoviridae	C2	P. 34	P. 34	65 × 44	10	24	Short tail collar with fibrils	18.1		
11. Podoviridae	C3	KSY1	KSY1	220–260	5	25–35	Elongated head, short tail Triple collar with spikes			

*Modified from Jarvis *et al.* (1991)

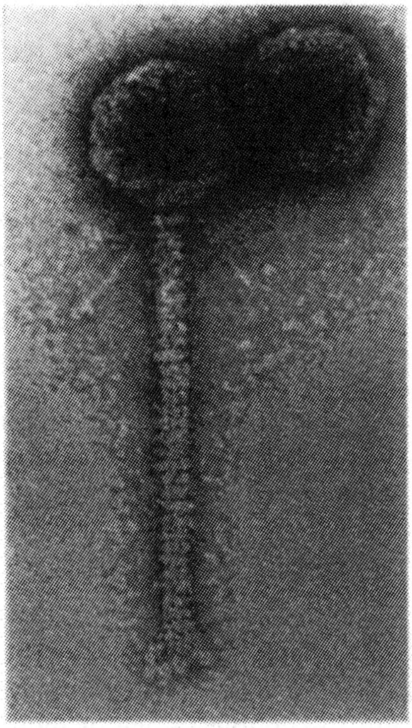

Figure 6.18 Electron micrograph of the common *Lactococcus lactis* bacteriophage species P008 (see also Table 6.9) (Loof *et al.*, 1983).

5×10^9 cfu/ml at the end of the fermentation. Nine cell divisions will yield this increase. If only half of the world's cheese production is assumed to be manufactured with lactococci (i.e. 7 million tons or 7×10^9 kg) this corresponds to 7×10^{10} litres of milk inoculated with about 10^{21} cfu proliferating to 10^{23} viable bacteria every year. Since cheese whey from mixed strain fermentation contains typically 10^6–10^8 phage particles per ml (even in the absence of a noticeable phage attack), up to at least 10^{21} phage particles are released during roughly 10^{22} cell divisions. This again provides ample opportunities for evolution of bacteria and phages.

Phage adsorption is usually not equally and densely packed over the whole cell surface, but to a few specific spots at the surface, possibly in the vicinity of the plane of cell division (Budde-Niekiel and Teuber, 1987). For that reason, the biochemical identification and isolation of phage receptor molecules has not yet been successful, although membrane lipoteichoic acid, cell surface galactose- and rhamnose-containing polysaccharides and murein complexes have been suggested to participate in phage adsorption. Latency periods after infection are around 90 min, burst sizes between 10 and 100 per infected cell (Neve and Teuber, 1991).

Most isolated virulent phages are very specific for specific strains, a few may even cross the subspecies line between '*lactis*' and '*cremoris*'. Especially the prolate headed phage species c2 (No. 9 in Table 6.9) may do so. In addition, bacteria infected and lysed by such phages produce a highly active bacteriophage-coded lysin, which is able to lyse lactococci from the outside even if the strain is resistant to phage adsoption. For temperate phage particles released by mitomycin or UV-induction from isolates from mixed strain starter culture, it is difficult to find sensitive host strains in that and in other starter cultures.

This may in part reflect the situation that prophage sensitive bacterial strains are immediately and constantly eliminated in a starter culture either by lysis or by lysogenization. *Lactococcus lactis* subsp. *cremoris* Wg2, a strain well known from investigations of the plasmid-coded cell-wall-protease system, is exceptionally susceptible to many prophage particles (due to lack of a restriction/modification system?).

6.6.2.2 Distribution of bacteriophage resistance. If single isolates (23 each) of three undefined mixed starter cultures (A, B, C) were tested against 375 phage strains from our collection, only 34 phages, that is less than 10%, showed lytic activities against one or more of the 69 bacterial strains. In all nine bacterial strains from culture A, 14 from culture B, and two from culture C were susceptible to one or more phages, the highest sensitivity being one bacterium from culture A against eight different phages (Budde-Niekiel *et al.*, 1985). This observation suggests that phage–host interaction seems to be rather specific. This is also evident from the observation showing the specificity of phages isolated during the application cycles of the three starter cultures in a cheese factory. The phages appearing during cycle A were mainly specific for bacterial strains from culture A, and likewise for the cycles B and C. The distribution of phage resistant strains of varying degrees in four different undefined mixed strains starter cultures is remarkable. A significant individual resistance level is evident which should be known to the culture producer, as well as to the user (see Figure 6.19). From a total of seven undefined mixed strain starters, 30 permanently and highly phage resistant strains were obtained by the Heap and Lawrence (1976) technique by challenging with a phage cocktail, containing all phages ever isolated in the author's laboratory (Möller and Teuber, 1988). This proves the high potential of lactococci to develop phage resistance mechanisms which actually allowed the successful use of such bacteria in spontaneous milk fermentations during the hundreds of years before the invention of starter cultures 100 years ago by Weigmann and others.

If the spatial distribution of phages in dairy plants is investigated, high numbers are found in whey and in the product (if it is not pasteurized). After enrichment, phages are detected in raw and pasteurized milk, on the

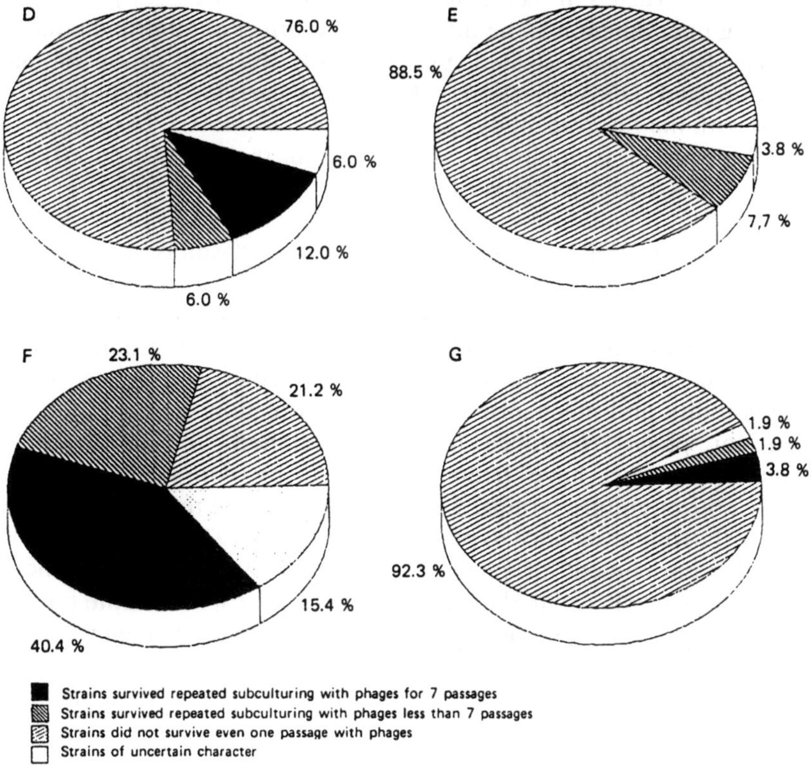

Figure 6.19 Distribution of phage resistant strains of lactococci in four different undefined mixed-strain starter cultures (D, E, F, G). Fifty strains each were randomly isolated and tested against a phage cocktail containing phages. (From Möller and Teuber, 1988).

surface of equipment, the clothes and skin of the cheesemakers, and in the air, especially in the neighbourhood of areas where whey-containing aerosols are produced, for example, close to whey separators. Cold disinfection of phage containing surfaces and material is most effective with hypochlorite and peracetic acid (Lembke and Teuber, 1981).

6.6.2.3 Mechanisms of bacteriophage resistance. Restriction of bacteriophages can be commonly demonstrated in lactococci, e.g. with host–virus systems isolated from cheese factories. The observed efficiencies of restriction in plaque numbers are around 10^3–10^7. In addition, inhibition of phage adsorption has quite regularly been reported. However, since phage adsorption to lactococcal cells due to the 'hot-spot' mechanism is difficult to determine and to distinguish from unspecific binding, adsorption studies can only be trusted if controlled by electron microscopy. The most interesting phenomenon is the quite common effect of 'abortive infection'. This has been demonstrated unequivocally by the proof of phage adsorption and DNA-injection with phages carrying a genomic DNA labelled with the fluorescent dye DAPI (Geis *et al.*, 1992). In some systems phage replication is inhibited as late as in the translation of phage messenger RNA into phage protein.

6.6.2.4 Plasmids and bacteriophage resistance. Plasmids are common in lactococci as discussed above. In ecological terms, lactococci are able to donate and receive genes necessary for better survival in milk between each other. Table 6.10 gives a list of important phage resistance plasmids reported in lactococci, including their molecular sizes, host species, transmissibility and mechanisms of phage protection. Some of the genes coding for phage resistance have been cloned. In addition, phage resistance plasmids have been transferred by conjugation to phage sensitive starter culture strains. Use of such strains in the dairy industry in the USA showing resistance to phage species c2 and 936 (see Table 6.9) has led to the enrichment of phages from the P335 species in American dairies. These newly isolated phages (known in Europe for some time) had high numbers of restriction sites in their genome in contrast to the traditional c2 and 936 phage species. This observation is suggested to result from a selection during evolution for phage strains with none or few restriction sites, which are therefore able to escape the restriction/ modification systems of lactococci (Moineau *et al.*, 1993).

6.6.2.5 Molecular biology of lactococcal bacteriophages. The molecular biology of the bacteriophages of lactococci is rudimentary and in an initial stage. All phages hitherto isolated contain double-stranded DNA in a closed circular, linear or permutated form. In terms of DNA sequences, we know the nucleotide sequences of a gene coding for a presumptive structural protein. Another example is the structural gene for lysin gene in prolate-headed lactococcal bacteriophages of the c2 species (for more detail see Table 6.11). Several more or less strong promotors have been isolated from lactococcal phages with the use of promotor probe vectors. Some work has been done regarding the integration of temperate phages into the lactococcal chromosome. Also, the characterization of restriction/ modification systems involved in phage resistance is in an early phase. An interesting approach is the use of antisense mRNA to induce bacteriophage resistance in lactococci (Kim and Batt, 1991).

6.6.2.6 Stability of bacteriophage resistance in lactococci. The fact that many resistance phenomena are coded for on plasmids suggests that plasmid loss could lead to loss of phage resistance, which is indeed the

Table 6.10 Bacteriophage resistance plasmids in *Lactococcus lactis**

Plasmid	Host	Size (kb)	Remarks
Adsorption inhibition			
pME0030	Lc. ME2	48	One of five resistances in ME2
pSK112	Lc. SK11	54	Encodes galactosylated LTA
p2520L	Lc. P25	38	Tra+
pC1528†	Lc. UC503	46	Tra+, also contains Abi.
pAH90†	Lc. DPC721	27	Co-integrate, also R/M
pNP40†	Lc. DRC3	65	Also encodes Abi1 and Abi2
Restriction and modification			
pME100	Lc. KH	17	Cloned
pLR1020	Lc. M12R	32	
pTN20†	Lc. ME2	28	Tra+, R/M and *AbiC*
pTR2030†	Lc. ME2	46	Tra+, *Lla*I and *AbiA*, cloned
pIL6	Lc. IL594	28	
pIL103	Lc. IL964	5.7	Cloned
pIL107	Lc. IL964	15	Cloned
pBF61†	Lc. KR5	42	Also encodes Abi
pKR223†	Lc. KR2	38	Also encodes Abi
pJW563	Lc. W56	12	
pJW566	Lc. W56	25	
pJW565	Lc. W56	14	
pFV1001	Lc. T912.2	13	
pFV1202	Lc. T2235	17	
pAH82	Lc. DPC220	20	
pAH90†	Lc. DPC721	27	'Silent' R/M activated
pTRK68	Lc. NCK202	46	
pTRK12	Lc. TDM1	30	One of three R/M in this strain
pTRK30	Lc. TDM1	28	One of three R/M in this strain
pTRK317	Lc. TDM1	16	One of three R/M in this strain
pHD131†	Lc. TDM1	131	May also possess Abi
Abortive infection			
pTR2030†	Lc. ME2	46	Tra+, R/M and *abiA*, cloned
pCI829	Lc. UC811	44	Tra+, *abiA*, cloned
pTN20†	Lc. ME2	28	Tra+, R/M and *abiC*, cloned
pIL611	Lc. IL416	–	Chomosomal *abi416*, cloned
pNP40†	Lc. DRC3	65	Tra+, two *abi* genes cloned
pBUI-8	Lc. BU1	64	Tra+
pIL105	Lc. IL964	9	*abi105*, cloned
pCI528†	Lc. UC503	46	Tra+, ads inhibition, cloned
pCI750	Lc. UC653	65	Tra+, cloned
pAJ1106	Lc. 4942	106	Tra+, co-integrate plasmid
pKR223†	Lc. KR2	38	R/M and Abi, cloned
pBF61†	Lc. KR5	42	R/M and Abi
pHD131†	Lc. HID600	131	Tra+, R/M and Abi
pCLP51R	Lc. 33–4	90	Tra+
pJS88	Lc. 11007	132	Tra+
pJS40	Lc. JS30	65	Tra+
pCC34	Lc. C3	51	Tra+
pNP2	Lc. WM4	132	Tra+
pEB56	Lc. EB7	84	Tra+

*Modified from Hill (1993).
†Those plasmids with more than one resistance mechanism.

Table 6.11 Genes/elements cloned from phage of *Lactococcus lactis**

Gene/element	Phage	Vector	Cloning host	Sequence determined
Lysin	ΦvML3	λgt10	*E. coli*	+
		pTG262	*B. subtilis*	
p35, p43	F4–1	pUC13	*E. coli*	+
			Lc. lactis	
MCP	F4–1	pUC13	*E. coli*	−
ORF1365	Φ7–9	pUC13	*E. coli*	+
*Lla*I methylase	F4-1	pBluescript	*E. coli*	+
		pSA3	*Lc. lactis*	
bpi	BK5-T	pACYC194	*E. coli*	+
per/rep	Φ50	pSA3	*E. coli*	+
			Lc. lactis	
Lysin	ΦU53	pLK	*E. coli*	+

*Data taken from Klaenhammer and Fitzgerald (1994).

case. In everyday cheesemaking; the loss of phage resistance of undefined mixed strain cultures was unequivocally shown by the experience of Stadhouders and Leenders (1984) in The Netherlands. As soon as the cultures were propagated in the laboratory in sterile skim milk in the absence of phages, the mixed strain cultures converted into a culture with just a few dominating strains. The previously high bacteriophage resistance known from cheese fermentations, was completely lost within a short time. This observation emphasizes the need of a constant challenge to such cultures under manufacturing conditions in order to keep them in an optimized state.

6.6.2.7 Practical phage control in dairies (for detailed treatment, see IDF, 1991). A variety of concepts has been elaborated in order to minimize the risk of a bacteriophage contamination.

(1) The necessity to work as far as possible under strictly hygienic conditions (e.g. by using closed fermentation utilities with air filter systems that retain phage particles). The dairy equipment has to be subjected to regular daily cold disinfection. Active chlorine and peracetic acid are suitable to totally inactivate bacteriophages.
(2) Propagations of mother and bulk cultures prior to the batch fermentations are frequently done in phage-inhibitory media, which usually contain phosphate and/or citrate chelating divalent cations (in particular Ca^{2+}), which are essential for lytic bacteriophage development.
(3) By using deep-frozen or lyophilized concentrated starter cultures, which are inoculated directly into the culture tank, prolonged exposure of the mother culture and the bulk culture to bacteriophages in the cheese factory can be avoided efficiently.

(4) Starter bacteria exhibiting different phage spectra are used as single strains or are blended for a defined multiple starter culture within a rotation scheme. By continuous phage monitoring in the plant, phage-sensitive strains are immediately replaced by alternative cultures.
(5) However, the use of starter bacteria with total or at least high phage insensitivity is desirable.

6.6.3 Bacteriocins of Lactococcus lactis *(for reviews see Klaenhammer, 1993; Dodd and Gasson, 1994)*

Bacteriocins are proteins that kill closely related bacteria. Of 280 *Lc. lactis* strains surveyed, 5% were shown to excrete proteinaceous bacteriocins (Geis *et al.*, 1983). These compounds can now be classified into several distinct classes based on structure and mechanism of action:

(i) Lantibiotics which are small, membrane-active peptides (<5 kDa) containing the unusual amino acids lanthionine, *β*-methyl lanthionine, and other dehydro residues: nisin, lacticin 481. Heat sensitive at pH 9.4; wide host range in Gram-positive bacteria.
(ii) Small heat-stable, non-lanthionine containing membrane-active peptides (<10 kDa) characterized by Gly-Gly^{-1**+1} × aa processing site in the bacteriocin precursor. The mature bacteriocins are predicted to form amphiphilic helices with varying amounts of hydrophobicity, *β*-sheet structure, and moderate (100°C) to high (121°C) heat stability: lactococcin A, lactococcin B, lactococcin G, lactococcin M, diplococcin. Narrow host range against lactococci.
(iii) Uncharacterized proteins active at low pH values only: lactostrepcins. Since these compounds have not been confirmed by all investigators, they will not be further discussed.

6.6.3.1 Nisin. Nisin, produced by many strains of *Lc. lactis* subsp. *lactis*, is the prototype lantibiotic which has achieved commercial application in the food industry and was discovered in 1928. It is now permitted as a food additive in at least 47 countries (Vandenbergh, 1993). Food commodities preserved by nisin include processed cheese, hard cheese, desserts, milk, yogurt, cottage cheese, fermented beverages, meat products like bacon and frankfurters, smoked fish, and canned vegetables. It is particularly useful to inhibit *Clostridium* species like *botulinum* and *tyrobutyricum*. It is the first bacteriocin granted GRAS (generally recognized as safe) status in the USA for use in processed cheese. The effectiveness of nisin in dairy products is demonstrated in Table 6.12: between 5 and 500 IU/g or ml of the product are sufficient for protection against *Bacillus* and *Clostridium* defects. The host range of nisin includes most Gram-positive bacteria, e.g.

some staphylococci, enterococci, pediococci, lactobacilli, leuconostocs, listerias, corynebacteria, *Mycobacterium tuberculosis*, and, especially, germinating spores of bacilli and clostridia. The mechanism of action involves the breakdown of the electrochemical potential of bacterial membranes. The chemical structure has been elucidated and is shown in Figure 6.20.

The pentacyclic peptide of 34 amino acid residues contains the three unusual amino acids dehydroalanine, lanthionine and β-methyl lanthionine. Nisin is ribosomally synthesized as a 57 amino acids containing prepeptide

Table 6.12 Effectiveness of nisin in dairy products*

Product	Nisin application (IU/g or ml)	Benefit
Processed cheese	250–500	Afforded greater flexibility in moisture content of cheese spreads.
Hard cheese	200	Reduce 'blowing' defect.
Chocolate dairy dessert	150	Increased shelf-life at 7°C.
Pasteurized milk	30–50	Increased shelf-life 6 days at 15°C and 2 days at 20°C.
Yogurt	50–200	Increased shelf-life of 7 days and no wheying off.
Cottage cheese	5	100% growth inhibition of *Bacillus* spp.

*Data taken from Vandenbergh (1993).

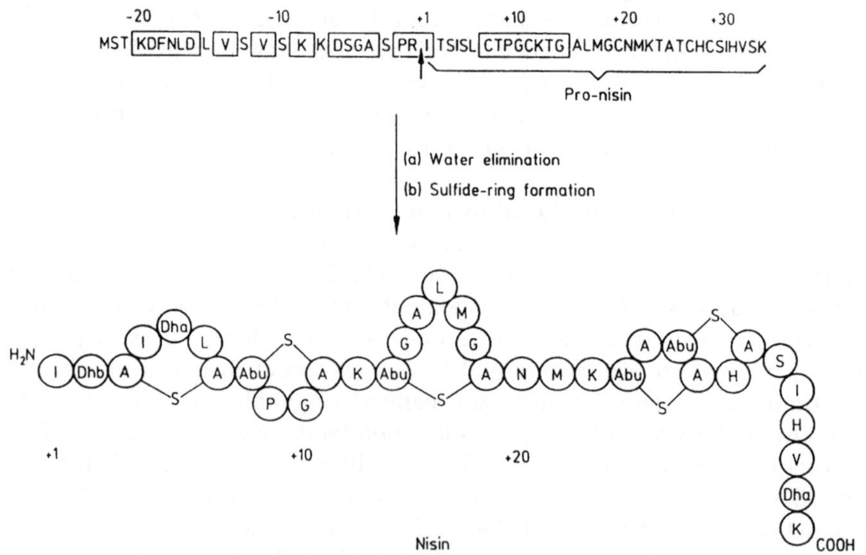

Figure 6.20 Processing of the pre-pro-nisin amino acid chain into the antibiotically active nisin. (Courtesy of Kaletta and Entian, 1989.)

```
AAAATAAATTATAAGGAGGCACTCAAA ATG AGT ACA AAA GAT TTT AAC TTG GAT
                             M   S   T   K   D   F   N   L   D
                                        -20

TTG GTA TCT GTT TCG AAG AAA GAT TCA GGT GCA TCA CCA CGC ATT ACA
 L   V   S   V   S   K   K   D   S   G   A   S   P   R   I   T
                -10                                 -1 ↑ +1

AGT ATT TCG CTA TGT ACA CCC GGT TGT AAA ACA GGA GCT CTG ATG GGT
 S   I   S   L   C   T   P   G   C   K   T   G   A   L   M   G
                            +10

TGT AAC ATG AAA ACA GCA ACT TGT CAT TGT AGT ATT CAC GTA AGT AAA
 C   N   M   K   T   A   T   C   H   C   S   I   H   V   S   K
    +20                                     +30

TAACCAAATCAAAGGATAGTATTTT
oc
```

Figure 6.21 Nucleotide sequence of the nisin structural gene (NISA) from *Lactococcus lactis*. (Courtesy of Kaletta and Entian, 1989.)

which is post-translationally processed. The meso-lanthionine, and the 3-methyl lanthionine bridges are formed by post-translational modifications at a cysteine, serine and threonine residue, respectively. The nucleotide sequence of the structural gene is shown in Figure 6.21. At least nine genes and sequences necessary for nisin synthesis and excretion are characterized in the nisin region which together with sucrose fermentation capacity is usually localized on a transposon (see Table 6.13). Many nisin producing strains can transfer nisin production, nisin immunity and sucrose fermentation to other lactococci by conjugation (see above). Genetic engineering of the nisin structural gene and the nisin operon is reality (Dodd and Gasson, 1994).

6.6.3.2 Lactococcins. The most intensively studied small heat stable bacteriocins are lactococcins A, B, and M. All three compounds are coded on the 60 kb plasmid p9B4-6 of *Lc. lactis* spbsp. *cremoris* 9B4 (Geis *et al.*, 1983). The plasmids contain in addition the genetic determinants for immunity, excretion and conjugal transfer (tra) to other lactococci (see Figure 6.22). The amino acid sequences including the leader peptides are given in Figure 6.23.

The host range of these compounds is rather narrow, being against lactococcal strains only. The mechanism of action is on the bacterial membrane of susceptible organisms. Liposomes, however, are only susceptible if they contain membrane proteins of a susceptible host, in contrast to nisin which destroys the permeability barrier of pure liposomes.

Table 6.13 Genes and sequence characteristics of the nisin region*

Order/Feature	Size		Description
1/*nisA*	171 bp	57aa	Structural gene for nisin precursor, no apparent upstream promoter.
2/*IR*	20 bp		Inverted repeat located between *nisA* and *nisB* proposed mRNA processing site.
3/*nisB*	2982 bp	993aa	117 kDa protein homologous to SpaB/EpiB, several C terminal amphipathic transmembrane helices and associated with membrane fractions.
3/*nisB*	2556 bp	851aa	110 kDa protein homologous to SpaB/EpiB, one C-terminal transmembrane spanning domain implicated as a membrane anchor. NisB provides an essential processing/maturation function; implicated in dehydration reactions.
4/*IR*	25 bp		Proposed rho-independent terminator (ΔG–10.6 kCal/mol) that defines the limit of a nisin operon, halts transcription into *nisT*.
5/Promoter			Two tandem consensus promoters to initiate transcription into *nisT*.
5/*nisT*	1803 bp	600aa	69 kDa ATP-dependent membrane translator with homology to HlyB.
6/*nisC*	1257 bp	418aa	43.7 kDA protein with homology to SpaC and EpiC, essential for biosynthesis of subtilin and epidermin.
7/*nisI*	738 bp	245aa	Lipoprotein that confers nisin immunity.
8/*nisP*	2049 bp	683aa	Subtilisin-like serine protease proposed to cleave the nisin precursor to form mature nisin.
9/*nisR*	690 bp	229aa	Transcriptional regulatory protein belonging to the family of two component regulators.

*Data taken from Klaenhammer (1993).

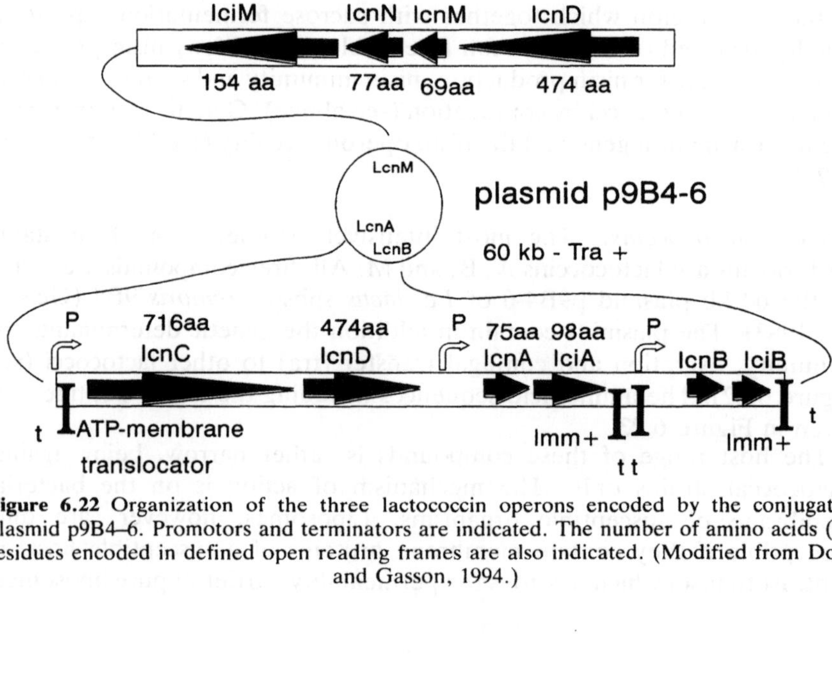

Figure 6.22 Organization of the three lactococcin operons encoded by the conjugative plasmid p9B4–6. Promotors and terminators are indicated. The number of amino acids (aa) residues encoded in defined open reading frames are also indicated. (Modified from Dodd and Gasson, 1994.)

```
                         processing site
                                 ▼
          NH2-MK......17aa.......GG  K.....................................................-COOH
              **                 **  *

Lactococcin A    MKNQLNFNIVSDEELSEANGG  KLTFIQSTAAGDLYYNTNTHKYVYQQTQNAFGAAANTIVNGWMGGAAGGFGLHH
Lactococcin B    MKNQLNFNIVSDEELAEVNGG  KLQYVMSAGPYTWYKDTRTGKTICKQTIDTASYTFGVMAEGWGKTFH
Lactococcin M    MKNQLNFEILSDEELQGINGG  KRGTGKGLAAAMVSGAAMGGAIGAFGGPVGAIMGAWGGAVGGAMKYSI
                 ******* * *****   ***  *
```

Figure 6.23 Primary amino acid sequences of small heat-stable bacteriocins of *Lactococcus lactis*. Homologies between sequences in the *N*-terminal leader region are indicated by asterisks. (Modified from Dodd and Gasson, 1994.)

6.6.4 Production and preservation of starter cultures

Liquid, dried and frozen starter cultures are in use. Starter cultures must have a high survival rate of microorganisms coupled with optimum activity for the desired technological performance; e.g. the fermentation of lactose to lactate, controlled proteolysis of casein, and production of aroma compounds like diacetyl. Since the genes for lactose and citrate fermentation as well as for certain proteases are located on plasmids, continuous culture has not been successful because fermentation-defective variants easily develop. In most instances, pasteurized or sterilized skim milk is the basic nutrient medium for the large-scale production of starter cultures because it ensures that only lactotococci fully adapted to the complex medium, milk, will develop. For liquid starter cultures, the basic milk medium may be supplemented with yeast extract, glucose, lactose, and calcium carbonate. To obtain optimum activity and survival, it may be necessary to neutralize the lactic acid that is produced, by addition of sodium or ammonium hydroxide. Since many strains of lactic streptococci produce hydrogen peroxide during growth under microaerophilic conditions, it has been beneficial to add catalase to the growth media, thus leading to cell densities of more than 10^{10} viable units per millilitre of culture.

Of course, many important details of the art of producing starter cultures are not being published for the protection of natural commercial interests. One example, for the manufacture of freeze-dried yogurt starters, has been described by Tamine and Robinson (1976). In contrast to this procedure, which uses reconstituted skimmed milk, the media for the production of concentrated starters are clarified by proteolytic digestion of skimmed milk with papain or bacterial enzymes to avoid precipitation of casein in the separators used to collect the lactococcal biomass.

6.6.4.1 Batch fermentation for concentrated starter cultures (Stanley, 1977). The (pilot) plant consists of 40 gallon and 100 gallon (1 gallon ≈ 4.5 litres) batch pasteurization tanks for use in medium preparation. Medium is pumped via a single, centrally placed peristaltic pump through a pasteurizer and into the 40 gallon fermenter. This fermenter is fitted with pH control equipment, a stirrer for efficient mixing and hot and cold water jackets for temperature control. The pasteurized medium is aseptically inoculated with appropriate starter and incubated for 16–18 h at 22°C. A positive air pressure is maintained in the fermenter throughout this time. When grown, the culture is rapidly cooled in the fermenter, and harvested in a selfcleaning clarifier. The concentrated cells are automatically recovered by the desludging action of the separator and are pumped into the CCV (cell collection vessel) where gentle agitation ensures homogeneity of the product, and a cooling device maintains a refrigeration temperature. From the CCV the concentrate passes to the microflow cabinet where it is packed semiautomatically in syringes via a syringe filler unit. These syringes are then frozen under vapour phase liquid nitrogen and stored until ready for use.

As an alternative, the bacterial sludge is dropped into liquid nitrogen to obtain frozen pellets suitable for direct vat inoculation. The classical liquid starter culture with about 10^9 viable cells/ml must be further propagated in the factory. The same principle applies to the traditional lyophilized cultures containing about 10^9 viable bacteria/g. However, they can be easily shipped and kept at ambient temperatures for several months. Concentrated lyophilized starters are also suited for direct vat inoculation, having a short lag phase, however, before growth is resumed. Direct vat inoculation may have a certain advantage regarding protection against bacteriophages which are common in the open dairy fermentation systems.

6.7 Habitats (Teuber and Geis, 1981)

The lactococci comprise the species *Lactococcus lactis*, *Lc. garviae*, *Lc. plantarum*, *Lc. raffinolactis* and *Lc. piscium* (Table 6.14). *Lactococcus lactis* subsp. *lactis* and *Lc. lactis* subsp. '*diacetylactis*' have commonly been detected directly or following enrichment in plant material, including fresh and frozen corn, corn silks, navy beans, cabbage, lettuce, peas, wheat middlings, grass, clover, potatoes, cucumbers, and cantaloupe (Sandine *et al.*, 1972). Lactococci are usually not found in faecal material or soil. Only small numbers occur on the surface of the cow and in its saliva. Since raw cow's milk consistently contains *Lc. lactis* subsp. *lactis*, and to a much lesser extent *Lc. lactis* subsp. '*diacetylactis*' and *Lc. lactis* subsp. *cremoris*, it is tempting to suggest that lactococci enter the milk from the exterior of the udder during milking and from the feed, which may be the primary source of inoculation. *Lactococcus lactis* subsp. *cremoris* has hitherto not been isolated with certainty from habitats other than milk, fermented milk, cheese, and starter cultures. According to Schleifer *et al.* (1985), the subspecies status 'diacetylactis' is no longer valid. Because of the

Table 6.14 Characteristics differentiating species and subspecies of the genus *Lactococcus**

Species	Source	Acid production from							Hydrolysis of arginine
		Galactose	Lactose	Maltose	Melibiose	Melizitose	Raffinose	Ribose	
Lc. lactis subsp. *lactis*	Raw milk and dairy products	+	+	+	–	–	–	+	+
Lc. lactis subsp. *cremoris*	Raw milk and dairy products	+	+	–	–	–	–	–	–
Lc. lactis subsp. *hordniae*	Leaf hopper	–	–	–	–	–	–	–	+
Lc. garviae	Bovine samples	+	+	v	v	–	–	+	+
Lc. plantarum	Frozen peas	–	–	+	–	+	–	–	–
Lc. raffinolactis	Raw milk	+	+	+	+	v	+	v	v
Lc. piscium	Diseased rainbow trout yearling	+	+		+	+	+	+	–

*+, positive; –, negative; v, variable.

importance of diacetyl-forming lactococci, especially in dairy fermentations, we still distinguish, for practical reasons, between *Lc. lactis* subsp. *lactis* and *Lc. lactis* subsp. '*diacetylactis*'. The most important habitats for lactococci, however, are in the dairy industry, and are shown in Table 6.1.

6.7.1 Other lactococci

Information on the habitats of the species *Lc. garviae*, *Lc. planatarum*, and *Lc. raffinolactis* is scarce. Regarding *Lc. garviae*, only a few strains have been isolated from samples of cows with mastitis. Koch's postulates for *Lc. garviae* as infective agent of mastitis have not been fulfilled. *Lactococcus piscium* is a single strain isolated from a diseased rainbow trout (see Table 6.14).

6.8 Enrichment and isolation

Lactic acid bacteria are nutritionally fastidious. They all require complex media for optimal growth. In synthetic media, all strains of lactococci require amino acids such as isoleucine, valine, leucine, histidine, methionine, arginine, and proline, and vitamins (niacin, Ca-pantothenate, and biotin).

Detailed procedures for isolating lactococci (and other lactic acid bacteria) are beyond the scope of the present book, but the interested reader is referred to Whittenbury (1965), silage, and Olsen *et al.*, (1978), a wide range of dairy materials.

Unfortunately, no satisfactory selective medium is available for the isolation of lactococci. Two media, both commercially available, are generally accepted to give reliable growth of these organisms. The medium proposed by Elliker *et al.* (1956) is widely used for the isolation and enumeration of lactococci. M 17 medium (Terzaghi and Sandine, 1975), a complex medium supplemented with 1.9% β-disodium glycerophosphate, resulted in improved growth of lactococci.

Elliker agar medium for isolation of lactococci (Elliker *et al.*, 1956); medium contains (per litre):

Tryptone	20.0 g
Yeast extract	5.0 g
Gelatin	2.5 g
Glucose	5.0 g
Lactose	5.0 g
Sucrose	5.0 g
Sodium chloride	4.0 g
Sodium acetate	1.5 g
Ascorbic acid	0.5 g
Agar	15.0 g

The medium has a pH of 6.8 before autoclaving

This medium is probably the most cited for the isolation and growth of lactococci, although it is unbuffered. This disadvantage can be overcome by the addition of suitable buffer substances. Addition of 0.4% (w/v) of diammonium phosphate improves the enumeration of lactic streptococci on Elliker agar. Colony counts were up to about eight times greater due to improved buffering capacity (Barach, 1979).

M17 Medium for isolation of lactococci (Terzaghi and Sandine, 1975); medium contains (per litre):

Phytone peptone	5.0 g
Polypeptone	5.0 g
Yeast extract	5.0 g
Beef extract	2.5 g
Lactose	5.0 g
Ascorbic acid	0.5 g
β-Disodium glycerophosphate	19.0 g
1.0 M $MgSO_4.7H_2O$	1.0 ml
Glass-distilled water	1.0 l

The medium is sterilized at 121°C for 12 min. The pH of the broth is 7.1. Solid medium contains 10 g agar/litre of medium.

This medium is useful for the isolation of all strains of *Lc. lactis* subsp. *cremoris*, *Lc. lactis* subsp. *lactis*, *Lc. lactis* subsp. '*diacetylactis*', and *Streptococcus thermophilus* and mutants of those strains lacking the ability to ferment lactose.

Addition of a pH indicator dye (bromcresol purple) and reduction of β-disodium glycerophosphate (to 5 g) allows an easy differentiation between lactose fermenting (large yellow colonies) and non-fermenting (small white colonies) strains.

The suppression of *Lactobacillus delbrueckii* subsp. *bulgaricus* in M17 medium can be demonstrated. The majority of these *Lactobacillus* strains failed to grow in this medium adjusted to pH 6.8. Since M17 medium supported good growth of *Streptococcus thermophilus*, it can be used for the selective isolation of this microorganism from yogurt.

In recent years, this medium has become the standard for genetic investigations of lactococci.

Lactococcal bacteriophages can be efficiently demonstrated and distinguished on M17 agar. Plaques larger than 6 mm in diameter could be observed as well as turbid plaques, indicating lysogeny (Terzaghi and Sandine, 1975).

6.8.1 *Enumeration of citrate-fermenting bacteria in lactic starter cultures and dairy products*

To control gas and aroma (diacetyl) production in the fermentation of various dairy products it is important to know the quantitative composition of the used starter cultures. *Leuconostoc* species and *Lactococcus lactis* subsp. '*diacetylactis*' are components of many mesophilic starter cultures (Table 6.1). These organisms are able to ferment citrate with concominant production of CO_2 and diacetyl.

For the collective enumeration of leuconostocs and *Lc. lactis* subsp. '*diacetylactis*' in starter and fermented dairy products, a whey agar containing calcium lactate and casamino acids (WACCA) has been introduced by Galesloot *et al.* (1961).

A different medium developed by Nickels and Leesment (1964) for the same purpose gives comparable results. A modified medium based on the different action of the lactose analogue 5-bromo-4-chlor-3-indolyl-β-D-galactopyranoside (Xgal) has been recently suggested (Vogensen *et al.*, 1987).

6.8.2 *Complex synthetic media for genetic and biochemical studies*

Two synthetic media which support logarithmic (exponential) growth of *Lactococcus lactis* have been developed and evaluated (Jensen and Hammer, 1993; see Table 6.15).

6.9 Identification and differentiation

Lactococci are Gram-positive, microaerophilic cocci which lack the cytochromes of the respiratory chain. They can be simply differentiated from pediococci and leuconostrocs by the main fermentation products from glucose (see Table 6.16).

The common morphology consists of spherical or ovoid cells, 0.5–1 μm in diameter, in pairs or more-or-less long chains (Figures 6.1–6.3). The lactic streptococci are in practice differentiated by their growth behaviour at different temperatures (Table 6.17) into mesophilic species *Lactococcus lactis* (Schleifer *et al.*, 1985) and the thermophilic species *Streptococcus thermophilus*. *Streptococcus thermophilus* is differentiated at first glance from the enterococci by its inability to grow in the presence of 6.5% NaCl. Problems may arise in the identification of *Lc. lactis* subsp. '*diacetylactis*' if the plasmid-coded fermentation of citrate is lost. The GC content of DNA ranges from 34 to 43 mol% (Schleifer *et al.*, 1985). The genome sizes of different lactococci were estimated to be 2300–2600 kb (Le Bourgeois *et al.*, 1993). Since the mesophilic dairy species differ only in a few properties

Table 6.15 Defined media which support exponential growth of *Lactococcus lactis**

Constituent	Conc. (mM) or presence in medium	
	SA	BL
L-Alanine	3.4	
L-Arginine	1.1	
L-Asparagine	0.8	0.8
L-Cysteine	0.8	
L-Glutamate	2.1	21
L-Glutamine	0.7	0.7
Glycine	2.7	
L-Histidine	0.3	0.3
L-Isoleucine	0.8	0.8
L-Leucine	0.8	1.5
L-Lysine-HCl	1.4	
L-Methionine	0.7	0.5
L-Phenylalanine	1.2	
L-Proline	2.6	
L-Serine	2.9	
L-Threonine	1.7	
L-Tryptophan	0.5	
L-Tryrosine	0.3	
L-Valine	0.9	2.6
NH_4Cl	9.5	9.5
K_2SO_4	0.28	0.28
KH_2PO_4	1.3	1.3
Na-acetate	15	15
Glucose	50	50
MOPS	40	190
Tricine	4	4
$CaCl_2$	0.0005	0.0005
$MgCl_2$	0.52	0.52
$FeSO_4$	0.01	0.01
NaCl	50	50
Vitamins†	+	+
Micronutrients‡	+	+

*Data taken from Jensen and Hammer (1993).
†Vitamins: 0.4 μM biotin, 10 μM pyridoxal-HCl, 2.3 μM folic acid, 2.6 μM riboflavin, 8 μM niacinamide, 3 μM thiamine-HCl, and 2 μM pantothenate.
‡Micronutrients: 0.003 μM $(NH_4)_6(MO_7)_{24}$, 0.4 μM H_3BO_3, 0.03 μM $CoCl_2$, 0.01 μM $CuSO_4$, 0.08 μM $MnCl_2$, and 0.01 μM $ZnSO_4$.

Table 6.16 Differentiation scheme for lactococci, pediococci, and leuconostocs*

Fermentation products of glucose	Genus
L(+)-Lactic acid	*Lactococcus*
D(−)-Lactic acid, CO_2, acetic acid, ethanol	*Leuconostoc*
DL-Lactic acid	*Pediococcus*

*Data taken from Teuber *et al.* (1991).

Table 6.17 Physiological and other properties of dairy lactococci used for identification and differentiation*

Properties	*Lc. lactis* subsp. *lactis*†	*Lc. lactis* subsp. '*diacetylactis*'†	*Lc. lactis* subsp. *cremoris*†
Growth at 10°C	+	+	+
Growth at 40°C	+	+	−
Growth at 45°C	−	−	−
Growth in 4% NaCl	+	+	−
Growth in 6.5% NaCl	−	−	−
Growth at pH 9.2	+	+	−
Growth with methylene blue (0.1% milk)	+	+	−
Growth in presence of bile (40%)	+	+	+
NH_3 from arginine	+	+	−
CO_2 from citrate	−	+	−
Diacetyl and acetoin	−	+	−
Fermentation of maltose	+	+	Rarely
Hydrolysis of starch	−	−	−
Heat resistance (30 min at 60°C)	v	v	v
Serological group	N	N	N
GC content of DNA (mol%)	33.8–36.9	33.6–34.8	35.0–36.2

*Data taken from Teuber *et al.* (1991).
†+, positive: −, negative; v, variable.

(Table 6.14), the GC data are meaningless for this differentiation. That the three mesophilic dairy species are closely related is also implied by the observation that many bacteriophages cross the 'subspecies' line and attack strains of all three subspecies (see section 6.6.2). Also, the plasmid patterns investigated so far do not allow a species differentiation (see above), but do allow the reidentification of strains. Another approach is differentiation on the basis of protein patterns after gel electrophoresis of soluble cell extracts (Jarvis and Wolff, 1979). By this method, classification of closely related strains seems possible. The lactococci contain menaquinones with nine isoprene units as the major component, in contrast to dimethylmenaquinones with nine isoprene units and menaquinones with eight isoprene units in enterococci. At the moment, it is not possible to assess the number of different lactococcal strains existing in dairies and starter cultures throughout the world.

The routine differentiation of the dairy lactococci is outlined in Table 6.17.

6.10 Maintainance and preservation

Since lactococcal cultures easily lose their viability in unbuffered acid media, the best routine method for maintainance is a stab culture in buffered M17-agar (see above) which keeps for months at 4°C. Freeze-drying is mentioned in section 6.6.4 on starter cultures. Mother cultures for the dairy industry must be kept in milk since dairy lactococci rapidly degenerate when grown, e.g. in Elliker medium. When cultivated in M17-medium, they may need some time to adapt back to milk as a medium, if they do so at all. The loss of plasmids with technologically important functions is a constant but strain dependent threat (see section 6.4.1). Milk grown cultures (log phase) keep very well when frozen in liquid nitrogen and stored at −70°C. Storage at higher temperatures may induce a decrease of viability in many strains.

6.11 Species of the genus *Lactococcus*

The following descriptions are taken from Schleifer *et al.* (1985) and Williams *et al.* (1990). For the isolation, characterization and description of the non-dairy lactococci, the original reports should be consulted (Garvie, 1978; Garvie *et al.*, 1981; Garvie and Farrow, 1982; Collins *et al.*, 1983). An excellent summary of the traditional description of the 'old' genus *Streptococcus* is given by Deibel and Seeley (1974).

Description of the genus *Lactococcus* (lac.to.co'cus, L.n. *lac*, lactis of milk, Gr.n. *coccus*, a grain or berry, M.L.masc.n. *Lactococcus* milk coccus).

Spheres or ovoid cells occur singly, in pairs or in chains, and are often elongated in the direction of the chain. Gram-positive. Endospores are not formed. Non-motile. Not *β*-haemolytic. Facultatively anaerobic, catalase negative. Growth at 10°C but not at 45°C. Usually grows in 4% (w/v) NaCl with the exception of *Lc. lactis* subsp. *cremoris* which only tolerates 2% (w/v) NaCl. Chemo-organotrophs. Metabolism fermentative. The predominant end product of glucose fermentation is L(+)-lactic acid. Most strains react with group N antisera (Lancefield, 1933).

Some strains possess low levels of menaquinones. The major glycolipid of all strains is Glc (α1–2) Glc(α1–3)acyl$_2$Gro. A constant minor component is Glc(α1–2), acyl-6Glc(α1–3)acyl$_2$Gro. All strains contain phophatidylglycerol and cardiolipin. Lipoteichoic acid structure, and occurrence of aminophospholipids, are species rather than genus-specific. Non-hydroxylated long-chain fatty acids are primarily of the straight-chain saturated and monounsaturated types; some strains produce cyclopropane-ring acids. The major fatty acids are hexadecanoic and *cis*-11,12-octadecenoic acids; *cis*-11,12-methylenoctadecanoic acid is also present in major amounts in most strains with the exception of *Lc. lactis* subsp. *hordniae* and *Lc. raffinolactis*. The G+C content of the DNA ranges from 34–43 mol%. Type species: *Lactococcus lactis*.

6.11.1 Lactococcus lactis *subsp.* lactis *(Schleifer* et al., *1985) Lactococcus* (lac.to.coc'cus, L.n. *lac*, lactis of milk., Gr.n. *coccus*, a grain or berry, M.L.masc.n. *Lactococcus* milk coccus).

Colonies on blood agar or nutrient agar are circular, smooth and entire. Non-pigmented. Non-haemolytic (some strains may produce a weak α-reaction). Ovoid cells elongated in direction of the chain. Mostly in pairs or short chains. Gram-positive, non-motile. Facultatively anaerobic. Catalase-negative. Growth at 10°C but not at 45°C. Grows in 4% (w/v) NaCl and 0.1% methylene blue milk. Chemo-organotroph: metabolism fermentative. All strains produced acid from galactose, glucose, fructose, lactose, maltose, mannose, N-actylglycosamine, ribose and trehalose. Most strains produced acid from arbutin, cellobiose, β-gentiobiose and salicin. Acid not produced from D-arabinose, L-arabinose, arabitol, adonitol, 2-keto-gluconate, 5-keto-gluconate, dulcitol, erythritol, D-fucose, L-fucose, glycerol, glycogen, inositol, gluconate, melibiose, melizitose, α-methyl-D-glucoside, α-methyl-D-mannoside, raffinose, rhamnose, L-sorbose, sorbitol, D-tagatose, D-turanose, xylitol, and L-xylose. Variable results may be obtained form amygdalin, inulin, mannitol, sucrose, starch and D-xylose. Aesculin and hippurate hydrolysed. Arginine dehydrolase and leucine arylamidase positive. Alkaline phosphatase, α-galactosidase, β-galactosidase and β-glucuronidase negative. A few strains are pyrrolidonyl-arylamidase positive. Some strains can utilize citrate (in conjunction with a fermentable carbohydrate) with the production of CO_2, acetoin and diacetyl. Reacts with Lancefield serological Group N antiserum. The peptidoglycan type is Lys-D-Asp. Low levels of menaquinones produced, with MK-9 predominating. Contains poly(glycerophosphate)-lipoteichoic acid partially substituted with α-galactosyl residues and D-alanine ester. The major glycolipid is Glc(α1–2)Glc(α1–3)-acyl$_2$Gro. Aminophospholipids are not present. Major non-hydroxylated long-chain fatty acids are hexadecanoic, *cis*-11,12-octadecenoic and *cis*-11,12-methylenoctadecanoic acids. The G+C content of DNA ranges from 34.4 to 36.3 mol% as determined by melting temperature. The type strain is NCDO 604 (DSM 20481, ATCC 19435). Source: milk.

6.11.2 Lactococcus lactis *subsp.* cremoris *(Schleifer* et al., *1985)* cre mo'ris. L.n. *cremor* juice or cream; L.gen.n. *cremoris* of cream)

In most respects the description of *Lc. lactis* subsp. *cremoris* corresponds to the description of *Lc. lactis* subsp. *lactis*. It differs in the following characteristics. Grows in 2% but not in 4% (w/v) NaCl. No growth at 40°C. Acid not produced from maltose and ribose; most strains do not produce acid from β-gentiobiose, salicin and trehalose. Arginine dehydrolase negative. The G+C content of DNA ranges from 34.8 to 35.6 mol% as determined by melting temperature. The type strain is NCDO 607 (DSM 20069, ATCC 19257). Source: raw milk and milk products.

6.11.3 Lactococcus lactis *subsp.* hordniae *(Schleifer* et al., *1985)* (hord'ni.ae M.L.fem.n. *Hordnia* generic name; M.L.gen.n. *hordniae* of *Hordnia circellata*, the name of the leafhopper from which the organism was isolated)

In most respects the description of *Lc. lactis* subsp. *hordniae* corresponds to the description of *Lc. lactis* subsp. *lactis*. It differs in the following characteristics. Grows in 2% but not 4% (w/v) NaCl. No growth at 40°C. Acid not produced from galactose, lactose, maltose or ribose. Hippurate is not hydrolysed. *cis*-11,12-Methylenoctadecanoic acid is absent. The G+C content of DNA of NCDO 2181 is 35.2. The type strain is ATCC 29071 (DSM 20450, NCDO 2181). Source: Leafhopper.

6.11.4 Lactococcus raffinolactis *(Schleifer* et al., *1985)* (raf.fi.no. lac'tis. M.L.adj. *raffinosus* of raffinose; L.gen.n. *lactis* of milk; L.gen.n. *raffinolactis*)

Colonies on blood agar are circular, smooth and entire. Non-pigmented. Not *β*-haemolytic. Spheres or ovoid cells elongated in the direction of the chain; mostly in pairs or short chains. Gram-positive. Non-motile. Facultatively anaerobic. Catalase negative. Growth at 10°C but not at 40°C. Does not grow in 4% (w/v) NaCl or 0.1% methylene blue milk. Chemoorganotroph: metabolism fermentative. Acid is usually produced from arbutin, D-fructose, galactose, lactose, maltose, D-mannose, melezitose, melibiose, *N*-acetylglucosamine, raffinose, salicin, starch, sucrose and trehalose. Most strains produce acid from D-xylose. Acid not produced from adonitol, arabitol, D-arabinose, dulcitol, erythritol, D-fucose, L-fucose, gluconate, glycerol, glycogen, inositol, 2-keto-gluconate, D-lyxose, α-methyl-D-mannoside, *β*-methylxyloside, rhamnose, sorbitol, D-tagatose, L-xylose and xylitol. Variable results may be obtained with amygdalin, L-arabinose, dextrin, *β*-gentiobiose, inulin, mannitol, melizitose, α-methyl-D-glucoside, ribose, L-sorbose and turanose. Aesculin hydrolysed. Casein, hippurate and gelatin not hydrolysed. α-Galactosidase and leucine arylamidase positive. Alkaline phosphatase, *β*-galactosidase, *β*-glucuronidase and pyrrolidonylarylamindase negative. Most strains are arginine dehydolase negative. Reacts with group N antisera.

The petpidoglycan type of strain DSM 20443 is Lys-Thr-Ala. Menaquinones are absent. Contains poly(glycerophosphate)-lipoteichoic acid substituted solely with D-alanine ester. The major glycolipid is Glc/(α1–2)Glc-/(α1–3)acyl$_2$Gro. Contains D-alanyl- and L-lysylphosphatidylglycerol.

Major non-hydroxylated long-chain fatty acids are hexadecanoic and *cis*-11,12-octadecenoic acids. *cis*-11,12-Methylenoctadecanoic acid is absent. The G+C content of DNA ranges from 40 to 43 mol% as determined by melting temperature (T_m). The type strain is NCDO 617 (DSM 20443). Source: raw milk.

6.11.5 Lactococcus plantarum *(Schleifer* et al., *1985)* (plan.ta'rum. M.L.n. *planta*; M.L.gen.pl.n. *plantarum* of plants)

Colonies on blood agar or nutrient agar are circular, smooth and entire. Non-pigmented. Not β-haemolytic. Spheres or ovoid cells elongated in the direction of the chain; mostly in pairs or short chains. Gram-positive, non-motile. Facultatively anaerobic. Catalase-negative. Growth at 10°C but not at 45°C. Grows in 4% (w/v) NaCl. Does not grow in 0.1% methylene blue milk. Chemo-organotroph: metabolism fermentative. Acid produced from amygdalin, arbutin, cellobiose, dextrin, fructose, glucose, maltose, D-mannose, mannitol, melizitose, *N*-acetyl-glucosamine, salicin, sorbitol, sucrose and trehalose. Acid not produced from adonitol, D-arabinose, L-arabinose, arabitol, 2-keto-gluconate, 5-keto-gluconate, dulcitol, erythritol, D-fucose, L-fucose, galactose, gluconate, glycogen, glycerol, inositol, inulin, lactose, D-xylose, melibiose, α-methyl-D-glucoside, α-methyl-D-mannoside, β-methylxyloside, raffinose, ribose, rhamnose, L-sorbose, D-tagatose, D-xylose, L-xylose and xylitol. Variable results may be obtained from β-gentiobiose and turanose. Aesculin hydrolysed. Starch, hippurate and gelatin not hydrolysed. Leucine arylamidase positive. β-Galactosidase, β-glucuronidase, alkaline phosphatase, arginine dehydrolase and pyrrolidonylarylamidase negative. Some strains are α-galactosidase positive. Reacts with group N antisera.

The peptidoglycan type of NCDO 1869 is Lys-Ser-Ala. Menaquinones are absent. Contains poly(glycerophosphate)-lipoteichoic acid substituted solely with D-alanine ester. The major glycolipid is Glc(α1–2)Glc-(α1–3)acyl$_2$Gro. Contains D-alanyl- and L-lysylphosphatidylglycerol. Major non-hydroxylated long-chain fatty acids are hexadecanoic, *cis*-11, 12-octadecenoic acids and *cis*-11,12-methyleneoctadecanoic acid. The G+C content of DNA ranges from 36.9 to 38.1 mol% as determined by melting temperature (T_m). The type strain is NCDO 1869. Source: frozen peas.

6.11.6 Lactococcus garvieae *(Schleifer* et al., *1985)* (M.L.gen.n. gar'vie.ae named for E.I. Garvie, a British microbiologist)

Colonies on blood agar or nutrient agar are circular, smooth and entire. Non-pigmented. Not β-haemolytic. Ovoid cells elongated in the direction of the chain; mostly in pairs or short chains. Gram-positive. Non-motile. Facultatively anaerobic. Catalase-negative. Grows at 10°C and 40°C but not at 45°C. Grows in 4% (w/v) NaCl. Grows in and reduces 0.1% methylene blue milk. Chemo-organotroph: metabolism fermentative. Acid produced from galactose, glucose, fructose, cellobiose, amygdalin, arbutin, mannose, ribose, trehalose, salicin, *β*-gentiobiose and *N*-acetylglucosamine. Acid not produced from D-arabinose, L-arabinose, D-arabitol, L-arabitol, adonitol, 2-keto-gluconate, 5-keto-gluconate, dulcitol, erythritol, D-fucose, L-fucose, glycogen, inulin, inositol, melibiose, melezitose, *β*-methylxyloside, α-methyl-D-glucoside, α-methyl-D-mannoside, D-lyxose, raffinose, rhamnose, L-sorbose, sorbitol, D-turanose, xylitol, D-xylose and L-xylose. Variable results may be obtained from maltose, mannitol, sucrose and D-tagatose. Aesculin hydrolysed. Hippurate hydrolysis variable. Starch hydrolysis negative. Arginine dehydrolase, leucine arylamidase and pyrrolidonylarylamidase positive. Some strains are *β*-glucuronidase positive. α-Galactosidase, *β*-galactosidase and alkaline phosphatase negative. Some strains react with Lancefield group N antisera.

The peptidoglycan type is Lys-Ala-Gly-Ala. Low levels of menaquinones produced with MK-9 predominating. Contains poly[Gal(α1–6)Gal(α1–3)Gro-1-P(2←1αGal)] lipoteichoic acid (Koch and Fischer, 1978). The major glycolipid is Glc(α1–2)Glc(α1–3)acyl$_2$Gro. Aminophospholipids are not detected. Major non-hydroxylated long-chain fatty acids are hexadecanoic, *cis*-11,12-octadecenoic and *cis*-11,12-methylene-octadecanoic acids. The G+C content of DNA ranges from, 38.3 to 38.7 mol% as determined by melting temperature (T_m). The type strain is NCDO 2155. Source: bovine mastitis.

6.11.7 Lactococcus pis'cium *(Williams* et al., *1990)* (pis'cium. M.L.n. *piscis*; M.L.gen.pl.n. *piscium* of fishes)

Cells are short rods to ovoid in shape, mostly in pairs or short chains. Gram-positive and non-motile. Facultatively anaerobic. Catalase-negative. Grows at 5°C and 30°C; no growth at 40°C. Acid is produced from amygdalin, L-arabinose, arbutin. *N*-acetylglucosamine, cellobiose, D-fructose, galactose, *β*-gentiobiose, gluconate, glucose, lactose, maltose, D-mannose, mannitol, melibiose, melezitose, D-raffinose, ribose, salicin, sucrose, trehalose, D-turanose and D-xylose. Acid is not produced from adonitol, D-arabinose, D-arabitol, L-arabitol, dulcitol, erythritol, D-fucose, L-fucose, glycogen, glycerol, inositol, inulin, 2-ketogluconate, 5-ketogluconate, D-lyxose, α-methyl-xyloside, rhamnose, L-sorbose, sorbitol, D-tagatose, xylitol, and L-xylose. Aesculin is hydrolysed. Starch hydrolysis is slow and weak. Arginine hydrolysis and urease negative. H_2S is not produced. The long-chain cellular fatty acids are of the straight chain saturated, monounsaturated and cyclopropanoic types. The major acids correspond to hexadecanoic acids, Δ11-octadecenoic acid and Δ11-methylene-octadecanoic acid. The G+C content of DNA is 38.5 mol%, as determined by melting temperature. Isolated from diseased rainbow trout yearling. The type strain is NCFB 2778 (=HR1A-68). Source: diseased rainbow trout yearling.

References

Aguilar, A. (1991) Biotechnology of lactic acid bacteria: an European perspective. *Food Biotechnology*, **5**(3), 323–330.

Aguilar, A. (1992) Community activities in biotechnology. The Bridge "T" project on the biotechnology of lactic acid bacteria. In *Lactic Acid Bacteria, Research and Industrial Applications in the Agro-food Industries*. Adria Normandie, Caen, France, pp. 333–341.

Anderson, A.W. and Elliker, P.R. (1953) Transduction in *Streptococcus lactis*. *Journal of Dairy Research*, **30**, 351–357

Andresen, A., Geis, A., Krusch, U. and Teuber, M. (1984) Plasmid profiles of mesophilic dairy starter cultures. *Milchwissenschaft*, **39**, 140–143.

Barach, J.T. (1979) Improved enumeration of lactic acid streptococci on Elliker agar containing phosphate. *Applied Environmental Microbiology*, **38**, 173–174.

Budde-Niekiel, A. and Teuber, M. (1987) Electron microscopy of the adsorption of bacteriophages to lactic acid streptococci. *Milchwissenschaft*, **42**, 551–554.

Budde-Niekiel, A., Möller, V., Lembke, J. and Teuber, M. (1985) Oekologie von Bakteriophagen in einer Frischkäserei. *Milchwissenschaft*, **40**, 477–481.

Chopin, A. (1993) Organization and regulation of genes for amino acid biosynthesis in lactic acid bacteria. *FEMS Microbiology Reviews*, **12**, 21–38.

Collins, M.D., Farrow, J.A.E., Phillips, B.A. and Kandler, O. (1983) *Streptococcus garvieae* sp. nov. and *Streptococcus plantarum* sp. nov. *Journal of General Microbiology*, **129**, 3427–3431.

Collins, M.D., Ash, C., Farrow, J.A.E., Wallbanks, S. and Williams, A.M. (1989) 16S Ribosomal ribonucleic acid sequence analyses of lactococci and related taxa. Description of *Vagococcus fluvialis* gen. nov., sp. nov. *Journal of Applied Bacteriology*, **47**, 453–460.

Condon, S. (1987) Responses of lactic acid bacteria to oxygen. *FEMS Microbiology Reviews*, **46**, 269–280.

Davies, F.L., Underwood, H.M. and Gasson, M.J. (1981) The value of plasmid profiles for strain identification in lactic streptococci and the relationship between *Streptococcus lactis* 712. ML3 and C2. *Journal of Applied Bacteriology*, **51**, 325–337.

Deibel, R.M. and Seeley Jr, H.W. (1974) Genus I. *Streptococcus*. In *Bergey's Manual of Determinative Bacteriology*, 8th edn (eds Buchanan, R.E. and Gibbons, N.E.). Baltimore, MD, USA, pp. 490–509.

de Vos, W.M. and Simons, G.F.M. (1994) Gene cloning and expression systems in *Lactococci*. In *Genetics and Biotechnology of Lactic Acid Bacteria* (eds Gasson, M.J. and de Vos, W.M.). Blackie Academic and Professional, Glasgow, UK, pp. 52–105.

de Vos, W.M., Kuiper, H., Lever, A. and Ventris, J. (1989) Heterogrammic replication of *Lactococcus lactis* plasmid PJH71 is regulated by a repressor–operator control circuit. *American Society for Microbiology, Annual Meeting, New Orleans, Abstract H276.*

Dodd, H.M. and Gasson, M.J. (1994). Bacteriocins of lactic acid bacteria. In *Genetics and Biotechnology of Lactic Acid Bacteria* (eds Gasson, M.J. and de Vos, W.M.). Blackie Academic and Professional, Glasgow, UK, pp. 211–251.

Elliker, P.R., Anderson, A.W. and Hannesson, G. (1956) An agar medium for lactic acid streptococci and lactobacilli. *Journal of Dairy Science*, **39**, 1611–1612.

Galesloot, T.E., Hassing, F. and Stadhouders, J. (1961) Agar media voor het isoleren en tellen van aromabacterien in zuursels. *Netherlands Milk and Dairy Journal*, **15**, 127–150.

Garvie, E.I. (1978) *Streptococcus raffinolactis* (Orla-Jensen and Hansen), a group N streptococcus found in raw milk. *International Journal of Systematic Bacteriology*, **28**, 190–193.

Garvie, E.I. and Farrow, J.A.E. (1982) *Streptococcus lactis* subsp. *cremoris* (Orla-Jensen) comb. nov. and *Streptococcus lactis* subsp. *diacetilactis* (Matuszewski *et al.*) nom. rev., comb. nov. *International Journal of Systematic Bacteriology*, **32**, 453–455.

Garvie, E.I., Farrow, J.A.E. and Phillips, B.A. (1981) A taxonomic study of some strains of streptococci which grow at 10°C but not at 45°C including *Streptococcus lactis* and *Streptococcus cremoris*. *Zentralblatt für Bakteriologie, Parasitenkunde, Infektionskrankheiten und Hygiene (1. Abteilung Originale)*, **C2**, 151–165.

Gasson, M.J. and Davies, F.L. (1980) High-frequency conjugation associated with *Streptococcus lactis* donor cell aggregation. *Journal of Bacteriology*, **143**, 1260–1264.

Gasson, M.J. and de Vos, W.M. (1994) *Genetics and Biotechnology of Lactic Acid Bacteria*. Blackie Academic and Professional, Glasgow, UK.

Gasson, M.J. and Fitzgerald, G.F. (1994) Gene transfer systems and transposition. In *Genetics and Biotechnology of Lactic Acid Bacteria*. (eds Gasson, M.J. and de Vos, W.M.). Blackie Academic and Professional, Glasgow, UK, pp. 1–51.

Gasson, M.J., Hill, S.H.A. and Anderson, P.H. (1987) Molecular genetics of metabolic traits in lactic streptococci. In *Streptococcal Genetics* (eds Ferretti, J.J. and Curtiss III, R.E.). American Society for Microbiology, Washington, DC, USA, pp. 242–245.

Geis, A. (1982) Transfection of protoplasts of *Streptococcus lactis* subsp. *diacetylactis*. *FEMS Microbiology Letters*, **15**, 119–122.

Geis, A., Singh, J. and Teuber, M. (1983) Potential of lactic streptococci to produce bacteriocin. *Applied and Environmental Microbiology*, **45**, 205–211.

Geis, A., Janzen, T., Teuber, M. and Wirsching, F. (1992) Mechanism of plasmid-mediated bacteriophage resistance in lactococci. *FEMS Microbiology Letters*, **94**, 7–14.

Harlander, S.K. (1987) Transformation of *Streptococcus lactis* by electroporation. In *Streptococcal Genetics* (eds Ferretti, J.J. and Curtiss III, R.E.). American Society for Microbiology, Washington, DC, USA, pp. 229–233.

Heap, H.A. and Lawrence, R.C. (1976) The selection of starter strains for cheesemaking. *New Zealand Journal of Dairy Science and Technology*, **11**, 16–20.

Hengstenberg, W., Kohlbrecher, D., Witt, E., Kruse, R., Christiansen, I., Peters, D., Pogge von Strandmann, R., Städtler, P., Koch, B. and Kalbitzer, H.-R. (1993) Structure and function of proteins of the phosphotransferase system and of 6-phospho-β-glycosidases in Gram-positive bacteria. *FEMS Microbiology Reviews*, **12**, 149–164.

Hill, C. (1993) Bacteriophage and bacteriophage resistance in lactic acid bacteria. *FEMS Microbiology Letters*, **12**, 87–108.

Hill, C., Daly, C. and Fitzgerald, G.F. (1985) Conjugative transfer of transposon Tn979 to lactic acid bacteria. *FEMS Microbiology Letters*, **30**, 115–119.

Hugenholtz, J. (1993) Citrate metabolism in lactic acid bacteria. *FEMS Microbiology Reviews*, **12**, 165–178.

IDF (1991) *Practical Phage Control* (Monograph Bulletin No. 263). International Dairy Federation, Brussels, Belgium.

Israelson, H. and Hansen, E.B. (1993) Insertion of transposon Tun917 derivatives into the *Lactococcus lactis* subsp. *lactis* chromosome. *Applied and Environmental Microbiology*, **59**, 21–26.

Jarvis, A.W. and Wolff, J.M. (1979) Grouping of lactic streptococci by gel electrophoresis of soluble cell extracts. *Applied and Environmental Microbiology*, **37**, 391–398.

Jarvis, A.W., Fitzgerald, G.F., Mata, M., Mercenier, A., Neve, H., Powell, I.B., Ronda, C., Saxelin, M. and Teuber, M. (1991) Species and type phages of lactococcal bacteriophages. *Intervirology*, **32**, 2–9.

Jensen, P.R. and Hammer, K. (1993) Minimal requirements for exponential growth of *Lactococcus lactis*. *Applied and Environmental Microbiology*, **59**, 4363–4366.

Kaletta, C. and Entian, K.D. (1989) Nisin, a peptide antibiotic: cloning and sequencing of the nisA gene and posttranslational processing of its peptide product. *Journal of Bacteriology*, **171**, 1597–1601.

Kim, S.G. and Batt, C. (1991) Antisense mRNA-mediated bacteriophage resistance in *Lactococcus lactis* subsp. *lactis*. *Applied and Environmental Microbiology*, **57**, 1109–1113.

Klaenhammer, T.R. (1993) Genetics of bacteriocins produced by lactic acid bacteria. *FEMS Microbiology Reviews*, **12**, 39–86.

Klaenhammer, T.R. and Fitzgerald, G.F. (1994) Bacteriophages and bacteriophage resistance. In *Genetics and Biotechnology of Lactic Acid Bacteria* (eds Gasson, M.J. and de Vos, W.M.). Blackie Academic and Professional, Glasgow, UK, pp. 106–168.

Koch, H.U. and Fischer, W. (1978) Acyldiglucosyldiacylglycerol-containing lipoteichoic acid with a poly(3-*O*-galabiosyl-2-*O*-galactose-su-glycero-l-phosphate) chain from *Streptococcus lactis* kiel 42172. *Biochemistry*, **17**, 5275–5281.

Kok, J. and de Vos, W.M. (1994). The proteolytic system of lactic acid bacteria. In *Genetics and Biotechnology of Lactic Acid Bacteria* (eds Grasson, M.J. and de Vos, W.M.). Blackie Academic and Professional, Glasgow, UK, pp. 169–210.

Konings, W.N. and Otto, R. (1983) Energy transduction and solute transport in streptococci. *Antonie van Leeuwenhoek*, **49**, 247–257.

Lancefield, R.C. (1933) A serological differentiation of human and other groups of hemolytic streptococci. *Journal of Experimental Medicine*, **57**, 571–595.

Lawrence, R.C., Thomas, T.D. and Terzaghi, B.E. (1976) Reviews of the progress of dairy science: cheese starters. *Journal of Dairy Research*, **43**, 141–193.

Le Bourgeois, P., Lautier, M. and Ritzenthaler, P. (1993) Chromosome mapping in lactic acid bacteria. *FEMS Microbiology Reviews*, **12**, 109–124.

Lembke, J. and Teuber, M. (1981) Inaktivierung von Bakteriophagen durch Desinfektionsmittel. *Deutsche Molkereizeitung*, **18**, 580–587.

Lister, J. (1873) A further contribution to the natural history of bacteria and the germ theory of fermentative changes. *Quarterly Microbiological Sciences*, **13**, 380–408.

Löhnis, F. (1909) Die Benennung der Milchsäurebakterien. *Zentralblatt für Bakteriologie, Parasitenkunde, Infektioskrankheiten und Hygiene (2. Abteilung Originale)*, **22**, 553–555.

Loof, M., Lembke, J. and Teuber, M. (1983) Characterization of the genome of the *Streptococcus lactis* subsp. *diacetylactis* bacteriophage P008 wide-spread in German cheese factories. *Systematic and Applied Microbiology*, **4**, 413–423.

Martinussen, J., Andersen, P.S. and Hammer, K. (1994) Nucleotide metabolism in *Lactococcus lactis*: Salvage pathways of exogenous pyrimidines. *Journal of Bacteriology*, **176**, 1514–1516.

McKay, L.L., Baldwin, K.A. and Zottola, E.A. (1972) Loss of lactose metabolism in lactic streptococci. *Applied Microbiology*, **23**, 1090–1096.

McKay, L.L., Cords, B.R. and Baldwin, K.A. (1973) Transduction of lactose metabolism in *Streptococcus lactis* C2. *Journal of Bacteriology*, **15**, 810–815.

McKay, L.L., Baldwin, K.A. and Efstathiou, J.D. (1976) Transductional evidence for plasmid linkage of lactose metabolism in *Streptococcus lactis* C2. *Appiied Environmental Microbiology*, **32**, 45–52.

Moineau, S., Pandian, S., Todd, R. and Klaenhammer, T.R. (1993) Restriction/modification systems and restriction endonucleases are more effective on lactococcal bacteriophages that have emerged recently in the dairy industry. *Applied and Environmental Microbiology*, **59**, 197–202.

Möller, V. and Teuber, M., (1988) Selection and characterization for phage-resistant mesophilic lactococci from mixed strain dairy starter cultures. *Milchwissenschaft*, **43**, 482–4867.

Neve, H. and Teuber, M. (1991) Basic microbiology and molecular biology of bacteriophages of lactic acid bacteria in dairies. *Bulletin of the International Dairy Federation*, **263**, 3–15.

Neve, H., Geis, A. and Teuber, M. (1984) Conjugal transfer and characterization of bacteriocin plasmids in group N (lactic acid) streptococci. *Journal of Bacteriology*, **157**, 833–383.

Neve, H., Geis, A. and Teuber, M. (1987) Conjugation, a common plasmid transfer mechanism in lactic acid streptococci of dairy starter cultures. *Systematic Applied Microbiology*, **9**, 151–157.

Nickels, C. and Leesment, H. (1964) Methode zur Differenzierung und quantitativen Bestimmung von Säureweckerbakterien. *Milchwissenschaft*, **19**, 374–378.

Olson, N.F., Anderson, R.F. and Sellars, R. (1978) Microbiological methods for cheese and other cultured products. In *Standard Methods for the Examination of Dairy Products*,

14th edn (ed Marth, E.H.). American Public Health Association, Washington, DC, USA, pp. 161–164.

Orla-Jensen, S. (1919) *The Lactic Acid Bacteria*. Host and Son, Copenhagen, Denmark.

Pasteur, L. (1857) Mémoire sur la fermentation appelée lactique. *CR Séances' Académiques de Sciences*, **45**, 913–916.

Pechmann, H. and Teuber, M. (1980) Plasmid pattern of group N (lactic) streptococci. *Zentralblatt für Bakteriologie Parasitenkunde, Infektionskrankheiten und Hygiene (1. Abeilung Originale)*, **C1**, 133–136.

Poolman, B. (1993) Energy transduction in lactic acid bacteria. *FEMS Microbiology Reviews*, **12**, 125–148.

Pritchard, G.G. and Coolbear, T. (1993) The physiology and biochemistry of the proteolytic system in lactic acid bacteria. *FEMS Microbiology Reviews*, **12**, 179–206.

Sandine, W.E., Radich, P.C. and Elliker, P.R. (1972) Ecology of the lactic streptococci. A review. *Journal of Milk and Food Technology*, **35**, 176–184.

Scheele, C.W. (1780) Von der Milch und ihrer Säure. In *Sämtliche Physische und Chemische Werke* (ed. Hermbstädt, D.W.F.). Vol. 1, (1793). Heinrich August Rottmann, Berlin, pp. 249–260.

Schleifer, K.H., Kraus, J., Dvorak, G., Kilpper-Bälz, R., Collins, M.D. and Fischer, W. (1985) Transfer of *Streptococcus lactis* and related streptococci to the genus *Lactococcus* gen.nov. *Systematic and Applied Microbiology*, **6**, 183–195.

Smid, E.J., Poolman, B. and Konings, W.N. (1991) Casein utilization by lactococci. *Applied Environmental Microbiology*, **57**, 2447–2452.

Stackebrandt, E. and Teuber, M. (1988). Molecular taxonomy and phylogenetic position of lactic acid bacteria. *Biochimie*, **70**, 317–324.

Stadhouders, J. and Leenders, G.J.M. (1984) Spontaneous developed mixed-strain cheese-starters. Their behaviour towards phages and their use in the Durch cheese industry. *Netherlands Milk Dairy Journal*, **38**, 157–181.

Stanley, G. (1977) The manufacture of starters by batch fermentation and centrifugation to produce concentrates. *Journal of the Society of Dairy Technology*, **30**, 36–39.

Tamine, A.Y. and Robinson, R.K. (1976) Recent developments in the production and preservation of starter cultures for yogurt. *Dairy Industries International*, **41**, 408–411.

Terzaghi, B.E. and Sandine, W.E. (1975) Improved medium for lactic streptococci and their bacteriophages. *Applied Microbiology*, **29**, 807–813.

Teuber, M. (1987) Milchsäurebakterien im Aufwind – Gentechnologie ermöglicht massgeschneiderte Lebensmittel. *Forschungsreport Ernährung, Landwirtschaft, Forsten* **2**, 9–11.

Teuber, M. (1993) Lactic acid bacteria. In *Biotechnology, Vol. 1*, 2nd edn (eds Rehm, H.R., Reed, G.A., Pühler, and Stadler, P.) Verlag Chemie, Weinheim, Germany , pp. 326–365.

Teuber, M. and Geis, A. (1981) The family Streptococcaceae (nonmedial aspects). In *The Prokaryotes*. Vol. 2 (eds. Starr, M.P., Stolp, H., Trüper, H.G., Balows, A. and Schlegel, H.G.) Springer-Verlag, New York, USA, pp. 1614–1630.

Teuber, M., Geis, A. and Neve, H. (1991) The genus *Lactococcus*. In *The Prokaryotes*, Vol. 2, 2nd edn (eds Balows, A., Trüper, H.G., Dworkin, M., Harder, W. and Schleifer, K.H.). Springer-Verlag, New York, USA, pp. 1482–1501.

Thompson, J. (1988) Lactic acid bacteria: model systems for in vivo studies of sugar transport and metabolism in Gram-positive organisms. *Biochimie*, **70**, 325–336.

Tulloch, D.L., Finch, L.R. Hillier, A.J. and Davidson, B.E. (1991) Physical map of the chromosome of *Lactococcus lactis* subsp. *lactis* DL11 and localization of six putative rRNA operons. *Journal of Bacteriology*, **173**, 2768–2775.

Van de Guchte, M., Kok, J. and Venema, G. (1992) Gene expression in *Lactococcus lactis*. *FEMS Microbiology Reviews*, **88**, 73–92.

Vandenbergh, P.A. (1993) Lactic acid bacteria, their metabolic products and interference with microbial growth. *FEMS Microbiology Reviews*, **12**, 221–238.

Vogensen, F.K., Karst, T., Larsen, J.J., Kringelum, B., Ellekjaer, D. and Waagner Nielsen, E. (1987) Improved direct differentiation between *Leuconostoc cremoris*, *Streptococcus lactis* and *Streptococus cremoris/Streptococcus lactis* on agar. *Milchwissenschaft*, **42**, 646–648.

von Milczewski, K.E. (1990) 100 Jahre technische Mikrobiologie in Kiel. In *100 Jahre Mikrobiologie an der Bundesanstalt für Milchforschung in Kiel 1889–1989* (eds Teuber, M.,

Von Milczewski, K.E.) Bundesanstalt für Milchforschung, Kiel, Germany, pp. 27–111.
Ward, L.J.H. and Jarvis, A.W. (1991) Rapid electroporation-mediated plasmid transfer between *Lactococcus lactis* and *Escherichia coli* without the need for plasmid preparation. *Letters in Applied Microbiology*, **13**, 278–280.
Weigmann, H. (1899) Versuch einer Einteilung der Milchsäurebakterien des Molkereigewerbes. *Zentralblatt für Bakteriologie*, *Parasitenkunde*, *Infektionskrankheiten und Hygiene* (2. Abteilung Originale), **5**, 825–831, 859–870.
Weigmann, H. (1905–1908). Das Reinzuchtsystem in der Butterbereitung und in der Käserei. In *Handbuch der Technischen Mykologie, Vol. 2* (ed. Lafar, F.) Gustav-Fischer-Verlag, Jena, Germany, pp. 2293–2309.
Whittenbury, R. (1965) The enrichment and isolation of lactic acid bacteria from plant material. *Zentralblatt für Bakteriologie, Parasitenkunde, Infektionskrankheiten und Hygiene, (1. Abteilung Originale)*, Suppl. **1**, 395–398.
Williams, A.M., Fryer, J.L. and Collins, M.D. (1990) *Lactococcus piscium* sp. nov. a new *Lactococcus* species from salmonid fish. *FEMS Microbiology Letters*, **68**, 109–114.
Zitzelsberger, W., Götz, F. and Schleifer, K.H. (1984) Distribution of superoxide dismutases, oxidases, and NADH peroxide in various streptococci. *FEMS Microbiology Letters*, **21**, 243–246.

7 The genus *Leuconostoc*

F. DELLAGLIO, L.M.T. DICKS and S. TORRIANI

7.1 Introduction

The genus *Leuconostoc* is phenotypically related to *Lactobacillus* and *Pediococcus* (Stackebrandt *et al.*, 1983; Stackebrandt and Teuber, 1988) and share many features with the heterofermentative lactobacilli. In a recent comparative study of the 16S rRNA sequences (Yang and Woese, 1989), it was shown that the leuconostocs form a natural phylogenetic group with *Lb. confusus*, *Lb. halotolerans*, *Lb. kandleri*, *Lb. minor*, and *Lb. viridescens*.

The leuconostocs have complex nutritional requirements (Garvie, 1967b, 1986) and are found in plants, dairy products, meat and various fermented food products (Holzapfel and Schillinger, 1992). In silage fermentation *Leuc. mesenteroides* subsp. *mesenteroides*, *Leuc. mesenteroides* subsp. *dextranicum*, and *Leuc. paramesenteroides* are well represented (Dellaglio *et al.*, 1984; Woolford, 1984; Dellaglio and Torriani, 1986; Daeschel *et al.*, 1987). *Leuconostoc mesenteroides* subsp. *mesenteroides* produces an excess of exopolysaccharides and is the major contaminant in sugar milling plants (Sharpe *et al.*, 1972; Tilbury, 1975; Pivnick, 1980). *Leuconostoc mesenteroides* subsp. *mesenteroides* and *Leuc. lactis* are the dominant leuconostocs in milk and fermented milk products (Daly, 1983; Keller *et al.*, 1987; Marshall, 1987). *Leuconostoc mesenteroides* subsp. *cremoris* and *Leuc. paramesenteroides* are less well represented in milk (El-Gendy *et al.*, 1983), probably due to their slow growth under psychrotrophic conditions. Recently, a new species (*Leuc. argentinum*) has been isolated from Argentinian raw milk (Dicks *et al.*, 1993). *Leuconostoc mesenteroides* subsp. *mesenteroides*, *Leuc. amelibiosum*, *Leuc. carnosum*, *Leuc. gelidum* and *Leuc. citreum* have been isolated from various meat products (Savell *et al.*, 1981; Holzapfel and Gerber, 1986; Korkeala *et al.*, 1988; Farrow *et al.*, 1989; Schillinger *et al.*, 1989; Shaw and Harding, 1989). *Leuconostoc oenos* plays a key role in the malolactic fermentation of wines (reviewed by Van Vuuren and Dicks, 1993).

In spite of the wide distribution and practical importance of leuconostocs, these bacteria have not been studied to the same extent as the genus *Lactobacillus* and the other lactic acid bacteria. Scientists only recently became more aware of the practical importance of leuconostocs, especially the role they play in changing the organoleptic quality and texture of fermented food products such as milk, butter, cheese and meat. Some leuconostocs might also play a role in the fermentation of dough.

In this chapter, the growth characteristics of *Leuconostoc* spp., taxonomy, genetics, plasmids, bacteriocins and bacteriophages are discussed. The practical importance of the genus *Leuconostoc* is summarized.

7.2 Growth characteristics

Leuconostoc spp. are facultative anaerobes (Garvie, 1986). All species require a rich medium with complex growth factors and amino acids (Reiter and Oram, 1962; Garvie, 1967b, 1986). The optimum growth of the non-acidophilic species is between pH 6 and 7, depending on the medium used. Growth of *Leuc. mesenteroides* stops when the internal pH reaches a value of 5.4–5.7 (McDonald *et al.*, 1990). *Leuconostoc oenos*, on the other hand, is acidophilic and grows best in acid medium with a pH of 4.2–4.8 (Garvie, 1986). The enrichment and isolation media are listed by Holzapfel and Schillinger (1992).

Growth is stimulated by the addition of 0.05% cysteine HCl. Growth in broth cultures is uniform, except when cells in long chains sediment. In stab cultures growth is more concentrated in the lower two thirds. Colonies are smooth, round, greyish white and seldom larger than 1 mm in diameter (Garvie, 1986). Growth on surface plates is poor (Garvie, 1986), but is stimulated when incubated under reducing conditions, namely in the presence of a gas mixture of 19.8% CO_2, 11.4% H_2, and nitrogen as balance (unpublished). Optimum growth is between 20 and 30°C. The minimum growth temperature for most species is 5°C (Garvie, 1986). However, growth at 1°C has been recorded for strains of *Leuc. gelidum* and *Leuc. carnosum* isolated from meat (Shaw and Harding, 1989). Strains of *Leuc. mesenteroides* subsp. *mesenteroides* have a relatively short generation time and good growth is obtained within 24 h of incubation at 30°C. *Leuconostoc mesenteroides* subsp. *cremoris* may require a 48 h incubation time, preferably at 22–30°C. Growth of *Leuc. oenos*, compared to that of non-acidophilic *Leuconostoc* spp., is slow and broth cultures can take 5–7 days at 22°C. However, most strains of *Leuc. oenos* we studied reached mid-logarithmic growth after 48 h in acidic grape broth at 30°C (Dicks *et al.*, 1990). The optimum growth temperature for *Leuc. oenos* is between 20 and 30°C (Garvie, 1986).

7.2.1 Morphology

Cells are Gram-positive, asporogenous and non-motile. Cell morphology varies with growth conditions (Garvie, 1986). Cells grown in a glucose medium are elongated and appear morphologically closer to lactobacilli than to streptococci. Most strains form coccoid cells when cultured in milk. Cells may occur singularly or in pairs, and form short to medium length chains. When grown on solid medium, cells are elongated and can be mistaken for rods. True cellular capsules are not formed (Garvie, 1986). Certain strains of *Leuc. mesenteroides* produce extracellular dextran which forms an electron-dense coat on the cell surface (Brooker, 1977).

7.2.2 Metabolism

Growth is dependent on the fermentation of a fermentable carbohydrate (Garvie, 1986). *Leuconostoc* spp. are heterofermentative. Under microaerophilic conditions, glucose is converted to equimolar amounts of D-lactate, ethanol and CO_2 via a combination of the hexose monophosphate (6-phosphate gluconate) and pentose phosphate pathways (DeMoss *et al.*, 1951; Garvie, 1986; Cogan, 1987; Schmitt *et al.*, 1992). Glucose 6-phosphate dehydrogenase and xylulose-5-phosphoketolase are the key enzymes present in all *Leuconostoc* spp. (Garvie, 1986). Reduced NAD (NADH) is regenerated to NAD by lactic dehydrogenase (LDH), acetaldehyde dehydrogenase, and alcohol dehydrogenase (Condon, 1987). However, in the presence of oxygen, strains of *Leuc. mesenteroides* use NADH oxidases and NADH peroxidases as alternative mechanisms to regenerate NAD (Condon, 1987). Acetate, instead of ethanol, and double the amount of ATP are produced (Johnson and McCleskey, 1957; Keenan, 1968; Ito *et al.*, 1983; Condon, 1987). This resulted in an almost doubling of the molar growth yield of *Leuc. mesenteroides* (Lucey and Condon, 1986). Similar results were recorded by Fitzgerald (1983), Johnson and McCleskey (1957), and Whittenbury (1963, 1966). The tendency to diverge from the ethanol branch has also been observed in the fermentation of fructose. In certain lactic acid bacteria fructose serves as hydrogen acceptor and mannitol is formed instead of ethanol (Kandler, 1983). It may be concluded from these results that the ethanol branch of the heterolactic pathway is a secondary route that wastes high-energy phosphate (Lucey and Condon, 1986).

The fermentation end products of *Leuc. oenos* are the same as the other *Leuconostoc* spp., suggesting that it follows the heterolactic fermentation pathway. However, NADP instead of NAD is used as coenzyme (Garvie, 1975). A faster growth rate was obtained for *Leuc. oenos* when oxygen was replaced by nitrogen (Kelly *et al.*, 1989). It might well be that a diverged heterolactic pathway, similar to that described for *Leuc. mesenteroides*, also exists in *Leuc. oenos*. The metabolic pathway for glucose fermentation by *Leuc. oenos* has not been fully confirmed. All evidence thus far obtained indicates that *Leuconostoc* spp. obtain their metabolic energy from substrate-level phosphorylation (Garvie, 1986).

Recent research indicates that intact cells of *Leuc. oenos* generate more ATP when grown in the presence of L-malate (Cox and Henick-Kling, 1989, 1990). This suggests that the malolactic enzyme system evolved as an energy-producing mechanism in *Leuc. oenos*, especially in habitats of low pH, such as wine. According to Britz and Tracey (1990), low concentrations of sulphur dioxide and ethanol stimulate *Leuc. oenos* to conduct the malolactic fermentation (MLF).

Leuconostoc spp. have complex nutritional requirements (Reiter and Oram, 1962; Garvie, 1967a, b, 1986; Weiller and Radler, 1972; Tracey and Britz, 1989; Holzapfel and Schillinger, 1992). Nicotinic acid, thiamine, biotin, and pantothenic acid (or its derivatives) are generally required for growth (Garvie, 1986). However, Kole *et al.* (1983) reported that *Leuc. oenos* strain 44.40, one of the malolactic starter cultures used in the wine industry, does not require nicotinic acid. Most strains of *Leuc. oenos* require a gluco-derivative of pantothenic acid (Amachi *et al.*, 1970), generally referred to as the tomato juice factor (Garvie and Mabbitt, 1967). Leuconostocs do not require cobalamin, or *p*-aminobenzoic acid (Garvie, 1986). *Leuconostoc lactis* does not require folic acid, whilst *Leuc. mesenteroides* subsp. *mesenteroides* requires only glutamic acid and valine for growth (Garvie, 1967b). The amino acid requirement of strains within a species differs. *Leuconostoc oenos* and *Leuc. paramesenteroides* are the most fastidious regarding their need for amino acids (Garvie, 1967b).

7.2.3 Carbohydrates

Chapter 1 carries schematics for the major carbohydrate fermentation routes active in lactic acid bacteria.

The carbohydrates fermented by *Leuconostoc* spp. are listed in Table 7.1. Glucose is fermented by all *Leuconostoc* spp., but fructose is preferred (Garvie, 1986). *Leuconostoc* spp. do not contain fructose 1,6-diphosphate aldolase. However, an active glucose 6-phosphate dehydrogenase is present (Garvie, 1986). Glucose is, therefore, fermented via a combination of the hexose monophosphate and pentose phosphate pathways (DeMoss *et al.*, 1951; Garvie, 1986; Cogan, 1987; Schmitt *et al.*, 1992). Glucose is phosphorylated and then oxidized to 6-phosphogluconate, followed by decarboxylation. The resulting pentose is converted to lactic acid and ethanol or acetate (Kandler, 1983). The acetate/ethanol ratio depends on the oxidation–reduction potential of the system (Kandler, 1983). The complete pathway of glucose fermentation in *Leuc. oenos* has not been confirmed (Garvie, 1986). Ito *et al.* (1983) and Lucey and Condon (1986) suggested that glucose metabolism is regulated by systems which are not yet fully understood.

Lactose is fermented by all strains of *Leuc. mesenteroides* subsp. *dextranicum*, *Leuc. lactis* (Garvie, 1986), and *Leuc. argentinum* (Dicks *et al.*, 1993). Variable results were recorded for *Leuc. mesenteroides* subspp. *mesenteroides* and *cremoris*, *Leuc. paramesenteroides*, and *Leuc. pseudomesenteroides* (Garvie, 1986). *Leuconostoc oenos*, *Leuc. gelidum*, *Leuc. carnosum*, *Leuc. citreum*, and *Leuc. amelibiosum* do not ferment lactose (Garvie, 1986). The metabolism of lactose by lactic acid bacteria has been studied in great detail, especially for some of the *Lactococcus* strains used as dairy starter cultures. However, little is known about the fermentation of lactose by *Leuconostoc* spp. Most lactobacilli transport lactose and galactose into the cell via specific permeases (Lawrence and Thomas, 1979). It is not yet known if *Leuconostoc* spp. contain such a transport system. Furthermore, it is not known if the Leloir pathway exists in *Leuconostoc* spp.

Most streptococci and a few lactobacilli transport lactose and galactose via the phosphoenolpyruvate (PEP)-dependent phosphotransferase (PTS) system (Thompson, 1979). Lactose phosphate is hydrolysed by phospho-β-galactosidase to glucose and galactose 6-phosphate (Thompson, 1980). Glucose and galactose 6-phosphate are then further fermented to pyruvate via glycolysis and the tagatose 6-phosphate pathways, respectively (Kandler, 1983). The genes coding for lactose uptake by the PTS system and the enzymes in the tagatose 6-phosphate pathway of *Lactococcus lactis* subspp. *lactis* and *cremoris* are plasmid-linked, whilst the genes coding for permease and β-galactosidase are cryptic in *Lactococcus* spp. (Crow *et al.*, 1983).

Fructose is fermented by all *Leuconostoc* spp., except *Leuc. mesenteroides* subsp. *cremoris* and some strains of *Leuc. argentinum* (Table 7.1). The reduction of fructose to mannitol results in less ethanol being produced (Kandler, 1983). The excess formation of mannitol in wine by *Leuc. oenos* leads to mannit spoilage (Amerine and Kunkee, 1968) which is usually accompanied by low levels of glycerol (Eltz and Vandemark, 1960).

All heterofermentative lactic acid bacteria have phosphoketolase and are thus theoretically able to ferment pentoses to lactate and acetate (Kandler, 1983). The obligate homofermentative lactic acid bacteria of *Lactobacillus* Group I do not have phosphoketolase and are thus not able to ferment pentoses (Kandler and Weiss, 1986), although the thermophilic *Lactobacillus* sp. isolated by Barre (1978) fermented L-arabinose and D-ribose without phosphoketolase. The facultatively homofermentative lactococci and pediococci, on the other hand, have a pentose-inducible phosphoketolase (Kandler, 1983). Pentoses are taken up by specific permeases and converted to D-xylose-5-phosphate by appropriate enzymes (see Figure 1.3).

Since phosphoketolase is one of the key enzymes in the heterolactic fermentation pathway (Kandler, 1983), it is assumed that pentoses are as a rule fermented by all *Leuconostoc* spp. However, the fermentation profiles of arabinose, ribose and xylose were different for each of the *Leuconostoc* spp. (Table 7.1). Certain strains of *Leuc. oenos*, for instance, do not

Table 7.1 Diagnostic characteristics of the subspecies and species belonging to the genus *Leuconostoc**

Characteristics	*Leuc. mesenteroides* subsp.			*Leuc. para-mesen-teroides*	*Leuc. lactis*	*Leuc. oenos*	*Leuc. gelidum*	*Leuc. carnosum*	*Leuc. pseudo-mesen-teroides*	*Leuc. citreum*	*Leuc. ameli-biosum*	*Leuc. argen-tinum*	*Leuc. fallax*
	mesen-teroides	*dex-tranicum*	*cremoris*										
Acid from													
Arabinose	+	−	−	d	−	d	+	−	d	+	+	d	−
Arbutin	d	−	−	−	−	ND	+	−	d	+	ND	−	−
Cellulose	d	d	−	(d)	−	d	ND	ND	ND	ND	ND	ND	ND
Cellobiose	d	−	−	d	−	d	+	d	d	d	+	d	−
Fructose	+	+	−	+	+	+	+	+	+	+	+	d	+
Galactose	+	d	+	+	+	d	−	−	d	−	ND	+	−
Lactose	d	+	d	d	+	−	−	−	d	−	−	+	−
Maltose	+	+	−	+	+	−	d	−	+	+	+	+	+
Mannitol	d	−	−	−	−	−	−	−	−	d	−	d	(d)
Mannose	d	d	−	+	d	d	+	d	+	+	ND	+	+
Melibiose	d	d	−	+	d	d	+	d	d	−	−	+	−
Raffinose	d	d	−	d	d	−	+	−	d	−	−	+	−
Ribose	d	ND	−	+	−	d	d	d	+	−	−	−	+
Salicin	d	−	−	−	d	d	+	d	d	+	+	−	−
Sucrose	+	+	−	+	+	−	+	+	d	+	+	+	+
Trehalose	+	+	−	+	−	+	+	+	+	+	+	d	(d)
Xylose	d	d	−	−	−	d	+	−	+	−	−	d	−

Hydrolysis of esculin	+	d	−	d	−	+	+	d	d	+	ND	−	ND
Dextran formation	+	+	−	−	−	−	+	+	ND	ND	+	−	ND
Growth at pH 4.8	−	−	−	−	−	+	ND	ND	ND	ND	−	ND	ND
Requirement for TJF	−	−	−	−	−	d	−	−	−	−	−	−	−
Growth in 10% ethanol	−	−	−	−	−	+	ND	ND	ND	ND	ND	ND	ND
NAD-dependent G6PDH present	+	+	+	+	+	−	ND	ND	ND	ND	ND	ND	ND
Growth at 37°C	d	+	−	d	+	d	−	−	+	d	ND	+	+
Peptidoglycan type	Lys-Ser-Ala_2	Lys-Ser-Ala_2	Lys-Ser-Ala_2	Lys-Ala_2	Lys-Ala_2	Lys-Ser_2 Lys-Ala-Ser	ND	ND	Lys-Ser-Ala_2	ND	Lys-Ala_2	ND	Lys-Ala_2

*+, 90% or more of the strains positive; −, 90% or more of the strains negative; d, 11–98% of the strains positive; (d), delayed reaction; ND, no data. Data for *Leuc. mesenteroides*, *Leuc. paramesenteroides*, *Leuc. lactis* and *Leuc. oenos* are from Garvie (1986). Data for *Leuc. gelidum* and *Leuc. carnosum* are from Shaw and Harding (1989), *Leuc. pseudomesenteroides* and *Leuc. citreum* from Farrow *et al.* (1989), *Leuc. amelibiosum* from Schillinger *et al.* (1989), *Leuc. argentinum* from Dicks *et al.* (1993), and *Leuc. fallax* from Martinez-Murcia and Collins (1991).

ferment xylose and arabinose (Garvie, 1986). Based on these differences, Peynaud and Domercq (1968) suggested that xylose and arabinose negative strains of *Leuc. oenos* be reclassified as *Leuc. gracile* and xylose and arabinose positive strains as *Leuc. oinos*. Detailed studies on the genetics of pentose fermentation in *Leuc. oenos* have not been published. It might be that the genes encoding pentose fermentation are plasmid-linked.

The formation of dextran from sucrose has been recorded for *Leuc. mesenteroides* subspp. *mesenteroides* and *dextranicum* (Garvie, 1986), *Leuc. carnosum*, certain strains of *Leuc. gelidum* (Shaw and Harding, 1989), and *Leuc. amelibiosum* (Schillinger *et al.*, 1989), but not *Leuc. mesenteroides* subsp. *cremoris*, *Leuc. paramesenteroides*, *Leuc. lactis*, *Leuc. oenos* (Garvie, 1986), and *Leuc. argentinum* (Dicks *et al.*, 1993). The ability to form dextran is often lost when serial transfers are made in media of increasing salt concentrations (Pederson and Albury, 1955). On the other hand, non-dextran-producing strains of *Leuconostoc* could be reverted to produce dextran when inoculated in media containing tomato or orange juice (Pederson and Albury, 1955). Dextran formation cannot, therefore, be used as a differential characteristic.

Dextrans, or polysaccharides produced by *Leuconostoc* spp. are responsible for the viscous texture of pulque, a fermentation product of agave juice (Sanchez-Marroquin and Hope, 1953; Steinkraus, 1983). *Leuconostoc* spp., specifically *Leuc. mesenteroides* subsp. *mesenteroides*, have been associated with slime formation on the surfaces of various meat products (Korkeala *et al.*, 1988; Von Holy and Holzapfel, 1989), in sauerkraut (Vaughn, 1985), and in sugar and beet sugar plants (Sharpe *et al.*, 1972; Pivnick, 1980). Vaughn (1955) reported the production of extracellular polysaccharides by non-acidophilic lactic acid bacteria in wine. Garvie and Farrow (1980) recorded levan production by strains of *Leuc. oenos*, previously classified as *Leuc. mesenteroides* subsp. *lactosum*, *Leuc. infrequens*, and *Leuc. blayaisense*. Twenty of the 53 strains of *Leuc. oenos* studied by Dicks *et al.* (1990) produced extracellular polysaccharides in a medium which contained 25% grape juice. The production of dextran from sucrose by *Leuc. mesenteroides* subsp. *mesenteroides*, *Leuc. mesenteroides* subsp. *dextranicum*, and *Leuc. amelibiosum* is well documented (Tilbury, 1975; Schillinger *et al.*, 1989). Dextrans produced by *Leuc. mesenteroides* subsp. *mesenteroides* are of industrial importance (Holzapfel and Schillinger, 1992).

Polysaccharides and most alcohols are not metabolized. The metabolism of malate and citrate is discussed elsewhere in this chapter.

7.2.4 Nitrogenous components

The amino acid requirements, as with many other properties of the leuconostocs, are variable between different strains of all species (Garvie,

1967b). *Leuconostoc* spp. can be divided in two groups according to their amino acid requirements, namely the dextran forming strains of *Leuc. mesenteroides* subsp. *dextranicum* and *Leuc. mesenteroides* subsp. *mesenteroides* which require up to eight amino acids, and the non-dextran-forming strains which require more than eight amino acids. Only valine and glutamic acid are required by all strains, and methionine is markedly a stimulator for most. Riboflavin and folic acid are often required and some strains do not grow in the absence of adenine, guanine, xanthine and uracil. None of the leuconostocs requires alanine. All strains require thiamine, panthothenic acid and biotin (Garvie, 1967b). *Leuconostoc* spp. are non-proteolytic.

7.2.5 *Organic acids*

The organic acids most frequently fermented by *Leuconostoc* spp. are citrate and malate (Figure 7.1). All strains of *Leuc. mesenteroides* subsp. *cremoris*, most of the other dairy leuconostocs (*Leuc. mesenteroides* subsp. *mesenteroides*, *Leuc. mesenteroides* subsp. *dextranicum*, *Leuc. lactis*, and *Leuc. paramesenteroides*), and *Leuc. oenos* ferment citrate in the presence of a fermentable carbohydrate (Garvie, 1986). The metabolism of citrate by starter cultures of *Lactococcus* spp., *Leuc. mesenteroides* subsp. *cremoris* and *Leuc. lactis* used in the production of cheese and yogurt, have

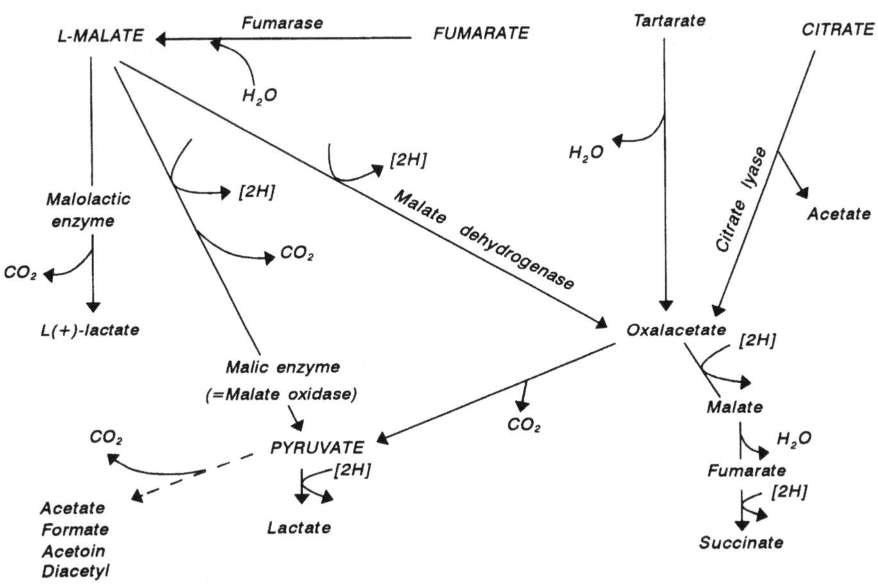

Figure 7.1 Metabolic pathways of carboxylic acids by lactic acid bacteria (Radler and Bröhl, 1984). Important substrates for *Leuc. oenos* are indicated in capital letters.

been well studied (Daly, 1983; Marshall, 1987). Malic acid is actively fermented by most strains of *Leuc. oenos* (Garvie, 1986). It is not known if *Leuc. gelidum*, *Leuc. carnosum*, (Shaw and Harding, 1989), *Leuc. citreum*, *Leuc. pseudomesenteroides* (Farrow *et al.*, 1989), *Leuc. amelibiosum* (Schillinger *et al.*, 1989), and *Leuc. argentinum* (Dicks *et al.*, 1993) are able to ferment citrate and malate.

Lactic acid bacteria metabolize citrate to acetate, acetoin, diacetyl, 2,3-butylene glycol, and CO_2 (Speckman and Collins, 1968). The ratio in which these compounds are formed relative to one another is dependent on the pH of the medium. At neutral pH, acetate, lactate, CO_2, and small amounts of formate and 2,3-butylene glycol are formed (Keenan, 1968). At low pH, on the other hand, acetoin and small quantities of diacetyl are produced (Cogan *et al.*, 1981). Although the production of acetoin and diacetyl from citrate has been recorded in some wines during MLF, little is known about their production by *Leuc. oenos* (Davis et *al.*, 1985b; Wibowo *et al.*, 1985).

A detailed study on strains of *Leuc. lactis*, *Leuc. mesenteroides*, and *Leuc. paramesenteroides* indicated that the additional pyruvic acid formed during the co-metabolism of citrate and glucose is used to re-oxidize NADH and NADPH (Cogan, 1987). Due to this, acetaldehyde is not reduced to ethanol, and acetyl-phosphate is converted to acetate and ATP by an active acetate kinase. The increased levels of ATP result in more rapid growth compared to cells grown on glucose without citrate. Furthermore, cells which fermented citrate showed a decrease in glucose metabolism and lactate dehydrogenase activity. In this study no diacetyl, acetoin or 2,3-butylene glycol were detected. One explanation for this could be that all the citrate was metabolized by the time the medium reached a pH of approximately 6. In a previous study (Cogan *et al.*, 1984) the lack of acetoin production was ascribed to the possible inhibition of acetolactate synthase by intermediate metabolites formed from glucose.

Apart from generating more ATP, citrate could serve as a carbon source for the synthesis of lipids and possibly other essential cell constituents. Experiments on *Leuc. mesenteroides* subsp. *mesenteroides* indicated that radioactively labelled citrate was incorporated into cell materials, whilst glucose was used primarily as an energy source (Schmitt *et al.*, 1992).

The conversion of L-malic acid to L(+)-lactic acid and CO_2, the so-called malolactic fermentation, is primarily conducted by *Leuc. oenos*. Unlike other malolactic bacteria, *Leuc. oenos* contains only the malolactic enzyme and not the malic enzyme or malate dehydrogenase (Schütz and Radler, 1973; Radler, 1986). *Leuconostoc oenos* is, therefore, unable to grow on L-malate as sole carbon source. However, intact cells of *Leuc. oenos* generated more ATP when grown in the presence of L-malate (Cox and Henick-Kling, 1989, 1990). Further research indicated that an electrochemical gradient formed when L-lactate, CO_2, and protons are transported out of the cell (Cox and Henick-Kling, 1989, 1990). The membrane potential formed by the efflux of protons is sufficient to drive the generation of ATP via the membrane-bound ATPase. The metabolism of malate and 2-oxoglutarate by *Leuc. oenos* is reviewed by Van Vuuren and Dicks (1993).

7.3 Taxonomy

7.3.1 Phylogenetic status

Leuconostoc, *Lactobacillus* and *Pediococcus* spp. are phylogenetically closely related and form a supercluster within the clostridia subbranch of the Gram-positive bacteria, as shown by 16S rRNA oligonucleotide cataloguing (Figure 7.2). Phenotypically, the leuconostocs, lactobacilli and pediococci share many characteristics and are often isolated from the same habitat (Sharpe *et al.*, 1972; Garvie, 1976). *Lactobacillus fructosus*, *Lb. viridescens* and *Lb. minor* are phenotypically more closely related to certain *Leuconostoc* spp. than to other lactobacilli (Kandler and Weiss, 1986). Based on 23S rRNA similarity studies, *Lb. fructosus* grouped with *Leuc. mesenteroides* subsp. *mesenteroides* in one cluster, while *Lb. viridescens*, *Lb. minor*, *Lb. confusus*, *Lb. halotolerans*, and *Lb. kandleri* grouped with *Leuc. paramesenteroides* in a separate cluster (Figure 7.3). This is contradictory to the results obtained from DNA–rRNA hybridization studies which showed that *Leuc. mesenteroides* (and its subspecies), *Leuc. lactis*, and *Leuc. paramesenteroides* form a natural group distinct from *Lb. confusus* and *Lb. viridescens* (Garvie, 1981). However, a comparative study of 16S rRNA sequences showed that *Lb. confusus*, *Lb. viridescens*, *Lb. halotolerans*, *Lb. kandleri*, *Lb. minor*, *Leuc. mesenteroides* subsp. *cremoris*, *Leuc. lactis*, *Leuc. mesenteroides*, *Leuc. oenos*, and *Leuc. paramesenteroides* form a natural phylogenetic group, referred to as the '*Leuconostoc* branch' of the lactobacilli (Yang and Woese, 1989). Included in the latter phylogenetic group are some of the other *Leuconostoc* spp. more recently described; *Leuc. carnosum*, *Leuc. citreum*, *Leuc. gelidum*, and *Leuc. pseudomesenteroides* (Figure 7.4). *Leuconostoc oenos* is phylogenetically different from the non-acidophilic leuconostocs and lactobacilli, as indicated by comparative 23S rRNA (Schillinger *et al.*, 1989), 16S rRNA (Yang and Woese, 1989; Martinez-Murcia and Collins, 1990), and DNA–rRNA (Garvie, 1981) studies.

Cross-reactivity studies with antisera against D-lactate dehydrogenase (D-LDH) and glucose 6-phosphate dehydrogenase (G6PDH) indicated that *Leuconostoc* spp. are phylogenetically unrelated to the heterofermentative lactobacilli (Hontebeyrie and Gasser, 1975).

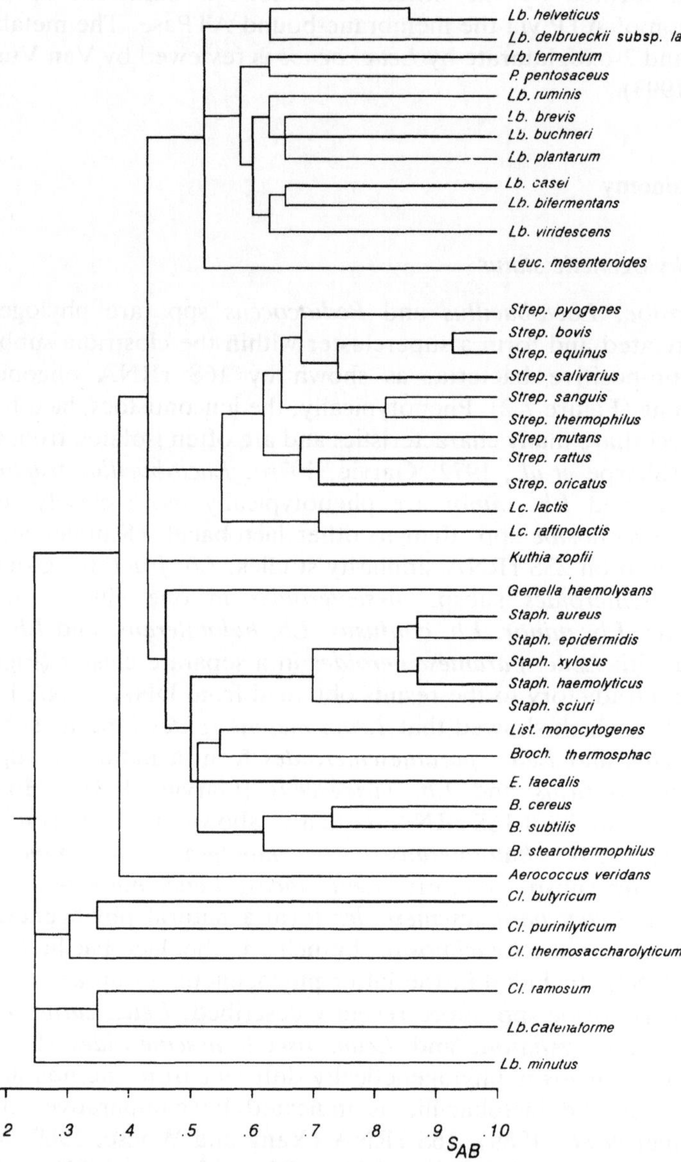

Figure 7.2 16S rRNA dendrogram indicating the phylogenetic position of *Leuconostoc* and the other facultative anaerobic and aerobic members of the *Clostridium* subdivision (Stackebrandt and Teuber, 1988). *Lb., Lactobacillus; P., Pediococcus; Leuc., Leuconostoc; Strep., Streptococcus; Lc., Lactococcus; Staph., Staphylococcus; List., Listeria; Broch., Brochothrix; E., Enterococcus; B., Bacillus;* and *Cl., Clostridium.*

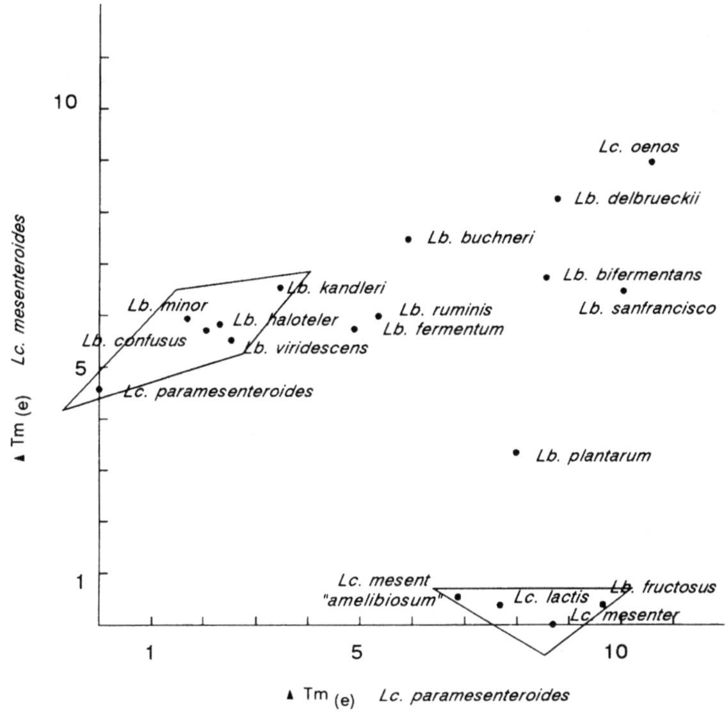

Figure 7.3 Similarity map of hybrids between 23S rRNA of *Leuconostoc mesenteroides* subsp. *mesenteroides* DSM 20343^{T} and *Leuconostoc paramesenteroides* DSM 20288^{T} and DNA of various leuconostocs and lactobacilli. Organisms sharing high rRNA similarity are surrounded by a line (Schillinger *et al.*, 1989). *Lb.*, *Lactobacillus*; and *Lc.*, *Leuconostoc*.

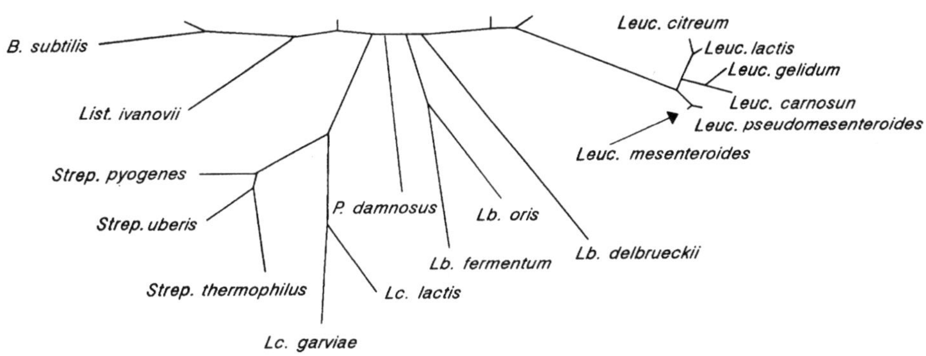

Figure 7.4 Unrooted phylogenetic tree showing the interrelationships of leuconostocs, heterofermentative lactobacilli and other lactic acid bacteria. The tree is based on a comparison of 1340 nucleotides. For abbreviations see Figure 7.2.

7.3.2 Species differentiation

The genus *Leuconostoc* consists of the following genospecies: *Leuc. mesenteroides* (subspp. *dextranicum* and *cremoris*), *Leuc. paramesenteroides*, *Leuc. lactis*, *Leuc. oenos* (Garvie, 1986), *Leuc. carnosum*, *Leuc. gelidum* (Shaw and Harding, 1989), *Leuc. amelibiosum* (Schillinger *et al.*, 1989), *Leuc. pseudomesenteroides*, *Leuc. citreum* (Farrow *et al.*, 1989) *Leuc. fallax* (Martinez-Murcia and Collins, 1991), and *Leuc. argentinum* (Dicks *et al.*, 1993). The type species of the genus is *Leuc. mesenteroides* (Tsenkovskii) Van Tiegham, 1878 (Garvie, 1974, 1986).

Leuconostoc spp. are phenotypically differentiated on the basis of sugar fermentations, hydrolysis of esculin, dextran formation, growth pH, growth temperature, and NAD-dependent G6PDH (Table 7.1). *Leuconostoc oenos* is the only species that prefers an initial growth pH of 4.8 and grows in the presence of 10% ethanol (Garvie, 1986).

7.3.3 Phenotypic relatedness

7.3.3.1 Sugar fermentations. Hucker and Pederson (1931) divided the genus *Leuconostoc* into the species *Leuc. mesenteroides*, *Leuc. dextranicum*, and *Leuc. citrovorum* based on the fermentation of arabinose, xylose, and dextran produced from sucrose. Garvie (1960) proposed a more detailed classification scheme and divided the leuconostocs into six groups based on the fermentation of 19 carbohydrates, the formation of dextran from sucrose, growth in various concentrations of NaCl, growth in litmus milk, and growth at different temperatures. In the latter classification, the three subspecies of *Leuc. mesenteroides* (subspp. *mesenteroides*, *dextranicum* and *cremoris*) were divided into three different groups. The more recently described leuconostocs (*Leuc. pseudomesenteroides*, *Leuc. citreum*, *Leuc. amelibiosum*, *Leuc. gelidum*, *Leuc. carnosum*, *Leuc. fallax*, and *Leuc. argentinum*) were differentiated from the 'classical' species (*Leuc. mesenteroides*, *Leuc. paramesenteroides*, *Leuc. lactis*, and *Leuc. oenos*) by a few key sugar fermentations (see Table 7.1). *Leuconostoc citreum* produces a lemon pigment and is distinguished from *Leuc. mesenteroides* subsp. *mesenteroides* and *Leuc. pseudomesenteroides* by the inability to utilize ribose and raffinose (Farrow *et al.*, 1989). Furthermore, strains of *Leuc. citreum* are α- and β-D-galactosidase and β-D-xylosidase negative (Farrow *et al.*, 1989). *Leuconostoc amelibiosum* is similar to *Leuc. mesenteroides* subsp. *mesenteroides*, but differs from this species by not fermenting melibiose or raffinose (Schillinger *et al.*, 1989). *Leuconostoc gelidum* and *Leuc. carnosum* are distinguished from each other by L-arabinose, arbutin and xylose fermentations, and from most non-acidophilic *Leuconostoc* spp. by not fermenting galactose (Shaw and Harding, 1989).

Leuconostoc oenos is distinguished from the other *Leuconostoc* spp. by its inability to ferment lactose, maltose and sucrose (Garvie, 1967a, 1986). All strains of *Leuc. oenos* ferment glucose, fructose and trehalose. Variable reactions were recorded for the fermentation of arabinose, cellulose, cellobiose, galactose, mannose, melibiose, ribose, salicin, and xylose (Garvie, 1967a, 1986). Nonomura and Ohara (1967) classified the wine leuconostocs into four species based on pentose fermentations: *Leuc. mesenteroides* (arabinose and xylose positive), *Leuc. infrequens* (arabinose positive and xylose negative), *Leuc. blayaisense* (arabinose negative and xylose positive), and *Leuc. dextranicum* (arabinose and xylose negative). These species were further divided based on hexose and disaccharide fermentations (Nonomura and Ohara, 1967). *Leuconostoc mesenteroides* var. *lactosum*, *Leuc. infrequens* and *Leuc. blayaisense* were distinguished from *Leuc. mesenteroides*, *Leuc. paramesenteroides*, and *Leuc. dextranicum* by sucrose fermentation and the inability to ferment maltose (Nonomura and Ohara, 1967). *Leuconostoc dextranicum* var. *vinarium* and *Leuc. dextranicum* var. *debile* were distinguished from *Leuc. dextranicum* in Garvie's group IV (Garvie, 1967a) by their inability to ferment lactose, maltose, salicin and melibiose (Nonomura and Ohara, 1967).

Tracey and Britz (1987) divided *Leuc. oenos* strains from wine into six clusters using numerical analysis of results recorded by the API 50 CHL system. However, these authors could not positively identify the strains within each of these clusters since none of the reference strains of *Leuc. oenos* included in their study grouped into any of the clusters.

Strains within the same species often display a different fermentation profile. Furthermore, the fermentation of sugars is often dependent on culture conditions. Certain strains of *Leuconostoc*, when grown under acidic conditions, are able to ferment sucrose (Garvie, 1967a). On the other hand, non-polysaccharide-producing strains of *Leuc. mesenteroides* formed dextran in media containing tomato juice or orange juice (Pederson and Albury, 1955; Langston and Bouma, 1959). Dicks *et al.* (1990) recorded the production of extracellular polysaccharides by *Leuc. oenos* in medium containing grape juice. Certain sugars are only fermented after prolonged incubation, as indicated for *Leuc. mesenteroides* subsp. *mesenteroides* and *Leuc. pseudomesenteroides* (Farrow *et al.*, 1989). Due to the variable results obtained, sugar fermentation patterns are of little value in the classification of *Leuconostoc* spp. and could lead to misclassifications. However, from an academic and industrial viewpoint, it is important to know which sugars are fermented.

7.3.3.2 Cell wall composition. The cell wall forms the basis for many classical taxonomic studies. Variations in the relative proportions of the sugars, and the sequence of amino acids in the interpeptide bridge of peptidoglycan, are used to differentiate Gram-positive species. The application of peptidoglycan (murein) type as a taxonomic criterion for Gram-positive bacteria was investigated by Cummins and Harris (1956) and reviewed by Schleifer and Kandler (1972). Antibodies prepared against cell wall polysaccharides, cell wall teichoic acids and membrane teichoic acids are used to differentiate bacteria into various antigenic groups (Sharpe, 1970; Knox and Wicken, 1976). Thus far no antibodies were raised against the cell wall components in *Leuconostoc* spp.

The peptidoglycan types can be used to differentiate among certain *Leuconostoc* spp. (Table 7.1) (Garvie, 1986; Holzapfel and Schillinger, 1992). Two peptidoglycan types were recorded for *Leuc. paramesenteroides* (Lys-Ala_2 and Lys-Ser-Ala_2). Based on these results, Kandler (1970) suggested that *Leuc. paramesenteroides* be divided into two subspecies: *paramesenteroides* and *lactophilum*. Two peptidoglycan types were recorded for *Leuc. mesenteroides*, viz. Lys-Ser-Ala_2 and Lys-Ala_2. The strains containing the Lys-Ala_2 type were later reclassified as *Leuc. amelibiosum* based on results obtained by DNA–DNA hybridizations (Schillinger *et al.*, 1989). *Leuconostoc pseudomesenteroides* is characterized by the Lys-Ser-Ala_2 type (Garvie, 1986; Farrow *et al.*, 1989). *Leuconostoc lactis* contains the peptidoglycan types Lys-Ala_2 and the less commonly found Lys-Ser-Ala_2 (Kandler, 1970; Garvie, 1986). *Leuconostoc fallax* contains the Lys-Ala_2 type (Martinez-Murcia and Collins, 1991). The interpeptide types of *Leuc. citreum*, *Leuc. gelidum*, *Leuc. carnosum*, and *Leuc. argentinum* were not recorded (Farrow *et al.*, 1989; Shaw and Harding, 1989; Dicks *et al.*, 1993). Two peptidoglycan types were reported for strains of *Leuc. oenos*; Lys-Ala-Ser and the less common Lys-Ser_2 which is also found in *Lb. viridescens* (Kandler, 1970; Schleifer and Kandler, 1972; Garvie, 1986). The variation in serine content in the peptidoglycan of *Leuc. oenos* could not be ascribed to the amino acid composition of the medium and is thus genetically controlled (Kandler, 1970). The unique peptidoglycan amino acid type of *Leuc. oenos* supports its taxonomic separation from the other *Leuconostoc* spp.

7.3.3.3 Electrophoretic mobilities of enzymes. Gasser (1970) and Williams and Sadler (1971) were the first to use electrophoresis of enzymes in the classification of lactic acid bacteria. Since then, the technique has been used extensively in the classification of leuconostocs (Garvie, 1969, 1975; Garvie and Farrow, 1980; Irwin *et al.*, 1983) and lactobacilli (London, 1976; Sharpe, 1981).

The number and nature of the lactic dehydrogenases and their respective electrophoretic mobilities are characteristic of a species (Gasser, 1970; London, 1976; Garvie, 1980). *Leuconostoc* spp. produce D(–)-lactate from glucose and have a single NAD-dependent lactate dehydrogenase [D(–)nLDH]. Several strains of *Leuconostoc* sp. have NAD-independent

L(+)- and D(−)-LDHs [L(+)iLDH and D(−)iLDH]. However, the D(−) iLDH seems to be more active *in vitro* (Garvie, 1969). Small concentrations of L(+)nLDH were recorded for *Leuc. mesenteroides* strain 39 and strains of *Leuc. oenos* (Doelle, 1971), but this has not been confirmed in subsequent publications.

The electrophoretic mobility of the D-LDHs of *Leuc. mesenteroides* and *Leuc. lactis* correspond well (Garvie, 1969; Dicks and Van Vuuren, 1990). Differences in the intensity of the D(–)nLDH bands of *Leuc. lactis* were observed on a stacked polyacrylamide gel with Tris-glycine (pH 8.9) used as upper electrode buffer (Dicks and Van Vuuren, 1990). The D(–)nLDH of *Leuc. paramesenteroides* is similar to that of *Leuc. mesenteroides* and *Leuc. lactis* (Dicks and Van Vuuren, 1990). However, different banding patterns of the D(–)nLDH of *Leuc. paramesenteroides* were obtained when a stacked polyacrylamide gel was used (Dicks and Van Vuuren, 1990). The different bands on the stacked gel might represent different subunits of LDH or different nLDHs (Dicks and Van Vuuren, 1990).

Leuconostoc oenos is the best distinguished from all the other leuconostocs by its slower migrating D(–)nLDH (Garvie, 1975). The species is divided into two subgroups based on the electrophoretic mobilities of its D(−)nLDH (Garvie, 1969; Dicks and Van Vuuren, 1990), confirming the results obtained by pentose fermentations (Peynaud and Domercq, 1968). Based on these results, and that obtained by Irwin *et al.* (1983), *Leuc. oenos* was considered a genetically heterogeneous species. Similar results were recorded for 6-phosphogluconate dehydrogenase (6-PGD) (Irwin *et al.*, 1983). Garvie and Farrow (1980) reclassified *Leuc. mesenteroides* subsp. *lactosum*, *Leuc. infrequens*, *Leuc. blayaisense*, *Leuc. dextranicum* subsp. *vinarium* and *Leuc. dextranicum* subsp. *debile* as *Leuc. oenos* based on results obtained by D(–)nLDH and G6PDH electrophoretic patterns.

The LDH patterns of *Leuc. pseudomesenteroides*, *Leuc. citreum*, *Leuc. gelidum*, *Leuc. carnosum*, *Leuc. fallax*, and *Leuc. argentinum* have not been reported.

Electrophoresis of 6PDH, G6PDH, alcohol dehydrogenase (ADH), and NADH indicated some relatedness between the non-acidophilic *Leuconostoc* spp., *Lb. confusus* and *Lb. viridescens* (Garvie, 1975).

7.3.3.4 Enzyme homology. Enzyme homology studies can be used to determine the evolution of bacteria. Hybridization studies with D(–)nLDH indicated that *Leuc. oenos*, the non-acidophilic *Leuconostoc* spp., *Lb. confusus*, and *Lb. viridescens* are closely related (Chilson *et al.*, 1965). Moreover, these results indicated that *Leuc. oenos* is not that different from the non-acidophilic *Leuconostoc* spp. (Garvie, 1975). However, antisera prepared from *Leuc. oenos* showed no cross-reaction with G6PDH and D(–)-LDH of *Leuc. lactis* (Hontebeyrie and Gasser, 1975) and

G6PDH of *Leuc. mesenteroides* and vice versa (Gasser and Hontebeyrie, 1977).

Electrophoretically identical enzymes of *Leuconostoc* spp. are serologically different (Hontebeyrie and Gasser, 1975; Gasser and Hontebeyrie, 1977). Strains of *Leuc. lactis* and *Leuc. paramesenteroides* grouped into two separate immunological groups (Hontebeyrie and Gasser, 1975; Gasser and Hontebeyrie, 1977). *Leuconostoc mesenteroides* grouped into five immunological groups with antisera prepared from G6PDH of *Leuc. mesenteroides* and *Leuc. lactis*. *Leuconostoc dextranicum* grouped into only one G6PDH immunological group, but with *Leuc. mesenteroides*. From these results, it can be concluded that all non-acidophilic *Leuconostoc* spp. stemmed from one ancestor (Hontebeyrie and Gasser, 1975; Gasser and Hontebeyrie, 1977). Enzyme homology data for *Leuc. pseudomesenteroides*. *Leuc. citreum*, *Leuc. gelidum*, *Leuc. carnosum*, *Leuc. fallax*, and *Leuc. argentinum* have not been reported.

7.3.3.5 Numerical analysis of total soluble cell protein patterns. Banding patterns of cellular proteins are characteristic for each bacterial strain and can be used as a 'fingerprint' (Kersters and De Ley, 1980). Each band on the gel represents structurally different protein species of the same electrophoretic mobility. The amino acid sequence, molecular weight, net electrical charge and amount of each protein species are genetically determined.

Numerical analysis of total soluble cell protein patterns proved valuable in the differentiation among strains within a genus. This technique has been used for a large number of different bacteria (Kersters and De Ley, 1980), including *Lactobacillus* (Dicks and Van Vuuren, 1987, 1988; Dellaglio *et al.*, 1991) and *Leuconostoc* (Dicks *et al.*, 1990; Fantuzzi *et al.*, 1992; Dicks *et al.*, 1993). The groupings obtained by numerical analysis of total soluble cell protein patterns corresponded well to groupings obtained by DNA–DNA hybridizations (Dicks *et al.*, 1990, 1993; Dellaglio *et al.*, 1991; Fantuzzi *et al.*, 1992).

By using numerical analysis of total soluble cell protein patterns, strains of *Leuc. oenos*, *Leuc. mesenteroides* subsp. *mesenteroides*, *Leuc. mesenteroides* subsp. *dextranicum*, *Leuc. paramesenteroides* and *Leuc. lactis* were grouped into four clusters (Figure 7.5). Strains of *Leuc. oenos* grouped in cluster 1 at $r \geqslant 0.83$, indicating it is a genotypically homogeneous species. The major differences observed in the protein patterns of *Leuc. oenos* and the non-acidophilic *Leuconostoc* spp. indicated a low genetic relatedness between these species. The type strain of *Leuc. paramesenteroides* grouped with the type strain of *Pediococcus acidilactici* (ATCC 12697) at $r \geqslant 0.94$ in cluster II, which places a question on the taxonomic status of *Leuc. paramesenteroides*. The remaining strains of *Leuc. paramesenteroides* grouped with unidentified *Leuconostoc* spp. in

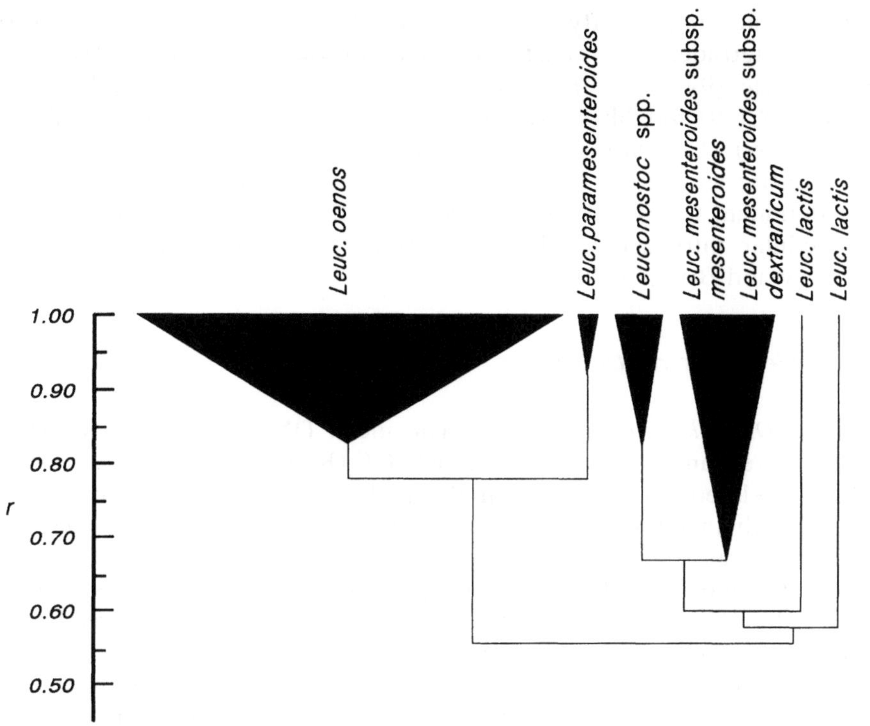

Figure 7.5 Simplified dendrogram showing the genotypic relatedness among *Leuc. oenos* (53 strains), *Leuc. paramesenteroides* (three strains), *Leuconostoc* spp. (three strains), *Leuc. mesenteroides* subsp. *mesenteroides* (seven strains), *Leuc. mesenteroides* subsp. *dextranicum* (five strains), and *Leuc. lactis* (three strains), based on computerized numerical analysis of total soluble cell protein patterns. Grouping was by the unweighted average pair-group method.

cluster III at $r \geqslant 0.84$. Most of the strains of *Leuc. mesenteroides* subsp. *mesenteroides*, *Leuc. mesenteroides* subsp. *dextranicum*, *Leuc. paramesenteroides* and *Leuc. lactis* grouped in cluster IV at $r \geqslant 0.68$, indicating they are genotypically related. Two strains from *Leuc. lactis* did not group into any cluster, indicating that it is a genotypically heterogenous collection of strains.

Numerical analysis of total soluble cell protein patterns was used to differentiate the leuconostocs isolated from Argentine raw milk from the other non-acidophilic *Leuconostoc* spp. (Fantuzzi *et al.*, 1992). A new species, *Leuc. argentinum*, was proposed (Dicks *et al.*, 1993).

7.3.3.6 Cellular fatty acids. Fatty acid profiles obtained by capillary gas chromatography were used to differentiate among *Leuconostoc* spp. (Schmitt *et al.*, 1989; Shaw and Harding, 1989; Tracey and Britz,

1989). The major fatty acids isolated from *Leuconostoc* spp. were myristic (tetradecanoic) acid (C14:0), palmitic (hexadecanoic) acid (C16:0), palmitoleic (9-hexadecenoic) acid (C16:1–9), oleic (*cis*-9-octadecenoic) acid (C18:1–9) and dihydrosterculic (*cis*-9,10-methyleneoctadecanoic) acid (C19-cyc-9). *Leuconostoc* spp., especially *Leuc. oenos*, differ from the other lactic acid bacteria in containing oleic acid, and not vaccenic acid, as the dominant C18:1 fatty acid (Tracey and Britz, 1989). Results obtained by cellular fatty acid analyses confirmed the phenotypic relationships (Tracey and Britz, 1989).

7.3.4 Genotypic relatedness

7.3.4.1 DNA base composition. The mean DNA base composition of prokaryotes ranges from 22 to 75 mol% GC (Kandler and Schleifer, 1980). Closely related bacteria have similar mol% G+C values. However, two strains with an identical DNA base composition are not necessarily closely related, because the linear arrangement of the nucleotides in the DNA is not accounted for in the technique (Johnson, 1986). Bacterial strains of which the GC content differs more than 20%, have no long stretches of nucleotide sequences in common (Kandler and Schleifer, 1980).

The GC content for most *Leuconostoc* spp. ranges from 37 to 40 mol%, calculated from thermal denaturation curves (Table 7.1). This is in the range of the GC content for *Lb. viridescens*, *Lb. trichodes* (now *Lb. fructivorans*), *Lb. vermiforme*, *Lb. brevis* and *Lb. buchneri* (Gasser and Mandel, 1968; Garvie *et al.*, 1974). The GC content of *Leuc. lactis* and *Leuc. amelibiosum* is in the range of 42–45 mol%. Due to the narrow GC range, DNA base composition cannot be used to differentiate between *Leuconostoc* spp.

7.3.4.2 DNA and RNA homology. Strains in a species which have a similar DNA base composition ($T_{m(e)}$ values of 0–3°C) are usually genetically closely related and share high DNA homology values (80–90%). On the other hand, the $T_{m(e)}$ value for strains with a 60–70% DNA homology, differs with 6–9°C (Johnson, 1986). The cut-off point for strains belonging to one species is 60% DNA homology. This value was merely chosen because it seems to be the transitional point between genetic characteristics determined by cistrons and changes in the base sequences. The degree of DNA homology above which strains are considered to belong to the same species is, however, arbitrary and should be confirmed by using other genetic taxonomic techniques.

Different DNA–DNA hybridization techniques have been used to determine the genetic relatedness of *Leuconostoc* spp. The techniques used include the hydroxyl apatide method of Brenner (Hontebeyrie and

Gasser, 1977), the membrane filter method of Denhardt (Farrow *et al.*, 1989; Dicks *et al.*, 1990), the thermal denaturation method of De Ley and co-workers (Schillinger *et al.*, 1989), and the S1 nuclease method of Crosa and co-workers (Shaw and Harding, 1989). The DNA homology values recorded for the different *Leuconostoc* spp. are listed in Table 7.2.

Hontebeyrie and Gasser (1977) grouped 45 strains of *Leuconostoc* spp. into six DNA homology groups and proposed the subdivision of the genus into six genospecies. *Leuconostoc mesenteroides*, *Leuc. dextranicum* and *Leuc. cremoris* are genetically closely related and were reclassified as subspecies of *Leuc. mesenteroides* (Garvie, 1986). DNA–DNA hybridizations indicated that *Leuc. lactis* and *Leuc. paramesenteroides* are genetically not closely related to the other *Leuconostoc* spp. (Garvie, 1976; Hontebeyrie and Gasser, 1977; Farrow *et al.*, 1989; Schillinger *et al.*, 1989; Shaw and Harding, 1989; Dicks *et al.*, 1990; Fantuzzi *et al.*, 1992). According to results obtained by DNA–rRNA hybridizations, *Leuc. paramesenteroides* is phylogenetically more related to *Lb. confusus*, *Lb. viridescens*, *Lb. halotolerans*, *Lb. minor*, and *Lb. kandleri* than to the non-acidophilic leuconostocs. *Leuconostoc lactis*, on the other hand, clustered with *Leuc. mesenteroides*, *Leuc. amelibiosum*, and *Lb. fructosus* (Garvie, 1981; Schillinger *et al.*, 1989). These results are in good agreement with the 16S rRNA sequence analyses which indicated that the non-acidophilic leuconostocs are phylogenetically related to *Lb. confusus*/*Lb. viridescens*, the so-called 'leuconostoc branch' of the lactobacilli (Yang and Woese, 1989).

According to the DNA homology values obtained by Farrow *et al.* (1989), *Leuc. pseudomesenteroides* is genetically not closely related to *Leuc. citreum*, *Leuc. lactis*, *Leuc. mesenteroides* subsp. *mesenteroides* and *Leuc. mesenteroides* subsp. *dextranicum*. However, Dicks *et al.* (1990) and Fantuzzi *et al.* (1992) recorded DNA homology values of 78 and 82% between *Leuc. pseudomesenteroides*, and *Leuc. mesenteroides* subsp. *mesenteroides* and *Leuc. mesenteroides* subsp. *dextranicum*, respectively. The taxonomic status of *Leuc. pseudomesenteroides* is, therefore, uncertain.

DNA–DNA hybridizations clearly differentiated the recently described species, *Leuc. carnosum*, *Leuc. gelidum* (Shaw and Harding, 1989), *Leuc. citreum* (Farrow *et al.*, 1989), *Leuc. fallax* (Martinez-Murcia and Collins, 1991), and *Leuc. argentinum* (Dicks *et al.*, 1993) from the other *Leuconostoc* spp.

Leuconostoc oenos is a genetically homogeneous species as indicated by the high DNA homology values (80–97%) with the type strain (NCDO 1674^T). However, *Leuc. oenos* shares a DNA homology of only 2–8% with the non-acidophilic *Leuconostoc* spp., indicating it is genetically not closely related to the other leuconostocs (Dicks *et al.*, 1990). This is not surprising, since *Leuc. oenos* is physiologically and biochemically different from the non-acidophilic *Leuconostoc* spp. In contrast to the other

Table 7.2 Levels of DNA–DNA homology for leuconostocs*

Species or subspecies	% Homology with ^{3}[H]DNA from				
	Leuc. mesenteroides subsp. *mesenteroides*	*Leuc. pseudo-mesenteroides*	*Leuc. para-mesenteroides*	*Leuc. lactis*	*Leuc. carnosum*
Leuc. mesenteroides subsp. *mesenteroides*	73–108	24–63	7–18	7–38	1–17
Leuc. mesenteroides subsp. *cremoris*	66–106	20–50	5–10	5–35	4–10
Leuc. mesenteroides subsp. *dextranicum*	84–110	38–49	5–19	6–35	5–8
Leuc. pseudomesenteroides	18–48	70–100	9–22	13–36	ND
Leuc. paramesenteroides	6–19	7–30	82–100	8–14	0–3
Leuc. lactis	16–49	23–61	0–25	74–100	7–22
Leuc. amelibiosum	39	34	22	32	ND
Leuc. carnosum	19–32	ND	0–6	0–25	78–116
Leuc. gelidum	9–31	ND	0–3	0–10	3–21
Leuc. citreum	21–30	27–56	10–20	23	ND
Leuc. oenos	8–15	5–10	5–7	3–11	ND
Leuc. fallax	28–41	ND	25	ND	ND
Leuc. argentinum	9	ND	ND	35–39	ND

*Data compiled from Hontebeyrie and Gasser (1977), Garvie (1986), Farrow *et al.* (1989), Schillinger *et al.* (1989), Shaw and Harding (1989), Martinez-Murcia and Collins (1991), and Dicks *et al.* (1993). ND, no data.

Leuconostoc spp., *Leuc. oenos* prefers an initial growth pH of 4.8 and grows in media supplemented with 10% (v/v) ethanol (Garvie, 1986). The electrophoretic mobility of the D(–)nLDH, 6PGDH and ADH of *Leuc. oenos* is different from that of the other *Leuconostoc* spp. (Garvie, 1967a, 1969, 1976, 1980, 1986; Dicks *et al.*, 1990). *Leuconostoc oenos* does not contain NAD-dependent G6PDH. Furthermore, antisera prepared from *Leuc. oenos* show no cross-reactivity with anti-G6PDH and anti-D(–)nLDH of *Leuc. lactis* and anti-G6PDH of *Leuc. mesenteroides* subsp. *mesenteroides* (Gasser and Hontebeyrie, 1977). The amino acid sequence of the cell wall interpeptide bridge of *Leuc. oenos* (Lys-Ala-Ser; Lys-Ser_2) is unusual for the genus (Kandler, 1970; Garvie, 1986; Schillinger *et al.*, 1989). Based on results obtained by numerical analysis of total soluble cell protein patterns (Dicks *et al.*, 1990), *Leuconostoc oenos* is not closely related to the other leuconostocs. Moreover, DNA–rRNA hybridizations (Garvie, 1981; Schillinger *et al.*, 1989) and 16S rRNA sequencing analyses (Yang and Woese, 1989; Martinez-Murcia and Collins, 1990) clearly indicated that *Leuc. oenos* is phylogenetically distant from other *Leuconostoc* spp. Yang and Woese (1989) argued that *Leuc. oenos* is a fast evolving organism. Martinez-Murcia and Collins (1990) indicated that *Leuc. oenos* is equidistant from the clusters *Leuc. paramesenteroides* and *Leuconostoc sensu stricto*. Based on all these characteristics, *Leuc. oenos* should be excluded from the genus *Leuconostoc*.

A new addition to the identification of *Leuc. oenos* and other malolactic bacteria is the use of DNA probes (Lonvaud-Funel *et al.*, 1991). The method proved to be genus specific. No cross-hybridization was obtained for most of the species belonging to the genera *Leuconostoc*, *Pediococcus* and *Lactobacillus*.

7.4 Genetics

7.4.1 Plasmids

Plasmids have been isolated from *Lactococcus* (Davies and Gasson, 1981; McKay, 1983), *Lactobacillus* (Chassy *et al.*, 1978; Klaenhammer, 1984), and *Pediococcus* (Gonzalez and Kunka, 1983). O'Sullivan and Daly (1982) were the first to isolate plasmids from *Leuconostoc* spp. Orberg and Sandine (1984) were the first to report plasmids in *Leuc. oenos*. Janse *et al.* (1987) developed a plasmid isolation procedure specifically for *Leuc. oenos* and identified 11 plasmids from eight strains. The plasmid size ranged from 2.47 to 4.61 kb pairs and were smaller than the 40 and 80 MDa plasmids previously isolated (O'Sullivan and Daly, 1982; Cavin *et al.*, 1988). Apart from indications that citrate metabolism in *Leuc. mesenteroides* subsp. *mesenteroides* strain 19D might be plasmid linked (Cavin *et al.*, 1988) no phenotypic characteristic could be linked to the plasmids isolated from

Leuconostoc spp. The plasmids isolated from *Lactobacillus* and *Pediococcus* spp. code for the utilization of sugars (Chassy *et al.*, 1978), production of proteinase (Davies and Gasson, 1981), nisin (Clewell, 1981), diplococcin, bacteriocins, restriction modification systems, drug resistance, slime formation, arginine hydrolysis and bacteriophage resistance (McKay, 1983; de Vos *et al.*, 1984).

With the increasing interest in genetic engineering of lactic acid bacteria, including *Leuconostoc* spp., more research is being done on plasmids to develop cloning vectors.

7.4.2 Bacteriocins

Bacteriocins are defined as proteins or protein complexes active against bacteria genetically closely related to the producer strain (Tagg *et al.*, 1976). The bacteriocins isolated from *Lactobacillus*, *Pediococcus* and *Lactococcus* spp. have been reviewed by Klaenhammer (1988). Little is known about bacteriocins produced by *Leuconostoc* spp.

Orberg and Sandine (1984) reported the production of antibacterial compounds by strains of *Leuc. mesenteroides* subsp. *dextranicum* and other *Leuconostoc* spp., but did not identify the antibacterial substance(s). Hastings and Stiles (1991) isolated a bacteriocin produced by *Leuc. gelidum* which inhibited lactobacilli, leuconostocs and *Listeria monocytogenes*. The bacteriocin is sensitive to proteases and trypsin and can be precipitated with ammonium sulphate, indicating that it is proteinaceous. Activity was not lost when the sample was heated to 62°C for 30 min. The bacteriocin is most likely bacteriostatic. Production of the bacteriocin is encoded by one of the plasmids.

Daba *et al.* (1991) isolated a bacteriocin from *Leuc. mesenteroides*. This bacteriocin differs from the bacteriocin produced by *Leuc. gelidum* in that it is produced during the late exponential phase of growth and is resistant to heat (100°C for 30 min). The concentrated supernatant had a bacteriostatic effect against strains of *Listeria*, *Streptococcus faecalis*, *Brevibacterium linens* and *Pediococcus pentosaceus*. The genes encoding the production and immunity of this bacteriocin are not genetically linked.

Recently, Lewus *et al.* (1992) isolated a bacteriocin from *Leuc. paramesenteroides*. Like the bacteriocin isolated from *Leuc. gelidum*, it is resistant to a 30 min heat treatment at 60°C. The bacteriocin is inactivated by α-amylase, trypsin, α-chymotrypsin, protease and proteinase K. It has a bacteriostatic effect on *Listeria monocytogenes*, *Staphylococcus aureus*, *Yersinia enterocolitica* and *Clostridium botulinum*.

7.4.3 Bacteriophages

Bacteriophages of the lactic acid bacteria have been studied extensively on a morphological and serological level (Teuber and Lembke, 1983; Mata and Ritzenthaler, 1988). Only recently these phages have been studied on a molecular level. DNA hybridization studies, restriction enzyme maps of phage genomes (Mata and Ritzenthaler, 1988), sequences of structural genes, and the location of sites for attachment and packaging initiation in temperate phage DNA (Klaenhammer, 1987; Jarvis, 1989; Raya *et al.*, 1992), have been published. These studies led to other genetic studies aimed at phage–host interactions and phage resistance.

Little is known about the phages of *Leuconostoc* spp. (Saxelin *et al.*, 1986; Jarvis, 1989). This is not surprising, since the leuconostocs represent only a small proportion of the lactic acid bacteria in mixed mesophilic starter cultures used in the dairy industry (Sozzi *et al.*, 1978). Furthermore, due to the slow growth of leuconostocs in these mixed cultures, their phages are not as easily detected. In addition, many *Leuconostoc* spp. produce exopolysaccharides which could protect the bacteria against phage adsorption. Even less is known about bacteriophages attacking *Leuc. oenos* and their effect on MLF.

Swiss researchers were the first to report the presence of bacteriophages in wine (Sozzi *et al.*, 1976, 1978; Cazelles and Gnaegi, 1982). These phages were shown to be lytic against *Leuc. oenos* (Sozzi *et al.*, 1982) and their presence in wine was associated with abnormal MLF. More recently, bacteriophages specific for *Leuc. oenos* have been isolated from Australian red wine (Davis *et al.*, 1985a; Henick-Kling *et al.*, 1986a, b) and South African red wines which underwent stuck MLF (Nel *et al.*, 1987).

Thus far, only two phages of *Leuc. oenos* (phages P58I and P58II) have been genetically characterized (Arendt *et al.*, 1991). Bacteriophage P58I is lytic and phage P58II lysogenic. Phage P58II was isolated after induction of *Leuc. oenos* with Mitomycin C. Electron microscopy studies showed that phages P58I and P58II are morphologically identical. Both phages have isometric heads (54 nm in diameter), tails (243 nm in length and 11 nm thick) and base plates (20 nm in diameter). Furthermore, phages P58I and P58II yielded identical DNA restriction enzyme fragment patterns and displayed similar structural proteins (Arendt *et al.*, 1991). It was concluded from these results that phage P58I originated from the lysogenic phage P58II which underwent a mutation (Arendt *et al.*, 1991).

The *Leuc. oenos* phages isolated from South African wines are divided into five genetic groups based on results obtained by restriction endonuclease digestion of their genomic DNA (Nel *et al.*, 1987). The genome of one of the phages (L5), was characterized by restriction enzyme mapping (Olwage, 1992). The middle two DNA fragments of the linear phage genome were cloned into plasmid vectors and transformed to *Escherichia coli*. One of these fragments, which coded for proteins of 2.8 kDa and 14.6 kDa, was sequenced and screened for DNA and protein homology in two separate data banks. The genome of *Leuc. oenos* phage L10, which is from a separate genetic group, has also been characterized, cloned and partially sequenced (Van der Meer, L.J. and Van Vuuren, H.J.J., 1993, pers. comm.). Phages L5 and L10 are lytic and have hexagonal heads and long flexible, non-contractile tails.

7.5 Practical importance

Leuconostoc spp. are often considered as spoilage bacteria. Because they are obligately heterofermentative, they produce carbon dioxide which causes blowing of certain cheeses and changes the texture of fermented foods. However, certain strains produce diacetyl and acetoin from citrate and contribute to the typical flavour and taste of many food and especially dairy products. These strains are frequently used as starter cultures for buttermilk, butter, cream cheese, and cheese (Sandine and Elliker, 1970; Collins and Speckman, 1974; Garvie, 1984; Sozzi and Pirovano, 1993).

In their natural habitat the leuconostocs represent a small percentage of the mesophilic microflora, which is largely composed of homofermentative lactobacilli, lactococci and pediococci. Even in a fermentation process the growth of *Leuconostoc* spp. is dominated by the other lactic acid bacteria. However, in a fermentation broth containing antibiotics, antibacterial substances, or bacteriophages, the leuconostocs can become the dominant lactic acid bacteria. Due to this, their slow growth and weak acidifying properties, the leuconostocs are often not included in mixed starter cultures. Exclusion of the leuconostocs often causes a reduction in flavour and aroma, and a general lowering in the typical characteristics of the product. However, in a well balanced starter culture, the leuconostocs play an important role in the reological and flavouring quality of the fermented product (Sozzi and Pirovano, 1993). It is therefore important to study the growth interactions between *Leuconostoc* and the other lactic acid bacteria, the volatile compounds they produce, and their resistance to bacteriophages.

At present most of the research programmes on *Leuconostoc* spp. are concentrated on the organism's antagonistic activity and ability to improve the organoleptic properties in the following main fields:

- fermented vegetables
- wine production
- dairy products.

7.5.1 Leuconostoc *in vegetables and fermented plant foods*

Several papers have been published on the initial cell concentration, available nutrients, and growth rate of lactic acid bacteria on plants. However, little is known about the interaction between leuconostocs and plants. Visser *et al.* (1986) suggested that lactic acid bacteria may protect the plant against opportunistic pathogenic microorganisms by producing antagonistic compounds such as organic acids and/or bacteriocin-like molecules. Daeschel *et al.* (1987) reported on the interaction between lactic acid bacteria and Gram-negative bacteria, and the role lactic acid bacteria play in controlling the microbial consortium in the phyllosphere. The success of this biological event depends on the initial microbial load, available nutrients, growth rate of specific species, and chemical and physical conditions (Daeschel *et al.*, 1987).

Lactic acid bacteria are present in low numbers on living plants, but increase when the plant tissue is damaged, for instance during ensiling (Stirling and Whittenbury, 1963; Dellaglio and Torriani, 1986; Daeschel *et al.*, 1987). From the data of Archibald and Fridovich (1981), it seems that the lactic acid bacteria growing on plants have developed defence mechanisms to the accumulation of Mn(II), whereas those associated with meat or dairy products have not.

Leuconostoc spp. are the most predominant lactic acid bacteria isolated from the aerial part of standing vegetables (Mundt *et al.*, 1967; Mundt, 1970). Of all leuconostocs isolated from plants, *Leuc. mesenteroides* subsp. *mesenteroides* is the most predominant (Mundt, 1970) and plays an important role in the fermentation of various plant and vegetable fermentations. *Leuconostoc paramesenteroides* represents one of the dominant species of lactic acid bacteria in fresh vegetables and play an important role in the first phase of silage fermentations (Dellaglio *et al.*, 1984; Dellaglio and Torriani, 1986).

Leuconostoc mesenteroides predominates in the initial phase of sauerkraut fermentation. However, as the fermentation proceeds, the cell numbers of *Pediococcus pentosaceus*, *Enterococcus faecalis* and *Lb. plantarum* increase and the growth of *Leuc. mesenteroides* is repressed (Mundt and Hammer, 1968; Vaughn, 1985). In carrot and cabbage fermentations *Leuc. mesenteroides* and *Lb. pentosus* are used as starter cultures. *Leuconostoc mesenteroides* produces acetic acid, which inhibits the growth of *Listeria monocytogenes* and most other spoilage bacteria, and increases the organoleptic quality of the product (Delclos, 1992). The organoleptic quality of jangsu, a Korean non-alcoholic beverage, and that of fermented salted Chinese cabbage (kimchi), improved when *Leuc. mesenteroides* was used in combination with other lactic acid bacteria starter cultures (Lee and Kim, 1988; Kim *et al.*, 1991). Beetroot fermented with starter cultures of *Leuc. mesenteroides* had a higher sensory quality and better consistency when compared to the uninoculated product (Buckenhueskes *et al.*, 1984). In the fermentation of sweet pepper *Leuc. paramesenteroides* and other leuconostocs are the dominant lactic acid bacteria (Valdez-de *et al.*, 1990). *Leuconostoc mesenteroides* is used as starter culture in the fermentation of coffee cherries for the production of Columbian coffee.

7.5.2 Leuconostoc *in oenology*

Leuconostoc oenos, *Leuc. mesenteroides* subsp. *mesenteroides*, *Lb.* plantarum, *Lb. casei* and *P. damnosus* have been isolated from fermented grape must during the alcoholic fermentation (Wibowo *et al.*, 1985). As the alcoholic fermentation commences, the growth of most lactic acid bacteria is repressed. *Leuconostoc oenos* is, however, more resistant to SO_2 and ethanol and is found in relatively high cell numbers at the end of the alcoholic fermentation (Wibowo *et al.*, 1985). During the secondary fermentation of wine, *Leuc. oenos* converts L-malic acid to L-lactic acid, increases the wine pH by 0.1–0.3 pH units, and renders the wine microbiologically stable (Davis *et al.*, 1985b). It is generally accepted that MLF improves the organoleptic quality of wines produced in cold viticultural regions. However, wines produced from grapes grown in a warm climate are less acidic and MLF could render the wine flat and insipid. Furthermore, high pH wines are more susceptible to spoilage by *Pediococcus* and *Lactobacillus* (Rankine and Bridson, 1971; Rankine, 1972).

The ecology of lactic acid bacteria in wines is influenced by the composition of the wine, interactions between the lactic acid bacteria (and yeast), and the technology of vinification. Recent research indicated that L-malic acid is utilized by *Leuc. oenos* during the initial growth phase, even before active growth commences (Hayman and Monk, 1982; Krieger *et al.*, 1986). Stuck MLF is often caused by bacteriophages which infect *Leuc. oenos* (Sozzi *et al.*, 1976, 1982; Davis *et al.*, 1985a, b; Henick-Kling *et al.*, 1986a, b; Nel *et al.*, 1987).

The practical importance of malolactic fermentation in wine is still a topic of dispute (Van Vuuren and Dicks, 1993). Van Wyk (1976) indicated that wines which underwent malolactic fermentation are not always superior to those which did not undergo the fermentation. However, according to recent data, wines which have been inoculated with malolactic bacteria improved in organoleptic quality (Giannakopoulous *et al.*, 1984; McDaniel *et al.*, 1987; Rodriguez *et al.*, 1990).

For a better understanding of the malolactic activity, strains of *Leuc. oenos* and *Lb. plantarum* have been studied under different environmental factors, such as temperature, pH, inoculum size, and different concentrations of ethanol and L-malic acid (Guerzoni *et al.*, 1993). Based on these results, the initial sugar concentration in the grape must is critical. Krieger *et al.* (1992) concluded that the growth rate, cell yield and malolactic activity are optimal when media contain fructose as the main carbohydrate source with added glucose and L-malate. The successful use of *Leuc. oenos* strains as starters for MLF is also related to the ability of the bacteria to adapt to a low pH and high alcohol content (Benda and Koehler, 1988).

By means of multivariate statistical analysis the growth response of several strains of *Leuc. oenos* to different pH, temperature values, and ethanol and sulphur dioxide concentrations were transformed into an informative plot showing the differences among the strains. The selected strain, *Leuc. oenos* 94, degraded L-malate completely in Aglianico wine and produced a good sensory quality (Rosi *et al.*, 1993).

7.5.3 Leuconostoc *in dairy products*

Leuconostoc mesenteroides subsp. *cremoris* and *Leuc. lactis* convert citrate to diacetyl and acetoin at low pH by using an inducible citrate lyase (Cogan *et al.*, 1981). The active metabolism of citrate and lactose, resistance to antibiotics, and growth interactions with other microorganisms are important criteria in the selection of a strain when incorporated in a dairy starter culture.

The organoleptic quality, consistency, texture, and eye formation of the following cheeses improved when selected strains of *Leuc. mesenteroides* and/or *Leuc. lactis* were included in the starter cultures: Manchego (Barneto and Ordonez, 1979; Ramos *et al.*, 1981), Danbo (Birkkjaer and Thompsen, 1981), Minas (Bonassi *et al.*, 1983), Sovetskii (Gudkov *et al.*, 1980), Fynbo (Barraquio *et al.*, 1983), Gouda (Toyoda and Kikuchi, 1983), Kefalotyri (Litopolou-Tzenetaki, 1990), water-buffalo Mozzarella (Coppola *et al.*, 1990), and Aracena (Garrido-Gomez *et al.*, 1991). Similarly, the organoleptic and keeping quality of fermented milks improved when fermented by leuconostocs (Kwan *et al.*, 1982; Abd-El-Gawad *et al.*, 1984; Taniuchi and Kiso, 1984; Marshall and Cole, 1985).

As with all the other leuconostocs, the starter cultures used in milk fermentations can be attacked by bacteriophages, causing a loss of specific aromatic properties. The acidification time of the fermented milk is, however, not necessarily influenced (Sozzi and Pirovano, 1993). *Leuconostoc* phages were isolated from Hungarian semi-hard cheese (Babella and Mike, 1977), Blue and cream cheese, Gouda, Edam, Camembert (Shin and Sato, 1979; Shin, 1983), Viili (Saxelin *et al.*, 1986), and from commercial starter cultures (Bannikova *et al.*, 1981; Lodics and Steenson, 1990). The phages isolated from cheese and butter dairy factories in Columbia had isometric heads and long non-contractile flexible tails and were identified as members of the Siphoviridae group of viruses (Sozzi and Pirovano, 1993).

7.5.4 Leuconostoc *in other foods*

The presence of *Leuconostoc* spp., associated with lactobacilli, is considered very important in the fermentation of dough, although the literature on this topic is scanty. According to Azar *et al.* (1977), the leuconostocs produce the desirable flavour compounds and sour taste in sangak dough.

However, further research is needed to determine the correct composition of these bacteria in the complex mixture of microorganisms. Dough fermentation could lead to exciting new products and merits further research.

Several papers have been published on the spoilage of meat by lactic acid bacteria. *Leuconostoc* spp. have been isolated from spoiled vacuum packed beef (Roxbeth *et al.*, 1991), dry sausages and raw hams (Hechelmann, 1986), and raw ripened sausage (Schillinger and Lücke, 1989). The authors suggested the use of starter culture metabolites, or the inclusion of bacteriocin-producing lactic acid bacteria in mixed cultures to control the negative effects of several microorganisms, including leuconostocs.

7.6 Conclusion

The leuconostocs are widespread in the natural environment and play an important role in several fermentations. Because of their physiological characteristics these bacteria are considered essential in maintaining the natural biological equilibrium. However, the inclusion of leuconostocs into starter cultures has to be carefully studied, since a wrong formula may lead to undesirable characteristics or spoilage of the product. It is important to study the growth interactions of leuconostocs and the other microorganisms in a mixed starter culture. Research should include a comparative study on specific strains of *Leuconostoc* spp. in combination with mesophilic and/or thermophilic starters. This will enable the microbiologists and technologists to obtain a real picture of the natural microbiological consortium responsible for a high-quality product and transfer this equilibrated potentiality in a starter for the production of modern hygienic and standardized foodstuffs.

7.7 List of species of the genus *Leuconostoc*

1 *Leuconostoc mesenteroides* (Tsenkovskii) (Van Tieghem 1879, p. 198[AL]; *Ascococcus mesenteroides* Tsenkovskii 1878, p. 159). me.sen.ter.oi'des. Gr.n. *mesenterium* the mesentery; Gr.n. *oides* form, shape; M.L.adj. *mesenteroides* mesentery-like. Cell morphology varies with growth conditions. Cells grown in a glucose medium are elongated and appear morphologically closer to lactobacilli than to streptococci. Most strains form coccoid cells when cultured in milk. Cells may occur singly or in pairs, and form short to medium length chains. When grown on solid medium, cells are elongated and can be mistaken for rods. True cellular capsules are not formed. Certain strains produce extracellular dextran which forms an electron-dense coat on the cell surface.

1a *Leuconostoc mesenteroides* subsp. *mesenteroides* (Tsenkovskii) (Van Tiegham 1879, p. 198[AL]). Dextran is produced from sucrose, especially when cells are grown at 20–25°C. Different colonial types are formed on sucrose agar, depending on the type of dextran produced. Cells which produce exopolysaccharides may withstand heating to 85°C. Temperature range: 10–37°C, optimum 20–30°C. Peptidoglycan type: Lys-Ser-Ala_2. The mol% G+C of the DNA is 37–39, with some strains 40–41 (T_m). Isolated from sugar milling plants, meat, milk, and various dairy products. Type strain: ATCC 8293 (NCFB 523, DSM 20343).

1b *Leuconostoc mesenteroides* subsp. *dextranicum* (Beijerinck) (Garvie, 1983, p. 118[VP]) (*Leuconostoc dextranicum* (Beijerinck) Hucker and Pederson 1930, p. 67; *Lactococcus dextranicus* Beijerinck 1912, p. 27). dex.tra'ni.cum. M.L.n. *dextranicum* dextran; M.L.neut.adj. *dextranicum* relating to dextran. Dextran is produced from sucrose. A number of sugars are fermented, but to a lesser extent than *Leuc. mesenteroides* subsp. *mesenteroides*. Requires more amino acids and vitamins for growth than *Leuc. mesenteroides* subsp. *mesenteroides*. Optimum growth temperatures and range are the same as for *Leuc. mesenteroides* subsp. *mesenteroides*. Peptidoglycan type: Lys-Ser-Ala_2. The mol% G+C of the DNA is 37–40 (T_m). Isolated from plants, meat, milk, and various dairy products. Type strain: ATCC 19255 (NCFB 529, DSM 20484).

1c *Leuconostoc mesenteroides* subsp. *cremoris* (Knudsen and Sorensen) (Garvie 1983, p. 118[VP]) (*Leuconostoc cremoris* (Knudsen and Sorensen) Garvie 1960, p. 288; *Betacoccus cremoris* Knudsen and Sorensen 1929, p. 81). Morphologically identical to subspecies *mesenteroides* and *dextranicum*, but long chains and flocculation in liquid medium are often observed. Metabolically less active than the other two subspecies of *Leuc. mesenteroides*. Fructose and sucrose are usually not fermented and dextran is not produced. Under certain growth conditions acetoin and diacetyl are formed from citrate. Requires a large number of amino acids and vitamins. Optimum growth temperature: 18–25°C. Peptidoglycan type: Lys-Ser-Ala_2. The mol% G+C of the DNA is 38–40 (T_m). All known strains have been isolated from milk and fermented milk products. Type strain: ATCC 19254 (NCFB 543, DSM 20346).

2 *Leuconostoc paramesenteroides* (Garvie 1967b, p. 446[AL]). pa.ra.me.sen.ter.oi'des. Gr.prep. *para* resembling; M.L. *mesenteroides* a specific epithet; M.L.adj. *paramesenteroides* resembling *Leuc. mesenteroides*. Morphologically similar to *Leuc. mesenteroides*. Dextran is not formed from sucrose. Amino acid requirements are complex and variable. Optimum growth temperature: 18–24°C, many strains grow at 30°C. Pseudocatalase positive when grown in a medium with low glucose concentrations. More tolerant to NaCl than *Leuc. mesenteroides*. Some strains will grow in medium with an initial pH below 5.0. Phenotypically very similar to non-dextran forming strains of *Leuc. mesenteroides*. Peptidoglycan type: Lys-Ala_2. The mol% G+C of the DNA is 37–38 (T_m). Type strain: ATCC 33313 (NCFB 803, DSM 20288).

Note: The type strain of *Leuc. paramesenteroides* (DSM 20288) shares a low DNA homology (6–19%) with *Leuc. mesenteroides* subsp. *mesenteroides*. The overall protein profile of strain DSM 20288 is also different from that recorded for all other *Leuconostoc* spp. (Dicks *et al.*, 1990). The latter technique enables the grouping of genetically closely related species. Based on DNA–rRNA hybridizations, *Leuc. paramesenteroides* is phylogenetically more related to *Lactobacillus confusus*, *Lb. viridescens*, *Lb. halotolerans*, *Lb. minor*, and *Lb. kandleri* than to the non-acidophilic leuconostocs (Garvie, 1981; Schillinger *et al.*, 1989).

3 *Leuconostoc lactis* (Garvie 1960, p. 290[AL]). lac.tis. L.n. *lac* milk; L.gen.n. *lactis* of milk. Morphologically similar to *Leuc. mesenteroides*. Lactose is usually fermented. Sucrose is not fermented and dextran is therefore not produced. Amino acid requirements are complex. Acetoin and diacetyl may be produced from citrate. The mol% G+C of the DNA is 43–45 (T_m). Mostly isolated from dairy products. Type strain: ATCC 19256 (NCFB 533, DSM 20202).

4 *Leuconostoc oenos* (Garvie 1967a, p. 431[AL]). oe.nos.′ Gr.n. *oinos* wine; Gr.gen.n. *oenos* of wine. Morphologically similar to *Leuc. mesenteroides*. Acidophilic and prefers an initial growth pH of 4.8. Requires a rich medium with complex growth factors and amino acids. Most strains will only grow in the presence of tomato juice, grape juice, apple juice, pantothenic acid, or 4′0-(β-glucopyranosyl) D-pantothenic acid. Grows in media supplemented with 10% (v/v) ethanol. Metabolically inactive and ferments only a few carbohydrates. Prefers fructose and usually ferments trehalose. In grape must or wine, pentose sugars (arabinose or xylose) are fermented before glucose, resulting in diauxic growth. Sucrose, lactose and maltose are not fermented. Although exopolysaccharides are usually not formed, certain strains are known to produce slime in medium containing grape juice (Dicks *et al.*, 1990). Growth in broth slow and usually uniform. Surface growth is enhanced by incubation in a 10% CO_2 atmosphere. Colonies develop usually only after five days and are less than 1 mm in diameter. Temperature range: 20–30°C. L-Malate is decarboxylated to L(+)-lactate in the presence of a fermentable carbohydrate. Nicotinamide adenine dinucleotide-dependent G6PDH is not present. Peptidoglycan type: L-Lys-L-Ala-L-Ser or L-Lys-L-Ser-L-Ser. The DNA base composition is 37–42 mol% G+C (T_m). Isolated from wine and related habitats. Type strain: ATCC 23279 (NCFB 1674, DSM 20252).

Note: *Leuc. oenos* is a genetically and phenotypically homogeneous species, as shown by DNA–DNA hybridizations and numerical analysis of total soluble cell protein patterns, respectively (Dicks *et al.*, 1990). The electrophoretic mobility of D(–)nLDH, 6PGDH and ADH, and the unique peptidoglycan amino acid type separates *Leuc. oenos* from the other *Leuconostoc* spp. (Kandler, 1970; Hontebeyrie and Gasser, 1977; Garvie, 1986; Dick *et al.*, 1990). Antisera prepared from *Leuc. oenos* show no cross-reactivity with anti-G6PDH and anti-D-nLDH of *Leuc. lactis* and anti-G6PDH of *Leuc. mesenteroides* subsp. *mesenteroides* (Garvie, 1986). Strains of *Leuc. oenos* share a DNA homology of 3–6% and 16–21% with the type strains of *Leuc. mesenteroides* subsp. *dextranicum* (NCFB 529) and *Leuc. mesenteroides* subsp. *mesenteroides* (NCFB 523), respectively. Furthermore, the type strain of *Leuc. oenos* (NCFB 1674) has a low DNA homology with strains of *Leuc. lactis* and *Leuc. paramesenteroides* (Dicks *et al.*, 1990). Groupings obtained by numerical analysis of total soluble cell protein patterns (Dicks *et al.*, 1990) corresponded well with those obtained by DNA–DNA hybridizations, confirming the low genotypic relatedness between *Leuc. oenos* and the other *Leuconostoc* spp. DNA–rRNA hybridizations and 16S rRNA sequencing analyses clearly indicated that members of *Leuc. oenos* are phylogenetically distant from other *Leuconostoc* spp. (Garvie, 1986). Changes in conserved positions of the 16S rRNA suggests that *Leuc. oenos* is a fast evolving organism (Yang and Woese, 1989). On the basis of results reported by Martinez-Murcia and Collins (1990), strains of *Leuc. oenos* are equidistant from the clusters *Leuc. paramesenteroides* and *Leuconostoc sensu stricto*. The taxonomic status of *Leuc. oenos* needs to be revised.

5 *Leuconostoc gelidum* (Shaw and Harding, 1989). ge'lidum. L.neut.adj. *gelidum* cold, referring to the ability to grow on chill-stored meat. Morphologically similar to *Leuc. mesenteroides*. Growth occurs at 1°C, most strains do not grow at 37°C. Most strains produce dextran from sucrose. Galactose is not fermented. The cellular fatty acids are straight-chain saturated, monounsaturated, and cyclopropane ring types, with tetradecanoic, hexadecanoic, *cis*-9,10-hexadecenoic, *cis*-11,12-octadecenoic, and *cis*-11,12-methyleneoctadecanoic acids predominating. Isolated from vacuum-packaged meat stored at low temperatures. The mol% G+C of the DNA of the type strain is 37 (T_m). The type strain of *Leuc. gelidum* shares a DNA homology of 9–31% with *Leuc. mesenteroides* subsp. *mesenteroides* (Shaw and Harding, 1989). Isolated from refrigerated meat. Type strain: NCFB 2775 (ATCC 49366).

6 *Leuconostoc carnosum* (Shaw and Harding, 1989). car.no'sum. L. neut.adj. *carnosum*, referring to flesh. Morphologically similar to *Leuc. mesenteroides*. Growth occurs at 1°C, most strains do not grow at 37°C. Most strains produce dextran from sucrose. The cellular fatty acids produced are identical to that recorded for *Leuc. gelidum*. The mol%

G+C of the DNA of the type strain is 39 (T_m). Isolated from vacuum-packaged meat stored at low temperatures. Type strain: NCFB 2776 (ATCC 49367).

Note: *Leuc. carnosum* is phenotypically closely related to *Leuc. gelidum*, except that most strains of *Leuc. carnosum* do not ferment L-arabinose, arbutin, salicin and xylose. The only reliable way to differentiate *Leuc. carnosum* from *Leuc. galidum* is by DNA–DNA hybridization. The type strain of *Leuc. carnosum* shares a DNA homology of 18% with the type strain of *Leuc. gelidum* and 23% with the type strain of *Leuc. mesenteroides* subsp. *mesenteroides* (Shaw and Harding, 1989).

7 *Leuconostoc pseudomesenteroides* (Farrow *et al.*, 1989). pseu.do.me.sen.ter.oi'des. Gr.adj. *pseudes*, false; Gr.n. *mesenterium*, the mesentery; Gr.n. *oides*, form, shape; M.L.adj. *pseudomesenteroides*, not the true *mesenteroides*. Morphologically similar to *Leuc. mesenteroides*. Grow at 10 and 37°C. Isolated from dairy products, food and clinical sources. The mol% G+C of the DNA ranges from 38.1–40.8 (T_m). Peptidoglycan type: Lys-Ser-Ala_2. Type strain: NCFB 768 (ATCC 49371).

Note: According to Farrow *et al.* (1989), *Leuc. pseudomesenteroides* shares a DNA homology of 18–48% with *Leuc. mesenteroides* subsp. *mesenteroides* and is genetically not closely related to *Leuc. citreum* and *Leuc. lactis*. However, Dicks *et al.* (1990) and Fantuzzi *et al.* (1992) recorded DNA homology values of 78 and 82% between *Leuc. pseudomesenteroides*, and *Leuc. mesenteroides* subsp. *mesenteroides* and *Leuc. mesenteroides* subsp. *dextrainicum*, respectively. The taxonomic status of *Leuc. pseudomesenteroides* needs to be revised.

8 *Leuconostoc citreum* (Farrow *et al.*, 1989). cit're.um. M.L.adj. *citreum*, lemon coloured. Morphology similar to *Leuc. mesenteroides*. Most strains produce a lemon yellow pigment. Differ from most strains of *Leuc. mesenteroides* subsp. *mesenteroides* by not fermenting ribose and raffinose. Grow at 10 and 30°C. Some strains grow at 37°C, but not at 40°C. Isolated from food and clinical sources. The mol% G+C of the DNA ranges from 38.0–40.3 (T_m). Shares a DNA homology of 21–30% with *Leuc. mesenteroides* subsp. *mesenteroides* and 27–56% with *Leuc. pseudomesenteroides* (Farrow *et al.*, 1989). Type strain: NCFB 1837 (ATCC 49370).

Note: Apart from some strains which produce a yellow pigment, *Leuc. citreum* is phenotypically not that clearly distinguished from *Leuc. pseudomesenteroides*. Sugar fermentations are not always consistent. However, all strains of *Leuc. citreum* thus far described do not ferment ribose and xylose.

9 *Leuconostoc amelibiosum* (Schillinger *et al.*, 1989). a'me.li.bi.o'sum. Gr.pref. a not; N.L.adj. *melibiosum* referring to disaccharide melibiose;

amelibiosum: not fermenting melibiose. Morphologically similar to *Leuc. mesenteroides*. Grows at 15°C, but not at 5°C or 45°C. Produces dextran from sucrose. Peptidoglycan type: Lys-Ala_2. The mol% G+C of the DNA of the type strain is 41.6 (T_m). The type strain shares a DNA homology of 40% with the type strain of *Leuc. mesenteroides* subsp. *mesenteroides* (Schillinger *et al.*, 1989). Isolated from sugar solutions and refineries. Type strain: DSM 20188 (ATCC 13146).

Note: Strain DSM 20188 is phenotypically closely related to *Leuc. mesenteroides* and was previously named *Leuc. mesenteroides* subsp. '*amelibiosum*' (Holzapfel, 1969; Kandler, 1970). However, strain DSM 20188 differs from *Leuc. mesenteroides* in its inability to ferment melibiose and raffinose and the absence of serine in its peptidoglycan interpeptide bridge.

10 *Leuconostoc argentinum* (Dicks *et al.*, 1993). ar.gen.ti'num. L.adj. argentinum, pertaining to Argentina. Morphologically similar to *Leuc. mesenteroides*. Growth temperature: 10–39°C, the optimum between 27 and 30°C. Fructose is not fermented and dextran is not produced from sucrose. The mol% G+C of the DNA of the type strain is 40.5 (T_m). The type strain shares a DNA homology of 9% with the type strain of *Leuc. mesenteroides* subsp. *mesenteroides*, and 35–39% with *Leuc. lactis* (Dicks *et al.*, 1993). The total soluble cell protein profiles of strains of *Leuc. argentinum* were different to those obtained for *Leuc. mesenteroides*, *Leuc. lactis* and *Leuc. paramesenteroides*, indicating that they are phenotypically not closely related. Isolated from Argentinian raw milk. Type strain: ATCC 51353 (LL 76).

11 *Leuconostoc fallax* (Martinez-Murcia and Collins, 1991). fal'lax. L.adj. *fallax* deceptive. Morphologically similar to *Leuc. mesenteroides*. Grows at 10 and 40°C, but not at 45°C. Peptidoglycan type: Lys-Ala_2. The mol% G+C of the DNA of the type strain is 40 (T_m). The type strain shares a DNA homology of 28–41% with strains of *Leuc. mesenteroides* subsp. *mesenteroides*, and 25% with *Leuc. paramesenteroides* (Schillinger *et al.*, 1989). Isolated from sauerkraut. Type strain: DSM 20189.

Note: Based on DNA–DNA hybridization studies, strain DSM 20189 is distantly related to *Leuc. mesenteroides*, *Leuc. amelibiosum*, *Leuc. lactis* and *Leuc. paramesenteroides* (Schillinger *et al.*, 1989). Data on 16S rRNA sequences confirm the latter findings and clearly show that *Leuc. fallax* is genetically distinct from other currently recognized *Leuconostoc* and *Lactobacillus* spp.

References

Abd-El-Gawad, I.A., Girgis, E.S., Mehriz, A.M., Anis, S.M.K. and Amer, S.N. (1984) Studies on the production of cultured buttermilk in Egypt. II. Type and ratio of starter culture. *Annals of Agricultural Science* (Moshtor), **21**, 739–747.

Amachi, T., Imamoto, S., Yoshizumi, H. and Senoh, S. (1970) Structure and synthesis of a novel pantothenic acid derivative, the microbial growth factor from tomato juice. *Tetrahedron Letters*, **56**, 4871–4874.

Amerine, M.A. and Kunkee, R.E. (1968) Microbiology of wine-making. *Annual Review of Microbiology*, **22**, 323–358.

Archibald, F.S. and Fridovich, I. (1981) Manganese, superoxide dismutase and oxygen tolerance in some lactic acid bacteria. *Journal of Bacteriology*, **146**, 928–936.

Arendt, E.K., Neve, H. and Hammes, W.P. (1991) Characterization of phage isolates from a phage-carrying culture of *Leuconostoc oenos* 58N. *Applied Microbiology and Biotechnology*, **34**, 220–224.

Azar, M., Ter-Sarkissian, N., Ghavifek, H., Ferguson, T. and Ghassemi, H. (1977) Microbiological aspects of sangak bread. *Journal of Food Science and Technology* (India), **14**, 251–254.

Babella, G. and Mike, Z.M. (1977) Bacteriophage contamination in semi-hard cheese factories and consequent variations in cheese quality. *Tejipar*, **26**, 1–16.

Bannikova, L.A., Mytnik, L.G. and Zadoyana, T. (1981) Role of test cultures detecting bacteriophages in standardization of technological process of manufacture of cultured milk products. *Molochnaya Promyshlennost*, **11**, 14–17, 42.

Barneto, R. and Ordonez, J.A. (1979) Preparation of a starter for industrial manufacture of Manchego cheese. *Alimentaria*, **107**, 39–44.

Barraquio, V.L., San Jose, M.T.D., Diongco, O.T. and Cruz, L.L. (1983) Fynbo cheese. I. Its acceptability and chemical quality as affected by the type of starter culture used and length of ripening period. *Philippine Journal of Veterinarian and Animal Sciences*, **6**, 301–306.

Barre, P. (1978) Identification of thermobacteria and homofermentative thermophilic pentose-utilizing lactobacilli from high temperature fermenting grape must. *Journal of Applied Bacteriology*, **44**, 125–129.

Benda, I. and Koehler, H.J (1988) Bacterienstarterculturen in der Keller wirtschaft – eine kritische betrachtung aus mikrobiologischer sicht. *Wein Wissenschaft*, **43**, 279–284.

Birkkjaer, H.E. and Thompsen, D. (1981) Deep-frozen starter cultures in cheesemaking and fermentation of lactose and citric acid by aroma-producing bacteria in fresh cheese. *Beretning Staten Forsogsmejeri*, **235**, 40.

Bonassi, I.A., Goldoni, J.S. and Kroll, B.L. (1983) Effect of the mesophilic lactic acid bacteria *Streptococcus cremoris*, *Streptococcus lactis*, *Streptococcus lactis* subsp. *diacetylactis* and *Leuconostoc cremoris* on the characteristics of Minas cheese. Organo-leptic properties. *Ciencia Tecnologie Alimentos*, **3**, 24–34.

Britz, T.J. and Tracey, R.P. (1990) The combination effect of pH, SO_2, ethanol and temperature on the growth of *Leuconostoc oenos*. *Journal of Applied Bacteriology*, **68**, 23–31.

Brooker, B.E. (1977) Ultrastructural surface changes associated with dextran synthesis by *Leuconostoc mesenteroides*. *Journal of Bacteriology*, **131**, 288–292.

Buckenhueskes, H., Schmidt, T. and Gierschner, K. (1984) Erste Ergebnisse über die milchsaure vergaerung von Roten Beeten. *Industrielle Obst- und Gemüseverwertung*, **69**, 367–372.

Cavin, J.F., Schmitt, P., Arias, A., Lin, J. and Divies, C. (1988) Plasmid profiles in *Leuconostoc* species. *Microbiologie Aliments Nutrition*, **6**, 55–62.

Cazelles, O. and Gnaegi, F. (1982) Enquête sur l'importance pratique du problème de bacteriophages dans le vin. *Revue Suisse de Viticulture, Arboriculture, Horticulture*, **14**, 267–270.

Chassy, B.M., Gibson, E.M. and Giuffrida, A. (1978) Evidence for plasmid-associated lactose metabolism in *Lactobacillus casei* subsp. *casei*. *Current Microbiology*, **1**, 141–144.

Chilson, P.O., Castello, L.A. and Kaplan, N.O. (1965) Studies on the mechanism of hybridization of lactic dehydrogenases *in vivo*. *Biochemistry*, (New York), **4**, 271–281.

Clewell, D.B. (1981) Plasmids, drug resistance, and gene transfer in the genus *Streptococcus*. *Microbiological Reviews*, **45**, 409–436.

Cogan, T.M. (1987) Co-metabolism of citrate and glucose by *Leuconostoc* spp.: effects on growth, substrates and products. *Journal of Applied Bacteriology*, **63**, 551–558.

Cogan, T.M., O'Dowd, M. and Mellerick, D. (1981) Effects of pH and sugar on acetoin

production from citrate by *Leuconostoc lactis*. *Applied and Environmental Microbiology*, **41**, 1–8.
Cogan, T.M., Fitzgerald, R.J. and Doonan, S. (1984) Acetolactate synthase of *Leuconostoc lactis* and its regulation of acetoin production. *Journal of Dairy Research*, **51**, 597–604.
Collins, E.B. and Speckman, R.A. (1974) Influence of acetaldehyde on growth and acetoin production by *Leuconostoc citrovorum*. *Journal of Dairy Science*, **57**, 1428–1431.
Condon, S. (1987) Responses of lactic acid bacteria to oxygen. *FEMS Microbiology Reviews*, **46**, 269–280.
Coppola, S., Villani, F., Coppola, R. and Parente, E. (1990) Comparison of different starter systems for water-buffalo Mozzarella cheese manufacture. *Le Lait*, **70**, 411–423.
Cox, D.J. and Henick-Kling, T. (1989) Chemiosmotic energy from malolactic fermentation. *Journal of Bacteriology*, **171**, 5750–5752.
Cox, D.J. and Henick-Kling, T. (1990) A comparison of lactic acid bacteria for energy-yielding (ATP) malolactic enzyme systems. *American Journal of Enology and Viticulture*, **41**, 215–218.
Crow, V.L., Davey, G.P., Pearce, L.E. and Thomas, T.D. (1983) Plasmid linkage of the D-tagatose 6-phosphate pathway in *Streptococcus lactis*: effect on lactose and galactose metabolism. *Journal of Bacteriology*, **153**, 76–83.
Cummins, C.S. and Harris, H. (1956) The chemical composition of the cell wall in some Gram-positive bacteria and its possible value as a taxonomic character. *Journal of General Microbiology*, **14**, 583–600.
Daba, H., Pandian, S., Gosselin, J.F., Simard, R.E., Huang, J. and Lacroix, C. (1991) Detection and activity of a bacteriocin produced by *Leuconostoc mesenteroides*. *Applied and Environmental Microbiology*, **57**, 3450–3455.
Daeschel, M.A., Andersson, R.E. and Fleming, H.P. (1987) Microbial ecology of fermenting plant materials. *FEMS Microbiology Reviews*, **46**, 357–367.
Daly, C. (1983) The use of mesophilic cultures in the dairy industry. *Antonie van Leeuwenhoek Journal of Microbiology*, **49**, 297–312.
Davies, F.L. and Gasson, M.J. (1981) Reviews of the progress of dairy science: genetics of lactic acid bacteria. *Journal of Dairy Research*, **48**, 363–376.
Davis, C.R., Silveira, N.F.A. and Fleet, G.H. (1985a) Occurrence and properties of bacteriophages of *Leuconostoc oenos* in Australian wines. Applied and Environmental Microbiology, **50**, 872–876.
Davis, C.R., Wibowo, D., Eschenbruch, R., Lee, T.H. and Fleet, G.H. (1985b) Practical implications of malolactic fermentation: A review. *American Journal of Enology and Viticulture*, **36**, 290–301.
Delclos, M. (1992) Vegetables preservation by a mixed organic acid fermentation. PhD thesis, University of Surrey, Guildford, UK.
Dellaglio, F. and Torriani, S. (1986) DNA–DNA homology, physiological characteristics and distribution of lactic acid bacteria from maize silage. *Journal of Applied Bacteriology*, **60**, 83–93.
Dellaglio, F., Vescovo, M., Morelli, L. and Torriani, S. (1984) Lactic acid bacteria in ensiled high-moisture corn grain: physiological and genetic characterization. *Systematic and Applied Microbiology*, **5**, 534–544.
Dellaglio, F., Dicks, L.M.T., Du Toit, M. and Torriani, S. (1991) Designation of ATCC 334 in place of ATCC 393 (NCDO 161) as the neotype strain of *Lactobacillus casei* and rejection of the name *Lactobacillus paracasei* (Collins *et al.*, 1989). Request for an opinion. *International Journal of Systematic Bacteriology*, **41**, 340–342.
DeMoss, R.D., Bard, R.C. and Gunsalus, I.C. (1951) The mechanism of heterolactic fermentation: a new route of ethanol formation. *Journal of Bacteriology*, **62**, 499–511.
de Vos, W.M., Underwood, H.M. and Davies, F.L. (1984) Plasmid encoded bacteriophage resistance in *Streptococcus cremoris* SK11. *FEMS Microbiology Letters*, **23**, 175–178.
Dicks, L.M.T. and Van Vuuren, H.J.J. (1987) Relatedness of heterofermentative *Lactobacillus* species revealed by numerical analysis of total soluble cell protein patterns. *International Journal of Systematic Bacteriology*, **37**, 437–440.
Dicks, L.M.T. and Van Vuuren, H.J.J. (1988) Identification and physiological characteristics of heterofermentative strains of *Lactobacillus* from South African red wines. *Journal of Applied Bacteriology*, **64**, 505–513.

Dicks, L.M.T. and Van Vuuren, H.J.J. (1990) Differentiation of *Leuconostoc* species by nicotinamide adenine dinucleotide-dependent D(–)-lactic dehydrogenase profiles. *FEMS Microbiology Letters*, **67**, 9–14.

Dicks, L.M.T., Van Vuuren, H.J.J. and Dellaglio, F. (1990) Taxonomy of *Leuconostoc* species, particularly *Leuconostoc oenos*, as revealed by numerical analysis of total soluble cell protein patterns, DNA base compositions, and DNA–DNA hybridizations. *International Journal of Systematic Bacteriology*, **40**, 83–91.

Dicks, L.M.T., Fantuzzi, L., Gonzalez, F.C., Du Toit, M. and Dellaglio, F. (1993) *Leuconostoc argentinum* sp. nov., isolated from Argentine raw milk. *International Journal of Systematic Bacteriology*, **43**, 347–351.

Doelle, H.W. (1971) Nicotinamide adenine dinucleotide-dependent and nicotinamide adenine dinucleotide-independent lactate dehydrogenases in homofermentative and heterofermentative lactic acid bacteria. *Journal of Bacteriology*, **108**, 1284–1289.

El-Gendy, S.M., Abdel-Galil, H., Shahin, Y. and Hegazi, F.Z. (1983) Characteristics of salt tolerant lactic-acid bacteria, in particular lactobacilli, leuconostocs and pediococci isolated from salted raw milk. *Journal of Food Protection*, **46**, 429–433.

Eltz, R.W. and Vandemark, P.J. (1960) Fructose dissimilation by *Lactobacillus brevis*. *Journal of Bacteriology*, **79**, 763–776.

Fantuzzi, L., Dicks. L.M.T., Du Toit, M., Reneiro, R., Bottazzi, V. and Dellaglio, F. (1992) Identification of *Leuconostoc* strains isolated from Argentine raw milk. *Systematic and Applied Microbiology*, **14**, 229–234.

Farrow, J.A.E., Facklam, R.R. and Collins, M.D. (1989) Nucleic acid homologies of some vancomycin-resistant leuconostocs and description of *Leuconostoc citreum* sp. nov. and *Leuconostoc pseudomesenteroides* sp. nov. *International Journal of Systematic Bacteriology*, **39**, 279–283.

Fitzgerald, F.M. (1983) Aerobic metabolism of group N streptococci and *Leuconostoc mesenteroides*. MSc thesis, University College, Cork, Republic of Ireland.

Garrido-Gomez, M.P., Barneto, R. and Quintana, M.A. (1991) Microbiological and physico-chemical study of Aracena cheese. *Chemie Mikrobiologie Technologie der Lebensmittel*, **13**, 173–177.

Garvie, E.I. (1960) The genus *Leuconostoc* and its nomenclature. *Journal of Dairy Research*, **27**, 283–292.

Garvie, E.I. (1967a) *Leuconostoc oenos* sp. nov. *Journal of General Microbiology*, **48**, 431–438.

Garvie, E.I. (1967b) The growth factor and amino acid requirements of species of the genus *Leuconostoc*, including *Leuconostoc paramesenteroides* (sp. nov.) and *Leuconostoc oenos*. *Journal of General Microbiology*, **48**, 439–447.

Garvie, E.I. (1969) Request for an opinion that the name *Leuconostoc citrovorum* be rejected and the name *Leuconostoc cremoris* be conserved. *International Journal of Systematic Bacteriology*, **19**, 283–290.

Garvie, E.I. (1974) Gram-positive cocci. In *Bergey's Manual of Determinative Bacteriology* (eds. Buchanan, R.E. and Gibbons, N.W.). Williams and Wilkins, Baltimore, MD, USA, pp. 510–512.

Garvie, E.I. (1975) Some properties of gas-forming lactic acid bacteria and their significance in classification. In *Lactic Acid Bacteria in Beverages and Food* (eds Carr, J.G., Cutting, C.V. and Whiting, G.C.). Academic Press, London, UK, pp. 339–349.

Garvie, E.I. (1976) Hybridization between the deoxyribonucleic acids of some strains of heterofermentative lactic acid bacteria. *International Journal of Systematic Bacteriology*, **26**, 116–122.

Garvie, E.I. (1980) Bacterial lactate dehydrogenases. *Microbiological Reviews*, **44**, 106–139.

Garvie, E.I. (1981) Sub-divisions within the genus *Leuconostoc* as shown by RNA/DNA hybridization. *Journal of General Microbiology*, **127**, 209–212.

Garvie, E.I. (1984) Separation of species of the genus *Leuconostoc* and differentiation of the leuconostocs from other lactic acid bacteria. *Methods in Microbiology*, **16**, 147–178.

Garvie, E.I. (1986) Genus *Leuconostoc* van Tieghem 1878, 198^{AL} emend mut. char. Hucker and Pederson 1930, 66^{AL}. In *Bergey's Manual of Systematic Bacteriology*, Vol. 2 (eds Sneath, P.H.A., Mair, N.S., Sharpe, M.E. and Holt, J.G.). Williams and Wilkins, Baltimore, MD, USA, pp. 1071–1075.

Garvie, E.I. and Farrow, J.A.E. (1980) The differentiation of *Leuconostoc oenos* from non-acidophilic species of *Leuconostoc*, and the identification of five strains from the American Type Culture Collection. *American Journal of Enology and Viticulture*, **31**, 154–157.

Garvie, E.I. and Mabbitt, L.A. (1967) Stimulation of the growth of *Leuconostoc oenos* by tomato juice. *Archives of Microbiology*, **55**, 398–407.

Garvie, E.I., Zezala, V. and Hill, V.A. (1974) Guanine plus cytosine content of the deoxyribonucleic acid of the leuconostocs and some heterofermentative lactobacilli. *International Journal of Systematic Bacteriology*, **24**, 248–251.

Gasser, F. (1970) Electrophoretic characterization of lactic dehydrogenases in the genus *Lactobacillus*. *Journal of General Microbiology*, **62**, 223–239.

Gasser, F. and Hontebeyrie, M. (1977) Immunological relationships of glucose-6-phosphate dehydrogenase of *Leuconostoc mesenteroides* NCDO 768 (=ATCC 12291). *International Journal of Systematic Bacteriology*, **27**, 6–8.

Gasser, F. and Mandel, M. (1968) Deoxyribonucleic acid base composition of the genus *Lactobacillus*. *Journal of Bacteriology*, **96**, 580–588.

Giannakopoulous, P.I., Markakis, P. and Howell, G.S. (1984) The influence of malolactic strain on the fermentation and wine quality of three eastern red wine grape cultivars. *American Journal of Enology and Viticulture*, **35**, 1–4.

Gonzalez, C.F. and Kunka, B.S. (1983) Plasmid transfer in *Pediococcus* spp.: Intergeneric and intrageneric transfer of pIP501. *Applied and Environmental Microbiology*, **46**, 81–89.

Gudkov, A.V., Anishchenko, I.P., Ostroumov, L.A. and Aleksoova, M.A. (1980) Characteristics of microbiological processes in Sovetskii cheese. *Molochnaya Promyshlennost*, **2**, 13–17, 47.

Guerzoni, M.E., Torriani, S., Sinigaglia, M. and Gardini, F. (1993) Modelling of the growth of *Lactobacillus plantarum* and *Leuconostoc oenos* in relation to the modulation of some chemico-physical factors. In *Proceedings of the Symposium on Biotechnology and Molecular Biology of Lactic Acid Bacteria for the Improvement of Foods and Feeds Quality* (eds Zamorani, A., Manachini, P.L., Bottazzi, V. and Coppola, S.). Istituto Poligrafico e Zecca dello Stato, Rome, Italy, pp. 280–288.

Hastings, J.W. and Stiles, M.E. (1991) Antibiosis of *Leuconostoc gelidum* isolated from meat. *Journal of Applied Bacteriology*, **70**, 127–134.

Hayman, D.C. and Monk, P.R. (1982) Starter culture preparation for the induction of malolactic fermentation in wine. *Food Technology in Australia*, **34**, 16–18.

Hechelmann, H. (1986) Mikrobiellverursachte Fehlfabrikate bei Rohwurst und Rohschinken. *Fleischwirtschaft*, **66**, 528.

Henick-Kling, T., Lee, T.H. and Nicholas, D.J.D. (1986a) Inhibition of bacterial growth and malolactic fermentation in wine by bacteriophage. *Journal of Applied Bacteriology*, **61**, 287–293.

Henick-Kling, T., Lee, T.H. and Nicholas, D.J.D. (1986b) Characterization of the lytic activity of bacteriophages of *Leuconostoc oenos* isolated from wine. *Journal of Applied Bacteriology*, **61**, 525–534.

Holzapfel, W.H. (1969) Aminosäuresequenz des Mureins und Taxonomie der Gattung *Leuconostoc*. Dissertation, Technischer Hochschule, Munich, Germany.

Holzapfel, W.H. and Gerber, E.S. (1986) Predominance of *Lactobacillus curvatus* and *Lactobacillus saki* in spoilage association of vacuum-packaged meat products. Paper presented at the 32nd European Meeting of the Meat Research Workers, Ghent, Belgium, 24–29 Aug.

Holzapfel, W.H. and Schillinger, U. (1992) The genus *Leuconostoc*. In *The Prokaryotes*, Vol. 2 (eds Balows, A., Trüper, H.G., Dworkin, M., Harder, W. and Schleifer, K.-H.). Springer-Verlag, Berlin, Germany, pp. 1508–1534.

Hontebeyrie, M. and Gasser, F. (1975) Comparative immunological relationships of two distinct sets of isofunctional dehydrogenases in the genus *Leuconostoc*. *International Journal of Systematic Bacteriology*, **25**, 1–6.

Hontebeyrie, M. and Gasser, F. (1977) Deoxyribonucleic acid homologies in the genus *Leuconostoc*. *International Journal of Systematic Bacteriology*, **27**, 9–14.

Hucker, G.J. and Pederson, C.S. (1931) A study of the physiology and classification of the genus *Leuconostoc*. *Zentralblatt für Bakteriologie, Parasitenkunde, Infektionskrankheiten und Hygiene (2. Abteilung Originale)*, **85**, 65–114.

Irwin, O.R., Subden, R., Lautensach, A. and Cunningham, J.P. (1983) Genetic heterogeneity in lactobacilli and leuconostocs of enological significance. *Canadian Institute of Food Science and Technology Journal*, **16**, 79–81.

Ito, S., Kobayashi, T., Ohta, Y. and Akiyama, Y. (1983) Inhibition of glucose catabolism by aeration in *Leuconostoc mesenteroides*. *Journal of Fermentation Technology*, **61**, 353–358.

Janse, B.J.H., Wingfield, B.D., Pretorius, I.S. and Van Vuuren, H.J.J. (1987) Plasmids in *Leuconostoc oenos*. *Plasmid*, **17**, 173–175.

Jarvis, A.W. (1989) Bacteriophages of lactic acid bacteria. *Journal of Dairy Science*, **72**, 3406–3428.

Johnson, L. (1986) Bacterial classification III. Nucleic acids in bacterial classification. In *Bergey's Manual of Systematic Bacteriology*, Vol. 2 (eds Sneath, P.H.A., Mair, N.S., Sharpe, M.E. and Holt, J.G.). Williams and Wilkins, Baltimore, MD, USA, pp. 972–975.

Johnson, M.K. and McCleskey, C.S. (1957) Studies on the aerobic carbohydrate metabolism of *Leuconostoc mesenteroides*. *Journal of Bacteriology*, **74**, 22–25.

Kandler, O. (1970) Amino acid sequence of the murein and taxonomy of the genera *Lactobacillus*, *Bifidobacterium*, *Leuconostoc* and *Pediococcus*. *International Journal of Systematic Bacteriology*, **20**, 491–507.

Kandler, O. (1983) Carbohydrate metabolism in lactic acid bacteria. *Antonie van Leeuwenhoek Journal of Microbiology*, **49**, 209–224.

Kandler, O. and Schleifer, K.-H. (1980) D. Taxonomy. I. Systematics of bacteria. In *Progress in Botany* (Fortschritte der Botanik 42) (eds Ellenberg, H., Esser, K., Kubitzki, K., Schnepf, E. and Ziegler, H.). Springer-Verlag, Berlin, Germany, pp. 234–252.

Kandler, O. and Weiss, N. (1986) Genus *Lactobacillus*. In *Bergey's Manual of Systematic Bacteriology*, Vol. 2 (eds Sneath, P.H.A., Mair, N.S., Sharpe, M.E. and Holt, J.G.). Williams and Wilkins, Baltimore, MD, USA, pp. 1209–1234.

Keenan, T.W. (1968) Production of acetic acid and other volatile compounds by *Leuconostoc citrovorum* and *Leuconostoc dextranicum*. *Applied Microbiology*, **16**, 1881–1885.

Keller, J.J., Holzapfel, W.H. and Steinman, M.A. (1987) The microbiological population differences between pasteurized and spoiled pasteurized milk. *South African Journal of Dairy Science*, **19**, 85–95.

Kelly, W.J., Asmundson, R.V. and Hopcroft, D.H. (1989) Growth of *Leuconostoc oenos* under anaerobic conditions. *American Journal of Enology and Viticulture*, **40**, 277–282.

Kersters, K. and De Ley, J. (1980) Classification and identification of bacteria by electrophoresis of their proteins. In *Microbiological classification and identification* (eds Goodfellow, M. and Board, R.G.). Academic Press, London, UK, pp. 273–297.

Kim, S.Y., Souane, M., Kim, G.E. and Lee, C.H. (1991) Microbial characterization of jangsu. *Korean Journal of Food Science and Technology*, **23**, 689–694.

Klaenhammer, T.R. (1984) Interactions of bacteriophages with lactic streptococci. *Advances in Applied Microbiology*, **30**, 1–29.

Klaenhammer, T.R. (1987) Plasmid-directed mechanisms for bacteriophage defence in lactic streptococci. *FEMS Microbiology Reviews*, **46**, 313–325.

Klaenhammer, T.R. (1988) Bacteriocins of the lactic acid bacteria. *Biochimie*, **70**, 337–349.

Knox, K.W. and Wicken, A.J. (1976) Grouping and cross-reacting antigens of oral lactic acid bacteria. *Journal of Dental Research*, **55**, A116–A122.

Kole, M., Altosaar, I. and Duck, P. (1983) Effect of vitamin supplements on growth of *Leuconostoc oenos*. *Journal of General Microbiology*, **48**, 1380–1381.

Korkeala, H., Suortti, T. and Makela, P. (1988) Ropy slime formation in vacuum-packed cooked meat products caused by homofermentative lactobacilli and a *Leuconostoc* species. *International Journal of Food Microbiology*, **7**, 339–347.

Krieger, S., De-Frenne, E. and Hammes, W.P. (1986) Ausfuehrung des biologischen Sauerabbaus im Wein mit *Leuconostoc oenos*. *Chemie Mikrobiologie Technologie der Lebensmittel*, **10**, 13–18.

Krieger, S., Hammes, W.P. and Henick-Kling, T. (1992) Effect of medium composition on growth rate, growth yield and malolactic activity of *Leuconostoc oenos* LOZH1-t7-1. *Food Microbiology*, **9**, 1–11.

Kwan, A.J., Kilara, A., Friend, B.A. and Shahani, K.M. (1982) Comparative B-vitamin content and organoleptic qualities of cultured and acidified sour cream. *Journal of Dairy Science*, **65**, 697–701.

Langston, C.W. and Bouma, C. (1959) A study of the microorganisms from grass silage. I. The cocci. *Applied Microbiology*, **8**, 212–222.

Lawrence, R.C. and Thomas, T.D. (1979) The fermentation of milk by lactic acid bacteria. In *Microbial Technology: Current State, Future Prospects* (eds Bull, A.T., Ellwood, D.C. and Ratledge, C.). University Press, Cambridge, UK, pp. 187–219.

Lee, S.H. and Kim, S.D. (1988) Effect of starters on fermentation of Kimchi. *Journal of the Korean Society of Food and Nutrition*, **17**, 342–347.

Lewus, C.B., Sun, S. and Montville, T.J. (1992) Production of an amylase-sensitive bacteriocin by an atypical *Leuconostoc paramesenteroides* strain. *Applied and Environmental Microbiology*, **58**, 143–149.

Litopolou-Tzanetaki, E. (1990) Changes in numbers and kinds of lactic acid bacteria during ripening of Kefalotyri cheese. *Journal of Food Science*, **55**, 111–113.

Lodics, T.A. and Steenson, L.R. (1990) Characterization of bacteriophages and bacteria indigenous to a mixed-strain cheese starter. *Journal of Dairy Science*, **73**, 2685–2696.

London, J. (1976) The ecology and taxonomic status of the lactobacilli. *Annual Review of Microbiology*, **30**, 279–301.

Lonvaud-Funel, A. and Strasser de Saad, A.M. (1982) Purification and properties of a malolactic enzyme from a strain of *Leuconostoc mesenteroides* isolated from grapes. *Applied and Environmental Microbiology*, **43**, 357–361.

Lonvaud-Funel, A., Joyeux, A. and Ledoux, O. (1991) Specific enumeration of lactic acid bacteria in fermenting grape must and wine by colony hybridization with non-isotopic DNA probes. *Journal of Applied Bacteriology*, **71**, 501–508.

Lucey, C.A. and Condon, S. (1986) Active role of oxygen and NADH oxidase in growth and energy metabolism of *Leuconostoc*. *Journal of General Microbiology*, **132**, 1789–1796.

Marshall, V.M. (1987) Lactic acid bacteria: starters for flavour. *FEMS Microbiology Reviews*, **46**, 327–336.

Marshall, V.M. and Cole, W.M. (1985) Methods for making kefir and fermented milks based on kefir. *Journal of Dairy Research*, **52**, 451–456.

Martinez-Murcia, A.J. and Collins, M.D. (1990) A phylogenetic analysis of the genus *Leuconostoc* based on reverse transcriptase sequencing of 16S rRNA. *FEMS Microbiology Letters*, **70**, 73–84.

Martinez-Murcia, A.J. and Collins, M.D. (1991) A phylogenetic analysis of an atypical *Leuconostoc*: description of *Leuconostoc fallax* sp. nov. *FEMS Microbiology Letters*, **82**, 55–60.

Mata, M. and Ritzenthaler, P. (1988) Present state of lactic acid bacteria phage taxonomy. *Biochimie*, **70**, 395–399.

McDaniel, M., Henderson, L.A., Watson Jr, B.T. and Heatherbell, D. (1987) Sensory panel training and screening for descriptive analysis of the aroma of Pino noir wine fermented by several strains of malolactic bacteria. *Journal of Sensory Studies*, **2**, 149–167.

McDonald, L.-C., Flemming, H.P. and Hanssen, H.M. (1990) Acid tolerance of *Leuconostoc mesenteroides* and *Lactobacillus plantarum*. *Applied and Environmental Microbiology*, **56**, 2120–2124.

McKay, L.L. (1983) Functional properties of plasmids in latic streptococci. *Antonie van Leeuwenhoek Journal of Microbiology*, **49**, 259–274.

Mundt, J.O. (1970) Lactic acid bacteria associated with raw plant food material. *Journal of Milk and Food Technology*, **33**, 550–553.

Mundt, J.O. and Hammer, J.L. (1968) Lactobacilli on plants. *Applied Microbiology*, **16**, 1326–1330.

Mundt, J.O., Graham, W.F. and McCarty, I.E. (1967) Spherical lactic acid-producing bacteria of southern-grown raw and processed vegetables. *Applied Microbiology*, **15**, 1303–1308.

Nel, L., Wingfield, B.D., Van der Meer, L.J. and Van Vuuren, H.J.J. (1987) Isolation and characterization of *Leuconostoc oenos* bacteriophages from wine and sugarcane. *FEMS Microbiology Letters*, **44**, 63–67.

Nonomura, H. and Ohara, Y. (1967) Die Klassifikation der Äpfelsäure-milchäure-bakterien. *Mitteilungsblatt in Klosterneuburg*, **17**, 449–466.

Olwage, M. (1992) Genetic characterisation of *Leuconostoc oenos* bacteriophage L5. MSc thesis, University of Stellenbosch, Stellenbosch, Republic of South Africa.

Orberg, P.K. and Sandine, W.E. (1984) Common occurrence of plasmid DNA and vancomycin resistance in *Leuconostoc* spp. *Applied and Environmental Microbiology*, **48**, 1129–1133.

O'Sullivan, T. and Daly, C. (1982) Plasmid DNA in *Leuconostoc* species. *Irish Journal of Food Science and Technology*, **6**, 206.

Pederson, C.S. and Albury, M.N. (1955) Variation among the heterofermentative lactic acid bacteria. *Journal of Bacteriology*, **70**, 702–708.

Peynaud, E. and Domercq, S. (1968) Étude de quatre cents souches de coques hétérolactiques isolés de vins. *Annales de l'Institut Pasteur* (Paris), **19**, 159–170.

Pivnick, H. (1980) Sugar, cocoa, chocolate, and confectioneries. In *Microbial Ecology of Foods*, Vol. 2 (eds Silliker, J.H., Elliot, R.P., Baird-Parker, A.C., Bryan, F.L., Christian, J.H.B., Clark, D.S., Olson Jr, J.C. and Roberts, T.A.). Academic Press, New York, USA, pp. 778–821.

Radler, F. (1986) Microbial biochemistry. *Experientia*, **42**, 884–893.

Radler, F. and Bröhl, K. (1984) The metabolism of several carboxylic acids by lactic acid bacteria. *Zeitschrift für Lebensmittel-Untersuchung und -Forschung*, **179**, 228–231.

Ramos, M., Barneto, R. and Ordonez, J.A. (1981) Evaluation of a specific starter for Manchego cheese production. *Milchwissenschaft*, **36**, 528–530.

Rankine, B.C. (1972) Influence of yeast strain and malo-lactic fermentation on composition and quality of table wines. *American Journal of Enology and Viticulture*, **23**, 152–158.

Rankine, B.C. and Bridson, D.A. (1971) Bacterial spoilage in dry red wine and its relationship to malo-lactic fermentation. *Australian Wine, Brewing and Spirit Review*, **90**, 44, 46, 48, 50.

Raya, R.R., Fremaux, G.L., De Antoni and Klaenhammer, T.R. (1992) Site-specific integration of the temperate bacteriophage øadh into the *Lactobacillus gasseri* chromosome and molecular characterization of the phage (*att*P) and bacterial (*att*B) attachment sites. *Journal of Bacteriology*, **174**, 5584–5592.

Reiter, B. and Oram, J.D. (1962) Nutritional studies on cheese starters. 1. Vitamin and amino acid requirements of single strain starters. *Journal of Dairy Research*, **29**, 63.

Rodriguez, S.B., Amberg, E., Thornton, R.J. and McLellan, M.R. (1990) Malolactic fermentation in Chardonnay: growth and sensory effects of commercial strains of *Leuconostoc oenos*. *Journal of Applied Bacteriology*, **68**, 139–144.

Rosi, I., Dinardo, E., Salicone, H. and Bertuccioli, H. (1993) Characteristics of lactic acid bacteria and wine quality. In *Proceedings of the Symposium on Biotechnology and Molecular Biology of Lactic Acid Bacteria for the Improvement of Foods and Feeds Quality* (eds Zamorani, A., Manachini, P.L., Bottazzi, V. and Coppola, S.). Istituto Poligrafico e Zecca dello Stato, Rome, Italy, pp. 264–273.

Roxbeth, M., Kalchayanand, N., Ray, B. and Field, R.A. (1991) Shelf-life extension of vacuum-packaged refrigerated beef using starter culture metabolites. *Proceedings of the Western Section of the American Society of Animal Science*, **42**, 50–53.

Sanchez-Marroquin, A. and Hope, P.H. (1953) Agave juice: Fermentation and chemical composition studies of some species. *Journal of Agricultural Food Chemistry*, **1**, 246–249.

Sandine, W.E. and Elliker, P.R. (1970) Microbiologically induced flavors and fermented foods flavor in fermented dairy products. *Journal of Agricultural Food Chemistry*, **18**, 557–562.

Savell, J.W., Hanna, M.O., Vanderzant, C. and Smith, G.C. (1981) An incident of predominance of *Leuconostoc* sp. in profile of steaks stored in O_2-CO_2–N_2 atmospheres. *Journal of Food Protection*, **44**, 742–745.

Saxelin, M.L., Nurmiaho-Lessila, E.L., Merilainen, V.T. and Forsen, R.I. (1986) Ultra-structure and host specificity of bacteriophages of *Streptococcus cremoris*, *Streptococcus lactis* subsp. *diacetylactis* and *Leuconostoc cremoris* from Finnish fermented milk 'Viili'. *Applied and Environmental Microbiology*, **52**, 771–774.

Schillinger, U. and Lücke, F.-K. (1989) Einsatz von Milchsäurebakterien als Schutzkulturen bei Fleischerzeugnissen. *Mitteilungsblatt der Bundesanstalt für Fleischforschung, Kulmbach*, **104**, 200–207.

Schillinger, U., Holzapfel, W.H. and Kandler, O. (1989) Nucleic acid hybridization studies on *Leuconostoc* and heterofermentative lactobacilli and description of *Leuconostoc amelibiosum* sp. nov. *Systematic and Applied Microbiology*, **12**, 48–55.

Schleifer, K.H. and Kandler, O. (1972) Peptidoglycan types of bacterial cell walls and their taxonomic implications. *Bacteriological Reviews*, **36**, 407–477.

Schmitt, P., Mathot, A.G. and Davies, C. (1989) Fatty acid composition of the genus *Leuconostoc*. *Milchwissenschaft*, **44**, 556–559.

Schmitt, P., Davies, C. and Cardona, A. (1992) Origin of end-products from the co-metabolism of glucose and citrate by *Leuconostoc mesenteroides* subsp. *cremoris*. *Applied Microbiology and Biotechnology*, **36**, 679–683.

Schütz, M. and Radler, F. (1973) Das 'Malatenzym' von *Lactobacillus plantarum* und *Leuconostoc mesenteroides*. *Archives of Microbiology*, **91**, 183–202.

Sharpe, M.E. (1970) Cell wall and cell membrane antigens used in the classification of lactobacilli. *International Journal of Systematic Bacteriology*, **20**, 509–518.

Sharpe, M.E. (1981) The genus *Lactobacillus*. In *The Prokaryotes*, Vol. 2 (eds Starr, M., Stolp, H., Trüper, H.G., Balows, A. and Schlegel, H.G.). Springer-Verlag, Berlin, Germany, pp. 1653–1679.

Sharpe, M.E., Garvie, E.I. and Tillbury, R.H. (1972) Some slime forming heterofermentative species of the genus *Lactobacillus*. *Applied Microbiology*, **23**, 389–397.

Shaw, B.G. and Harding, C.D. (1989) *Leuconostoc gelidum* sp. nov. and *Leuconostoc carnosum* sp. nov. from chill-stored meats. *International Journal of Systematic Bacteriology*, **39**, 217–223.

Shin, C. (1983) Some characteristics of *Leuconostoc cremoris* bacteriophages isolated from blue cheese. *Japanese Journal of Zootechnical Science*, **54**, 481–486.

Shin, C. and Sato, Y. (1979) Isolation of leuconostoc bacteriophages from dairy products. *Japanese Journal of Zootechnical Science*, **50**, 419–422.

Sozzi, T. and Pirovano, F. (1993) Bacteriophages of *Leuconostoc* spp. In *Proceedings of the Symposium on Biotechnology and Molecular Biology of Lactic Acid Bacteria for the Improvement of Foods and Feeds Quality* (eds Zamorani, A., Manachini, P.L., Bottazzi, V. and Coppola, S.). Istituto Poligrafico e Zecca dello Stato, Rome, Italy, pp. 252–263.

Sozzi, T., Maret, R. and Poulin, J.M. (1976) Mise en evidence de bactériophages dan le vin. *Experimentia*, **32**, 568–569.

Sozzi, T., Poulin, J.M., Maret, R. and Pousaz, R. (1978) Isolation of bacteriophage of *Leuconostoc mesenteroides* from dairy products. *Journal of Applied Bacteriology*, **44**, 159–161.

Sozzi, T., Gnaegi, F., D'Amico, N. and Hose, H. (1982) Difficultées de fermentation malolactique du vin dues à des bacteriophages de *Leuconostoc oenos*. *Revue Suisse de Viticulture Arboriculture et Horticulture*, **14**, 17–23.

Speckman, R.A. and Collins, E.B. (1968) Diacetyl biosynthesis in *Streptococcus diacetylactis* and *Leuconostoc citrovorum*. *Journal of Bacteriology*, **95**, 174–180.

Stackebrandt, E. and Teuber, M. (1988) Molecular taxonomy and phylogenetic position of lactic acid bacteria. *Biochimie*, **70**, 317–324.

Stackebrandt, E., Fowler, V.J. and Woese, C.R. (1983) A phylogenetic analysis of lactobacilli. *Pediococcus pentosaceus* and *Leuconostoc mesenteroides*. *Systematic and Applied Microbiology*, **4**, 326–337.

Steinkraus, K.H. (1983) Lactic acid fermentation in the production of foods from vegetables, cereals and legumes. *Antonie van Leeuwenhoek Journal of Microbiology*, **49**, 337–348.

Stirling, A.C. and Whittenbury, R. (1963) Sources of the lactic acid bacteria occurring in silage. *Journal of Applied Bacteriology*, **26**, 86–90.

Tagg, J.R., Dajani, A.S. and Wannamaker, L.W. (1976) Bacteriocins of Gram-positive bacteria. *Bacteriological Reviews*, **40**, 722–756.

Taniuchi, Y. and Kiso, C. (1984) Investigations into the manufacture of cultured cream. *Japanese Journal of Dairy and Food Science*, **33**, A93–A103.

Teuber, M. and Lembke, J. (1983) The bacteriophages of lactic acid bacteria with emphasis on genetic aspects of group N lactic streptococci. *Antonie van Leeuwenhoek Journal of Microbiology*, **49**, 283–295.

Thompson, J. (1979) Lactose metabolism in *Streptococcus lactis*: phosphorylation of galactose and glucose moieties in vivo. *Journal of Bacteriology*, **140**, 774–785.

Thompson, J. (1980) Galactose transport systems in *Streptococcus lactis*. *Journal of Bacteriology*, **144**, 683–691.

Tilbury, R.H. (1975) Occurrence and effects of lactic acid bacteria in the sugar industry. In

Lactic Acid Bacteria in Beverages and Food (eds Carr, J.G., Cutting, C.V. and Whiting, G.C.). Academic Press, London, UK, pp. 177–191.

Toyoda, S. and Kikuchi, T. (1983) Accelerated ripening of cheese. *Japanese Journal of Dairy Food Science*, **32**, A279–A286.

Tracey, R.P. and Britz, T.J. (1987) A numerical taxonomic study of *Leuconostoc oenos* strains from wine. *Journal of Applied Bacteriology*, **63**, 523–532.

Tracey, R.P. and Britz, T.J. (1989) The effect of amino acids on malolactic fermentation by *Leuconostoc oenos*. *Journal of Applied Bacteriology*, **67**, 589–595.

Valdez-de, G.F., Giori, G.S., Garro, M., Mozzi, F. and Oliver, G. (1990) Lactic acid bacteria from naturally fermented vegetables. *Microbiologie Aliments Nutrition*, **8**, 175–179.

Van Vuuren, H.J.J. and Dicks, L.M.T. (1993) *Leuconostoc oenos*: A review. *American Journal of Enology and Viticulture*, **44**, 99–112.

Van Wyk, C.J. (1976) Malo-lactic fermentation in South African table wines. *American Journal of Enology and Viticulture*, **27**, 181–185.

Vaughn, R.J. (1955) Bacterial spoilage of wines with special reference to California conditions. *Advances in Food Research*, **6**, 67–108.

Vaughn, R.H. (1985) The microbiology of vegetable fermentation. In *Microbiology of Fermented Foods*, Vol. 2 (ed. Wood, B.J.B.). Elsevier, New York, USA, pp. 49–109.

Visser, R., Holzapfel, W.H., Bezuidenhout, J.J. and Kotze, J.M. (1986) Antagonism of lactic acid bacteria against phytopathogenic bacteria. *Applied and Environmental Microbiology*, **52**, 552–555.

Von Holy, A. and Holzapfel, W.H. (1989) Spoilage of vacuum packaged processed meats by lactic acid bacteria, and economic consequences. Proceedings of the Xth WAFVH International Symposium, Stockholm, Sweden, 6–9 July 1989.

Weiller, H.G. and Radler, F. (1972) Vitamin- und Aminosäurebedarf von Milchsäurebakterien aus Wein und von Rebenblättern. *Mitteilungen der Hoeheren Bundeslehr und Versuchsanstalt für Wein und Obstbau Klosterneuburg Serie B Obst und Garten*, **22**, 4–18.

Whittenbury, R. (1963) The use of soft agar in the study of conditions affecting the utilization of fermentable substrates by lactic acid bacteria. *Journal of General Microbiology*, **32**, 375–385.

Whittenbury, R. (1966) A study of the genus *Leuconostoc*. *Archives of Microbiology*, **53**, 317–327.

Wibowo, D., Eschenbruch, R., Davis, C.R., Fleet, G.H. and Lee, T.H. (1985) occurrence and growth of lactic acid bacteria in wine: A review. *American Journal of Enology and Viticulture*, **36**, 302–313.

Williams, R.A.D. and Sadler, S. (1971) Electrophoresis of glucose-6-phosphate dehydrogenase, cell wall composition and the taxonomy of heterofermentative lactobacilli. *Journal of General Microbiology*, **65**, 351–358.

Woolford, M.K. (1984) The microbiology of silage. In *The Silage Fermentation*, Vol. 14 (eds Laskin, A.I. and Mateles, R.I.). Marcel Dekker, New York, USA, pp. 23–70.

Yang, D. and Woese, C.R. (1989) Phylogenetic structure of the 'leuconostocs': an interesting case of a rapidly evolving organism. *Systematic and Applied Microbiology*, **12**, 145–149.

8 The genus *Bifidobacterium*

B. SGORBATI, B. BIAVATI and D. PALENZONA

8.1 General description of the genus

The first recorded mention in the annals of science of the name '*bifidus*' as applied to a cell dates to 1900, when Tissier (1900) discovered in the faeces of breast-fed infants a rod-shaped, Gram-positive, non-gas-producing, anaerobic bacterium with bifid morphology which he termed *Bacillus bifidus*. At the beginning of the 20th century, Orla-Jensen, in a detailed paper on bacteria that produce lactic acid, classified the *Bacillus bifidus* as part of the Lactobacteriaceae family, and in 1924 he tried to propose it as a separate species, explaining that the various species of bifidobacteria "doubtless constitute a separate genus, possibly forming a connecting link between lactic acid bacteria and the propionic acid bacteria". Although studies of this bacterial group gradually declined therafter, since 1950 there has been a flourish of new research that has brought the initial listing of *Lactobacillus bifidus* in the seventh edition of Bergey's *Manual of Determinative Bacteriology* (Breed *et al.*, 1957) to the present 24 species noted in its latest edition (Scardovi, 1986). Since then an additional five species have been described (*B. gallicum*, Lauer, 1990; *B. gallinarum*, Watabe *et al.*, 1983; *B. ruminantium* and *B. merycicum*, Biavati and Mattarelli, 1991; *B. saeculare*, Biavati *et al.*, 1991a).

One of the initial studies that led to the separation of bifidobacteria from lactobacilli was a 1957 paper by Dehnert, who proposed dividing bifidobacteria into five groups on the basis of their sugar fermentation: either pentose negative or pentose positive. A more detailed study was published in 1963 by Reuter, who isolated seven new species of bifidobacteria in the faeces of children, adults and adolescents on the basis of fermentation and serological characteristics: *B. infantis*, *B. parvulorum*, *B. breve*, *B. liberorum*, *B. lactentis*, *B. adolescentis* and *B. longum*. Confirmation of Reuter's work was provided in 1969 by the Japanese researcher Mitsuoka, who added two new biotypes to the *B. longum* species (*B. longum* subsp. *animalis a* and *b*) and isolated for the first time from the faeces of pig, chicken, calf and rat the two new species *B. thermophilum* and *B. pseudolongum*. Simultaneously in Italy Scardovi *et al.* (1969) isolated *B. ruminale* and *B. globosum* from the rumen of cattle and *B. asteroides*, *B. indicum* and *B. coryneforme* from the intestine of the honey bee.

The species described to date can be regrouped in four different ecological niches: the human intestine, vagina and oral cavity, the animal intestine, the insect intestine and sewage. The intestinal microflora, one of the most complex ecological niches, is still not fully understood. Many types of facultative anaerobic bacteria live together, influencing each other. Host development age, the differing kinds of nourishment and the adaptability of each type influence the composition of microflora in the intestine, which can thus be considered as one of the most complex and variable of habitats. Only after the first week of life, or even later, is a comparatively permanent presence of microflora in the colon established.

While in the first day of life the faeces of both bottle- and breast-fed infants contain several facultative type groups, it takes another 5–7 days for flora bifida to be established: the bifidobacteria become the predominant bacteria in the faeces of breast-fed infants (Mitsuoka *et al.*, 1973; Biavati *et al.*, 1984) whereas no such predominance has ever been noted in the faeces of bottle-fed infants. This compositional diversity of the microflora in infants nourished with mother's or artificial milk is due to the fact that the former has specific growth factors for the development of bifidobacteria: for the *B. bifidum* var. *pennsylvanicus* the growth factor is an oligosaccharide mixture containing *N*-acetylglucosamine, glucose, galactose and fructose (Gyorgy, 1953; Gauche *et al.*, 1954). In the adult intestine the most frequently found bifid species are *B. pseudocatenulatum* and *B. longum* (Biavati *et al.*, 1986). Bifidobacteria have also been located in the oral cavity; the most common is *B. dentium* (Scardovi and Crociani, 1974), which seems to be involved in dental plaque formation.

Many types of bifidobacteria have been found in animal faeces (rabbit, chicken, cattle, mice and piglets), some of which seem to be 'host specific': *B. magnum* and *B. cuniculi* only in rabbit faeces, *B. pullorum* and *B. gallinarum* in the intestine of chicken, and *B. suis* only in piglet faeces. From a phylogenetically antique ecological niche two species have been isolated: *B. asteroides* is the only species found in *Apis mellifera* intestine and *B. indicum* in that of *Apis cerana* and *Apis dorsata* (from the Philippines and Malaysia) (Scardovi and Trovatelli, 1969). While in the early years of research it was widely believed that the bifidobacteria were typically confined to the ecological nich of the intestine, twelve species of this genus have been found in sewage, two of them, *B. minimum* and *B. subtile*, being found nowhere else (Biavati *et al.*, 1982): a fact that might be a sign of faecal pollution of water.

Many media have been studied for the isolation and the growth of bifidobacteria, essential components including sheep or horse blood (Dehnert, 1957; Reuter, 1963; Mitsuoka *et al.*, 1965), human milk (Dehnert, 1957), trypticase (BBL) and proteose peptone No. 3 (DIFCO), phytone (BBL), yeast extract (DIFCO), carbon source, mineral salts like Mg^{2+}, Ca^{2+}, Zn^{2+}, Mn^{2+}, Fe^{2+}, and vitamins like biotin and calcium pantothenate. One of the most frequently used media for isolation is TPY (Scardovi, 1986): 10 g trypticase (BBL); 5 g phytone (BBL); 15 g glucose; 2.5 g yeast extract (DIFCO); 1 ml Tween 80; 0.5 g cystein hydrochloride; 2 g K_2HPO_4; 0.5 g $MgCl_2.6H_2O$; 0.25 g $ZnSO_4{\cdot}7H_2O$; 0.15 g $CaCl_2$; $FeCl_3$ trace; and 15 g agar to 1 litre water at pH 6.5. Since on this non-selective medium other bacteria can develop, such as streptococci and lactobacilli, whose colony morphology is indistinguishable from that of bifidobacteria, it is necessary to isolate on selective ground. Among the latest media with a considerable degree of selectivity are those of Muñoa and Pares (1988) and Beerens (1990).

Bifidobacteria have cell, bifid or multiple-branching rods and exhibit pleomorphism: in adverse growth conditions its morphology undergoes changes resulting in a cell with greater branching. In a medium deficient in β-methyl-D-glucosamine the bifid cell takes on a more 'branched shape' (Glick *et al.*, 1960), and by adding certain amino acids (e.g. serine, alanine, aspartic acid) to a medium minimum and highly branched bifida cells become curved rods. The genus *Bifidobacterium* can generally be characterized as gram-positive, non-spore-forming, non-motile anaerobes (some species can tolerate O_2 only in the presence of CO_2) that are catalase-negative (excepting *B. asteroides* and *B. indicum* which are catalase positive when grown in the presence of air with or without the addition of hemin) and saccharoclastic (acetic and lactic acids are formed in the molar ratio of 3:2 without CO_2 production except in the degradation of gluconate). Small amounts of formic acid, ethanol and succinic acid are produced, while butyric and propionic acid never are. Glucose is degraded exclusively and characteristically by fructose-6-phosphoketolase (F6PPK-EC 4.1.2.22), which cleaves fructose 6-phosphate into acetylphosphate and erythose 4-phosphate. The G+C content of DNA (Bd or Tm) varies from 55–67 mol%.

Like all the Gram-positive bacteria, the cell wall of bifidobacteria consists of three macromolecular components: peptidoglycan (murein), polysaccharides and lipoteichoic acid. Kandler and Lauer (1974) and Lauer and Kandler (1983), in detailed studies on the peptidoglycan structures, showed that peptidoglycan is made up of *N*-acetylmuramic acid and *N*-acetylglucosamine, which form a polymer bound via oligopeptide chains whose interpeptide bridges differ from species to species, and proposed them as taxonomic characteristics. The same authors also report that the cell wall contains signficiant amounts of polysaccharides which usually consist of glucose and galactose and are often associated with rhamnose.

Two research groups, one from The Netherlands and the other from Germany, have taken an interest in the structure of the lipoteichoic acid, leading to controversial results. Veerkamp *et al.* (1983) in The Netherlands classified it chemically as the 1,2-linked phosphoglycerol oligomer containing either a galactose or glucose polymer chain at its hydrophilic end and the glycerol 1-phosphoryl-diacyl-galactofuranosyl-diacylglyceride component at its hydrophobic end. For Fischer (1987) in Germany the chemical structure of the lipotheicoic acid is a galactofuranose-glucopyranose polymer, with glycerophosphate residues attached to the galactose residues.

The bifidobacteria ferment various types of sugars. Some, like lactose, galactose and sucrose, are metabolized by a large number of species while others, like mannitol or sorbitol, are fermented by a restricted number of species. The fermentative characteristics of the species of the genus *Bifidobacterium* are reported in Table 8.1. Bifidobacteria also have extracellular or surface enzymes that catalyse the hydrolysis of complex polysaccharides: amylose, amylopectin, xylan, gum arabic (Salyers *et al.*, 1978); the gastric mucin of pigs is degraded in the species *B. bifidum*, *B. infantis* and *B. longum* (Hoskins *et al.*, 1985) by enzymes that are partially extracellular and partially associated with the cells; dextran is metabolized by a dextranase that seems to be an extracellular endolytic enzyme (α-1,6 glucosidase).

In 1965 Scardovi and Trovatelli were the first to discover the 'fructose-6-phosphate shunt' (Figure 8.1). This pathway, often referred to as the bifid pathway, is utilized by many authors as a marker for the genus *Bifidobacterium*. The key enzyme of this pathway is a fructose-6-phosphate phosphoketolase which splits the hexose phosphate to erythrose-4-phosphate and acetyl phosphate. From tetrose and hexose phosphates, through the subsequent action of transaldolase and transketolase, are formed pentose phosphates that, via the usual 2–3 cleavage, give rise to lactic acid and additional amounts of acetic acid so that lactic and acetic acids are formed in the theoretical ratio 1.0:1.5. Phosphoroclastic cleavage of some pyruvate to formic and acetic acids and the reduction of acetate to ethanol can often alter the fermentation balance (De Vries and Stouthamer, 1968) to a highly variable extent (Lauer and Kandler, 1976).

The enzyme fructose-6-phosphate phosphoketolase, the key enzyme of glycolytic fermentation, also serves as a taxonomic character in the identification of the genus but does not enable the distinguishing of one species from another. In order to identify one species from another biochemically, electrophoretic studies were carried out on the transaldolase enzymes (EC2.2.I.2) and 6-phosphogluconate dehydrogenase (ECI.I.I.44) (Scardovi *et al.*, 1979a). Fourteen isoenzymes of transaldolase and nineteen 6PGD have been identified: their distribution among the various bifidobacterial species is shown in Table 8.2. The urease activity found in nearly all the strains of the species *B. suis* and in a few strains of *B. breve*, *B. magnum* and *B. subtile*, strains grown in the absence of the urea, forces us to consider that this enzyme can not be induced (Craciani and Matteuzzi, 1982). The enzymes glutamate dehydrogenase and glutamine synthetase, which are involved in the assimilation of ammonia, were found

Figure 8.1 The fermentative characteristics of the species of the genus *Bifidobacterium**

	Sorbitol	L-Arabinose	Raffinose	D-Ribose	Starch	Lactose	Inulin	Cellobiose	Melezitose	Gluconate	Xylose	Mannose	Fructose	Galactose	Sucrose	Maltose	Threalose	Meliniose	Mannitol	Salicin
1. *B. bifidum*	−	−	−	−	−	+	−	−	−	−	−	−	+	+	d	−	−	±	−	−
2. *B. longum*	−	+	+	+	−	+	−	−	+	−	d	d	+	+	+	+	−	+	−	−
3. *B. infantis*	−	−	+	+	−	+	d	−	−	−	d	d	+	+	+	+	−	+	−	−
4. *B. breve*	d	−	+	+	−	+	d	d	d	−	−	+	+	+	+	+	d	−	d	+
5. *B. adolescentis*	d	+	+	+	+	+	d	+	+	+	+	d	+	+	+	+	d	−	d	+
6. *B. angulatum*	d	+	+	+	+	+	+	−	−	d	+	−	+	+	+	+	−	−	−	−
7. *B. catenulatum*	+	+	+	+	−	+	d	+	−	d	+	−	+	+	+	+	d	+	d	−
8. *B. pseudocatenulatum*	d	+	+	+	+	+	−	d	−	d	+	+	+	+	+	+	d	−	+	−
9. *B. dentium*	−	+	+	+	+	+	−	+	+	+	+	+	+	+	+	+	+	−	+	−
10. *B. globosom*	−	d	+	+	+	+	−	−	−	−	d	−	+	+	+	+	−	+	−	−
11. *B. pseudolongum*	−	+	+	+	+	d	−	d	d	−	+	+	+	+	+	+	−	+	−	−
12. *B. cuniculi*	−	+	−	−	+	−	−	−	−	−	+	−	−	+	+	+	−	+	−	−
13. *B. choerinum*	−	−	+	−	+	+	−	−	−	−	−	−	−	+	+	+	−	+	−	−
14. *B. animalis*	−	+	+	+	+	+	−	d	d	−	+	d	+	+	+	+	d	+	−	+
15. *B. thermophilum*	−	−	+	−	+	d	d	d	d	−	−	−	+	+	+	+	d	+	−	d
16. *B. boum*	−	−	+	−	+	d	+	−	−	−	−	−	+	+	+	+	−	+	−	−
17. *B. magnum*	−	+	+	+	−	+	−	−	−	−	+	−	+	+	+	+	−	+	−	−
18. *B. pullorum*	−	+	+	+	−	−	+	−	−	−	+	+	+	+	+	+	+	+	−	+
19. *B. gallinarum*	−	+	+	+	−	+	+	d	d	ND	+	d	+	+	+	+	+	+	−	+
20. *B. suis*	−	+	+	−	−	+	−	−	−	−	+	d	d	+	+	+	−	+	−	−
21. *B. minimum*	−	−	−	−	+	−	−	−	−	−	−	−	+	−	+	+	−	−	−	−
22. *B. subtile*	+	−	+	+	+	−	d	−	+	+	−	−	+	+	+	+	d	+	−	d
23. *B. coryneforme*	−	−	+	+	−	−	ND	+	−	+	+	−	+	ND	+	+	−	+	−	+
24. *B. asteroides*	−	−	+	+	−	−	−	+	−	d	+	−	+	d	+	d	−	+	−	+
25. *B. indicum*	−	−	+	+	−	−	−	+	−	+	−	d	+	d	+	d	−	+	−	+
26. *B. gallicum*	−	+	−	+	+	−	−	−	−	−	+	−	+	+	+	+	−	−	−	+
27. *B. ruminantium*	−	−	+	+	+	+	−	−	−	−	−	−	+	+	+	+	−	−	+	+
28. *B. merycicum*	−	+	+	+	+	+	∓	∓	−	−	+	−	∓	+	+	+	−	−	−	∓
29. *B. seaculare*	−	+	+	+	−	∓	+	−	+	−	+	+	+	∓	+	+	+	+	−	−

*Symbols: d, 11–89% of strains are positive; ND, not determined; ∓, when positive it is weakly fermented.

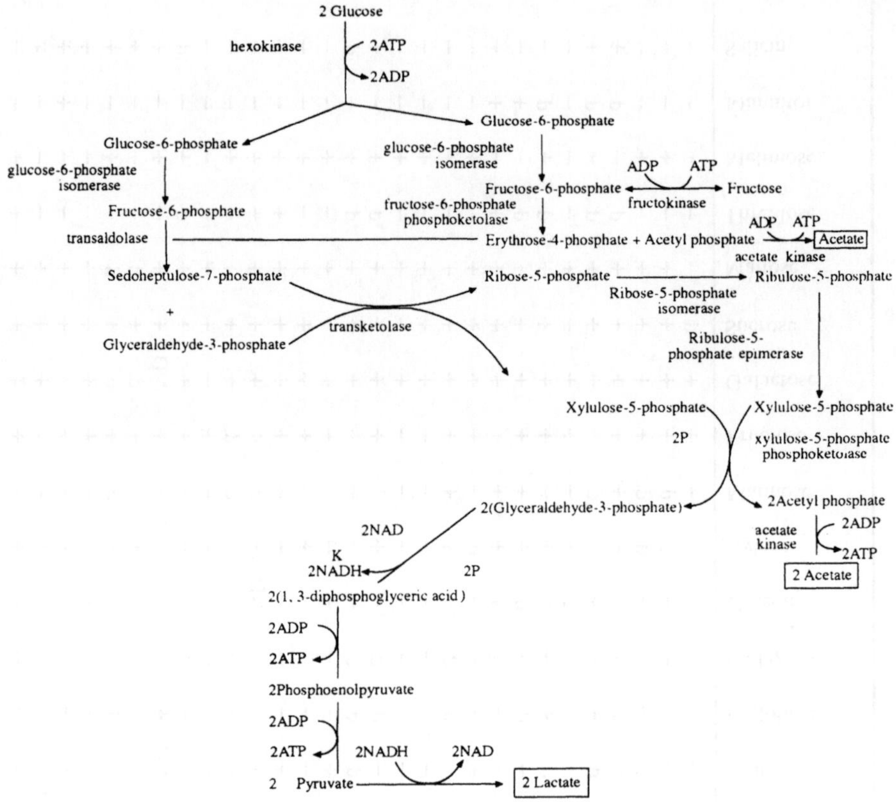

Figure 8.1 The fructose-6-phosphate shunt.

only in the species *B. bifidum*, *B. infantis*, *B. breve*, *B. adolescentis*, *B. thermophilum*, *B. longum* and *B. pseudolongum* (Hatanaka *et al.*, 1987).

The research carried out on the secretion activity of the bifidobacteria has highlighted a production of amino acids, the most frequent being alanine, valine and aspartic acid in strains of the species *B. thermophilum*, *B. adolescentis*, *B. dentium*, *B. animalis* and *B. infantis*. Exceptionally, *B. bifidum* can produce up to 150 mg/litre of threonine (Matteuzzi *et al.*, 1978). In 1977 Crociani *et al.* pointed out that the production of isoleucine and valine in mutants of *B. thermophilum* increased when the cell was grown in the presence of aminoisobutyric acid, a precursor of isoleucine.

Molecular-structure studies were initiated by Scardovi *et al.* (1971). Methodologies of DNA–DNA hybridization and the value of mol% G+C have reconfirmed species that had initially been proposed only on the basis of phenotypic characteristics (see Table 8.2). Subsequently, a total of 1500 bacteria isolates representing all the species of the genus were examined for the presence of plasmids. Approximately 20% contained detectable

Table 8.2 Mol% G+C values and electrophoretic patterns of transaldolase and 6-phosphogluconate dehydrogenase enzymes*

Species	% G+C (average values)	6-PGD isoenzyme	Transaldolase isoenzyme
1. *B. bifidum*	60.8	7	7
2. *B. longum*	60.8	5	6–8
3. *B. infantis*	60.5	4	3
4. *B. breve*	58.4	6–7	6
5. *B. adolescentis*	58.9	5	8
6. *B. angulatum*	59.0	5	5
7. *B. catenulatum*	54.0	6–8	5
8. *B. pseudocatenulatum*	57.5	1–3	4
9. *B. dentium*	61.2	–	4
10. *B. globosom*	63.8	6	2
11. *B. pseudolongum*	59.5	7	2
12. *B. cuniculi*	64.1	4	1
13. *B. choerinum*	66.3	4	3
14. *B. animalis*	60.0	8–9	5
15. *B. thermophilum*	60.0	7–8–9	8
16. *B. boum*	60.0	8–9	6
17. *B. magnum*	60.0	7	5
18. *B. pullorum*	67.5	Absent	2
19. *B. gallinarum*	65.7	Absent	Present
20. *B. suis*	62.0	5–8	6
21. *B. minimum*	61.5	6	10
22. *B. subtile*	61.5	2	3
23. *B. coryneforme*	—	6	6
24. *B. asteroides*	59.0	10	8
25. *B. indicum*	60.0	6–8	7–8–9
26. *B. gallicum*	61.0	ND	ND
27. *B. ruminantium*	57.0	Present	Present
28. *B. merycicum*	59.0	Present	Present
29. *B. saeculare*	63.0	Absent	Present

*Symbols: ND, not determined. Present, determined by other method.

plasmids but only four species were represented: *B. longum*, *B. globosum*, *B. asteroides* and *B. indicum*. The plasmid profile is multiple in *B. longum* and *B. asteroides* while the strains of the species *B. globosum* and *B. indicum* exhibit only one profile (Sgorbati *et al.*, 1982).

8.2 Phylogenetic relationships

The identification of bifidobacteria as a phylogenetically separate type and its relatedness to similar bacterial types have been the subject of studies and controversies for years. Indeed, bacterial phylogeny is no simple matter as it lacks the corroborating evidence of historical remnants that enabled evolutionists to trace so many differentiation paths for eukaryotes. As a matter of fact, since their discovery by Tissier, bifidobacteria had long been considered very similar to lactobacilli: almost a century after Tissier (1900), Poupard *et al.* (1973) and others proposed on the basis of the morphology of cells and colonies that bifidobacteria be considered phylogenetically related to *Actinomyces*. Of course, it is utterly inadequate to infer phylogenetic differentiation merely on the basis of a few different morphological traits, though it may be a starting point, a working hypothesis against which many other different criteria should be tested in order to reach something like a stochastic function of the stability of the differentiated state in time.

As a bold approximation, one might say that in bacteria the stability of the differentiated state in time must be inferred from the right combination of morphological, functional and genetically determined traits, and that the number of such traits should be as high as possible since the probability of a large number being random is small. Again, it should be considered that the discriminating value for a phylogenetic character is higher with the increasing conservation of its genetic basis. More simply, it may be stated that the more a trait is stable, the more it appears unchanged in a high number of different strains, and the higher is the probable length of its evolutionary life. Added to the general difficulty of bacterial evolution is the scarce development of the genetics of the Gram-positive bacteria. With reference to *Lactobacillus*, *Bifidobacterium*, *Leuconostoc* and related genera, this means very few characters are known to be genetically determined, which in turn means that our interference in the time function is even more questionable or that the stability of the differentiated state is even less proven.

Of the different criteria proposed to establish the differentiated state of the *Bifidobacterium*, some rely on the 'discrimination level' of a single trait while others try to accumulate many traits, each one not very discriminating in itself. An example of the first criterion is the study of the enzyme fructose-6-phosphate phosphoketolase (Scardovi and Trovatelli, 1965), whose discriminating value is supposed to be high since it is a key enzyme in the carbohydrate metabolism of the genus. Cellular extracts of *Bifidobacterium* isolates from a number of various habitats show the presence of the F6PPK activity, suggesting a high degree of conservation for this trait. However, it should be borne in mind that there is no convincing experimental proof of the genetic determination of this trait and, moreover, the genetic structure of the strains that do not show F6PPK activity is unknown.

Sgorbati and London (1982) examined transaldolase, an enzyme in carbohydrate metabolism, which is supposed to be very significant in bacterial evolution as it controls the source of energy for the bacterial cell. The double-diffusion method of Oucherlony and microcomplement fixation techniques were used to study the intercurrent immunological relationships between the transaldolases present in several *Bifidobacterium* species against eight antitransaldolase sera from *B. angulatum*, *B. asteroides*, *B. cuniculi*, *B. globosum*, *B. minimum*, *B. infantis*, *B. suis* and *B. thermophilum*. Species were grouped and immunological distances were estimated on the basis of similarity of amino acid sequences among enzyme molecules in different species (Sgorbati and Scardovi, 1979; Sgorbati and London, 1982). The degree of conservation of the primary structure of the enzymes was employed to develop a taxonomic index and an evolution dendrogram (Figure 8.2) for the grouping of species according to their habitat: human, animal, insect and sewage. The dendrogram suggests that isolates from insects and sewage habitats are at an extreme position in the space defined by antisera reactivity, and this might be interpreted as evidence of a longer evolutionary history of the species as represented by the isolates considered.

Interferences from these results should be considered as of the recurrently additive type, since the trait used to collect experimental evidence, even if important, cannot be taken as representative of a taxonomic entity, much less of a stable differentiated state of a bacterium. For this to be inferred one must be convincingly informed of the stability of the character on a genetic basis (and possibly with additional evidence of a different kind as well): of course, the evidence proposed by Poupard *et al.* (1973) and by Holdeman *et al.* (1977), based on the quantity of acetic acid produced (in *Bifidobacterium* it is expected to be larger than that of lactic acid but many genetic and environmental factors could hinder the expected result) needs an even broader basis of investigation since the specific informative value of the trait about *Bifidobacterium* evolution is very scarce. The same considerations hold too for the findings of Kandler (1970), who on the basis of the murein structure of the cell wall suggests that bifidobacteria are more closely related to lactobacilli than to *Actinomyces*.

The lack of genetic information can partly be overcome by studying the nucleic acid structure, but the phenotypes corresponding to the base sequences are not known in bacteria from these genera and therefore the interferences resulting from information on nucleic acid structure are still weak. Even the nature of the evidence is sometimes not fully informative. For example, the relative proportion of GC bases: without clear-cut information of its genetic value, it should be considered as a tentative differentiation with a possibly ambiguous meaning. Fox *et al.* (1977) report studies of rRNA sequences taken as 'evolutionary markers' on the assumption that these sequences are extensively conserved during the course of evolution. Woese *et al.* (1978) divide prokaryotes into *Archaebacteria* (including the methane-producing bacteria) and *Eubacteria* (including the bacteria associated with human physiology and pathology), bifidobacteria being included in the family Actinomycetaceae. Yet it must be noted that conservation alone is not very informative on the

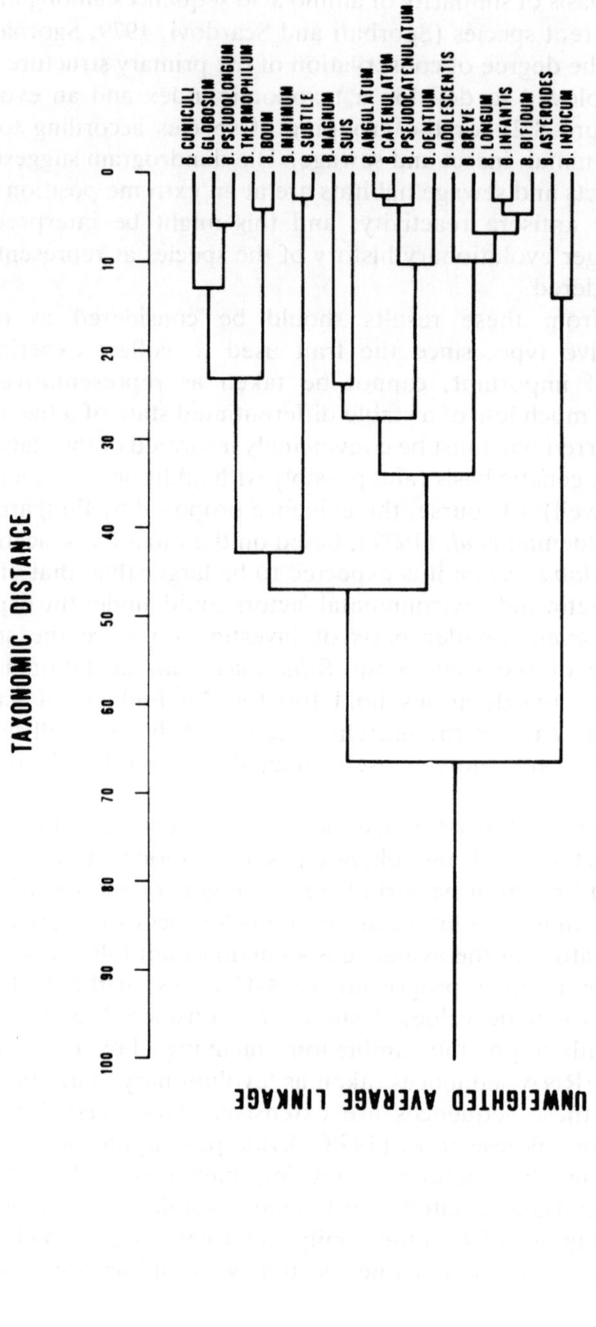

Figure 8.2 Dendrogram of the evalutionary relationships among the species of the genus *Bifidobacterium*.

evolutionary value of the trait studied unless the trait itself is shown to be subjected to selective pressure; otherwise, conservation may only mean not evolved.

8.3 The importance of the genus

The bifidobacteria that normally constitute part of the microflora in the human body are distributed in various niches: in the mouth, the ileum, the colon (where they are present in counts of 10^9/g and 10^{10}/g), the vagina and the cervix. Only the species *B. dentium* seems to cause dental caries, while all the other species, especially the intestinal ones, have a beneficial effect by maintaining a proper balance in the human intestinal flora. Given the therapeutic action of the intestinal bacterial group, both the food industry and pharmaceutical manufacturers have started adding cultures of bifidobacteria to milk and its by-products (yogurt in particular) and to pharmaceutical preparations (Rasìc and Kurmann, 1983; Bezkorovainy and Miller-Catchpole, 1989). The most common species used are *B. bifidum*, *B. longum* and *B. breve*, often combined with cultures of *Lactobacillus acidophilus* and *Streptococcus thermophilum* to facilitate acidification. These pharmaceutical preparations have been used in correcting the balance of intestinal microflora after antibiotic therapy for cirrhosis of the liver, in enterocolitis and in the case of intestinal stypsis.

The high levels of ammonia in the intestine, which enters the blood circulation in patients with portal systemic encephalopathy, can induce neurological disorders and the eventual development of coma (Podolsky and Isselbacher, 1987). Among the various therapies for the reduction of intestinal ammonia is a treatment which contains a strain of *B. bifidum* (Eugalanforte) or lactulose (Cephulac) (Kosman, 1976). The lactulose is not metabolized gastrically but in the colon, and it is degraded to lactic acid and acetic acid by the bifidobacteria, lactobacilli and streptococci. Lactulose has a beneficial effect for various reasons: it increases, albeit moderately, the growth of Gram-positive anaerobic bacteria, inhibits the growth of aerobic bacteria which are the major producers of ammonia, the ammonia produced in minor quantities lowers the pH and transforms itself into non-diffuse ammonia and therefore it is not absorbed by the intestine. The final result is that the acid produced lowers the pH, so transforming the ammonia to ammonium ion, which cannot diffuse through the intestinal wall into the bloodstream, with a consequent clinical improvement in the patient.

A clinical application of bifidobacteria was studied by Sekine *et al.* (1985): they injected 10^9 bifidobacteria (*B. infantis*) into a Meth-A sarcomatous tumour that had been subcutaneously transplanted into a BALB/c mouse and, after five injections, noticed a regression in tumour mass. To determine which components of the cellular bacteria had an antitumoural action, three types of cell-wall preparations were then tested: whole peptidoglycan (WPG), cell-wall skeleton (CSW) and sonicated peptidoglycan (WPG). The impact of tumour regression after five injections was 70% (WPG), 40% (sonicated WPG) and 20% (CWS). From these results it can be inferred that it is the integrity of the cell wall (peptidoglycan in particular) that produces an antineoplastic action (Old *et al.*, 1959). These findings, recorded in experiments on mice only and not on man, suggest that bifidobacteria act like 'immunomodulators' that stimulate the reticuloendothelial system and activate natural killer cells, thereby strengthening the immunological defences against the neoplastic cells (Kohwi *et al.*, 1978).

8.4 List of species of the genus *Bifidobacterium*

The Approved Lists of Bacterial Names (Skerman *et al.*, 1980), which contain only the names of previously adequately described taxa, report 20 species under the genus *Bifidobacterium*. Another nine species have since been described and their names validated either because published in the *International Journal of Systematic Bacteriology* or validated under the procedure adopted in 1977 (Lessel, 1977). The following list reports the main information concerning each species together with the strains available at the ATCC (American Type Culture Collection), DSM (Deutsche Sammlung von Mikroorganismen) and JCM (Japan Collection of Microorganisms). Since certain morphological traits are characteristic of many *Bifidobacterium* species, the morphologies in Figure 8.3 are useful for species differentiation. Those features of the genus shared by a large part of the various species are reported in the general description and certain peculiar characteristics are discussed in a subsequent section of this chapter.

1 *Bifidobacterium adolescentis* (Reuter, 1963). In 1963 Reuter proposed four biovars (*a*, *b*, *c*, *d*) based on serological reactions and differences in the fermentation of mannitol and sorbitol. A study of DNA–DNA homology by Scardovi *et al.* (1971) indicates there is often little genetic relatedness between strains having the characteristics of Reuter's biovar groups. For example, biovars *b* and *d* of *B. adolescentis*, i.e. those which do not ferment sorbitol, cannot be distinguished phenotypically from *B. dentium*. Difficulties in distinguishing *B. adolescentis* on the basis of phenotypical characteristics from other bifidobacteria isolated from the faeces of human adults are reported by Yaeshima *et al.* (1992a), who used DNA base compositions and DNA–DNA homologies to assign phenotypically similar strains to different species; the PAGE procedure can also

be employed as an alternative (Biavati *et al.*, 1982, 1986). The cells exhibit a morphology very similar to that of many other species of the genus (Figure 8.3(1)). Isolated from faeces of human adult, bovine rumen and sewage. Type strain: E194a = ATCC 15703 = DSM 20083 = JCM 1275. Other strains: ATCC 11146; ATCC 15704 = JCM 7042; ATCC 15705 = DSM 20086 = JCM 7046; ATCC 15706; DSM 20087 = JCM 1251.

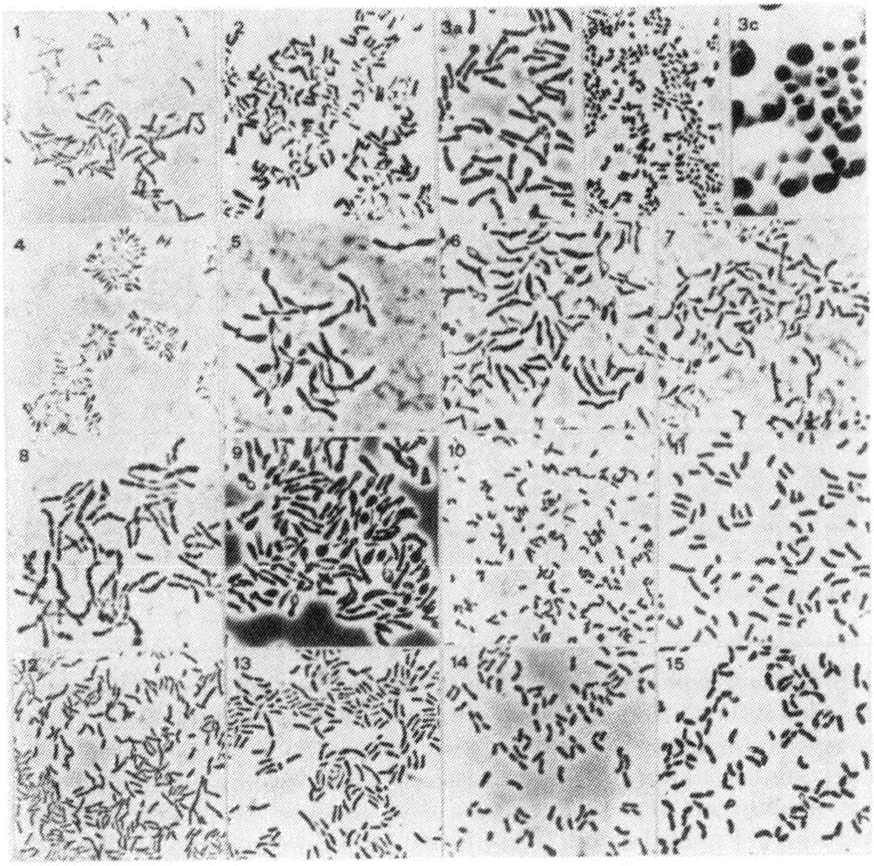

Figure 8.3 (cont'd on p. 292) Cellar morphology in the genus *Bifidobacterium*. Cells of the type strain were grown in anaerobic TPY medium stabs. Phase contrast photomicrographs × 825. (1) *B. adolescentis*; (2) *B. angulatum*; (3) *B. animalis* strain P23: a and b, from O and T colonies respectively; c, O and T colonies; (4) *B. asteroides*; (5) *B. bifidum*; (6) *B. boum*; (7) *B. breve*; (8) *B. catenulatum*; (9) *B. choerinum*; (10) *B. coryneforme*; (11) *B. cuniculi*; (12) *B. dentium*; (13) *B. gallicum*; (14) *B. gallinarum*; (15) *B. globosum*; (16) *B. indicum*; (17) *B. infantis*; (18) *B. longum*; (19) *B. magnum*; (20) *B. merycicum*; (21) *B. minimum*; (22) *B. pseudocatenulatum*; (23) *B. pseudolongum*; (24) *B. pullorum*; (25) *B. ruminantium*; (26) *B. saeculare*; (27) *B. subtile*; (28) *B. suis*; (29) *B. thermophilum*.

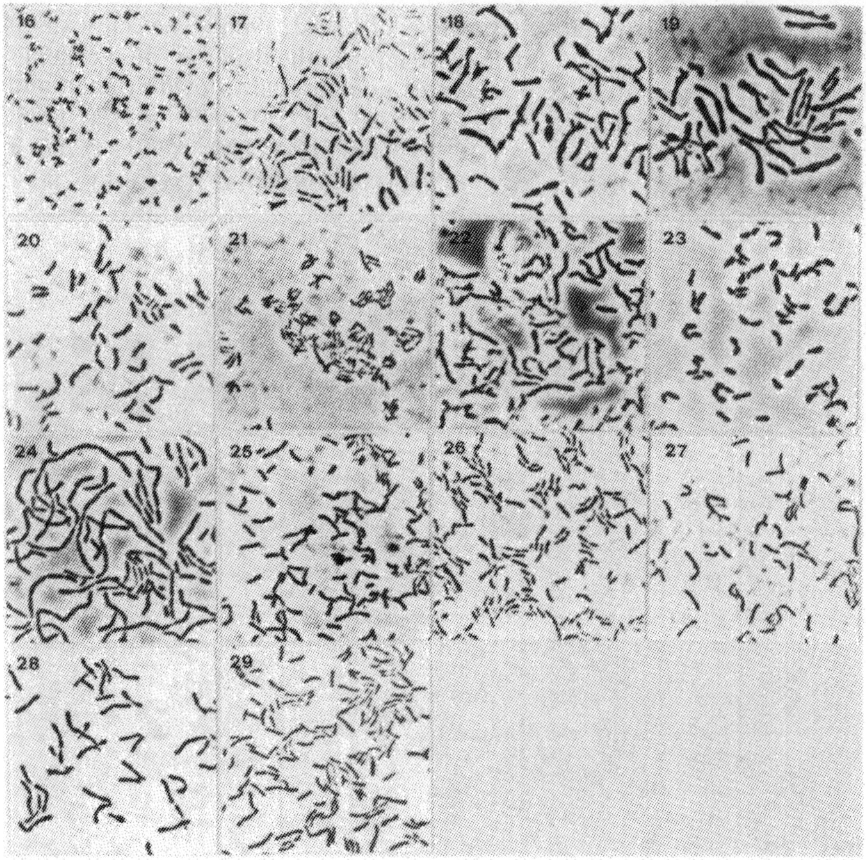

Figure 8.3 *cont'd*

2 *Bifidobacterium angulatum* (Scardovi and Crociani, 1974). This species was first isolated from human adults (only the type strain) and then found in sewage. Since no strains have been assigned to *B. angulatum* in many other studies concerning the presence of bifidobacteria in human faeces, the assumption that this species is a member of human intestinal microbiota is questionable. Its sensitivity to O_2 as measured by the depth of growth in stab is greater than that of most bifidobacteria; and CO_2 strongly enhances anaerobic growth. The morphology of cells grown in TPY agar stabs is unique among bifidobacteria (Figure 8.3(2)); characteristic are the arrangements of cells either in a V-shape or in palisade form similar to corynebacteria. Isolated from adult human faeces and sewage. Type strain: B 677 = ATCC 27535 = DSM 20098 = JCM 7096. Other strains. ATCC 27669; ATCC 27670 = DSM 20225 = JCM 1252; ATCC 27671.

3 *Bifidobacterium animalis* (Mitsuoka, 1969; Scardovi and Trovatelli, 1974). In 1969 Mitsuoka isolated from the faeces of various animals strains phenotypically very similar to *B. longum* and referred to as *B. longum* subsp. *animalis*; two biovars (*a*, *b*) were described. Two strains, one for each biovar, used as reference were found unrelated to *B. longum* upon DNA homology: one of them (biovar *b*) was genetically *B. pseudolongum* (Scardovi *et al.*, 1971) and the other (biovar *a*) was assigned to a group of isolates from animal faeces and sewage and proposed as *Bifidobacterium animalis* (Scardovi and Trovatelli, 1974). Recently, Biavati *et al.* (1992b) have identified the bifidobacteria from six samples of fermented milk products as *B. animalis* which, contrary to label information, was the only species of bifidobacteria present. It is generally held that bifidobacteria help to maintain a proper balance in human intestinal flora, and at the same time it is intuitively logical to assume that the bifidobacterial strains used as probiotic elements in the preparation of milk products should be from the human habitat. Strains of *B. animalis* isolated from commercial fermented milk products, faeces of rabbit, chicken and rat and from sewage have shown phase variations in colony appearance and in cellular morphology (Biavati *et al.*, 1992a). Some findings are documented in Figure 8.3(3a)–(3c). Most significant is the fact that the transition to colony morphotype transparent (T) and opaque (O) is accompanied by a dramatic change in cell dimension: minute and mostly spherical from T colonies while those from O colonies are large and show the species-specific shape, i.e. the central portion slightly enlarged. Isolated from faeces of calf, chicken, guinea pig, rabbit and rat; sewage; fermented milk products. Type strains: R 101–8 = ATCC 25527 = DSM 20104 = JCM 1190. Other strains: ATCC 27536 = DSM 20105 = JCM 1253; ATCC 27672; ATCC 27673; ATCC 27674 = JCM 7117.

4 *Bifidobacterium asteroides* (Scardovi and Trovatelli, 1969). Synonym: *Bacillus constellatus. Bifidobacterium asteroides*, one of three species isolated from honey bee gut, is the only one found in the intestine of *Apis mellifera* irrespective of geographical provenance (Scardovi and Trovatelli, 1969). Although unknown at present, the significance and the origin of bifidobacteria in the honey bee gut would make an interesting chapter of insect microbiology if they were found elsewhere in this class of animals. Cells grown anaerobically in fresh rich media are 2–2.5 μm long, pear-shaped or slightly curved, tend to have pointed ends and are usually arranged radially around a mass of common hold-fast material (Figure 8.3(4)). Colonies are circular, smooth and convex and feature a uniformly entire edge and glistening surface; their density is such that a colony can be removed by needle and can hardly be dispersed in water. Growth in static fluid culture tends to adhere to glass walls and to leave the liquid clear. A study of the presence of isozymes of transaldolase and 6-phosphogluconate dehydrogenase (6-PGD) in 85 strains detected eight transaldolases and nine 6-PGDs (Scardovi *et al.*, 1979a). Of the 224 strains tested for the presence of plasmids 33% contained a large variety of extra chromosomal elements of varying molecular weight (Sgorbati *et al.*, 1982). Preliminary data on the structural relatedness among plasmids were collected by means of blot hybridization using many selected unrestricted plasmic probes. Thirteen plasmids were found in *B. asteroides* and their frequencies and distribution reported (Sgorbati *et al.*, 1986), although the functions coded are still unknown. Isolated from the hind-gut of western honey bees, occasionally found in *Apis cerana*, an Asiatic honey bee. Type strain: C51 = ATCC 25910 = DSM 20089. Other strain: ATCC 25909 = DSM 20431.

5 *Bifidobacterium bifidum* (Tissier, 1900; Orla-Jensen, 1924). Synonyms: *Bacillus bifidus communis*, *Bacillus bifidus*, *Bacteroides bifidus*, *Bacterium bifidum*, *Tissieria bifida*, *Nocardia bifida*, *Actinomyces bifidus*, *Actinobacterium bifidum*, *Lactobacillus bifidus*, *Lactobacillus parabifidus*, *Cohnistreptotrix bifidus*. Type species of the genus. Strains of this species have most often been used for the production of foods containing bifidobacteria, such as cultured milk and other dairy products, and in many therapeutic preparations for the treatment of digestive disturbances in infants that are not breast fed, of enterocolitis, constipation, cirrhosis of the liver, imbalance of intestinal flora following antibiotic therapy and for the promotion of intestinal peristalsis. Some cells grown in TPY agar stabs have an 'amphora-like' cell shape (Figure 8.3(5)). They are anaerobic, dying rapidly in aerobic subcultures. *Bifidobacterium bifidum* is among the most easily recognized species in the genus because of its characteristic fermentation pattern. Isolated from faeces of human adults and infants, suckling calf and human vagina. Type strain: TI = ATCC 29521 = DSM 20456 = JMC 1255. Other strains: ATCC 11863 = DSM 20239 = JMC 1209; ATCC 15696 = JCM 7004; ATCC 35914; DSM 20082 = JCM 1254; DSM 20215.

6 *Bifidobacterium boum* (Scardovi *et al.*, 1979b). From a large number of animal strains surveyed by DNA–DNA hybridization, 36 strains from the rumen and five from piglet faeces were assigned to the new species *B. boum* (Scardovi *et al.*, 1979b). Two other species found in the same habitat, *B. choerinum* and *B. thermophilum*, cannot be distinguished on common phenotypical grounds, the differences in their morphology can be used for their distinction (Figure 8.3(6)) as can transaldolase electrophoresis (Scardovi *et al.*, 1979a) and PAGE protein electrophoresis (Biavati *et al.*, 1982), which also permit clear-cut identification. Isolated from bovine rumen, piglet faeces. Type strain: Ru 917 = ATCC 27917 = DSM 20432 = JCM 1211. No other strains available in culture collections.

7 *Bifidobacterium breve* (Reuter, 1963). Synonym. *Bifidobacterium parvulorum*. Cells are short, slender or thick and with or without bifurcations, a morphology that suggests the specific epithet (Figure 8.3(7)). The two biovars *a* and *b* were identified by Reuter (1963), and serologically related strains were referred to a separate species, i.e. *B. parvulorum*. DNA homology tests on type strains and other strains previously designated as *B. breve* and *B. parvulorum* proved the genetic identity of Reuter's two species (Scardovi *et al.*, 1971). Isolated from the faeces of infant and suckling calf; human vagina; sewage. Type strains: S1 = ATCC 15700 = DSM 20213 = JCM 1192. Other strains: ATCC 11147; ATCC 15698 = DSM 20091 = JCM 1273; ATCC 15701 = JCM 7016.

8 *Bifidobacterium catenulatum* (Scardovi and Crociani, 1974). This species is reported as one of the most frequently isolated from adult and infant faeces (Biavati *et al.*, 1982, 1986). Lauer and Kandler (1983), in a study on DNA–DNA homology, murein types and enzyme patterns of the type strains of the genus *Bifidobacterium*, suggest that *B. catenulatum* and *B. pseudocatenulatum* may be subspecies of *B. catenulatum*, i.e. *B. catenulatum* subsp. *catenulatum* and *B. catenulatum* subsp. *pseudocatenulatum*. The same suggestion is posited by Biavati *et al.* (1982) concerning the use of a PAGE procedure to distinguish genetically related and unrelated species of bifidobacteria. However, the G+C content of their DNA differs by 3 mol%, and the GC% of *B. catenulatum* is the lowest value found in bifidobacteria (see Table 8.2); none of the strains genetically identified as *B. catenulatum* ferments starch or mannose, whereas strains of *B. pseudocatenulatum* ferment these compounds. The isozymes of transaldolase and 6PGD of the two species migrated very differently (see Table 8.2); furthermore *B. catenulatum* as yet has never been isolated from a faeces of suckling calves. Cells grown in TPY agar stabs are generally arranged in chains of three or more globular elements (Figure 8.3(8)). Isolated from faeces of infant and human adult; human vagina; sewage. Type strain: B669 = ATCC 27539 = DSM 20103 = JCM 1194. Other strains: ATCC 27675; ATCC 27676; ATCC 27677; DSM 20224 = JCM 7130.

9 *Bifidobacterium choerinum* (Scardovi *et al.*, 1979b). *Bifidobacterium choerinum* exhibits a sugar fermentation pattern very similar to that of *B. thermophilum* and *B. boum*, two other species frequently found in piglet faeces. These three species can be distinguished by PAGE patterns of soluble proteins or by transaldolase electrophoresis. Cells grown in TPY agar stabs show a morphology similar to that of *B. globosum* (Figure 8.3(9)). Isolated from faeces of piglet and sewage. Type strain: SU 806 = ATCC 27686 = DSM 20434. No other strains available in culture collections.

10 *Bifidobacterium coryneforme* (Scardovi and Trovatelli, 1969; Biavati *et al.*, 1982). *Bifidobacterium coryneforme* was inadvertently omitted from the Approved Lists (Skerman *et al.*, 1980) and is therefore without taxonomic standing. Biavati *et al.* (1982) proposed reinstatement of the name *B. coryneforme* on the basis of its different electrophoretic pattern of cellular proteins and the previously confirmed differential characters. The growth is more abundant in MRS than in PYG. Cells grown in MRS agar stabs very rarely form radial groupings of cells. Their morphological traits are shown in Figure 8.3(10). Isolated from the intestine of honey bee, *Apis mellifera* subsp. *mellifera* and subsp. *caucasica*. Type strain: C 215 = ATCC 25911 = DSM 20216 = JCM 5819. No other strains available in culture collections.

11 *Bifidobacterium cuniculi* (Scardovi *et al.*, 1979). *Bifidobacterium cuniculi* inhabits rabbit intestinal tracts and its morphology (Figure 8.3(11)) is similar to that of *B. globosum*, *B. pseudocatenulatum* and *B. animalis*, which are also found in the same habitat. The fermentation patterns can be used to distinguish this species from the other three because unlike the latter, *B. cuniculi* does not ferment lactose, ribose or raffinose. *Bifidobacterium magnum*, another species isolated from the same source, can be distinguished via morphology (Figure 8.3(19)), in addition to its different fermentation pattern. Isolated from faeces of rabbit. Type strain: Ra 93 = ATCC 27916 = DSM 20435. No other strains available in culture collections.

12 *Bifidobacterium dentium* (Scardovi and Crociani, 1974). Synonyms: *Bifidobacterium appendicitis*, *Actinomyces eriksonii*, *Bifidobacterium eriksonii*. Scardovi *et al.* (1971), in a DNA–DNA homology study, first identified bifid strains isolated from human dental caries, faeces of human adult and human vagina as a distinct group from other strains, also identifiable phenotypically as *B. adolescentis*, and suggested considering them as 'dentium' group distinct from *B. adolescentis*. The phenotypic distinction from biovars *b* and *d* of *B. adolescentis* and *B. dentium* should be based either on transaldolase isozymes (Scardovi *et al.*, 1979a) or on electrophoretic patterns of soluble cellular proteins (Biavati *et al.*, 1982). The morphology of these two species is quite similar (Figure 8.3(1) and (12)). *Bifidobacterium dentium*, as well as *B. angulatum* and *B. catenulatum*, requires riboflavin and pantothenate for growth. That this species is frequently isolated from dental caries and often found in human clinical samples, supports the view that its definition by Georg *et al.* (1965) as a pathogenic anaerobic species may be appropriate. Isolated from human dental caries and oral cavities; faeces of human adult, human vagina, human clinical samples. Type strain: B 764 = ATCC 27534 = DSM 20436 = JCM 1195. Other strains: ATCC 15423 = DSM 20084; ATCC 15424; ATCC 27678; ATCC 27679; ATCC 27680; DSM 20221 = JCM 7135.

13 *Bifidobacterium gallicum* (Lauer, 1990). The description of this species is based on phenotypic and genotypic characterization of a strain found by

H. Beerens (Institut Pasteur de Lille, France) in 1967 as a human faecal isolate. This isolate proved to have very low genetic relatedness to the other species of the genus and to contain a type of peptidoglycan unique among the previously described species. Provided that the characteristics of the monotype strain are confirmed in further isolates, the identification of *B. gallicum* is possible from its carbohydrate fermentations: pentoses and starch are fermented whereas lactose, melibiose and raffinose are not. *Bifidobacterium gallicum* has been isolated from human faeces but its true habitat remains uncertain because no additional strains have been isolated by researchers in intestinal ecology. Its morphological appearance is shown in Figure 8.3(13). Monotype strain P 6 = DSM 20093.

14 *Bifidobacterium gallinarum* (Watabe *et al.*, 1983). This species, described by Watabe *et al.* (1983) as having very low or no genetic relatedness with the previously described species of the genus *Bifidobacterium*, was subsequently found to be closely related (61%) to a new species, *B. saeculare*, described by Biavati *et al.* (1991a). Carbohydrate fermentation patterns are characterized by the ability of *B. gallinarum* to ferment inulin and dextrin, which enables this species to be distinguished from the other starch-negative bifidobacteria. It has been reported (Watabe *et al.*, 1983) that there is a strain – Barnes EBF 92/68 – in this species that failed to ferment glucose; this is the only case known previously of the bifidobacteria strain that does not ferment glucose. The cell morphology is shown in Figure 8.3(14). Isolated from chicken caeca. Type strain: Ch 206–5 = ATCC 33777 = DSM 20670. Other strains ATCC 33778 = JCM 6291.

15 *Bifidobacterium globosum* (Scardovi *et al.*, 1969; Biavati *et al.*, 1982). This species was considered as a synonym of *B. pseudolongum* by Rogosa (1974) and not reported in the Approved Lists of Bacterial Names (Skerman *et al.*, 1980). These two species show about 70% DNA–DNA homology and have similar fermentation patterns and morphology (Figure 8.3(15) and (23)). However, the reported G+C contents of their DNA differ by 3 mol% (Scardovi *et al.*, 1971). The classification as an independent species is also sustained by differences in 6-phosphogluconate dehydrogenase (6PGD) zymogram, by their DNA similarity to other related species and by the presence of plasmids of larger molecular weight found only in *B. globosum* (Sgorbati *et al.*, 1982). Further indications on the validity of the species *B. globosum*, the revival of which was proposed by Biavati *et al.* (1982), include the identical PAGE protein patterns of selected strains from various sources, which are quite distinct from those of the most closely related species, *B. pseudolongum*. Recently, Yaeshima *et al.* (1992b), in a study comparing the DNA base composition, DNA similarities and phenotypic characteristics of *B. pseudolongum* and *B. globosum* strains, have proposed that *B. pseudolongum* and *B. globosum* should be unified into two subspecies: *B. pseudolongum* subsp. *pseudolongum* and *B. pseudolongum* subsp. *globosum*. *Bifidobacterium globosum* is undoubtedly the predominant bifidal species in a large number of animal digestive tracts. Isolated from faeces of piglet, suckling calf, rat, rabbit and lambs; sewage; bovine rumen and in a single specimen of human infant faeces. Type strain: Ru 224 = ATCC 25865 = DSM 20092 = JCM 5820. Other strain: ATCC 25864.

16 *Bifidobacterium indicum* (Scardovi and Trovatelli, 1969). *Bifidobacterium indicum*, as well as *B. asteroides* and *B. coryneforme*, are the bifids inhabiting the intestine of honey bees. Cells grown in TPY agar stabs are short, generally paired and coccoid; star-like clusters are not observed (Figure 8.3(16)). Colonies are less dense than those of *B. asteroides*. In liquid media the cells do not adhere to the walls and have an even turbidity and dispersible sediment. One-hundred and twenty-two strains were examined for their transaldolase and 6PGD zymograms; different isozymes obtained by starch gel electrophoresis were reported (Scardovi *et al.*, 1979a). Sixty-seven percent of the 106 strains of *B. indicum* surveyed for the presence of plasmids were positive (Sgorbati *et al.*, 1982). The structural relatedness among plasmids obtained by means of blot hybridization using selected unrestricted plasmid probes points to the presence of three differing plasmids (Sgorbati *et al.*, 1986). Unrelated by DNA homology to any other species of the genus. Isolated from intestine of *Apis cerana* and *Apis dorsata*. Type strain: C 410 = ATCC 25912 = DSM 20214 = JCM 1302. Other strain: ATCC 25913.

17 *Bifidobacterium infantis* (Reuter, 1963). Synonyms: *Bididobacterium liberorum*, *Bifidobacterium lactentis*, *Actinomyces parabifidus*. Cellular morphology is similar to that of many other species of the genus (Figure 8.3(17)). Most pertinent literature indicates that this species is predominant in the faeces of breast-fed infants and is apparently host-specific given that no strains of *B. infantis* have been found in the faeces of human adults (Biavati *et al.*, 1982, 1986). The close genetic relationships between *B. infantis* and *B. longum* are reflected in the similar fermentation activity and electrophoretic patterns of their soluble cellular proteins. Isozyme patterns of transaldolase and 6PGD were determined in 63 and 12 strains which were allotted via DNA homology to *B. infantis* and *B. longum*, respectively; 90% of *B. infantis* strains had a more anodal transaldolase (isozyme 5, migration 100) than that possessed by 72% of *B. longum* strains (isozyme 8, migration 90) (Scardovi *et al.*, 1979a). Identification of strains belonging to one of the two species remains a difficult task. The rules suggested by Scardovi (1986) can be adopted: strains not fermenting arabinose should be held as *B. infantis* while strains fermenting both arabinose and melezitose should regarded as *B. longum*; strains fermenting arabinose (and xylose) but not melezitose are *B. infantis* if their transaldolase migrates more anodically (isozyme 5, migration 100), and *B. longum* if their isozyme is less anodal (isozyme 8, migration 90). Isolated from faeces of infant and suckling calk; human vagina. Type strain: S12 = ATCC 15697 = DSM 20088 = JCM 1222. Other strains: ATCC 15702 = DSM 20090 = JCM 1272; ATCC 17930 = DSM 20218 = JCM 1260; ATCC 25962 = DSM 20223 = JCM 1210; ATCC 27920.

18 *Bifidobacterium longum* (Reuter, 1963). The cellular morphology of strains belonging to this species is characterized by very elongated and relatively thin cells with rare branchings (Figure 8.3(18)). *Bifidobacterium longum* may be considered the most common species of bifidobacteria, being found both in infant and adult faeces (Biavati *et al.*, 1984, 1986). Accordingly, the importance of this species in the production of food containing bifidobacteria such as cultured milk, other milk products or added preparations made by pharmaceutical industries for therapeutic purposes is obvious. Methods for the proper identification of the bifidobacteria employed in fermented milks are reported by Biavati *et al.* (1992b). Reuter (1963), in the description of the species, points out the presence of biovar *a*, which slowly ferments mannose in human adults, and biovar *b*, which is mannose-negative in infants. Mitsuoka (1969) proposed the *B. longum* subsp. *animalis* biovars *a* and *b*, which were subsequently recognized on the basis of DNA homology as *B. animalis* and *B. pseudolongum*, respectively (Scardovi *et al.*, 1971; Scardovi and Trovatelli, 1974). The DNA of this species is 50–76% related to that of *B. infantis*; comments about identification problems are reported under *B. infantis*. Furthermore, a group of 180 strains isolated form the faeces of suckling calves could not be identified because of their high degree (over 80%) of DNA homology with both *B. infantis* and *B. longum*, and hence could conceivably constitute a 'continuum' between the two species (Scardovi *et al.*, 1979a). *Bifidobacterium longum* is apparently the only species among those usually found in human faeces which possess a large variety of plasmids (Sgorbati *et al.*, 1982). *Bifidobacterium infantis* does not carry plasmids, although strains of both species, generally isolated from the same species, have been studied; the same was true for the 45 strains studied as *B. infantis–B. longum* 'intermediate' found in calf faeces. Isolated from faeces of human adult, infant and suckling calves; human vagina; sewage. Type strain: E 194b = ATCC 15707 = DSM 20219 = JCM 1217. Other strains: ATCC 15708 = JCM 7054.

19 *Bifidobacterium magnum* (Scardovi and Zani, 1974). This species is characterized by the unusually large dimension of its cell, especially when grown in the absence of Tween 80. Cells grown in TPY agar are 2 μm in diameter, 10–20 μm in length and are frequently arranged in aggregates (Figure 8.3(19)). Other features of the species are: the high stimulative effect of Tween 80, the optimum pH for growth (5.3–5.5), which characterize this species as the only acidophilic one within the genus, the unique isozyme pattern among bifidobacteria (Scardovi *et al.*, 1979a) and the PAGE protein pattern (Biavati *et al.*, 1982). Isolated from rabbit faeces. Type strain: RA 3 = ATCC 27540 = DSM 20222 = JCM 1218. Other strains: ATCC 27681 = JCM 7120; ATCC 27682 = DSM 20220 = JCM 7132.

20 *Bifidobacterium merycicum* (Biavati and Mattarelli, 1991). Four strains among several hundred bifidobacteria isolated from bovine rumen were recognized primarily on the basis of DNA–DNA hybridization results as members of a new homology group and posited as a new species. The morphological traits of the strains belonging to *B. merycicum* are not as distinctive as those of many other bifidobacteria (Figure 8.3(20)). This species is more sensitive to oxygen than most bifidobacteria; its fermentation characteristics, when compared to those of *Bifidobacterium* species that have been isolated from bovine rumen, may be helpful in identification (Biavati and Mattarelli, 1991). Isolated from bovine rumen. Type strain: Ru 915B = ATCC 49391 = DSM 6492 = JCM 8219. Other strains: DSM 6493 = JCM 8220; DSM 6494.

21 *Bifidobacterium minimum* (Scardovi and Trovatelli, 1974; Biavati *et al.*, 1982). The specific epithet refers to the very small size of the cells, a morphology resembling that of *B. asteroides* without star-like aggregates (Figure 8–3(21)). In addition to morphology, this species possesses two other features that are unique among bifidobacteria: the lys-ser murein type (Lauer and Kandler, 1983) and the last anodal isozyme of transaldolase (Scardovi *et al.*, 1979a). This taxon was called a 'minimum' DNA homology group by Scardovi and Trovatelli (1974) and consists of only two strains. Isolated from sewage. Type strain: F 392 = ATCC 27538 = DSM 20102 = JCM 5821. No other strains in culture collections.

22 *Bifidobacterium pseudocatenulatum* (Scardovi *et al.*, 1979b). According to Biavati *et al.* (1986), this species is the one most frequently found in human adult faeces. This fact may be influenced by difficulties in differentiating *B. pseudocatenulatum* from *B. catenulatum* (for comments on identification problems, see *B. catenulatum*); in any case both species are very common members of the intestinal microbiota. Strain-to-strain variability in cell morphology is common in this species. The morphological traits of the type strains grown on TPY agar stabs are shown in Figure 8.3(22). Riboflavin, pantothenate and nicotinic acid required for growth. The DNA of this species is closely related to that of *B. catenulatum* and practically unrelated to that of any other species of the genus. Isolated from faeces of human adult, infant and suckling calf and sewage. Type strain: B 1279 = ATCC 27919 = DSM 20438 = JCM 1200. Other strain: DSM 20439.

23 *Bifidobacterium pseudolongum* (Mitsuoka, 1969). Bifid strains isolated from a variety of animals were recognized as *B. pseudolongum* and the four biovars *a*, *b*, *c* and *d* were proposed on the basis of differences in the fermentation of mannose, lactose, cellobiose and melezitose (Mitsuoka, 1969). The strain Mo-2-10 representative of biovar *d*, was found completely homologous to strain C10-46 of *B. longum* subsp. *animalis* biotype *b* (Scardovi *et al.*, 1971). *Bifidobacterium pseudolongum* has about 70% DNA–DNA homology with *B. globosum*. Moreover, these two species are phenotypically very similar having the same murein type (Lauer and Kandler, 1983) and identical morphology (Figure 8.3(15) and (23)). They have recently been proposed as subspecies: *B. pseudolongum* subsp. *pseudolongum* and *B. pseudolongum* subsp. *globosum* (Yaeshima *et al.*, 1992b). Further comments on the differentiation of *B. globosum* and *B. pseudolongum* are reported under *B. globosum*. Isolated from faeces of pig, chicken, dog, bull, calf, rat and guinea pig. Type strain: PNC-2-9G = ATCC 25526 = DSM 20099 = JCM 1205. Other strains: JCM 1264; DSM 20094 = JCM 1266; DSM 20095 = JCM 1267.

24 *Bifidobacterium pullorum* (Trovatelli *et al.*, 1974). Morphology is of help in recognizing this species, whose cells are mostly arranged in irregular chains often of great length (Figure 8.3(24)). For good growth, nicotinic acid, pyridoxin, thiamin, folic acid, *p*-aminobenzoic acid and Tween 80 are required. The G+C content of the DNA is 67 mol% (as determined by thermal denaturation) the highest value reported for bifidobacteria. Unlike the other species of the genus, the isomeric type of lactic acid formed is DL. Isolated from chicken faeces. Type strain: P 145 = ATCC 27685 = DSM 20433. No other strains available in culture collections.

25 *Bifidobacterium ruminantium* (Biavati and Mattarelli, 1991). Bifidobacteria are present in rumen in large numbers, especially when animals are fed starch-rich diets. The species commonly found in this habitat are *B. globosum*, *B. thermophilum* and *B. boum*. Two other species, *B. ruminantium* and *B. merycicum*, have recently been described as inhabitants of cattle rumen; they have similar morphology (Figure 8.3(20) and (25)) but very low levels of DNA relatedness. The fermentation characteristics distinguishing the *Bifidobacterium* species isolated from bovine rumen are reported by Biavati and Mattarelli (1991). *Bifidobacterium ruminantium* DNA is 54% homologous (average value) to the DNA of *B. adolescentis*. Additional experiments (electrophoretograms of soluble cellular proteins and electrophoretic patterns of isozymes), performed to verify the taxonomic position of these two species confirmed that they should be recognized as distinct. *Bifidobacterium ruminantium* is more sensitive to oxygen than most bifidobacteria, carbon dioxide did not affect this sensitivity but it strongly enhanced anaerobic growth. Isolated from rumen of cattle. Type strain: Ru 687 = ATCC 49390 = DSM 6489 = JCM 8222. Other strains: DSM 6490 = JCM 8509; DSM 6491.

26 *Bifidobacterium saeculare* (Biavati *et al.*, 1991a). Trovatelli *et al.* (1974) found that four strains of bifidobacteria isolated from faeces of rabbit were characterized as 'unassigned homology group I'. Subsequently, on the basis of DNA–DNA hybridization, phenotypic characters and electrophoretic patterns of proteins and of isozymes the 'unassigned homology group I' was described as a new species and named *B. saeculare*. The specific epithet is in honour of the Ninth Centenary of the Foundation of Bologna University. A 60–64% range of intraspecific DNA homologies was observed between *B. saeculare* and both *B. pullorum* and *B. gallinarum*, two species isolated from chicken faeces. Additional experiments on fermentation analysis (19 complex carbohydrates and mucins were tested), isozyme electrophoresis and penicillin-binding protein detection were performed to verify the taxonomic position of *B. saeculare* with respect to *B. pullorum* and *B. gallinarum*. The various characteristics distinguishing the three species are listed in a table by Biavati and Mattarelli (1991). The PAGE procedure can be used alternatively. Cells grown in TPY agar stabs are short, slightly curved rods, isolated or in pairs (Figure 8.3(26)). Isolated from rabbit faeces. Type strain: Ra 161 = ATCC 49392 = DSM 6531 = JCM 8508. Other strains: DSM 6532; DSM 6533.

27 *Bifidobacterium subtile* (Biavati *et al.*, 1982). This taxon was referred by Scardovi and Trovatelli (1974) as a 'subtile' DNA homology group. Its cell morphology (Figure 8.3(27)) is similar to that of *B. breve* (Figure 8.3(7)), although the cells of the latter are usually shorter and thicker, swollen and branched. The two species have very similar fermentation patterns although *B. subtile* ferments starch and gluconate whereas *B. breve* does not and lactose is fermented only by *B. breve*. Its DNA is not related to that of any of the other species. The optimum temperature for growth (34–35°C) is the lowest among bifidobacteria (37–41°C range). Isolated from sewage. Type strain: F 395 = ATCC 27537 = DSM 20096 = JCM 5822. Other strains: ATCC 27683 = JCM 7109; ATCC 27684.

28 *Bifidobacterium suis* (Matteuzzi *et al.*, 1971). *Bifidobacterium suis*, commonly found in the faeces of piglets, share this habitat with other bifid species, i.e. *B. globosum*, *B. pseudolongum*, *B. thermophilum*, *B. boum* and *B. choerinum*, and can be distinguished from the latter species on the basis of fermentation characters: *B. globosum* and *B. pseudolongum* ferment starch whereas *B. suis* does not, and the last three species do not ferment arabinose and xylose whereas *B. suis* does. Cells are slender, elongated, with rare terminal bifurcations or clumps (Figure 8.3(28)). Riboflavin is the only demonstrable vitamin required for growth. Most strains posses a strong ureolytic activity. High levels of urease activity were present in the cells grown in the absence of urea, suggesting that this enzyme is not inducible (Crociani and Matteuzzi, 1982). Matteuzzi *et al.* (1971) did not find any DNA–DNA relationship between their reference strain of *B. suis* and any strain of the other species tested. In a subsequent study of DNA–DNA homology in the type strains of the genus *Bifidobacterium* (Lauer and Kandler, 1983) it was found that *B. suis* and *B. longum* are genetically closely related to each other (70% DNA–DNA homology) and to *B. infantis* (about 65% DNA–DNA homology). Isolated from piglet faeces. Type strain: Su 859 = ATCC 27533 = DSM 20211 = JCM 1269. Other strains: ATCC 27531; ATCC 27532 = JCM 7139.

29 *Bifidobacterium thermophilum* (Mitsuoka, 1969). Synonym: *Bifidobacterium ruminale.* Mitsuoka applied this specific epithet because of the ability of the strains belonging to this species to grow at 46.5°C. Cells are slender, slightly curved, often with tapered ends, protuberances or irregularities near the junction of paired cells, rare branchings and arranged singly or in pairs never in clumps or in angular disposition (Figure 8.3(29)). The four biovars *a*, *b*, *c* and *d* were distinguished by Mitsuoka (1969) according to differences in the fermentation of melezitose and lactose. For comments on the identification of *B. thermophilum*, *B. boum* and *B. choerinum* see *B. choerinum.* Isolated from faeces of pig, chicken and suckling calf; bovine rumen; sewage. Type strain: P2–91 = ATCC 25525 = DSM 20210 = JCM 1207. Other strains: ATCC 25866 = DSM 20212 = JCM 1268; ATCC 25867; DSM 20209.

References

Beerens, H. (1990) An elective and selective isolation medium for *Bifidobacterium* spp. *Letters in Applied Microbiology*, **11**, 155–157.

Bezkorovainy, A. and Miller-Catchpole, R. (1989) *Biochemistry and Physiology of Bifidobacteria.* CRC Press, Boca Raton, FL, USA.

Biavati, B. and Mattarelli, P. (1991) *Bifidobacterium ruminantium* sp. nov. and *Bifidobacterium merycicum* sp.nov. from the rumens of cattle. *International Journal of Systematic Bacteriology*, **41**, 163–168.

Biavati, B., Scardovi, V. and Moore W.E.C. (1982) Electrophoretic patterns of proteins in the genus *Bifidobacterium* and proposal of four new species. *International Journal of Systematic Bacteriology*, **32**, 358–373.

Biavati, B., Catagnoli, P., Crociani, F. and Trovatelli, L.D. (1984) Species of the *Bifidobacterium* in the faeces of infants. *Microbiologica*, **7**, 341–346.

Biavati, B., Castagnoli, P. and Trovatelli, L.D. (1986) Species of the genus *Bifidobacterium* in the faeces of human adults. *Microbiologica*, **9**, 39–45.

Biavati, B., Matterelli, P. and Crociani, F. (1991a) *Bifidobacterium saeculare* a new species isolated from faeces of rabbit. *Systematic and Applied Microbiology*, **14**, 389–392.

Biavati, B., Sgorbati, B. and Scardovi, V. (1991b) *The Genus Bifidobacterium. The Prokaryotes*, 2nd edn (eds Balows, A., *et al.*). Springer-Verlag, Berlin, Germany, pp. 816–833.

Biavati, B., Crociani, F., Mattarelli, P. and Scardovi, V. (1992a) Phase variations in *Bifidobacterium animalis. Current Microbiology*, **25**, 51–55.

Biavati, B., Mattarelli, P. and Crociani, F. (1992b) Identification of bifidobacteria from fermented milk products. *Microbiologica*, **15**, 7–13.

Breed, R.S., Murray, E.G.D. and Smith, N.R. (eds) (1957) *Bergey's Manual of Determinative Bacteriology*, 7th edn. Williams and Wilkins, Baltimore, MD, USA.

Crociani, F. and Matteuzzi, D. (1982) Urease activity in the genus *Bifidobacterium. Annales de Microbiologie* (Paris), **133A**, 417–423.

Crociani, F., Emaldi, O. and Matteuzzi, D. (1977) Increase in isoleucine accumulation by α-aminobutyric acid-resistant mutants of *Bifidobacterium ruminale. European Journal of Applied Microbiology*, **4**, 177–179.

Dehnert, J. (1957) Untersuchungen über die Gram-positive Stuhlflora des Brustmilchkinder. *Zentralblatt für Bakteriologie, Parasitenkunde, Infektionskrankheiten und Hygiene (1. Abteilung Originale)*, **A169**, 66–79.

De Vries, W. and Stouthamer, A.H. (1968) Fermentation of glucose, lactose, galactose, mannitol and xylose by bifidobacteria. *Journal of Bacteriology*, **96**, 472–478.

Fischer, W. (1987) Lipoteichoic acid of *Bifidobacteium bifidum* subspecies pennsylvanicum DSM 20239. *European Journal of Biochemistry*, **165**, 639–646.

Fischer, W., Bauer, W. and Finegold, M. (1987) Analysis of the lipoichoic-acid like macroamphiphile from *Bifidobacterium bifidum* subspecies *pennsylvanicum* by one and two-dimensional H- and C-NMR spectroscopy. *European Journal of Biochemistry*, **165**, 647–652.

Fox, G.E., Pechman, K. and Woese, C.R. (1977) Comparative cataloging of 16 S ribosomal ribonucleic acid; molecular approach to procaryotic systematics. *International Journal of Systematic Bacteriology*, **27**, 44–57.

Gauche, A., Gyorgy, P., Hoover, J.R., Rose, R., Ruelius, H.W. and Zilliken, F. (1954) Bifidus factor IV. Preparation obtained from human milk. *Archives of Biochemistry and Biophysics*, **48**, 214–220.

Glick, M.C., Sall, T., Zilliken, F. and Mudd, S. (1960) Morphological changes in *Lactobacillus bifidus* var. *pennsylvanicus* produced by a cell wall precursor. *Bichemica et Biophysica Acta*, **37**, 361–368.

Georg, L.K., Robertstad, G.W., Brinkman, S.A. and Hicklin, M.D. (1965) A new anerobic *Actinomyces* species. *Journal of Infectious Disease*, **115**, 88–99.

Gyorgy, P. (1953) A hitherto unrecognized biochemical difference between human milk and cow's milk. *Pediatrics*, **11**, 98–103.

Hatanaka, M., Tachiki, T., Kumagai, H. and Tochikura, T. (1987) Distribution and some properties of glutamine synthetase and glutamate dehydrogenase in bifidobacteria. *Agricultural Biological Chemistry*, **51**, 251–257.

Holdeman, L.V., Cato, E.P. and Moore, W.E.C. (1977) *Anaerobe Laboratory Manual*, 4th edn. Virginia Polytechnic Institute and State University, Blackburg, VA, USA.

Hoskins, L., Agustines, M., McCkee, W., Boulding, E., Kriaris, M. and Niedermeyer, G. (1985) Mucin degradation in human colon ecosystems. Isolation and properties of fecal strains that degrade ABH blood group antigens and oligosaccharides from mucin glycoproteins. *Journal of Clinical Investigation*, **75**, 944–947.

Kandler, O. (1970) Amino acid sequence of the murein and taxonomy of the genera *Lactobacillus, Bifidobacterium, Leuconostoc* and *Pediococcus. International Journal of Systematic Bacteriology*, **20**, 491–500.

Kandler, O. and Lauer, E. (1974) Neuere Vortstellungen zur Taxonomic der *Bifidobacterien. Zentralblatt für Bakteriologie, Parasitenkunde, Infektionskrankheiten und Hygiene. (1. Abteilung Originale)*, **A228**, 29–45.

Kohwi, Y., Imai, K., Tamura, Z. and Hashimoto, Y. (1978) Antitumor effect of *Bifidobacterium infantis* in mice. *Gann*, **69**, 613–616.

Kosman, M.E. (1976) Lactulose (Cephulac) in portosystemic encaphalopaty. *Journal of the American Medical Association*, **236**, 2444.

Lauer, E. (1990) *Bifidobacterium gallicum* sp.nov. isolated from human faeces. *International Journal of Systematic Bacteriology*, **40**, 100–102.

Lauer, E. and Kandler, O. (1976) Mechanismus der Variation des Verhaltnisses Aceta/lactat bei der Vergarung von Glucose durch Bifidobakterien. *Archives of Microbiology*, **110**, 271–277.

Lauer, E. and Kandler, O. (1983) DNA–DNA homology, murein types and enzyme patterns in the strains of the genus *Bifidobacterium. Systematic and Applied Microbiology*, **4**, 42–64.

Lessel, E.F. (1977) Concerning the valid publication of the new names and combinations effectively published on or after 1 January 1976 in publications other than the IJSB. *International Journal of Systematic Bacteriology*, **27**(3), iv.

Matteuzzi, D., Crociani, F., Zani, G. and Trovatelli, L.D. (1971) *Bifidobacterium suis* n.sp.: a new species of the genus *Bifidobacterium* isolated from pig faeces. *Zeitschrift für Allgemeine Mikrobiologie*, **11**, 387–395.

Matteuzzi, D., Crociani, F. and Emaldi, O. (1978) Amino acids produced by bifidobacteria and some clostridia. *Annales de Microbiologie Institut Pasteur* (Paris), **129b**, 175–181.

Mitsuoka, T. (1969) Vergleichende Untersuchungen über die Bifidobakterien aus dem Verdauungstrakt von Menschen und Tieren. *Zentralblatt für Bakteriologie, Parasitenkunde, Infektionskrankheiten und Hygiene (1. Abteilung Originale)*, **A210**, 52–64.

Mitsuoka, T. (1984) Taxonomy and ecology of bifidobacteria. *Bifidobacteria Microflora*, **3**, 11–28.

Mitsuoka, T., Sega, T. and Yamamoto, S. (1965) Eine verbesserte methodik der qulitativen und quantitative Analyse der Darmflora von Menschen und Tieren. *Zentralblatt für Bakteriologie, Parasitenkunde, Infektionskrankheiten und Hygiene (1. Abteilung Originale)*, **A195**, 455–469.

Mitsuoka, T., Hayakama, K. and Kimura, N. (1973) Die Faekalflora bei Mensche II Mitteilung. Die Zusammensetuzung der Bifidobakteriaenflora der verschiedenen Altersgruppen. *Zentralblatt für Bakteriologie, Parasitenkunde, Infektionskrankheiten und Hygiene*, **A226**, 469–478.

Muñoa, F.J. and Pares, R. (1988) Selective medium for isolation and enumeration of *Bifidobacterium* spp. *Applied and Environmental Microbiology*, **54**, 1715–1718.

Old, L.J., Clarke, D.A., and Benaauof, B. (1959) Effect of bacillus Calmette-Guerin infection on transplanted tumors in the mouse. *Nature*, **184**, 291–293.

Orla-Jensen, S. (1924) La classification des bacteries lactiques. *Lait*, **4**, 468–474.

Podolsky, D.K. and Isselbacher, K.J. (1987) Cirrhosis. In *Harrison's Principles of Internal Medicine*, IIth edn (eds Braunwald, E.K., *et al.*). McGraw-Hill, New York, USA, pp. 1349–1399.

Poupard, J., Husain, I. and Norris, R.F. (1973) Biology of the bifidobacteria. *Bacteriological Review*, **37**, 136–165.

Rasic, J.L. and Kurmann, J.A. (1983) *Bifidobacteria and their Role*. Birkhauser Verlag, Basel, Switzerland.

Reuter, G. (1963) Vergleichende Untersuchunge uber die Bifidus-flora im Sauglings und Erwashenenstuhlz. *Zentralblatt für Bakteriologie, Parasitenkunde, Infektionskrankheiten und Hygiene (1.Abteilung Originale)*, **A191**, 486–507.

Rogosa, M. (1974) Genus III, *Bifidobacterium* Orla-Jensen. In: Buchanan, R.E. & *Bergey's Manual of Determinative Bacteriology*, 8th edn (eds Buchanan, R.E. and Gibbons, N.S.). Williams and Wilkins, Baltimore, MD, USA, pp. 669–676.

Salyers, A., Palmer, J.K. and Wilkins, T.D. (1978) Degradation of polysaccharides by intestinal bacterial enzymes. *American Journal of Clinical Nutrition*, **31**, 128–132.

Scardovi, V. (1986) Genus *Bifidobacterium* Orla-Jensen 1924,472[al]. In: Sneath, *Bergey's Manual of Systematic Bacteriology*, Vol. 2 (eds Sneath, P.H.A., Mair, N.S., Sharpe, M.E. and Holt, J.G.). Williams and Wilkins, Baltimore, MD, USA, pp. 1418–1434.

Scardovi, V. and Crociani, F., (1974) *Bifidobacterium catenulatum*, *Bifidobacterium dentium* and *Bifidobacterium angulatum*: three new species and their deoxyribonucleic acid homology relationships. *International Journal of Systematic Bacteriology*, **24**, 6–20.

Scardovi, V. and Trovatelli, L.D. (1965) The fructose-6-phosphate shunt as peculiar pattern of hexose degradation in the genus *Bifidobacterium. Annali di Microbiologia ed Enzimologia*, **15**, 19–29.

Scardovi, V. and Trovatelli, L.D. (1969) New species of bifidobacteria from *Apis mellifica* and *Apis indica*. A contribution to the taxonomy and biochemistry of the genus *Bifidobacterium, Zentralblatt für Bakteriologie, Parasitenkunde, Infektionskrankheiten und Hygiene (2. Abteilung Originale)*, **123**, 64–88.

Scardovi, V. and Trovatelli, L.D. (1974) *Bifidobacterium animalis* (Mitsuoka) comb. nov.

and the "*minimum*" and "*subtile*" groups of new bifidobacteria found in sewage. *International Journal of Systematic Bacteriology*, **24**, 21–28.

Scardovi, V. and Zani, G. (1974). *Bifidobacterium magnum* sp. nov., a large acidophilic bifidobacterium isolated from rabbit faeces. *International Journal of Systematic Bacteriology*, **24**, 29–34.

Scardovi, V., Trovatelli, L.D., Crociani, F. and Sgorbati, B. (1969) Bifidobacteria in bovine rumen. New species of the genus *Bifidobacterium: B. globosum* sp. nov. and *B. ruminale* sp. nov. *Archiv für Mikrobiologie*, **68**, 278–294.

Scardovi, V., Trovatelli, L.D., Zani, G., Crociani, F. and Matteuzzi, D. (1971) Deoxyribonucleic acid homology relationships among species of the genus *Bifidobacterium. International Journal of Systematic Bacteriology*, **21**, 276–294.

Scardovi, V., Casalicchio, F. and Vincenzi, N. (1979a) Multiple electrophoretic forms of transaldolase and 6-phosphogluconate dehydrogenase and their relationships to the taxonomy and ecology of bifidobacteria. *International Journal of Systematic Bacteriology*, **29**, 312–327.

Scardovi, V., Trovatelli, L.D., Biavati, B. and Zani, G. (1979b) *Bifidobacterium cuniculi, Bifidobacterium choerinum, Bifidobacterium boum* and *Bifidobacterium pseudocatenulatum*: four new species and their deoxyribonucleic acid homology relationships. *International Journal of Systematic Bacteriology*, **29**, 291–311.

Sekine, K., Toida, T., Saito, M., Kuboyama, M., Kawashima, T. and Hasimoto, Y. (1985) A new morphologically characterized cell wall preparation (while peptidoglycan) from *Bifidobacterium infantis* with a higher efficacy on the regression of an established tumor in mice. *Cancer Research*, **45**, 1300–1308.

Sgorbati, B. and London, J. (1982) Demonstration of phylogenetic relatedness among members of the genus *Bifidobacterium* using the enzyme transaldolase as an evolutionary marker. *International Journal of Systematic Bacteriology*, **32**, 37–42.

Sgorbati, B. and Scardovi, V. (1979) Immunological relationships among transaldolases in the genus *Bifidobacterium. Antonie van Leeuwenhoek Journal of Microbiology and Serology*, **45**, 129–140.

Sgorbati, B., Scadovi, V. and Leblanc, D.J. (1982) Plasmids in the genus *Bifidobacterium. Journal of General Microbiology*, **128**, 2121–2131.

Sgorbati, B., Scardovi, V. and Leblanc, D.J. (1986) Related structures in the plasmid profiles of *Bifidobacterium asteroides, B. indicum* and *B. globosum. Microbiologica*, **9**, 443–456.

Skerman, V.B.D., McGowan, V. and Sneath, P.H.A. (1980) Approved lists of bacterial names. *International Journal of Systematic Bacteriology*, **30**, 225–420.

Tissier, H. (1900) Recherches sur la flore intestinale normale et pathologique du nourisson. Thesis, University of Paris, Paris, France.

Trovatelli, L.D., Crociani, F., Pedinotti, M. and Scardovi, V. (1974) *Bifidobacterium pullorum* sp. nov.: a new species isolated from chicken feces and a related group of bifidobacteria isolated from rabbit feces and a related group of bifidobacteria isolated from rabbit feces. *Archiv für Mikrobiologie*, **98**, 187–198.

Veerkamp, J.H., Hoelen, G.E.J.M. and Op den Camp, H.J.M. (1983). The structure of a mannitol teichoic acid from *Bifidobacterium bifidum* sub spp. *pensylvanicum. Biochimica et Biophysica Acta*, **755**, 439–451.

Watabe, J., Benno, Y. and Mitsuoka, T. (1983) *Bifidobacterium gallinarum* sp. nov.: a new species isolated from the ceca of chicken. *International Journal of Systematic Bacteriology*, **33**, 127–132.

Woese, C.R., Magrum, L.J. and Fox, G.E. (1978) Archaebacteria. *Journal of Molecular Evolution*, **11**, 245–252.

Yaeshima, T., Fujisawa, T. and Mitsuoka, T. (1992a) *Bifidobacterium* species expressing phenotypical similarity to *Bifidobacterium adolescentis* isolated from the feces of human adults. *Bifidobacteria Microflora*, **11**, 25–32.

Yaeshima, T., Fujisawa, T. and Mitsuoka, T. (1992b) *Bifidobacterium globosum* subjective synonym of *Bifidobacterium pseudolongum*, and description of *Bifidobacterium pseudolongum* subsp. *pseudolongum* comb. nov. and *Bifidobacterium pseudolongum* subsp. *globosum* comb. nov. *Systematic and Applied Microbiology*, **15**, 380–385.

9 The genus *Carnobacterium*

U. SCHILLINGER and W.H. HOLZAPFEL

9.1 History

Resembling the lactobacilli in several respects, including morphology and the absence of a cytochrome system, the carnobacteria have originally been considered as slightly 'unusual' representatives of the genus *Lactobacillus*. Probably the first mention of such unusual lactobacilli was made in the 1950s, and referred to strains isolated from poultry meat (Thornley and Sharpe, 1959).

During investigations into the bacterial associations of fresh and vacuum-packaged meats, the occurrence of 'atypical' lactobacilli not growing on acetate agar (Rogosa *et al.*, 1951) was reported by several authors (Thornley and Sharpe, 1959; Hitchener *et al.*, 1982; Von Holy, 1983; Holzapfel and Gerber, 1983; Shaw and Harding, 1984). As these isolates showed some similarities to heterofermentative lactobacilli, including (inconsistent) gas production from glucose, they were placed in the genus *Lactobacillus* but were not identifiable to species level. They differed from the described heterofermentative species in their production of virtually pure L(+)-lactic acid from glucose, in their sugar fermentation pattern, and from most other species in the presence of *meso*-diaminopimelic acid (*m*Dpm) in the cell wall. These unusual 'lactobacilli' were unable to grow on acetate agar at pH 5.6 and most of them fermented glycerol and also mannitol, properties which are very rare in lactobacilli (Holzapfel and Gerber, 1983; Shaw and Harding, 1984). Gas production was mostly weak, inconsistent and dependent on growth medium (Shaw and Harding, 1985), explaining why some authors described them as homofermentative lactobacilli (Shaw and Harding, 1984).

In 1983, Holzapfel and Gerber investigated a group of atypical lactobacilli isolated from vacuum-packaged minced beef during a shelf-life study, and proposed a new species, *Lactobacillus divergens*, on the basis of their ability to produce L(+)-lactic acid, their low mol% G+C (32–36) of the DNA, the presence of *m*Dpm in the cell wall and their presumed CO_2 production from glucose. Shaw and Harding (1985) included three strains of *Lb. divergens* in their study of non-aciduric lactobacilli isolated from vacuum-packaged beef, pork, lamb and bacon, and DNA–DNA hybridizations revealed that one group of their isolates was highly homologous to

Lb. divergens whereas the second group represented a new species for which the name *Lb. carnis* was proposed.

Collins *et al.* (1987) demonstrated a very high DNA–DNA homology between *Lb. carnis* and *Lb. piscicola*, a species which was isolated from diseased fish by Hiu *et al.* (1984). They suggested *Lb. carnis* as a synonym to *Lb. piscicola.*

Shaw and Harding (1984) assumed a certain relationship of their isolates from meat with the strains isolated from poultry meat by Thornley and Sharpe (1959), on the basis of the phenotypic descriptions of the latter. Collins *et al.* (1987) and Ferusu and Jones (1988) included strains of the three groups of isolates from Thornley and Sharpe (1959) in their taxonomic studies on atypical lactobacilli from meat and chicken. In the numerical taxonomic study of Ferusu and Jones (1988), using phenetic characteristics, the organisms of group 2 of Thornley and Sharpe were found to form three clusters, two of which resembled *Lb. carnis* (*piscicola*) (phenon 9) and *Lb. divergens* (phenon 11), respectively. In the study of Collins *et al.* (1987) the majority of the unidentified poultry strains were classified as *Lb. divergens* and *Lb. piscicola*. The remainder could not be allocated to any described species. As a result, Collins *et al.* (1987) introduced the name *Carnobacterium* and proposed to reclassify *Lb. divergens* and *Lb. piscicola* in this new genus as *Carnobacterium divergens* and *Carnobacterium piscicola* and to incorporate the remaining poultry isolates into the new species *Carnobacterium gallinarum* and *Carnobacterium mobile*.

The separate taxonomic position of this genus was confirmed by the results of DNA–RNA hybridizations by Champomier *et al.* (1989). The four species formed a group of high rRNA similarity and did not show any close genetic relatedness to other lactic acid bacteria.

More recently Franzmann *et al.* (1991) described two new *Carnobacterium* species, *C. funditum* and *C. alterfunditum* which were isolated from the anaerobic waters of an Antarctic lake. The allocation of these isolates to this genus as separate species was demonstrated by DNA–DNA hybridization and rRNA sequence comparison (Franzmann *et al.*, 1991).

9.2 Morphology

Young cultures of carnobacteria are Gram-positive, short to medium length, straight, slender rods which usually occur singly or in pairs and more rarely in short chains (see Figure 9.1). Cell size varies between 0.5 and 0.7 μm in diameter and 1.1 and 3.0 μm in length. Older cultures of some strains may become Gram variable (Hiu *et al.*, 1984) and elongated cells may be observed (Franzmann *et al.*, 1991).

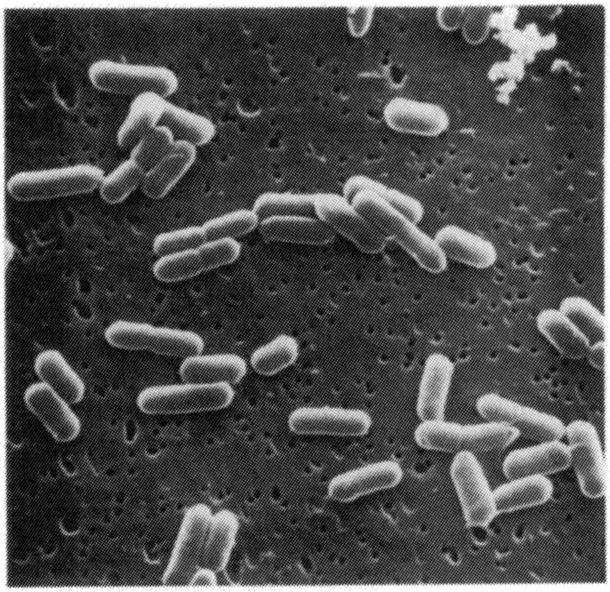

Figure 9.1 Electron micrograph of cells of *Carnobacterium piscicola* DSM 20730 after 24 h in D-MRS broth, pH 6.0, at 30°C. Magnification: × 7500.

9.3 Biochemistry/physiology

9.3.1 Carbohydrate metabolism

Carnobacteria metabolize hexoses to L(+)-lactic acid, CO_2, acetate and ethanol. In addition, varying amounts of acetoin and/or formic acid may be produced depending on the aeration conditions. Formic acid was found to be produced only in anaerobic cultures (Borch and Molin, 1989). Pentoses are metabolized to L(+)-lactic acid, acetate and ethanol (Holzapfel and Gerber, 1983).

Based on these observations it was assumed that these organisms are heterofermentative lactic acid bacteria. However, metabolic studies on the type strain of *C. divergens* revealed that glucose is fermented via the glycolytic pathway (De Bruyn *et al.*, 1987, 1988). Acetate, formate and CO_2 are the end-products of some secondary decarboxylation/dissimilation reactions of pyruvate. Experiments with radioactively labelled glucose precursors showed that about 75% of the lactate and less than 10% each of the formate and acetate were produced from glucose (De Bruyn *et al.*, 1988). The remainder of these products appears to be derived from endogenous substrates other than radioactive glucose such as intracellular amino acids or fructose.

9.3.2 Nitrogen metabolism

Carnobacteria possess an arginine deiminase or dihydrolase and produce ammonia from arginine. Studies on their specific nitrogen requirements have not been conducted thus far.

9.3.3 Vitamin and mineral requirements

Carnobacteria show nutritional requirements similar to other lactic acid bacteria. However, they seem to be somewhat less fastidious than many lactobacilli. For instance, Tween 80 is not needed for their growth and in anaerobic cultures growth of *C. divergens* and *C. piscicola* was found not to be stimulated by the addition of Tween 80 (Borch and Molin, 1989).

The exact vitamin and mineral requirements of all *Carnobacterium* spp. are not known. Strains of *C. piscicola* isolated from diseased fish were reported to require folic acid, riboflavin, pantothenate, and niacin for growth, but not vitamin B_{12}, biotin, thiamine or pyridoxal (Hiu *et al.*, 1984).

9.3.4 Reaction to oxygen

Comparative studies on the effect of oxygen on growth and product formation of different lactic acid bacteria in batch cultures showed that strains of *C. divergens* and *C. piscicola* are more oxygen tolerant than are homofermentative lactobacilli (Borch and Molin, 1989). Oxygen consumption was at maximum during the late exponential phase of growth. Aerobic conditions stimulated the formation of acetate from glucose.

9.3.5 Cell wall chemistry and cellular fatty acids

The cell wall of carnobacteria contains peptidoglycan and the *m*Dpm direct type.

Analysis of the cellular fatty acids revealed the dominance of straight chain saturated and monounsaturated types. The $C_{18:1}$ compound is a 9,10-isomer (Collins *et al.*, 1987). Cyclopropane-ring containing fatty acids may also be produced.

9.4 Genetics

Present knowledge on the genetics of carnobacteria is very limited and is restricted to the species *C. divergens* and *C. piscicola*. Plasmids seem to be present in most strains of these species. For instance, all eight strains of *C. divergens* screened by Ahrne *et al.* (1989) contained one to four plasmids.

The molecular size of these plasmids ranged from 1.4 MDa to 26.9 MDa (Ahrne *et al.*, 1989). Strains of *C. piscicola* also harbour plasmids of different size and in some cases, the genes responsible for bacteriocin production were shown to be plasmid linked (Ahn and Stiles, 1990a; Schillinger *et al.*, 1993). The carnobacteriocin A gene could be transferred and expressed in heterologous hosts using an 8.5 kb chloramphenicol-resistance plasmid from *Lb. plantarum* (Worobo *et al.*, 1993).

It is not known what other physiological or metabolic functions are encoded by the plasmids present in carnobacteria.

9.5 Phylogeny

Studies on RNA relatedness (Champomier *et al.*, 1989; Wallbanks *et al.*, 1990) supported the recognition of the genus *Carnobacterium* as a distinct taxon. The four species – *C. divergens*, *C. piscicola*, *C. gallinarum*, and *C. mobile* – exhibited a high degree of sequence similarity of 16S rRNA with each other, but only a very low sequence relatedness with other lactic acid bacteria (Wallbanks *et al.*, 1990) (see Figure 9.2).

The very distant relationship of the carnobacteria with lactobacilli was also shown by DNA–rRNA hybridizations (Champomier *et al.*, 1989). The 16S rRNA sequence data indicated a closer genetic relatedness of carnobacteria to the genera *Enterococcus* and *Vagococcus* than to the members of the genus *Lactobacillus* (Wallbanks *et al.*, 1990; Collins *et al.*, 1991). In the phylogenetic study of Collins *et al.* (1991), a close relatedness of the species *Lb. maltaromicus* with the carnobacteria was also detected. Comparing a 1340-nucleotide region of 16S rRNA, 100% sequence homology was found between a strain of *Lb. maltaromicus* and the type strain of *C. piscicola* (Collins *et al.*, 1991). Franzmann *et al.* (1991) compared the 16S rRNA sequences of the type strain of *C. funditum* with those of the type strains of other carnobacteria and *Vagococcus salmoninarum* and found a closer relationship of *C. funditum* to *C. mobile* than to other species of *Carnobacterium* ($K_{nuc} = 0.0214$). However, the results of Franzmann *et al.* (1991) also indicated that *C. funditum* and *C. mobile* are not as closely related to each other as other species of *Carnobacterium*.

9.6 Importance

9.6.1 Habitats

Carnobacteria have thus far been isolated from four habitats: meat (and meat products), poultry, fish and seawater.

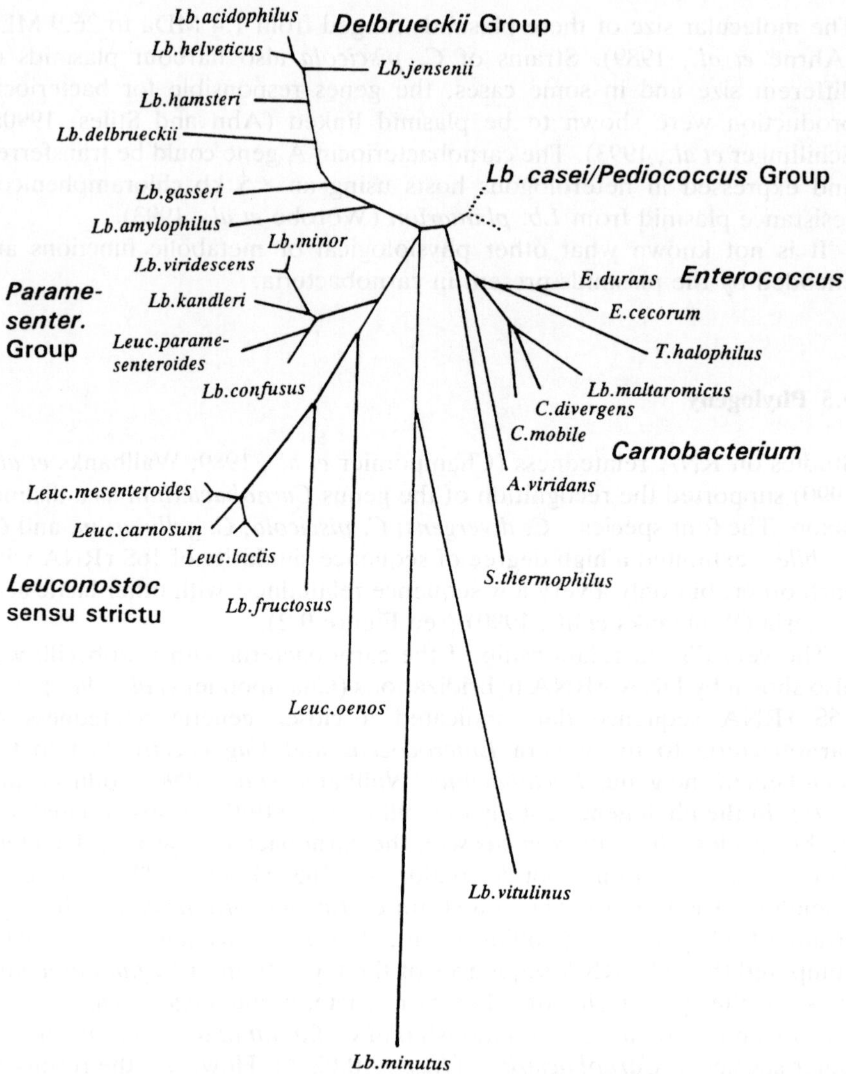

Figure 9.2 Evolutionary distance tree showing the phylogenetic relationships of the genus *Carnobacterium* to other lactic acid bacteria. K_{NUC} values were based on a comparison of 1340 nucleotides of 16S rRNA as described by Collins *et al.* (1991).

9.6.1.1 Meat. Unprocessed meat as well as some meat products represent one of the main habitats of carnobacteria. *Carnobacterium divergens* and *C. piscicola* belong to the microorganisms dominating on vacuum-packaged beef and pork stored at low temperatures (Shaw and Harding, 1984; Schillinger and Lücke, 1986, 1987; Borch and Molin, 1988). The association of atypical or non-aciduric lactobacilli with meat and meat products was reported in several studies on the microbial population of meats (Mol *et al.*, 1971; Kitchell and Shaw, 1975; Reuter, 1975; Hitchener *et al.*, 1982). Many of these unusual lactobacilli, not identifiable at that time, phenotypically resembled the species *C. divergens* and *C. piscicola*. Shaw and Harding (1984) detected carnobacteria only on uncured meat (vacuum-packaged beef, pork and lamb) but not on bacon, and they suggest that this may be due to the relatively low salt tolerance of *C. divergens* and *C. piscicola*. Von Holy *et al.* (1991) also did not detect any carnobacteria among their isolates from spoiled and unspoiled vacuum-packaged Vienna sausages. On the other hand, Borch and Molin (1988) isolated strains of *C. divergens* from cured CO_2-packed pork and from sliced vacuum-packed ham.

In the study of Schillinger and Lücke (1986) on the influence of lactic acid bacteria on shelf life of vacuum-packaged meat, *C. divergens* and *C. piscicola* (formerly *Lb. carnis*) were found to belong to the dominating lactic acid bacteria at 2 and 5°C. Carnobacteria were more frequently isolated from meat stored for 2 weeks than from 4-week-old samples. Lactobacilli (mostly strains of *Lb. sake*) were more competitive than carnobacteria and predominated after prolonged storage of vacuum-packaged meats. This is also indicated by the results of experiments using meats inoculated with selected strains of *Lb. sake* and *C. divergens* (Schillinger and Lücke, 1986). *Lactobacillus sake* presented more than 80% of the total bacterial population after 30 days, whereas in meat inoculated with a strain of *C. divergens* only 37% of the lactic acid bacteria were found to be carnobacteria after 30 days.

Ahn and Stiles (1990b) suggest that the early competitive advantage of these organisms may be the result of antibacterial substances produced by these organisms. Indeed, the investigation of 37 *Carnobacterium* strains isolated from meats (Schillinger and Holzapfel, 1990) revealed the frequent occurrence of bacteriocin-producing strains of *C. piscicola* and *C. divergens* (18 out of 37 strains). The finding that *C. piscicola* LV 17 produces its bacteriocin early in the growth phase supports the hypothesis of Ahn and Stiles (1990b) that bacteriocins may be an important factor contributing to the early dominance of carnobacteria in meats containing a mixed natural population of lactic acid bacteria.

9.6.1.2 Poultry. Relatively few reports have documented the occurrence of carnobacteria in poultry. First mention was probably made by Thornley (1957), describing various unusual lactic acid bacteria isolated from chicken meat during irradiation experiments. These strains, not identifiable at that time, reached high viable counts ($>10^8$/g) during storage in a nitrogen atmosphere at 5°C.

Collins *et al.* (1987) studied the taxonomic position of 59 of the isolates of Thornley (1957) 20 years later, using DNA–DNA hybridization techniques, and could allocate 83% to *Brochothrix thermosphacta*, *C. divergens* and *C. piscicola*. The remainder were shown to be new species of *Carnobacterium* and were designated to the new species *C. gallinarum* and *C. mobile*. The latter have apparently not been isolated from other habitats thus far.

In a numerical taxonomic study, Grant and Patterson (1991) reported on strains isolated from irradiated chicken packaged under various gas atmospheres, and which mainly resembled *Lb. sake*. Only one of the chicken isolates clustered with the type strain of *C. divergens* and *C. piscicola* and probably belonged to the genus *Carnobacterium*.

9.6.1.3 Fish and seawater. The isolation of two species of *Carnobacterium* from anaerobic waters of Ace Lake, Antarctica, indicates a potential role of these bacteria in the ecology of Antarctic meromictic lakes. Franzmann *et al.* (1991) suggest that *C. funditum* and *C. alterfunditum* may contribute to the initial production of a reduced environment in antarctic sulphurous waters of marine origin and that they may provide electron donors for exploitation by sulphate-reducing bacteria in such waters. Temperature (1·2°C), pH (6.6) and sodium concentration (*c.* 0.5 M) (Masuda *et al.*, 1988) allow growth of these psychrotrophic *Carnobacterium* species.

Whilst *C. funditum* and *C. alterfunditum* seem to live freely in sea water, other species of *Carnobacterium* were found to be associated with poikilotherm animal hosts.

Strains of *C. piscicola* were isolated from different fish species such as rainbow trout (*Oncorhynchus mykiss*) (Hiu *et al.*, 1984; Starliper *et al.*, 1992), carp (*Cyprinus carpio*) (Bausewein, 1988), spring chinook salmon (*Oncorhynchus tshawytscha*) (Hiu *et al.*, 1984), channel catfish (*Ictalurus punctatus*) (Baya *et al.*, 1991) and striped bass (*Morone saxatilis*) (Baya *et al.*, 1991). The occurrence of *C. piscicola* was first reported in fish from hatcheries in the Pacific Northwest of America (Hiu *et al*, 1984). Subsequent reports confirmed that carnobacteria are commonly found in farmed and natural populations of salmonids in Europe (Michel *et al.*, 1986) and Australia (Humphrey *et al.*, 1987). Carnobacteria can be considered part of the normal gut microbial population of many fish species. On the other hand, numerous reports refer to the association of *C. piscicola* with fish diseases, such as pseudo-kidney disease (Ross and Toth, 1974; Hiu *et al.*, 1984) and clinical or subclinical peritonitis (Humphrey *et al.*, 1987). Some authors attempted to infect salmonids and other fish with strains of *C. piscicola* in order to determine the virulence/pathogenicity of this species for fish. Most of these infection trials failed (Ahmed, 1987; Baya *et al.*, 1991), probably on account of an extremely low mortality rate (Michel *et al.*, 1986; Starliper *et al.*, 1992). In the study of Baya *et al.* (1991) only rainbow trout was susceptible to infection with some isolates of

Carnobacterium and no mortalities were recorded in striped bass. Nevetherless, *C. piscicola* could be re-isolated from fish inoculated with these strains. *Carnobacterium piscicola* seems to have a low virulence for fish; on the other hand, fish seem to be most susceptible when exposed to stress conditions such as spawning and handling (Starliper *et al.*, 1992). Baya *et al.* (1991) suggest a carrier state of *C. piscicola* in fish populations; however, so far the importance of this species as fish pathogen cannot be accurately assessed.

9.6.2 Bacteriocins

The characterization of several bacteriocin-producing strains of carnobacteria has been reported in recent years (Ahn and Stiles, 1990a,b; Schillinger and Holzapfel, 1990; Lewus *et al.*, 1991; Buchanan and Klawitter, 1992a,b; Stoffels *et al.*, 1992a,b; Schillinger *et al.*, 1993). The application of such strains for biological preservation, e.g. for extending the shelf-life of some foods such as vacuum-packaged meats (Stiles and Hastings, 1991) or fresh fish and lightly preserved fish products (Stoffels *et al.*, 1992a) has been suggested. The fact that carnobacteria are non-aciduric and will acidify the product to a lesser extent than other lactic acid bacteria, serves as an important consideration. Ahn and Stiles (1990a) suggest that these bacteria may be used in the development of preservative packaging systems for meat.

All bacteriocin-producing strains of carnobacteria studied to some detail so far were identified as *C. piscicola* and were isolates from meat (Ahn and Stiles, 1990a; Schillinger and Holzapfel, 1990; Lewus *et al.*, 1991; Buchanan and Klawitter, 1992a) or fish (Stoffels *et al.*, 1992a) (Table 9.1). Bacteriocinogenic strains of *C. divergens* (Schillinger and Holzapfel, 1990; Schillinger *et al.*, 1993) have been reported more recently, and deserve more intensive studies.

Like most bacteriocins of lactic acid bacteria, the bacteriocins produced by *C. piscicola* are small hydrophobic peptides. All of them are inactivated by a range of proteolytic enzymes; however, some are not sensitive to all types of proteases. Trypsin, α-chymotrypsin and papain are effective against most of them, whereas pepsin does not degrade the carnobacteriocin(s) from strain LV 17 (Ahn and Stiles, 1990b) and causes only partial inactivation of carnocin UI 49 produced by *C. piscicola* UI 49 (Stoffels *et al.*, 1992a). The bateriocins from carnobacteria are moderately heat stable and only a small loss of activity is observed after heating to 100°C for 60 min (Stoffels *et al.*, 1992a). Their activity is not affected by exposure to pH values from 2 to 8. However, at pH >8, carnocin UI 49 was rapidly inactivated (Stoffels *et al.*, 1992b) and piscicolin 61 also lost some of its activity (Schillinger *et al.*, 1993).

Carnobacterium piscicola LV 17 is able to produce bacteriocins of

Table 9.1 Properties of bacteriocins produced by strains of *Carnobacterium piscicola*

Bacteriocin	Producer strain	Origin	Heat stability at 100°C	pH stability	Sensitivity to proteases	Sensitive bacteria	Plasmids involved in production and immunity	Molecular mass (Da)
Carnobacteriocins A1, A2, A3, B1, B2 (Ahn and Stiles, 1990b)	LV 17	Pork	No inactivation after 30 min	2–11	Trypsin: + α-Chymotrypsin: + Pepsin: − Papain: +	carnobacteria enterococci lactobacilli pediococci *Listeria*	40 MDa 49 MDa	4541–5052
Piscicolin 61 (Schillinger *et al.*, 1993)	LV 61	Lamb	No inactivation after 20 min	2–8	Trypsin: + α-Chymotrypsin: + Pepsin: + Papain: +	carnobacteria enterococci lactobacilli pediococci leuconostocs *Listeria*	22 kb	5052
Carnocin UI49 (Stoffels *et al.*, 1992a,b)	UI 49	Fresh fish	Partial inactivation after 60 min	2–10	Trypsin: + α-Chymotrypsin: + Pepsin: ±	carnobacteria lactobacilli pediococci lactococci	ND	4635
Unnamed (Buchanan and Klawitter, 1992a)	LK 5	Raw ground beef	Partial inactivation after 5 min	ND	Trypsin: + Pepsin: − Papain: +	*Listeria*	probably no plasmid	ND

ND, no data.

different size and amino acid sequence and which were called carnobacteriocin A (obviously existing in three different forms A1, A2, A3) and carnobacteriocins B1 and B2 (Nettles and Barefoot, 1993). The molecular mass of carnobacteriocin A is identical to that of piscicolin 61, whereas the carnobacteriocins B are smaller peptides with molecular masses of 4541 and 4969 Da, respectively.

Carnocin UI 49 was found to have a molecular mass of 4635 Da and obviously belongs to the class of bacteriocins termed lantibiotics (Stoffels *et al.*, 1992b) whereas the bacteriocin molecules produced by *C. piscicola* LV 17 and *C. piscicola* LV 61 do not contain lanthionine.

The antibacterial spectrum of the bacteriocins produced by *C. piscicola* comprises different genera of lactic acid bacteria and generally also includes *Listeria monocytogenes*. Gram-negative bacteria are not inhibited. Within one species not all strains are equally susceptible to the bacteriocin and differences are observed in sensitivity between different species. Schillinger *et al.* (1993) observed that carnobacteria were the most sensitive organisms to piscicolin 61. Enterococci and species of *Listeria* were less sensitive and most other lactic acid bacteria such as lactobacilli, pediococci and leuconostocs were more or less resistant to piscicolin 61.

The bacteriocins described by Ahn and Stiles (1990a,b), Stoffels *et al.* (1992a,b) and Schillinger *et al.* (1993) were produced during the logarithmic phase of growth and maximum production was observed at the beginning of the stationary phase of growth. Several bacteriocins are also produced at refrigeration temperatures, although Stoffels *et al.* (1992a) report a very low yield of carnocin UI 49 at 4°C. By contrast, *C. piscicola* LK5 produced about the same amounts of bacteriocin at 5, 12, and 19°C, whilst the bacteriocin activity was reduced substantially at 28°C (Buchanan and Klawitter, 1992a). The increased effectiveness of this bacteriocin-producer at refrigeration temperatures seems to be caused by an increase in susceptibility of *Listeria monocytogenes* to the bacteriocin at lower temperatures and by a decrease in bacteriocin production at higher temperatures (Buchanan and Klawitter, 1992a). Temperatures above 25°C seem to affect bacteriocin production of *C. piscicola* adversely. No production of piscicolin 61 was observed above 30°C although growth was evident at 34°C and 37°C (Schillinger *et al.*, 1993). On the other hand, Stoffels *et al.* (1992a) report a faster destruction rate of carnocin UI 49 at 34°C.

Bacteriocin production is also influenced by the initial pH of the medium. Optimal growth of carnobacteria occurs at a higher pH than other lactic acid bacteria. At pH 8.0–9.0, higher amounts of piscicolin are produced than at pH 6.0 or 7.0. However, about the same yield of the bacteriocin can be achieved at pH 6.0 and 7.0 when the pH is maintained constant during growth of *C. piscicola* LV 61 (Schillinger *et al.*, 1993).

To date, only sparse information is available on the effectiveness of these bacteriocins in controlling pathogens in food systems and their ecological role in microbial interactions. In a study by Buchanan and Klawitter (1992b), *C. piscicola* LK5 was effective in suppressing growth of *Listeria monocytogenes* in a range of refrigerated foods such as UHT milk, crabmeat, creamed corn, and frankfurters. The bacteriocin-producer was more inhibitory at 5°C than at 19°C. The effectiveness of other bacteriocinogenic strains of *C. piscicola* to prevent undesirable bacteria from growth in foods has still to be studied.

9.7 Identification

Most phenotypic properties of the genus *Carnobacterium* are similar to those of lactobacilli. Carnobacteria are Gram-positive, catalase-negative, non-spore-forming rods or coccobacilli usually occurring singly, in pairs or short chains. They possess a fermentative metabolism and do not reduce nitrate to nitrite. Differences in some characteristics from lactobacilli, however, justify the phenetic separation of these bacteria from the genus *Lactobacillus* (see Table 9.2).

Carnobacteria do not grow on acetate agar at pH 5.4 (Rogosa *et al.*, 1951) commonly used for the selective isolation of lactobacilli. Their sensitivity to acetate allows only restricted or no growth in MRS broth (De Man *et al.*, 1960; Holzapfel and Gerber, 1983). Growth of carnobacteria is inhibited at pH 4.5 whereas most lactobacilli are able to grow at this low pH. On the other hand, carnobacteria show good growth at pH 9.0 which is unusual for lactobacilli.

The profile of cellular fatty acids of carnobacteria differs from that of lactobacilli. Carnobacteria possess oleic acid ($\Delta 9,10$) as major $C_{18:1}$ isomer (Collins *et al.*, 1987) whereas *cis*-vaccenic acid ($\Delta 11,12$) constitutes the predominant $C_{18:1}$ isomer in *Lactobacillus* species (Hofmann *et al.*,

Table 9.2 Major properties differentiating carnobacteria from lactobacilli

Property	*Carnobacterium*	*Lactobacillus*
Growth on acetate agar (pH 5.4)	−	+
Growth at pH 4.5	−	+
Growth at pH 9.0	+	−
Lactic acid isomers produced	L(+)	L(+), D(−), DL
Type of diamino acid in peptidoglycan	*m*Dpm	Lys, *m*Dpm, Orn
Major $C_{18:1}$ cellular fatty acid	Oleic acid ($\Delta 9,10$)	*cis*-Vaccenic acid ($\Delta 11,12$)
Fermentation type (glucose)	Atypical homo	Homo or hetero
G+C content (mol %)	33–37	32–55

1952; Hofmann and Sax, 1953; Veerkamp, 1971; Uchida and Mogi, 1973; Turujman *et al.*, 1974). Moreover, carnobacteria seem not to produce lactobacillic acid (Shaw and Harding, 1985; Collins *et al.*, 1987). *Carnobacterium divergens* is able to synthesize cyclopropane-ring-containing fatty acids which correspond to 9,10-methylenoctadecanoic acid (Collins *et al.*, 1987).

In their identification scheme, Montel *et al.* (1991) chose the presence of *m*Dpm in the cell wall, failure to grow on acetate agar, formation of only L(+)-lactate from glucose and the ability to produce citrulline from arginine as key properties to differentiate *C. divergens* and *C. piscicola* from meat lactobacilli.

As possibly suggested by the relatively close phylogenetic relationship to the genus *Enterococcus*, carnobacteria share several physiological characteristics with the enterococci. Strains of *C. divergens* and *C. piscicola* show a similar heavy metal resistance pattern to that of strains of *Enterococcus faecium* and *Enterococcus faecalis* (Bosch and Holzapfel, 1985, unpublished observations) and practically identical resistance spectra to about 60 different antibiotics. Moreover, in contrast to other lactic acid bacteria, carnobacteria and enterococci both grow well in media with a high pH of up to 9.5.

9.8 Isolation and enumeration

A summary of methods for the isolation and enumeration of carnobacteria is given by Hammes *et al.* (1992) and by Holzapfel (1992). The problem of selective differentiation from enterococci has not been completely solved, and explains the elective character of media suggested thus far. As indicated above, carnobacteria and enterococci share common resistance spectra to, for example, heavy metal salts, antibiotics and other physicochemical factors.

9.9 Maintenance and preservation

No special precautions are necessary for the maintenance and preservation of carnobacteria. To maintain cultures for several weeks, stab cultures in D-MRS agar (Schillinger and Holzapfel, 1990) may be used. They should be kept at 1–4°C and subcultured every 2–3 weeks (Hammes *et al.*, 1992). For longer storage, cultures may be lyophilized using cryoprotective agents such as milk solids, lactose or horse serum. Alternatively, fresh cultures may be frozen at −80°C after addition of 20% glycerol to the cultivation broth.

9.10 Species of the genus *Carnobacterium*

Species of *Carnobacterium* share the following common properties:

- *Morphology*: Microscopically the cells appear as straight, slender rods occurring singly or in short chains (see Figure 9.1). Macroscopically surface colonies on agar media are white, convex, shiny and circular.
- *Metabolism*: D-Glucose is fermented predominantly to L(+)-lactic acid via the glycolytic pathway (De Bruyn *et al.*, 1987, 1988). Small amounts of formic acid, acetic acid, ethanol, acetoin and CO_2 may be produced. Gas production is slow, inconsistent and often not detectable. In aerobic cultures the formation of acetic acid is stimulated (Borch and Molin, 1989).
- *General physiological features*: Growth may occur at low temperatures (0–2°C). Facultatively aerobic, carnobacteria are more tolerant to oxygen than homofermentative lactobacilli (Borch and Molin, 1989). Major cellular fatty acids are of the straight-chain saturated and monounsaturated types, the $C_{18:1}$ compound being a 9,10-isomer (Collins *et al.*, 1987).
  - The peptidoglycan is of the *m*Dpm direct type.
  - The G+C content of DNA ranges from 33–37 mol%.
  - All strains produce acid from cellobiose, D-fructose, D-glucose, D-mannose, ribose, salicin, and sucrose. Adonitol, arabinose, dulcitol, erythritol, glycogen, inositol, raffinose, rhamnose, sorbitol and xylose are not fermented.

To date the genus comprises six species: *C. divergens*, *C. piscicola*, *C. gallinarum*, *C. mobile*, *C. funditum*, and *C. alterfunditum*. The type species of *Carnobacterium* is *C. divergens*.

Characteristics used for species differentiation are presented in Table 9.3.

9.10.1 Carnobacterium divergens

Originally described as *Lactobacillus divergens* (Holzapfel and Gerber, 1983), the features of this new 'deviating *Lactobacillus* sp.' clearly differed from other 'atypical' lactobacilli, especially regarding CO_2 and L(+)-lactic acid production, and the presence of *m*Dpm in the cell wall.

The cells are non-motile. Growth in acetate-containing media is suppressed in the presence of glucose and citrate but is stimulated when ribose or fructose are added (Holzapfel and Gerber, 1983). Growth occurs between 0 and 40°C, but not at 45°C, pH 3.9 or in 10% NaCl.

The pattern of sugars fermented is given in Table 9.3. Arginine is

Table 9.3 Physiological and biochemical characteristics of species of *Carnobacterium* (All strains produce acid from cellobiose, fructose, glucose, maltose, mannose, ribose, salicin and sucrose, but not from adonitol, arabinose, dulcitol, erythritol, glycogen, inositol, raffinose, rhamnose, sorbitol and xylose)*

Characteristic	*C. alterfunditum*	*C. divergens*	*C. funditum*	*C. gallinarum*	*C. mobile*	*C. piscicola*
Growth at						
0°C	+	+	+	+	+	+
30°C	–	+	–	+	+	+
45°C	–	–	–	–	–	–
Acid produced from						
Amygdalin	W	+	–	+	–	+
Galactose	W	–(+)	W	+	+	+(–)
Gluconate	NT	+	NT	+	–	+
Glycerol	W	–	W	–		+(–)
Inulin	–	–	–	–	+	+
Lactose	–	–	–	+	–	+(–)
Mannitol	–	–	+	–	–	+
Melezitose	–	+(–)	–	+	–	+
Melibiose	–	–	–	–	–	+/–
α-Methyl-D-glucoside	NT	–	NT	+	–	+
α-Methyl-D-mannoside	NT	–	NT	–	–	+
D-Tagatose	NT	–	NT	+	–	–
Trehalose	–	+	+	+	+	+
D-Turanose	NT	–	NT	+	–	+
Motility	+	–	+	–	+	–
Voges–Proskauer Test	NT	+	NT	+	–	+
Gas production in arginine MRS	–	+	–	–	+(–)	+
G+C content (mol%)	33–34	33–36.4	32–34	34–36.4	35.5–37.2	33.7–36.4

*Symbols: –(+), occasional strain positive; +(–), occasional strain negative; W, weak; NT, not tested.

deaminated, starch and urea are not degraded (Ferusu and Jones, 1988). No slime is produced from sucrose.

They are oxidase-negative. No production of H_2O_2 (Borch and Molin, 1988). Production of pseudo-catalase is observed on haem-containing media. Major cellular fatty acids are of the straight-chain saturated, monounsaturated and cyclopropane-ring types with tetradecanoic, 9,10-octadecanoic and 9,10-methylenoctadecanoic acids predominating (Collins *et al.*, 1987). G+C content of the DNA is 33.0–36.4 mol%. Many strains contain plasmids, ranging from 2.5 to 26.9 MDa (Ahrne *et al.*, 1989). Some strains produce bacteriocins active against closely related species (Schillinger, unpublished observations).

Most strains are sensitive to chloramphenicol (50 μg), erythromycin (10 μg), gentamycin (10 μg), and tetracyclin (50 μg), and resistant to methicillin (10 μg) and nalidixic acid (30 μg). Source: Fresh and vacuum-packaged, refrigerated meat. Type strain: DSM 20623 = ATCC 35677. Other strains: DSM 20589, DSM 20624, DSM 20625.

9.10.2 Carnobacterium piscicola

Synonyms: *Lactobacillus piscicola, Lactobacillus carnis.*

Cells are non-motile. Growth occurs at temperatures from 1 to 40°C. Optimum pH ranges from 7.0 to 9.0.

Folic acid, riboflavin, panthothenate, and niacin are required for growth, whereas vitamin B_{12}, biotin, thiamine, and pyridoxal are not required. Pattern of sugars fermented is given in Table 9.3. Arginine is deaminated. Starch and urea are not degraded.

They are oxidase-negative. Major cellular fatty acids are of the straight-chain saturated and monounsaturated types with tetradecanoic, hexadecanoic, hexadecenoic, and 9,10-octadecenoic acids predominating. G+C content of DNA is 33.7–36.4 mol%.

Sensitivity to chloramphenicol (50 μg), erythromycin (10 μg), furazolidone (100 μg). Resistance to cloxacillin (5 μg), gentamycin (10 μg), nalidixic acid (30 μg), neomycin (5 μg) and steptomycin (25 μg).

Some strains produce bacteriocins active against other carnobacteria, some enterococci and lactobacilli, and strains of *Listeria monocytogenes* (Ahn and Stiles, 1990a,b; Schillinger and Holzapfel, 1990; Buchanan and Klawitter, 1992a; Stoffels *et al.*, 1992a). Sources: fresh and vacuum-packaged refrigerated meat; salmonid fish. Type strain: DSM 20730 = ATCC 35586. Other strains: DSM 20590, DSM 20722 = ATCC 43225.

9.10.3 Carnobacterium gallinarum

Cells are non-motile. Growth occurs from 0 to 35°C but not at 40°C.

Pattern of sugars fermented is given in Table 9.3. Arginine dihydrolase positive and lysine decarboxylase, ornithine decarboxylase, tryptophan desaminase, and oxidase negative. Starch and urea are not degraded. Voges–Proskauer positive. Tetradecanoic, hexadecanoic, and 9,10-octadecenoic acids are the predominating cellular fatty acids. G+C content of the DNA ranges from 34.3 to 36.4 mol%.

Sensitivity to ampicillin (25 μg), chloramphenicol (50 μg), erythromycin (10 μg), novobiocin (30 μg), and tetracycline (50 μg). Resistance to cloxacillin (5 μg), gentamycin (10 μg), nalidixic acid (30 μg), and streptomycin (25 μg). The type strain was isolated from ice slush around chicken carcasses. Sources: refrigerated poultry meat. Type strain: DSM 4847 = NCFB 2766

9.10.4 Carnobacterium mobile

Cells are motile. Growth occurs at 0–35°C, but not at 40°C.

Pattern of fermented sugars is given in Table 9.3. Arginine dihydrolase positive and lysine decarboxylase, ornithine decarboxylase, tryptophan desaminase and urease negative. They are Voges–Proskauer negative. Hexadecanoic, hexadecenoic and 9,10-octadecenoic acids are the cellular fatty acids predominating.

The G+C content of the DNA ranges from 35.5 to 37.2 mol%.

Sensitivity to ampicillin (25 μg), chloramphenicol (50 μg), erythromycin (10 μg), methicillin (10 μg), novobiocin (30 μg), and tetracycline (50 μg). Resistance of most strains to cloxacillin (5 μg) and nalidixic acid (30 μg). Sources: Type strain was isolated from irradiated chicken meat. Type strain: DSM 4848 = NCFB 2765.

9.10.5 Carnobacterium funditum

Cells are motile. Old cells are often non-motile and stain Gram-negative. Growth occurs at 0–23°C, but not at 30°C. Optimum initial pH: 7.0–7.4.

Sodium is required for growth (optimal growth at 1.7% NaCl; it also appears to be dependent on a component present in yeast extract (at least 0.1%).

Pattern of sugars fermented is given in Table 9.3. Gelatinase negative. Oxidase negative. C18:1*cis*9 is a major component of cellular fatty acids. The G+C content of DNA ranges from 31.6 to 34.0 mol%.

The type strain was isolated from the anaerobic monilimnion of an Antarctic lake. Sources: Antarctic lake water. Type strain: DSM 5970 = ATCC 49836. Other strains: DSM 5971.

9.10.6 Carnobacterium alterfunditum

Cells are motile by a single subpolar flagellum. Old cells usually stain Gram-negative and are non-motile. Growth occurs from 0 to 23°C, but not at 30°C. Optimum initial pH: 7.0–7.4.

Growth occurs in media containing 0.1% yeast extract without added sodium salts, and the optimum chloride concentration for growth is about 0.1 M.

Pattern of sugars fermented is given Table 9.3. C18:1*cis*9 is a major component of cellular fatty acids. The G+C content of DNA ranges from 33.3 to 34.4 mol%.

The type strain was isolated from the anaerobic waters of an Antarctic lake. Sources: Antarctic lake water. Type strain: DSM 5972 = ATCC 49837. Other strain: DSM 5973.

References

Ahmed, S.M. (1987) Untersuchungen zum kulturellen und fluoreszenz-mikroskopischen Nachweis und zur Pathogenität von *Renibacterium salmoninarum* und *Lactobacillus piscicola* bei Fischen in Norddeutschland unter Berücksichtigung der Abgrenzung zwischen beiden Keimarten. MS thesis, Tierärztliche Hochschule Hannover, Germany.

Ahn, C. and Stiles, M.J. (1990a) Antibacterial activity of lactic acid bacteria isolated from vacuum-packaged meats. *Journal of Applied Bacteriology*, **69**, 302–310.

Ahn, C. and Stiles, M.J. (1990b) Plasmid-associated bacteriocin production by a strain of *Carnobacterium piscicola* from meat. *Applied and Environmental Bacteriology*, **56**, 2503–2510.

Ahrne, S., Molin, G. and Stahl, S. (1989) Plasmids in *Lactobacillus* strains isolated from meat and meat products. *Systematic and Applied Microbiology*, **11**, 320–325.

Bausewein, G. (1988) Untersuchungen zur aeroben und fakultativ anaeroben bakteriellen Flora des Intestinaltrakts von Karpfen (*Cyprinus carpio*). MS thesis, Universität München, Munich, Germany.

Baya, A.M., Toranzo, A.E., Lupiani, B., Li, T., Roberson, B.S. and Hetrick, F.M. (1991) Biochemical and serological characterization of *Carnobacterium* spp. isolated from farmed and natural populations of striped bass and catfish. *Applied and Environmental Microbiology*, **57**, 3114–3120.

Borch, E. and Molin, G. (1988) Numerical taxonomy of psychotrophic lactic acid bacteria from prepacked meat and meat products. *Antonie van Leeuwenhoek*, **54**, 301–323.

Borch, E. and Molin, G. (1989) The aerobic growth and product formation of *Lactobacillus*, *Leuconostoc*, *Brochothrix*, and *Carnobacterium* in batch cultures. *Applied and Microbiological Biotechnology*, **30**, 81–88.

Buchanan, R.L. and Klawitter, L.A. (1992a) Characterization of a lactic acid bacterium, *Carnobacterium piscicola* LK5, with activity against *Listeria monocytogenes* at refrigeration temperatures. *Journal of Food Safety*, **12**, 199–217.

Buchanan, R.L. and Klawitter, L.A. (1992b) Effectiveness of *Carnobacterium piscicola* LK5 for controlling the growth of *Listeria monocytogenes* Scott A in refrigerated foods. *Journal of Food Safety*, **12**, 219–236.

Champomier, M.C., Montel, M.C. and Talon, R. (1989) Nucleic acid relatedness studies on the genus *Carnobacterium* and related taxa. *Journal of General Microbiology*, **135**, 1391–1394.

Collins, M.D., Ash, C., Farrow, J.A.E., Phillips, B.A., Ferusu, S. and Jones, J. (1987) Classification of *Lactobacillus divergens*, *Lactobacillus piscicola* and some catalase-negative, asporogenous, rod-shaped bacteria from poultry in a new genus, *Carnobacterium*. *International Journal of Systematic Bacteriology*, **37**, 310–317.

Collins, M.D., Rodrigues, U., Ash, C., Aguirre, M., Farrow, J.A.E., Martinez-Murcia, A., Phillips, B.A., Williams, A.M. and Wallbanks, S. (1991) Phylogenetic analysis of the genus *Lactobacillus* and related lactic acid bacteria as determined by reverse transcriptase sequencing of 16S rRNA. *FEMS Microbiology Letters*, **77**, 5–12.

De Bruyn, I.N., Louw, A.I., Visser, L. and Holzapfel, W.H. (1987) *Lactobacillus divergens* is a homofermentative organism. *Systematic and Applied Microbiology*, **9**, 173–175.

De Bruyn, I.N., Holzapfel, W.H., Visser, L. and Louw, A.I. (1988) Glucose metabolism by *Lactobacillus divergens*. *Journal of General Microbiology*, **134**, 2130–2109.

De Man, J.R., Rogosa, M. and Sharpe, M.E. (1960) A medium for the cultivation of lactobacilli. *Journal of Applied Bacteriology*, **23**, 130–135.

Ferusu, S.B. and Jones, B. (1988) Taxonomic studies on *Brochothrix*, *Erysipelothrix*, *Listeria* and atypical lactobacilli. *Journal of General Microbiology*, **134**, 1165–1183.

Franzmann, P.D., Höpfl, P., Weiss, N. and Tindall, B.J. (1991) Psychrotrophic, lactic acid-producing bacteria from anoxic waters in Ace Lake, Antarctica; *Carnobacterium funditum* sp. nov. and *Carnobacterium alterfunditum* sp. nov. *Archives of Microbiology*, **156**, 255–262.

Grant, I.R. and Patterson, M.F. (1991) A numerical taxonomic study of lactic acid bacteria isolated from irradiated pork and chicken packaged under various gas atmospheres. *Journal of Applied Bacteriology*, **70**, 302–307.

Hammes, W.P., Weiss, N. and Holzapfel, W.H. (1992) *Lactobacillus* and *Carnobacterium*. In *The Prokaryotes*, Vol. II, 2nd edn (eds Balows, A., Trüper, H.G., Dworkin, M., Harder, W. and Schleifer K.H.). Springer-Verlag, New York, USA, pp. 1535–1594.

Hitchener, J.B., Egan, A.F. and Rogers, P.J. (1982) Characteristics of lactic acid bacteria isolated from vacuum-packaged beef. *Journal of Applied Bacteriology*, **52**, 31–37.

Hiu, S.F., Holt, R.A., Sriranganathan, N., Seidler, R.J. and Fryer, J.L. (1984) *Lactobacillus piscicola*, a new species from salmonid fish. *International Journal of Systematic Bacteriology*, **34**, 393–400.

Hofmann, K. and Sax, S.M. (1953) The chemical nature of the fatty acids of *Lactobacillus casei*. *Journal of Biological Chemistry*, **205**, 55–63.

Hofmann, K., Lucas, R.A. and Sax, S.M. (1952) The chemical nature of the fatty acids of *Lactobacillus arabinosus*. *Journal of Biological Chemistry*, **195**, 473–485.

Holzapfel, W.H. (1992) Culture media for non-sporulating Gram-positive food spoilage bacteria. *International Journal of Food Microbiology*, **17**, 113–133.

Holzapfel, W.H. and Gerber, E.S. (1983). *Lactobacillus divergens* sp. nov., a new heterofermentative *Lactobacillus* species producing L(+)-lactate. *Systematic and Applied Microbiology*, **4**, 522–534.

Humphrey, J.D., Lancaster, C.E., Gudkovs, N. and Copland, J.W. (1987) The disease status of Australian salmonids: bacteria and bacterial diseases. *Journal of Fish Diseases*, **10**, 403–410.

Kitchell, A.G. and Shaw, B.G. (1975) Lactic acid bacteria in fresh and cured meat. In *Lactic Acid Bacteria in Beverages and Food* (eds Carr, J.G., Cutting, C.V. and Whiting, G.C.). Academic Press, London, UK, pp. 209–220.

Lewus, C.B., Kaiser, A. and Montville, T.J. (1991) Inhibition of foodborne bacterial pathogens by bacteriocins from lactic acid bacteria isolated from meat. *Applied and Environmental Microbiology*, **57**, 1683–1688.

Masuda, N., Nakaya, S., Burton, H.R. and Torii, T. (1988) Trace element distributions in some saline lakes of the Vestfold hills, Antarctica. *Hydrobiologia*, **165**, 103–114.

Michel, C., Faivre, B. and Kerouault, B. (1986). Biochemical identification of *Lactobacillus piscicola* strains from France and Belgium. *Diseases of Aquatic Organisms*, **2**, 27–30.

Mol, J.H.H., Hietbrink, J.E.A., Mollen, H.W.M. and Van Tinteren, J. (1971) Observations on the microflora of vacuum packed sliced cooked meat products. *Journal of Applied Bacteriology*, **34**, 377–397.

Montel, M.C., Talon, R., Fournaud, J. and Champomier, M.C. (1991) A simplified key for identifying homofermentative *Lactobacillus* and *Carnobacterium* spp. from meat. *Journal of Applied Bacteriology*, **70**, 469–472.

Nettles, C.G. and Barefoot, S. (1993) Biochemical and genetic characteristics of bacteriocins of food-associated lactic acid bacteria. *Journal of Food Protection*, **56**, 338–356.

Reuter, G. (1975) Classification problems, ecology and some biochemical activities of lactobacilli of meat products. In *Lactic Acid Bacteria in Beverages and Food* (eds Carr, J.G., Cutting, C.V. and Whiting, G.C.). Academic Press, London, UK, pp. 221–229.

Rogosa, M., Mitchell, J.A. and Wiseman, R.F. (1951) A selective medium for the isolation of oral and faecal lactobacilli. *Journal of Bacteriology*, **62**, 132–133.

Ross, A.J. and Toth, R.J. (1974) *Lactobacillus* – a new fish pathogen? *Progressive Fish Culturist*, **36**, 191.

Schillinger, U. and Holzapfel, W.H. (1990) Antibacterial activity of carnobacteria. *Food Microbiology*, **7**, 305–310.

Schillinger, U. and Lücke, F.-K. (1986) Milchsäurebakterien-Flora auf vakuumverpacktem Fleisch und ihr Einfluß auf die Haltbarkeit. *Fleischwirtschaft*, **66**, 1515–1520.

Schillinger, U. and Lücke, F.-K. (1987) Identification of lactobacilli from meat and meat products. *Food Microbiology*, **4**, 199–208.

Schillinger, U., Stilles, M.E. and Holzapfel, W.H. (1993) Bacteriocin production by *Carnobacterium piscicola* LV 61. *International Journal of Food Microbiology*, **20**, 131–147.

Shaw, B.G. and Harding, C.D. (1984) A numerical taxonomic study of lactic acid bacteria from vacuum-packed beef, pork, lamb and bacon. *Journal of Applied Bacteriology*, **56**, 25–40.

Shaw, B.G. and Harding, C.D. (1985) Atypical lactobacilli from vacuum-packaged meats; comparison by DNA hybridization, cell composition, and biochemical tests with a description of *Lactobacillus carnis* sp. nov. *Systematic and Applied Microbiology*, **6**, 291–297.

Starliper, C.E., Shotts, E.B. and Brown, J. (1992) Isolation of *Carnobacterium piscicola* and an unidentified Gram-positive *Bacillus* from sexually mature and post-spawning rainbow trout *Oncorhynchus mykiss*. *Diseases of Aquatic Organisms*, **13**, 181–187.

Stiles, M.E. and Hastings, J.W. (1991) Bacteriocin production by lactic acid bacteria: potential for use in meat preservation. *Trends in Food Science and Technology*, **2**, 247–251.

Stoffels, G., Nes, I. and Gudmundsdottir, A. (1992a) Isolation and properties of a bacteriocin-producing *Carnobacterium piscicola* isolated from fish. *Journal of Applied Bacteriology*, **73**, 309–316.

Stoffels, G., Nissen-Meyer, J., Gudmundsdottir, A., Sletten, K., Holo, H. and Nes, I.F. (1992b) Purification and characterization of a new bacteriocin isolated from a *Carnobacterium* sp. *Applied and Environmental Microbiology*, **58**, 1417–1422.

Thornley, M.J. (1957) Observations on the microflora of minced chicken meat irradiated with 4 MeV cathode rays. *Journal of Applied Bacteriology*, **20**, 286–298.

Thornley, M.J. and Sharpe, M.E. (1959) Microorganisms from chicken meat related to both lactobacilli and aerobic sporeformers. *Journal of Applied Bacteriology*, **22**, 368–376.

Turujman, N., Jabr, I. and Durr, I.F. (1974) Amino acid and fatty acid of *Lactobacillus* plantarum. *International Journal of Biochemistry*, **5**, 791–793.

Uchida, K. and Mogi, K. (1973) Cellular fatty acid spectra of hiochi bacteria, alcohol-tolerant lactobacilli, and their group separation. *Journal of General and Applied Microbiology*, **19**, 233–249.

Veerkamp, J.H. (1971) Fatty acid composition of *Bifidobacterium* and *Lactobacillus* strains. *Journal of Bacteriology*, **108**, 861–867.

Von Holy, A. (1983) Bacteriological studies on the extension of shelf life of raw minced beef. M.Sc. (microbiol.) Thesis, University of Pretoria, Republic of South Africa.

Von Holy, A., Cloete, T.E. and Holzapfel, W.H. (1991) Quantification and characterization of microbial populations associated with spoiled, vacuum-packaged Vienna sausages. *Food Microbiology*, **8**, 95–104.

Wallbanks, S., Martinez-Murcia, A.J., Fryer, J.L., Phillips, B.A. and Collins, M.D. (1990) 16S ribosomal ribonucleic acid sequence determination of members of the genus *Carnobacterium* and related lactic acid bacteria. Description of *Vagococcus salmoninarum* sp. nov. *International Journal of Systematic Bacteriology*, **40**, 224–230.

Worobo, R.W., Roy, K.L., Sailer, M., Vederas, J.C. and Stiles, M.E. (1993) Genetic determinants of carnobacteriocin A and the development of a food grade vector for multiple bacteriocin expression. *FEMS Microbiology Reviews*, **12**, P133.

10 The genus *Enterococcus*

L.A. DEVRIESE and B. POT

10.1 History

The history of the enterococci starts in 1899 with the description by Thiercelin of a new Gram-positive diplococcus, which was later included in the new genus *Enterococcus* with the type species *Enterococcus proteiformis* (Thiercelin and Jouhaud, 1903). In 1906, however, Andrewes and Horder renamed Thiercelin's 'entérocoque' as *Streptococcus faecalis* based on its ability to form short or long chains. Because of these early links, the history of the enterococci cannot be considered separately from that of the genus *Streptococcus*.

Identification and classification of streptococci (and enterococci) relied for a very long time on the serological groups introduced by Lancefield (1933). In this system, in which organisms are designated by letters of the alphabet, the enterococci known at that time all possessed the group D antigen. *Streptococcus bovis* and *Strep. equinus* also belong to group D, but have other characteristics which allow one to group them apart from the enterococci. Sherman (1937) divided the streptococci into four groups: (1) the enterococci, (2) the 'lactic-', (3) the 'viridans-', and (4) the 'pyogenic' streptococci. The terms 'enterococci' and 'faecal streptococci' have been used commonly in a loose sense during the past decades to designate streptococci of faecal origin, resembling more or less *Ent. faecalis*.

In a 'request for an opinion', Kalina (1970) proposed *Streptococcus faecalis* as the type species of the new genus *Enterococcus* as *Enterococcus faecalis* with the subspecies *Ent. faecalis* subsp. *liquifaciens* (previously *Strep. faecalis* subsp. *liquifaciens*) and *Ent. faecalis* subsp. *zymogenes* (previously *Strep. faecalis* subsp. *zymogenes*). Kalina also proposed a second species *Ent. faecium*, originally described by Orla-Jensen in 1919 as *Streptococcus faecium*. He added *Streptococcus durans* (Sherman and Wing, 1937) to *Ent. faecium* as *Ent. faecium* subsp. *durans*. The propositions of Kalina were never officially recognized, and in 1980 the genus *Enterococcus* did not appear in the Approved Lists of Bacterial Names (Skerman *et al.*, 1980).

The original proposals of Sherman (1937) were extended and modified after investigation of new physiological and biochemical data. Jones (1978) maintained the 'pyogenic' and the 'lactic' streptococci, replaced the names 'viridans' and 'enterococci' by 'oral' and 'faecal', respectively, and added the groupings 'pneumococci', 'anaerobic' and 'other' streptococci. "All these groupings are purely artificial . . . and do not necessarily imply any close relationship between the streptococci included in any group" (Jones, 1978). A further attempt was made by Bridge and Sneath (1982) who performed a numerical taxonomic study and divided the 'pyogenic' and the 'oral' into two or three further groups. The genus *Enterococcus* was finally resurrected in 1984 by Schleifer and Kilpper-Bälz, as will be discussed below.

10.2 Phylogeny

Based on 16S rRNA cataloguing (Ludwig *et al.*, 1985), DNA–DNA and DNA–rRNA hybridizations (Garvie and Farrow, 1981; Kilpper-Bälz and Schleifer, 1981, 1984; Kilpper-Bälz *et al.*, 1982; Schleifer and Kilpper-Bälz, 1984; Schleifer *et al.*, 1985) and serological studies using superoxide dismutase (SOD) antisera (Schleifer *et al.*, 1985), the streptococci *sensu lato* were subdivided into three genera (Schleifer and Kilpper-Bälz, 1984): (i) *Streptococcus sensu stricto* comprising the majority of the known species, (ii) *Enterococcus* for the enterococcal group, and (iii) *Lactococcus* for the lactic streptococci. Consequently '*Streptococcus*' *faecalis* and '*Streptococcus*' *faecium* were transferred to the revived genus *Enterococcus* (Thiercelin and Jouhaud, 1903) as *Ent. faecalis* and *Ent. faecium* respectively, as had been proposed by Kalina (1970). 16S rRNA sequence analysis performed on lactococci, enterococci and other Gram-positive reference strains (Collins *et al.*, 1989a) completely supported the proposed subdivision.

Since the revival of the genus *Enterococcus* in 1984 (Schleifer and Kilpper-Bälz), chemotaxonomic and phylogenetic studies have resulted in the addition of 17 other species to the genus *Enterococcus*. Several of these species have been transferred from *Streptococcus*: *Ent. avium* (Collins *et al.*, 1984) for '*Strep. avium*' (Nowlan and Deibel, 1967), *Ent. casseliflavus* (Collins *et al.*, 1984) for '*Strep. casseliflavus*' (Vaughn *et al.*, 1979), *Ent. durans* (Collins *et al.*, 1984) for '*Strep. durans*' (Sherman and Wing, 1937), *Ent. gallinarum* (Collins *et al.*, 1984) for '*Strep. gallinarum*' (Bridge and Sneath, 1982), *Ent. malodoratus* (Collins *et al.*, 1984) for '*Strep. faecalis* subsp. *malodoratus*' (Pette, 1955), *Ent. cecorum* (Williams *et al.*, 1989) for '*Strep. cecorum*' (Devriese *et al.*, 1983) and *Ent. saccharolyticus* (Rodrigues and Collins, 1990) for '*Strep. saccharolyticus*' (Farrow *et al.*, 1984). New valid descriptions were given for *Ent. columbae* (Devriese *et al.*, 1990), *Ent. dispar* (Collins *et al.*, 1991), *Ent. flavescens* (Pompei *et al.*, 1992a), *Ent. hirae* (Farrow and Collins, 1985), *Ent. mundtii* (Collins *et al.*, 1986), *Ent. pseudoavium* (Collins *et al.*, 1989b), *Ent. raffinosus* (Collins *et al.*, 1989b), *Ent. seriolicida* (Kusuda *et al.*, 1991), *Ent. solitarius* (Collins *et al.*, 1989b) and *Ent. sulfureus* (Martinez-Murcia and Collins, 1991).

DNA–rRNA hybridizations (Kilpper-Bälz *et al.*, 1982) and 16S rRNA cataloguing data (Ludwig *et al.*, 1985) have shown that the enterococci form a separate genus. This was confirmed with the more comprehensive 16S rRNA sequencing studies which elucidated the precise phylogenetic position of the genus. Based on almost complete 16S rRNA sequences the enterococci have been shown to belong to the clostridial subdivision of the Gram-positive bacteria, comprising also the lactic acid bacterial genera *Aerococcus*, *Carnobacterium*, *Globicatella*, *Lactobacillus*, *Lactococcus*, *Leuconostoc*, *Pediococcus*, *Streptococcus*, *Tetragenococcus*, and *Vagococcus*. The enterococci form a distinct cluster with the genera *Vagococcus*, *Tetragenococcus* and *Carnobacterium* as their closest neighbours (Collins *et al.*, 1989; Aguirre and Collins, 1992). This phylogenetic group of four genera is genealogically somewhat more closely related to *Aerococcus* and *Dolosigranulum* than to the streptococci or the lactococci, to which *Enterococcus* has historically been linked.

Within the genus *Enterococcus* 16S rRNA sequences revealed the presence of species groups (Williams *et al.*, 1991), as shown in Figure 10.1. The first group contains *Ent. durans*, *Ent. faecium*, *Ent. hirae* and *Ent. mundtii* (between 98.7 and 99.7% 16S rRNA sequence similarity). The second contains *Ent. avium*, *Ent. raffinosus*, *Ent. malodoratus* and *Ent. pseudoavium* (between 99.3 and 99.7% 16S rRNA sequence similarity). The third consists of the species pair *Ent. casseliflavus* and *Ent. gallinarum* (99.8% 16S rRNA sequence similarity) which is peripherally linked to the *avium* species group. *Enterococcus saccharolyticus*, *Ent. sulfureus*, *Ent. faecalis*, *Ent. cecorum* and *Ent. columbae* form individual lines of descent (Martinez-Murcia and Collins, 1991) but *Ent. cecorum* was found to be more related to *Ent. columbae* than to any other *Enterococcus* species.

Enterococcus solitarius appears to be phylogenetically more closely related to the genus *Tetragenococcus* (*c.* 98% 16S rRNA sequence similarity) than to the enterococci (*c.* 92–94% 16S rRNA sequence similarity; Collins *et al.*, 1990; Williams *et al.*, 1991), whereas the type strain (ATCC 49156) of *Ent. seriolicida* has been shown to be genealogically identical with *Lactococcus garvieae* (100% 16S rRNA similarity; Collins, M.D., unpublished). At present no 16S rRNA sequencing data are available for *Ent. flavescens*. By SDS-PAGE of whole-cell proteins, this species cannot be discriminated from *Ent. casseliflavus* (Pot *et al.*, unpublished). Therefore, its taxonomic status is uncertain.

10.3 Growth and isolation of enterococci

Because of their complex nutrient requirements, enterococci cannot be grown and studied in defined media. They require several vitamins, biotin,

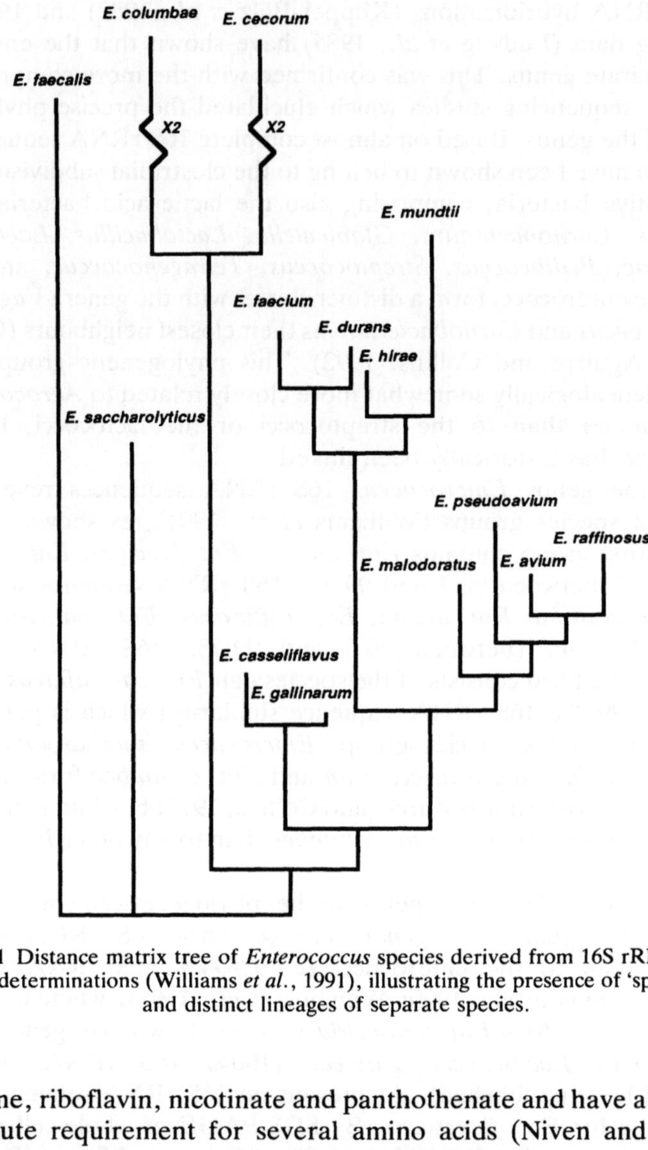

Figure 10.1 Distance matrix tree of *Enterococcus* species derived from 16S rRNA sequence homology determinations (Williams *et al.*, 1991), illustrating the presence of 'species groups' and distinct lineages of separate species.

pyridoxine, riboflavin, nicotinate and panthothenate and have a relative or an absolute requirement for several amino acids (Niven and Sherman, 1944; McCoy and Wender, 1953; Deibel, 1964). Profuse and rapid growth is only achieved in rich complex media such as brain heart infusion broth or Todd–Hewitt broth.

Growth occurs over a wide range of temperatures for most species. Ability to grow at 10°C as well as at 45°C has been used in the differentiation of the genus, but as described above, several of the new species fail to grow at these temperatures. Usually 35–37°C is preferred as incubation temperature.

As discussed extensively by Whittenbury (1978) the enterococci, and especially *Ent. faecalis* strains, are well equipped to survive and to grow in aerobic as well as in anaerobic conditions. Two species from animals, *Ent. cecorum* and *Ent. columbae*, show enhanced growth when incubated in 3–10% CO_2 (Devriese *et al.*, 1983, 1990).

The selective isolation of enterococci is widely used in food and drinking water bacteriology. A wide variety of media has been described, but none of them is completely or nearly completely selective for enterococci (Reuter, 1985; Garg and Mittal, 1991). The more selective types of media do not allow growth of several of the newly described species (Devriese *et al.*, 1991b) and may inhibit recovery and growth of stressed cells (Payne, 1978).

10.4 Phenotypic characteristics of the genus

Certain characteristics of the genus as described by Schleifer and Kilpper-Bälz in 1984 and 1987 are valid for all enterococcal species described since. These are as follows.

- Cells are ovoid, Gram-positive, occur singly, in pairs or in short chains. Within the chains the cells are frequently arranged in pairs and elongated in the direction of the chain.
- Endospores are absent.
- May be motile.
- Facultatively anaerobic.
- Chemo-organotrophs. Metabolism is fermentative. Enterococci have a homofermentative lactic acid fermentation. The predominant end product of glucose fermentation is L(+)-lactic acid.
- Long-chain fatty acids are predominantly of the straight-chain saturated or monounsaturated types; several species produce cyclopropane ring acids.
- Pepdidoglycan type: lysine-D-asparagine (with D-isoasparagine as cross-bridge), in all species described, except in *Ent. faecalis* which has a peptidoglycan of the lysine-alanine$_{2-3}$ type (Schleifer and Kilpper-Bälz, 1987).
- The minimal nutritional requirements are generally complex.
- They are benzidine negative and usually catalase negative, but some strains may produce pseudo-catalase.

A list of characteristics seen, with few exceptions, in all species described to date is shown in Table 10.1.

Other characteristics listed in the genus description of *Enterococcus* (Schleifer and Kilpper-Bälz, 1984) do not apply to all new enterococcal species described subsequently.

- Lancefield group D antigen: with many strains of the *avium* species group (see below), *Ent. cecorum*, *Ent. columbae*, *Ent. dispar* and *Ent. saccharolyticus*, this antigen cannot be demonstrated.
- Growth at 10 and 45°C: *Ent. dispar* and *Ent. sulfureus* do not grow at 45°C, while *Ent. cecorum* and *Ent. columbae* fail to grow at 10°C.
- Growth in 6.5% NaCl: *Ent. cecorum*, *Ent. columbae*, *Ent. avium* and related species are often negative.
- Pyrrolidonylarylamidase: negative in *Ent. saccharolyticus*, *Ent. cecorum* and *Ent. columbae*.

Table 10.1 Characteristics generally or with few exceptions found in all enterococci

Characteristic	Result
Resistance to 40% (v/v) bile	+
β-Glucosidase	+
Urease	−
VP	+*
Leucinearylamidase	+
β-Glucuronidase	−†
Aesculin hydrolysis	+
Acid from	
N-Acetyl glucosamine	+
Amygdalin	+
D-Arabinose	−
Arbutin	+
Cellobiose	+
Erythritol	−
D-Fructose	+
Galactose	+
β-Gentiobiose	+
Glucose	+
Glycogen	−‡
Inositol	−§
D-Fucose	−
L-Fucose	−
Lactose	+¶
Maltose	+
D-Mannose	+
Methyl-β-D-glucopyranoside	+‖
α-Methyl-D-xyloside	−
Pullulan	−‖
Ribose	+**
Salicin	+
Trehalose	+¶
L-Xylose	−

*Negative only in *Ent. saccharolyticus*.
†Positive only in most *Ent. cecorum* strains.
‡Positive only in some *Ent. gallinarum*, *Ent. cecorum* and *Ent. columbae* strains.
§Delayed positive in *Ent. raffinosus*.
¶*Enterococcus faecalis* (asaccharolytic variant) strains from human infections may be lactose and trehalose negative (Facklam and Collins, 1989; Ruoff *et al.*, 1990).
‖Not reported in some of the newer species.
**Negative only in *Ent. flavescens*.

10.5 Identification of the genus

No phenotypic criteria are hitherto available by which the genus *Enterococcus* can be separated unequivocally from others. The classical well known species *Ent. faecalis* and *Ent. faecium* on which the genus description (Schleifer and Kilpper-Bälz, 1984) was based, have a number of characteristics in common which distinguish them from other catalase-negative, Gram-positive, facultatively anaerobic cocci: their ability to grow both at 10°C and 45°C, in 6.5% NaCl broth and at pH 9.6 (Schleifer and Kilpper-Bälz, 1987). But other species, added later to the genus, do not show one or more of these characteristics (see above).

The test results in Table 10.1, obtained when studying enterococci representatives of all species described to date, are not specific for enterococci. Also, none of the possible combinations of these tests appears to be unique to the enterococci.

VP (Voges–Proskauer reaction, acetoin production) and acid from ribose have a high differentiating value because positive results are seen with nearly all enterococcal species, while few streptococci (only *Strep. agalactiae*, *Strep. uberis* and *Strep. porcinus*) react positive in both tests. Also *Lactococcus*, *Leuconostoc* and *Pediococcus* species may be VP as well as ribose positive (Facklam *et al.*, 1989). These two reactions may be helpful in screening procedures. However, it should be borne in mind that at least one species, *Ent. saccharolyticus*, is known to be VP negative and another species, *Ent. flavescens*, is ribose negative (Table 10.1). *Enterococcus saccharolyticus* strains have been isolated mainly from straw bedding in cattle stables; *Ent. flavescens* is a yellow pigmented and motile species which has been described in rare cases of human infection; its taxonomic status is uncertain. The low rates of positive VP results reported by Facklam and Collins (1989) in several enterococcal species, are probably due to the use of a less sensitive test method (see comparisons of VP test results in Facklam *et al.*, 1989).

Aesculin hydrolysis in combination with resistance to 40% (v/v) bile, or the equivalent bile salts, a most useful test described in 1926 by Meyer and Schonfeld, and revived in 1970 by Facklam and Moody, is still valuable for all new enterococcal species. Unfortunately, certain streptococci also react positive. The test has to be incubated in a CO_2-enriched atmosphere with the carboxyphylic species *Ent. cecorum* and *Ent. columbae*. Other selective agents combined with aesculin in certain *Enterococcus* selective media, such as kanamycin aesculin azide agar, inhibit aesculin hydrolysis of species belonging to the *Ent. avium* and *Ent. cecorum* groups.

Certain carbohydrate acidification tests, other than ribose, can also be helpful. Glycogen is usually, but not always, negative in the genera *Enterococcus*, *Lactococcus*, *Leuconostoc* and *Gemella* and positive in many streptococci. L-Arabinose is positive in several enterococcal species and negative in streptococci, except in some rare *Strep. bovis* strains.

Also *β*-glucosidase and leucinearylamidase are valuable. All enterococci react positive in both tests. *Gemella*, many *Streptococcus* and certain *Leuconostoc* species are *β*-glucosidase negative. *Aerococcus*, *Leuconostoc* and *Helcococcus* do not possess leucinearylamidase.

In cases where all enterococci are looked for, it is only possible to identify the genus by its component species. This means that genus identification necessarily follows species identification: e.g. when a strain shows the characteristics of one of the species described below, the strain is an *Enterococcus*.

For practical reasons the following approach can be used: catalase negative, Gram-positive cocci, able to grow in 6.5% (w/v) NaCl broth and showing good growth on media containing 0.04% sodium azide, commonly used in the selective isolation of enterococci, can be identified presumptively as belonging to the genus *Enterococcus*. In case of doubt, tests for ribose fermentation and VP can be added. Ribose and VP testing can also replace to some extent the 6.5% NaCl test. Typically, only *Strep. bovis* (*Strep. equinus*) shows colony characteristics similar to those of the 'classical' enterococci on selective media. *Streptococcus bovis* is always ribose negative and does not grow in 6.5% NaCl. VP cannot be used in the differentiation of enterococci from *Strep. bovis* since all *Strep. bovis* are VP positive.

It should be kept in mind that this procedure excludes several enterococcal species, as explained above. When only the 'classical' species are looked for, this approach is valid and useful, e.g. when working with samples from humans, or in faecal contamination control procedures for hygienic monitoring of drinking water.

10.6 Identification of enterococcal species

A useful approach to species identification within the genus *Enterococcus* is to use those characteristics common to the species groups or clusters described by Williams *et al.* (1991) based on 16S rRNA sequence studies (see section 10.2). It can be seen in Tables 10.3, 10.4, 10.7, 10.9 and 10.11 that each of the *Enterococcus* species groups has a remarkably long list of phenotypical characteristics in common, which can be used in differentiation tests. Some characteristics (Table 10.2) appear to be unique to phylogenetically distinct species groups and species.

Facklam and Collins (1989) used a similar procedure for the identification of enterococci from human clinical specimens with a limited number of conventional tests. They placed, for practical reasons, the species in three groups. Group I is composed exclusively of the species of the *Ent. avium*

Table 10.2 Characteristics typical of phylogenetically distinct lines of enterococcal species and species groups

Characteristic	Typical in
Lys-Ala$_{2-3}$ peptidoglycan type	*Ent. faecalis*
Motile	*Ent. gallinarum** species group
Carboxyphylic growth	*Ent. cecorum* species group
Respiratory quinones present	*Ent. gallinarum* species group
Major fatty acid: tetradecanoic	*Ent. avium* species group
Vancomycin resistant	*Ent. gallinarum* species group
VP negative	*Ent. saccharolyticus*
Alkaline phosphatase positive	*Ent. cecorum** species group
APPA†	*Ent. avium* species group
PYRA‡ negative	*Ent. cecorum* species group
	Ent. saccharolyticus
Arginine not hydrolysed	*Ent. avium* species group
	Ent. cecorum species group
	Ent. sulfureus
	Ent. saccharolyticus
Acid produced from	
Adonitol	*Ent. avium* species group
L-Sorbose	*Ent. avium* species group§
Acid not produced from ribose	*Ent. flavescens*

*Some strains are negative.
†Alanyl-phenylalanyl-prolinearylamidase (not yet reported for the newer species).
‡Pyrolidonylarylamidase.
§One L-sorbose positive *Ent. saccharolyticus* strain has been described (Farrow *et al.*, 1984).

group, but the others are mixtures of species from different phylogenetic groups.

In recent years alternative identification procedures have been described for certain enterococcal species, based on biochemical test galleries, with the Rapid ID 32 Strep system (Freney *et al.*, 1992) and the Biolog system (not yet evaluated with enterococci) probably as the most ambitious and most encompassing analysis of penicillin binding proteins (Williamson *et al.*, 1986), colony hybridization with 23S rRNA-targeted oligonucleotide probes (Betzl *et al.*, 1990), DNA and rRNA restriction fragment length polymorphism (Hall *et al.*, 1992; Jayaro *et al.*, 1992) and bacteriolytic activity patterns (Pompei *et al.*, 1992b).

10.7 Description of species

The different enterococcal species are grouped below and described as far as possible in the phylogenetically derived species groups (the *Ent. faecium*, *Ent. avium*, *Ent. gallinarum* and *Ent. cecorum* groups), together with the separate *Ent. faecalis* and the as yet ungrouped less well known species. As will be evident from the descriptions given below, the species groups differ considerably in physiology, growth, composition and biochemical characteristics. Table 10.3 presents tests for differentiation of the enterococcal species groups. The most important single phylogenetically different species, *Ent. faecalis*, is included in this table for comparison.

Table 10.3 Tests differentiating between *Ent. faecalis* and the *Ent. faecium* (*Ent. faecium*, *Ent. durans*, *Ent. hirae*, *Ent. mundtii*), the *Ent. avium* (*Ent. avium*, *Ent. pseudoavium*, *Ent. malodoratus*, *Ent. raffinosus*), the *Ent. gallinarum* (*Ent. gallinarum*, *Ent. casseliflavus*, *Ent. flavescens*) and the *Ent. cecorum* (*Ent. cecorum*, *Ent. columbae*) species groups

	Enterococcus				
Test	*faecalis*	*faecium* group	*avium* group	*gallinarum* group	*cecorum* group
Motility	−	−	−	+	−
Group D antigen	+	D+*	D	+	−
APPA†	−	−	+	−	−
PYRA‡	+	+	+	+	−
Alkaline phosphatase	−	−	−	−	D+
β-Galactosidase	−	D	D	+	+
Arginine dihydrolase	+§	+	−	+	−
Acid production from					
Adonitol	−	−	D+	−	−
L-Arabinose	−	D	D	+	D
D-Arabitol	−	−	+	−	D
L-Arabitol	−	−	D	−	W
D-Cyclodextrin	+	D+	D−	D	+
Dulcitol	−	−	D	−	−
Gluconate	D+	D	D+	+	−
Glycerol	+¶	−	D	D	−
Inulin	−	−(D−?)‖	−	+(D?)‖	+
2-Ketogluconate	D	−	+	−	D
D-Lyxose	−	−	D	−	−
Mannitol	+	D	+	+	D
Melezitose	D+	−	D	D−	D
Melibiose	−	D	D	+	+
α-Methyl-D-glucoside	−	−	D	+	D
D-raffinose	−	D	D	D	+
Sorbitol	D+	D−	+	D−	D
L-Sorbose	−	−	+	−	−
Xylitol	−	−	D+	−	D−
D-Xylose	−	D	D−	+	D
Pyruvate fermentation	+	−	D	−	?

*D, different or variable; D+, usually positive; D−, usually negative; W, weak.
†Alanyl-phenylalanyl-prolinearylamidase (not yet reported for all species).
‡Pyrolidonylarylamidase.
§Rare *Ent. faecalis* strains may be arginine dihydrolase negative.
¶In anaerobic conditions (under paraffin cover).
‖Unclear in the species description of *Ent. mundtii* (Collins *et al.*, 1986); stated to be positive in the species description of *Ent. gallinarum* and *Ent. casseliflavus* but listed as negative in *Ent. gallinarum* and usually positive in *Ent. casseliflavus* in the identification schemes of Facklam and Collins (1989).

10.7.1 Enterococcus faecalis

10.7.1.1 Enterococcus faecalis *(Schleifer and Kilpper-Bälz, 1984).* Formerly *Streptococcus faecalis* (Andrewes & Horder, 1906). Type strain: Strain Tissier (LMG 7937; NCTC 775; ATCC 19433; NCFB 581; DSM 20478; FIRDI 66; CCUG 19916; CECT 481; NCIMB 775; RIMD 3116001).

10.7.1.1.1 Genomic characterization. The size of chromosome may differ considerably: OG1X, a strain widely used in genetic studies was 2.75–2.76 Mb, while the size of two other strains ranged from 2.0–2.28 Mb (Miranda *et al.*, 1992). The G+C content of the DNA is 37–40 mol%. Sequence studies of its 16S rRNA showed that it forms a distinct line of descent within the genus (Figure 10.1).

Except for the restriction enzyme analyses for epidemiological purposes, the chromosome of *Ent. faecalis* has received little interest. Its transposons and plasmids, however, have been the subject of intensive study in recent years, because in *Ent. faecalis* two special genetic systems were discovered: sex pheromone plasmids and conjugative transposons.

Conjugative transposons are transferred between different cells of the same or of different species, in a DNAse-resistant, conjugation-like manner. The transposons are thought to play a major role in the spread of antibiotic resistance-determining genes among the enterococci and related genera.

Sex pheromone plasmids seem to occur in *Ent. faecalis* only. They act as follows: plasmid carrying donor cells are stimulated by the excretion of pheromones by plasmid free potentially recipient cells, to synthesize a corresponding adhesive protein ('aggregation substance'). This results in a tight aggregation of both types of cells, thus making the conjugative transfer of the sex pheromone plasmid possible. The system is remarkably regulated and versatile: inhibitor peptides are excreted by donor cells which neutralize the effects of the corresponding sex pheromones, and donor cells may produce sex pheromones, not related to the sex pheromone plasmids they harbour. The reader is referred to the reviews by Clewell and Gawron-Burke (1986); Clewell and Weaver (1989); Dunny (1990) and by Wirth in the reference by Devriese *et al.* (1991b) for further information.

10.7.1.1.2 Chemical composition. The cell wall peptidoglycan differs from that of other enterococci in being of the Lys-Ala_{2-3} type. The glycerol teichoic acid group D antigen is characterized by the glucose disaccharide kojibiose esterifying position 2 of glycerol, and a D-alanine residue linked to glucose. The major fatty acids are hexadecanoic, octadecenoic and *cis*-11,12-methylenoctadecanoic. Most strains contain demethylmenaquinones with nine isoprene units (Collins and Jones, 1979).

10.7.1.1.3 Biochemical characteristics. Pyruvate, serine, citrate, gluconate, malate and arginine can be utilized as energy sources (Deibel, 1964). End-products of glucose fermentation at pH above neutral, are mainly ethanol, formic acid and acetic acid; if the pH is allowed to drop below 5, mainly lactic acid is produced. Vigorous aeration results in a much higher cell yield in glucose-containing media, than anaerobic glucose fermentation. NADH oxidase, NADH peroxidase, superoxide dismutase and pyruvate oxidase activity has been demonstrated in *Ent. faecalis* (Zitzelsberger *et al.*, 1984). The metabolism of glycerol, gluconate, malate and lactose has been described by Jacobs and Van Demark (1960), Bernsmann *et al.* (1982), London and Meyer (1970) and Heller and Röschenthaler (1978). *Enterococcus faecalis* appears to be unique in being able to perform aerobic respiration when supplied with haemin. The presence of this compound in the growth medium induces cytochrome formation and stimulates oxidative dissimilation (Sijpesteijn, 1970; Whittenbury, 1978).

The production of extracellular products such as proteinase was studied by Defernando *et al.* (1991), cytolysin (see pathogenicity section, below), DNAse by Batish *et al.* (1982), hyaluronidase by Rosan and William (1966) and lipase by Chander *et al.* (1979).

Characteristics which *Ent. faecalis* shares with other enterococci are listed in Table 10.1. Other test results which differ between *Ent. faecalis* and the most important other species, notably the *Ent. avium*, *Ent. cecorum*, *Ent. faecium* and *Ent. gallinarum* species groups, are given in Table 10.3.

10.7.1.1.4 Morphology and growth characteristics. Cells are mostly ovoid and occur singly, in pairs or in short chains. Colonies are circular, smooth and entire; in broth, turbidity with a deposit is produced. Non-motile; motile strains mentioned in the species description were probably erroneously identified. Some strains carrying a cytolysin-coding plasmid are β-haemolytic on horse blood, but not on sheep blood.

10.7.1.1.5 Identification. A fairly reliable presumptive diagnosis is often made in busy routine work, based on the typically red colonies formed on tetrazolium (TTC) containing selective media such as Slanetz and Bartley agar or on the resistance to and reduction of 0.004% potassium tellurite. Other strongly tetrazolium reducing bacteria rarely belong to the genus *Enterococcus* and are often more or less inhibited on *Enterococcus* selective media. Tellurite reduction is less reliable than TTC reduction. Sorbitol and L-arabinose testing can be useful as described in the section on *Ent. faecium* identification. Other differentiation tests are listed in Table 10.3.

10.7.1.1.6 Occurrence and habitat. Enterococcus faecalis is often the dominating *Enterococcus* in the human bowel, although *Ent. faecium* may outnumber *Ent. faecalis* in some individuals, especially in some countries (see section 10.7.2.1.6). In animals the occurrence of *Ent. faecalis* is strongly dependent on age: it is predominant among the enterococci in preruminant calves and equally frequent as *Ent. faecium* in 1-day-old chicks, but it is found in less than 10% of animals a few weeks older (Devriese *et al.*, 1991d, 1992a). It was also isolated from the gut and the tonsils of about 50% of dogs and cats examined by Devriese *et al.* (1992b). *Enterococcus faecalis* was found less often than *Ent. faecium* in wild mammals and reptiles (Mundt, 1963). Stiles *et al.* (1978) described *Ent. faecalis* as the predominant *Enterococcus* in beef and pork cuts. Difficulties may be experienced with the identification of *Ent. faecalis* from plants or vegetable foods: Mundt (1975, 1976) attempted to make a distinction between a 'human' and a 'plant type' *Ent. faecalis*, but the definite identification of the latter remains uncertain. *Enterococcus faecalis* is also frequent in many types of dairy products, but its prevalence in different products varies greatly between different countries (e.g. Rao *et al.*, 1988; Wessels *et al.*, 1988).

10.7.1.1.7 Pathogenicity. Enterococcus faecalis accounts for the larger part of enterococcal infections in humans, usually representing over 80% or up to 90% of the cases in different surveys. A true increase in the incidence in these infections has been noted in the 1980s (George and Uttley, 1989). This is attributed to the intrinsic resistance to β-lactams, trimethoprim/sulphamethoxazole (*in vivo*), aminoglycosides (low level), fluoroquinolones (low level), lincosamides (low level) and acquired resistances against many of the antibiotics widely used especially in hospitals (Moellering, 1991). *Enterococcus faecalis* is mainly involved in endocarditis, urinary tract infections, soft tissue infections, biliary and abdominal sepsis. A cytolysin (haemolysin) encoded by large transmisible plasmids such as the sex pheromone plasmid pAD1, contributes to the virulence of some strains (Ike *et al.*, 1987).

Enterococcus faecalis has been reported but rarely in animal infections.

10.7.2 The Enterococcus faecium *species group*

Characteristics which are common to all species of this group (*Ent. faecium*, *Ent. durans*, *Ent. hirae* and *Ent. mundtii*) are listed in Table 10.4. Tests useful in the differentiation of this group from all others are shown in Table 10.3 and characteristics distinguishing the species of the *Ent. faecium* group are shown in Table 10.5.

10.7.2.1 Enterococcus faecium *(Schleifer and Kilpper-Bälz, 1984).* This species was known formerly as *Streptococcus faecium* (Orla-Jensen, 1919).

Table 10.4 Common characteristics of the *Enterococcus faecium* species group (*Ent. faecium*, *Ent. durans*, *Ent. hirae*, *Ent. mundtii*) other than those listed in the general and differentiating Tables 10.1, 10.2 and 10.3

Good growth in 6.5% NaCl, at 10 and 45°C
Resistant to 0.04% NaN_3
Less resistant than *Ent. faecalis* to tellurite
Triphenyl tetrazolium chloride not reduced
Possess neither menaquinones nor ubiquinones
Main fatty acids: hexadecanoic, octadecanoic and *cis*-11,12-methylenoctadecanoic
Hippurate variable
N-Acetyl-*β*-glucosaminidase variable
Glycyl-tryptophane arylamidase variable
β-Mannosidase variable
Acid not produced from
5-Ketogluconate
D-Turanose
Acid variable with
Amidon
α-Methyl-D-mannoside
Rhamnose
Sucrose
D-Tagatose

Type strain: LMG 8149; NCTC 7171; ATCC 19434; NCFB 942; DSM 20477; CCUG 542.

10.7.2.1.1 Genomic characterization. The G+C content of the DNA ranges from 37 to 40 mol%. The size of the chromosome of the plasmid-free strain GE-1 ranges from 2170 to 2155 kb (Miranda *et al.*, 1992). The distribution of certain plasmid sequences in human faecal *Ent. faecium* strains was reported by Watanabe *et al.* (1992). Restriction enzyme analysis of chromosomal DNA has been described as a useful tool to characterize strains and to detect strain relationships in epidemiological studies (Lacoux *et al.*, 1992).

10.7.2.1.2 Chemical composition. Peptidoglycan is of the Lys-D-Asp type. Group D antigen contains koijitriose which esterifies position 2 of glycerol in the teichoic acid backbone. D-Alanine and L-lysine are in an ester linkage to the hydroxyl groups of glucose in this polymer. Neither menaquinones nor ubiquinones, which act as non-cytochrome electron carriers in certain enterococci, are found in *Ent. faecium* (Collins and Jones, 1979). The main fatty acids are hexadecanoic, octadecenoic and *cis*-11,12-methylenoctadecanoic.

10.7.2.1.3 Biochemical characteristics. Enterococcus faecium differs in certain aspects from *Ent. faecalis* in its growth requirements and metabolism. It requires folic acid for growth and is unable to derive energy from pyruvate, citrate, malate, gluconate and serine. The species possesses a superoxide dismutase, NADH oxidase and L-lactate oxidase, but no NADH peroxidase or pyruvate oxidase (Zitzelsberger *et al.*, 1984). Synthetic ability for amino acids was studied genetically by Deguchi and Morishita (1992). *Enterococcus faecium* shows the more general characteristics of the genus listed in Table 10.1 as well as the characteristics of the *Ent. faecium* species group listed in Tables 10.3 and 10.4. Species-specific traits are given in Table 10.5. Certain host species-associated characters (ecovars) are known: *Ent. faecium* strains from poultry are usually raffinose positive; most bovine and canine strains produce acid from D-xylose, while strains from other origins are D-xylose negative (Devriese *et al.*, 1987). Most canine strains are sorbitol-positive (Devriese *et al.*, 1992b).

10.7.2.1.4 Morphology and growth characteristics. Ovoid cells in pairs or short chains, elongated in the direction of the chain. Colonies are smooth, circular and entire. They are nonpigmented and may be haemolytic. They have homogenous turbidity in broth and are non-motile. Schleifer and Kilpper-Bälz (1984) state in their species description that some strains are motile, but such strains may belong to one of the motile species such as *Ent. gallinarum*, which were not yet described at that time.

Table 10.5 Characteristics differentiating between members of the *Enterococcus faecium* species group

Characteristic	*Ent. faecium*	*Ent. durans*	*Ent. hirae*	*Ent. mundtii*
Yellow pigment	−	−	−	+
Hippurate	D+*	D	D	−
Acid from				
L-Arabinose	+	−	−	+
Gluconate	D	−	−	−
Mannitol	D+	−	−	+
Melibiose	D+	−†	+	+
α-Methyl-D-mannoside	D−	−	−	D+
D-Raffinose	−‡	D−	D−	D+
Rhamnose	D−	−	−	D+
Sorbitol	−§	−	−	D
Sucrose	D+	−¶	+	+
D-Xylose	D−	−	−	+

*D, different; D+, usually positive; D−, usually negative.
†Melibiose positive strains were identified by Facklam and Collins (1989) as *Ent. durans*. Their differentiation from *Ent. hirae* is uncertain.
‡*Enterococcus faecium* strains from poultry are usually raffinose positive.
§Rare *Ent. faecium* strains from humans and most strains from dogs are sorbitol positive.
¶Sucrose positive *Ent. durans* are described in the RAPID ID 20 STREP system (API-BioMérieux). Their differentiation from *Ent. hirae* is uncertain.

10.7.2.1.5 Identification. *Enterococcus faecium* produces pale pink colonies on tetrazolium-containing selective media, such as the medium of Slanetz and Bartley (1957). It shares this colony type with the related species *Ent. durans*, *Ent. hirae*, and *Ent. mundtii*, and also with *Streptococcus bovis*. It is more easily differentiated from *Ent. faecalis* which produces reddish colonies because of its ability to reduce triphenyl tetrazolium chloride. Positive arabinose and negative sorbitol reactions are useful confirmatory tests in specimens from humans, where species other than *Ent. faecalis* and *Ent. faecium* are infrequent. It should be noted that *Ent. faecalis* may be sorbitol negative and that rare *Ent. faecium* strains from humans and most strains from dogs may be sorbitol positive (Facklam and Collins, 1989; Devriese *et al.*, 1992b). Difficulties may also occur in the differentiation of raffinose positive poultry strains from *Ent. gallinarum*. The positive motility of the latter species is the major differentiating character here. Other useful identification characteristics can be found in Tables 10.3 and 10.5.

10.7.2.1.6 Occurrence and habitat. In humans *Ent. faecium* is, together with *Ent. faecalis*, the most frequently occurring enterococcal species in the gastrointestinal tract, but some geographical differences are evident (Mead, 1978; Noble, 1978). Diet is supposed to influence this, but the underlying mechanisms are not clear. In poultry and cattle the incidence of *Ent. faecium* decreases with age (Devriese *et al.*, 1991d, 1992a). It is frequent in pigs (Buttiaux, 1958) and occurs also in dogs and cats (Devriese *et al.*, 1992b), as well as in several wild mammal species and reptiles (Mundt, 1963).

Mundt (1976) isolated *Ent. faecium*-like strains from plants and frozen or dried foods, but many had atypical characteristics and their identification was tentative. *Enterococcus faecium* appears to be relatively frequent in certain types of processed foods (Stiles *et al.*, 1978). It is also dominant in raw milk (Kielwein, 1978; Wessels *et al.*, 1988) and several types of milk products.

10.7.2.1.7 Pathogenicity. Although *Ent. faecium* is far less frequent than *Ent. faecalis* (usually less than 10%; see Ruoff *et al.*, 1990), this species has received much attention recently. Reasons for this are its multiple antibiotic resistance which may cause hospital (nosocomial) infections and epidemics (George and Uttley, 1989; Moellering, 1991). Disease caused in animals by well-characterized and identified *Ent. faecium* has not been described to date (see also the section on *Ent. hirae*: section 10.7.2.3).

10.7.2.1.8 Applications. In recent years feeding of *Ent. faecium* has received much attention as a possible means to improve growth and feed conversion of farm animals and to prevent or to cure disease in animals and humans. This complex field of probiotic research is not covered here.

10.7.2.2 Enterococcus durans *(Collins* et al., *1984).* This species was known formerly as *Streptococcus durans* (Sherman and Wing, 1937). Type strain: Sherman 98D (LMG 12691; NCFB 596; ATCC 19432; DSM 20633; NCTC 8307; CCM 5612).

10.7.2.2.1 Genomic characterization. The G+C content of the DNA ranges from 38 to 40 mol%. Approx 40% DNA homology with the *Ent. faecium* type strain (20% under stringent conditions). The 16S rRNA primary sequences, determined by reverse transcription of small subunit rRNA, cluster with those of the *Ent. faecium*, *Ent. hirae* and *Ent. mundtii* (intragroup homology: 98.7–99.5%). Characterization of *Ent. durans* isolates by restriction enzyme analysis of DNA for epidemiological purposes was studied by Lacoux *et al.* (1992).

10.7.2.2.2 Chemical composition. As described for *Ent. faecium.*

10.7.2.2.3 Morphology and growth characteristics. Ovoid cells in pairs, short chains and small groups. Colonies are circular smooth and entire. Strains may be haemolytic, producing α- or *β*-haemolysis. They are non-pigmented, have turbid growth in broth and are non-motile.

10.7.2.2.4 Biochemical characteristics. Enterococcus durans shows the general or common enterococcal characteristics listed in Table 10.1, and shares many characteristics with the other species of the *Ent. faecium* species group as demonstrated in Tables 10.3 and 10.4. Typical reactions are listed in Table 10.5.

10.7.2.2.5 Identification. The species can be distinguished from its closest relatives mainly by its remarkable inertness on carbohydrates (Table 10.5). Difficulties may be experienced in its differentiation from *Ent. hirae*. Testing for acid production from melibiose and sucrose is usually sufficient. Raffinose may be useful, possibly more so for human (Facklam and Collins, 1989) than for animal strains (Devriese *et al.*, 1987). The melibiose and sucrose reactions of human strains pose some problems: *Ent. durans* is stated to be sucrose and melibiose negative in the species description (Collins *et al.*, 1984) and so were animal strains described by Devriese *et al.*, 1987). In the commercial RAPID ID 20 STREP system (API-Bio-Mérieux), however, certain sucrose positive strains are identified as *Ent. durans* and in the series of Facklam and Collins (1989) most human strains were melibiose-positive.

10.7.2.2.6 Occurrence and habitat. Enterococcus durans was originally described from milk and dairy products (Sherman and Wing, 1937) and has subsequently repeatedly been shown to be associated with these products, though less frequently than *Ent. faecium* and *Ent. faecalis* (Batish and Ranganathan, 1984). Wessels *et al.* (1988) isolated this species most frequently in butter. Insufficiently cleaned equipment is thought to play a major role in its persistence in the dairy environment (Kielwein, 1978). *Enterococcus durans* has also been shown to occur, infrequently, in beef meat (Stiles *et al.*, 1978). Probably some of the *Ent. durans* strains in the older reports may have been incorrectly identified as *Ent. hirae*.

The original habitat of *Ent. durans* is probably the gut, but this species seems to occur only in a minority of humans (Cooper and Ramadan, 1955; Finegold *et al.*, 1974). Among various domestic animals it was only found in the intestines of some preruminant calves by Devriese *et al.* (1992a) and in young chicks (Devriese *et al.*, 1991d).

10.7.2.2.7 Pathogenicity. No reports are available on the pathogenicity of *Ent. durans*. Animal pathogenic strains, originally identified as *Ent. durans*, were reclassified as *Ent. hirae* by Devriese and Haesebrouck (1991). Ruoff *et al.* (1990) identified only one *Ent. durans* in a collection of 302 consecutive cultures from routine clinical samples obtained from humans.

10.7.2.3 Enterococcus hirae *(Farrow and Collins, 1985).* Based on strains formerly known as '*Streptococcus faecium*' or '*Streptococcus durans*' homology group 2 (Knight *et al.*, 1984). Type strain: E.E. Snell strain R (LMG 6399; NCFB 1258; ATCC 8043; CCM 2423; NCIMB 6459; DSM 20160; NCTC 6469).

10.7.2.3.1 Genomic characterization. The G+C content of the *Ent. hirae* DNA ranges from 37 to 38 mol%. The species contains some strains from pig intestines which hybridize in DNA–DNA tests less efficiently than others with the type strain, and which can be considered as peripheral members of the species (Farrow and Collins, 1985).

rRNA sequencing studies demonstrated that *Ent. hirae* belongs to the *Ent. faecium* species cluster.

10.7.2.3.2 Chemical composition. The chemical composition of *Ent. hirae* is as described for *Ent. faecium*.

10.7.2.3.3 Biochemical characteristics. Enterococcus hirae (in earlier physiological reports usually named *Ent. faecium*) has been used in studies on muramidase. It possesses two distinct peptidoglycan hydrolases (*N*-acetylmuramoylhydrolases) called *Ent. hirae* muramidase-1 and -2

(Kariyama and Shockman, 1992). Other fields of study cover penicillin-binding proteins (Piraz *et al.*, 1990) and ion porters (Reizer *et al.*, 1992). *Enterococcus hirae* shows the characteristics listed in Tables 10.1, 10.3 and 10.4, and differs from its closest relatives in the test results indicated in Table 10.5.

10.7.2.3.4 Morphology and growth characteristics. Morphology and growth are as described for *Ent. faecium.*

10.7.2.3.5 Identification. The species is most easily confused with *Ent. durans*. It can be differentiated by its positive melibiose reaction and usually also by its acid production from sucrose. Probably all strains produce acid from at least one of these two carbohydrates (see also section on *Ent. durans*). Knight and Shlaes (1986) reported that *Ent. hirae* also differs from *Ent. durans* in its inability to form clots in litmus milk and in its resistance to tobramycin and clindamycin.

10.7.2.3.6 Occurrence and habitat. Enterococcus hirae was found to occur most frequently in the gut of dogs (Devriese *et al.*, 1992b), but it was also found in all important domestic animal species investigated by Devriese *et al.* (1987). Knight and Shlaes (1986) identified a strain from human sputum as *Ent. hirae*, but the species is apparently rare in humans.

10.7.2.3.7 Pathogenicity. The species description of Farrow and Collins (1985) was based on strains, originally described as '*Ent. faecium*', causing growth depression in chickens (Fuller *et al.*, 1979). The mechanisms underlying this growth depression are still unknown. Also in chicks *Ent. hirae* was found to cause focal necrosis of the brain (Devriese *et al.*, 1991c) and in psittacine birds it was involved in septicaemia cases (Devriese *et al.*, 1992c). *Enterococcus hirae* is an enteropathogen with a characteristic adhesion to the brush border of the small intestine of suckling animals such as rats (Etheridge *et al.*, 1988), foals (Tzipori *et al.*, 1984), piglets (Johnson *et al.*, 1983) and dog pups (Collins *et al.*, 1988). The causative organisms were identified as *Ent. durans* in the last three references, but Devriese and Haesebrouck (1991) demonstrated that they were *Ent. hirae*. The actual importance of *Ent. hirae* as an enteropathogen in animals is as yet unknown. In piglets enterococcal infection appears to be rare (Drolet *et al.*, 1990). Facklam and Collins (1989) described eight *Ent. hirae* strains from human infections, but the role of this *Enterococcus* in human disease has not received any closer attention.

10.7.2.3.8 Applications. Enterococcus hirae has been used frequently in recent years in physiological studies (see the Biochemical characteristics section: section 10.7.2.3.3).

10.7.2.4 Enterococcus mundtii *(Collins* et al., *1986).* Type strain: Mundt MUTK 559 (LMG 10748; NCFB 2375; ATCC 43186; DSM 4838).

10.7.2.4.1 Genomic characterization. The DNA G+C content was determined as 38 to 39 mol%. Sequencing of its 16S ribosomal RNA showed that *Ent. mundtii* belongs to the *Ent. faecium* species group (Williams *et al.*, 1991).

10.7.2.4.2 Chemical composition. Peptidoglycan, group D carbohydrate and long-chain fatty acids are composed as described for *Ent. faecium.*

10.7.2.4.3 Morphology and growth characteristics. Enterococcus mundtii differs only in its yellow pigmentation from *Ent. faecium* and the other species of this group.

10.7.2.4.4 Biochemical characteristics. Characteristics which *Ent. mundtii* shares with other species are listed in Tables 10.1, 10.3 and 10.4. Tests which differentiate it from the others are shown in Table 10.5.

10.7.2.4.5 Identification. The pigmentation of *Ent. mundtii* is usually visible after 1 day, but more pronounced after 2 days. The major trait which differentiates it from the other yellow-pigmented species is its lack of motility, but several other tests can confirm this differentiation (Table 10.6). Other useful tests can be found in Tables 10.3 and 10.5.

Table 10.6 Tests useful in the differentiation of yellow pigmented enterococci

Characteristic	*Ent. casseliflavus* (*Ent. flavescens**)	*Ent. mundtii*	*Ent. sulfureus*
Vancomycin resistance	+	−	?
Motility	D+†	−	−
Arginine dihydrolase	+	+	−
Group D antigen	+	+	−
Acid produced from			
L-Arabinose	+	+	−
Glycerol	−	D	−
Gluconate	+	−	+
Inulin	+	− (D−?)	−
α-Methyl-D-glucoside	+	−	+
Rhamnose	D+	D+	−
Sorbitol	−	D	−
D-Turanose	D+	−	?
D-Xylose	+	+	−

**Enterococcus flavescens* differs from *Ent. casseliflavus* only in its negative ribose reaction and its failure to produce α-haemolysis on sheep blood agar.
†D, different or variable; D+, usually positive; D−, usually negative.

10.7.2.4.6 Occurrence and habitat. Enterococcus mundtii appears to be typically associated with plants. Isolates from animals and humans are probably indirectly derived from this source.

10.7.2.4.7 Pathogenicity. The species has been reported from human infections (Kaufhold and Ferrieri, 1991), but appears to be rare.

10.7.3 The Enterococcus avium *species group*

Characteristics which are typical for all species (*Ent. avium*, *Ent. malodoratus*, *Ent. pseudoavium* and *Ent. raffinosus*) composing this group, are listed in Table 10.7. The *Ent. avium* species group has several characteristics which are unique among the enterococci (Table 10.2). Table 10.3 shows other tests useful in the differentiation of this group from others.

10.7.3.1 Enterococcus avium *(Collins* et al., *1984). Formerly Streptococcus avium* (Nowlan and Deibel, 1967). Type strain: Guthof E6844 (LMG 10744; ATCC 14025; DSM 20679; NCFB 2369; NCTC 9938). Isolated from human faeces.

10.7.3.1.1 Genomic characterization. The G+C content of the DNA ranges from 39 to 40 mol%. Forms a distinct species cluster within the genus, together with *Ent. raffinosus*, *Ent. malodoratus* and *Ent.*

Table 10.7 Characteristics common to the *Enterococcus avium* species group (*Ent. avium*, *Ent. pseudoavium*, *Ent. malodoratus*, and *Ent. raffinosus*) other than those listed in the general and differentiating Tables 10.1 and 10.2

Relatively small colonies
Strong greening haemolysis
Often poor growth on enterococcal selective media
Growth in 6.5% NaCl may be negative
Major fatty acids tetradecanoic, hexadecanoic and *cis*-11,12,methylenoctadecanoic
H_2S produced (lead acetate paper test)
Hippurate negative
N-Acetyl-*β*-glucosaminidase rarely positive
Glycyl-tryptophane arylamidase variable
β-Mannosidase negative
Acid usually produced from D-tagatose*
Acid production is variable with
5-Ketogluconate
α-Methyl-D-mannoside
Rhamnose
Sucrose
D-Turanose
Acid not produced from amidon

**Enterococcus pseudoavium* is negative.

pseudoavium (intragroup homology of 99.3–99.7% as determined by reverse transcriptase sequencing of small-subunit rRNA).

10.7.3.1.2 Chemical composition. Group A peptidoglycan is based on lysine (Lys-D-Asp). Many, but not all, strains contain a serologically demonstrable group D glycerol teichoic acid. They may also react with Lancefield group Q antiserum. The major fatty acids are non-hydroxylated long chain hexadecanoic, and *cis*-11,12-methylenoctadecanoic, as with *Ent. faecalis* and the *Ent. faecium* group, but *Ent. avium* contains additionally relatively large amounts of tetradecanoic ($C_{14:0}$) acid. Respiratory quinones are absent.

10.7.3.1.3 Biochemical characteristics. Folinic acid, but not riboflavin or pyrodoxal, are required for growth. Most strains can use gluconate as an energy source, but not pyruvate, serine, citrate, malate and arginine (Nowlan and Deibel, 1967). *Enterococcus avium* has NADH peroxidase, NADH oxidase, superoxide dismutase, and L-lactate dehydrogenase activity (Zitzelsberger *et al.*, 1984). Biochemical characteristics that are similar to those of other enterococci are shown in Table 10.1. Typical characteristics of the *Ent. avium* species group are listed in Tables 10.3 and 10.7. The species can be differentiated from *Ent. pseudoavium*, *Ent. malodoratus* and *Ent. raffinosus* by the carbohydrate fermentation tests listed in Table 10.8.

10.7.3.1.4 Morphology and growth characteristics. Cells are ovoid, usually in pairs or in short chains. They are non-motile. Colonies are

Table 10.8 Tests differentiating between members of the *Enterococcus avium* species group (*Ent. avium*, *Ent. pseudoavium*, *Ent. malodoratus*, and *Ent. raffinosus*)

Test	*Ent. avium*	*Ent. pseudoavium*	*Ent. malodoratus*	*Ent. raffinosus*
Acid from				
L-Arabinose	+	−	−	+
L-Arabitol	+	−	+	+
Dulcitol	D*	−	D+	−
Glycerol	+	−	D	+
Melezitose	+	−	−	+
Melibiose	−†	−	+	+
Raffinose	−	−	+	+
Rhamnose	+	−	+	+
D-Tagatose	+	−	+	+
Inositol	−	−	−	+‡

*D, different; D+, usually positive; D−, usually negative.
†Melibiose is variable in *Ent. avium* according to Facklam and Collins (1989).
‡Delayed positive (after 2 days).

smooth, circular and entire, somewhat smaller than *Ent. faecalis* or *Ent. faecium* strains, and are non-pigmented. A broad, sharply demarcated zone of greening double zone haemolysis is typically produced.

10.7.3.1.5 Identification. The *Ent. avium* species group is easily differentiated from all known enterococci in its production of alanyl-phenylalanyl-prolinearylamidase (APPA), and in its acid production from adonitol, D-lyxose and L-sorbose (Tables 10.2 and 10.3). Within this group problems may occur in the differentiation of *Ent. avium* from *Ent. raffinosus* (Table 10.8; see also section on *Ent. raffinosus* identification).

10.7.3.1.6 Occurrence and habitat. Guthof (1955) described 'group Q streptococci' first from human faeces, but Nowlan and Deibel (1967) coined the name *Ent. avium* for it because they found these bacteria mainly in chicken faeces. In modern poultry this species is absent or rare (Devriese *et al.*, 1991d). It is rather frequent in preruminant calves (Devriese *et al.*, 1992a) and occurs also in dog and pig intestines. In Japan *Ent. avium* was found to occur frequently in children, but was absent from adults (Watanabe *et al.*, 1981).

10.7.3.1.7 Pathogenicity. Enterococcus avium may be involved in human infections, but is apparently infrequent (Facklam and Collins, 1989; Ruoff *et al.*, 1990).

10.7.3.2 Enterococcus malodoratus *(Collins* et al., *1984).* Formerly the unofficial species *Streptococcus malodoratus* or *Streptococcus faecalis* var. *malodoratus* (Pette, 1955). Type strain: Galesloot 6 (LMG 10747; NCFB 864; ATCC 43197; DSM 20681). From Gouda cheese.

10.7.3.2.1 Genomic characterization. Enterococcus malodoratus clusters in the *Ent. avium* species group, as described above.

10.7.3.2.2 Chemical composition. Cell wall and fatty acids are as described for *Ent. avium*.

10.7.3.2.3 Biochemical characteristics. Enterococcus malodoratus possesses the characteristics of the *Ent. avium* species group (Tables 10.2, 10.3 and 10.7), but differs from the other species of the *Ent. avium* species cluster in the carbohydrate reactions listed in Table 10.8. It differs notably from *Ent. avium* in its negative L-arabinose and melezitose, positive raffinose and melibiose, and to some extent also in its positive glycyl-tryptophane arylamidase reaction.

10.7.3.2.4 Morphology and growth characteristics. Similar to *Ent. avium*.

10.7.3.2.5 Identification. Enterococcus malodoratus identification is mainly based on the *Ent. avium* group-specific characteristics and the carbohydrate reactions listed in Tables 10.3 and 10.8.

10.7.3.2.6 Occurrence and habitat. The species seems to be very rarely, if at all, associated with humans. Strains isolated relatively frequently from cat tonsils, which were described originally as *Ent. raffinosus* by Devriese *et al.* (1992b), were later found to be *Ent. malodoratus* (Pot, B. and Devriese, L.A., unpublished observations). The species is apparently infrequent in other domestic animals, in vegetation, in foods and in dairy products, though *Ent. malodoratus* was originally described from Gouda cheese.

10.7.3.2.7 Pathogenicity. No data are available.

10.7.3.3 Enterococcus pseudoavium *(Collins* et al., *1989b).* Type strain: Roguinsky 47–16 (LMG 11426; ATCC 49372; NCFB 2138). From bovine mastitis.

10.7.3.3.1 Genomic characterization. The species belongs to the *Ent. avium* species cluster (see *Ent. avium* – section 10.7.3.1).

10.7.3.3.2 Chemical composition, morphology and growth characteristics. These are as described for *Ent. avium* – section 10.7.3.1).

10.7.3.3.3 Biochemical characteristics. Only one strain has been described to date (Collins *et al.*, 1989b). *Enterococcus pseudoavium* does not grow in 6.5% NaCl and is less active on carbohydrates than other members of the *Ent. avium* species group (Table 10.8). Other reactions are as shown in Tables 10.1, 10.3 and 10.7.

10.7.3.3.4 Identification. The species can be identified using the group specific characters listed in Tables 10.3 and 10.7, together with the species specific tests shown in Table 10.8.

10.7.3.3.5 Occurrence and habitat. The single *Ent. pseudoavium* strain described was isolated originally from cow mastitis, but the species was not found in bovine intestines and faeces (Devriese *et al.*, 1992a) and is probably very rare also in bovine mastitis.

10.7.3.3.6 Pathogenicity. Unknown.

10.7.3.4 Enterococcus raffinosus *(Collins* et al., *1989b).* Type strain: Facklam 1789/79 (LMG 12888; ATCC 49427; NCTC 12192). From a human blood culture.

10.7.3.4.1 Genomic characterization. Enterococcus raffinosus belongs to the *Ent. avium* species cluster (see *Ent. avium* – section 10.7.3.1).

10.7.3.4.2 Chemical composition, morphology and growth characteristics. Results described to date are similar to those obtained for *Ent. avium*.

10.7.3.4.3 Biochemical characteristics. Enterococcus raffinosus grows weakly or not at all at 10°C. Other test results are as shown in Tables 10.1, 10.3, 10.7 and 10.8.

10.7.3.4.4 Identification. The species may be difficult to distinguish from *Ent. avium*. This is based (Collins *et al.*, 1989b) on raffinose, negative in *Ent. avium*. L-Arabitol and melibiose may also be helpful (Table 10.8). Grayson *et al.* (1991) reported that *Ent. raffinosus* is more resistant than *Ent. avium* to penicillin, but the difference was only marginal in some strains. Analysis of penicillin-binding proteins (Collins *et al.*, 1989b; Grayson *et al.*, 1991) confirmed the identifications based only on the raffinose-melibiose trait. The *Ent. raffinosus* type strain was raffinose negative using commercial API galleries, but raffinose positive using conventional methods (Devriese, L.A., unpublished).

10.7.3.4.5 Occurrence and habitat. Enterococcus raffinosus was isolated from human clinical sources (Collins *et al.*, 1989b) but its actual habitat is unknown. It appears to be rare in animals. Strains from cat tonsils described as *Ent. raffinosus* by Devriese *et al.* (1992b) were later shown to be *Ent. malodoratus* (see *Ent. malodoratus* – section 10.7.3.2.6).

10.7.3.4.6 Pathogenicity. Enterococcus raffinosus parallels somewhat *Ent. avium* in human infections. Facklam and Collins (1989) and Grayson *et al.* (1991) described strains from a wide variety of clinical sources, including wounds, biliary infections and haemocultures. Increased frequencies in certain hospitals may be due to antibiotic selective pressure effects (Chirurgi *et al.*, 1991).

10.7.4 The Enterococcus gallinarum *species group*

This group comprises *Ent. gallinarum* and *Ent. casseliflavus* and most probably also the species *Ent. flavescens* which was described in 1992 (Pompei *et al.*, 1992a). The group is unique (Table 10.2) in being motile and having intrinsic (natural) low level resistance to vancomycin. Other characteristics which are common to the species *Ent. gallinarum*, *Ent. casseliflavus* and *Ent. flavescens* are listed in Tables 10.3 and 10.9. Tests differentiating species of this group are shown in Table 10.10.

Table 10.9 Characteristics common to the *Enterococcus gallinarum* species group (*Ent. gallinarum*, *Ent. casseliflavus* and probably also *Ent. flavescens*), other than those listed in Tables 10.1, 10.2 and 10.3

Good growth in 6.5% NaCl, at 10 and 45°C
Intermediately resistant to 0.04% NaN_3 and tellurite
Triphenyl tetrazolium chloride reduced
Main fatty acids: hexadecanoic and octadecanoic
Menaquinones present
Low level resistance to vancomycin
Hippurate variable
N-Acetyl-*β*-glucosaminidase usually positive
Glycyl-tryptophane arylamidase usually positive
β-Mannosidase usually positive
Acid not produced from 5-ketogluconate
Acid may be produced from
Glycogen
α-Methyl-D-mannoside
Acid usually produced from
Rhamnose
Sucrose
D-Turanose
Acid variable with
Amidon
D-Tagatose

Table 10.10 Tests useful in the differentiation of members of the *Enterococcus gallinarum* species group

Test	*Ent. gallinarum*	*Ent. casseliflavus* (*Ent. flavescens**)
Yellow pigment	–	+†
Hippurate	D+‡	–
Acid from		
Sorbitol	D	–
Glycerol	–	D
Glycogen	D	–
D-Cyclodextrin	+	D–
D-Tagatose	+	D–
β-Haemolysis on horse blood agar	D+	–
α-Haemolysis on sheep blood agar	–	D+

**Enterococcus flavescens* differs from *Ent. casseliflavus* only in its negative ribose reaction and its failure to produce α-haemolysis on sheep blood (Pompei *et al.*, 1992a).
†Some strains are not pigmented.
‡D, different; D+, usually positive; D–, usually negative.

10.7.4.1 Enterococcus gallinarum *(Collins* et al.*, 1984)*. Formerly *Streptococcus gallinarum* (Bridge & Sneath, 1982). Type strain: Barnes F87/276 (LMG 11207; ATCC 49372; DSM 20628; NCTC 11428; NCFB 2313). From a chicken.

10.7.4.1.1 Genomic characterization. The G+C content of the DNA is 39–40 mol%. Reverse transcriptase sequencing of 16S rRNA demonstrated that *Ent. gallinarum* forms a species cluster with *Ent. casseliflavus* (Williams *et al.*, 1991).

10.7.4.1.2 Chemical composition. Group A peptidoglycan is of the Lys-D-Asp type. Glycerol teichoic acid of the group D antigen type is present. The major non-hydroxylated long-chain fatty acids are hexadecanoic and octadecenoic, with low levels of *cis*-11,12-methylenoctadecanoic acid. Small amounts of menaquinones are produced.

10.7.4.1.3 Morphology and growth characteristics. Cells are coccoid, mostly in pairs or short chains. Originally described as not motile, but subsequently shown to be typically motile (Devriese *et al.*, 1987). Colonies are circular, smooth, entire and not pigmented. *Enterococcus gallinarum* is often α-haemolytic on horse blood agar, but not on sheep blood (Williamson *et al.*, 1986).

10.7.4.1.4 Biochemical characteristics. Enterococcus gallinarum shows the common characteristics of the enterococci as listed in Table 10.1. Other test results which are typical for the species group are shown in Tables 10.3 and 10.9, and results typical for the species itself are in Table 10.10.

10.7.4.1.5 Identification. The low level resistance to vancomycin is a useful diagnostic characteristic of the *Ent. gallinarum* species group, as are also positive motility, and production of acid from inulin and D-xylose (Table 10.3). Difficulties may be experienced in the differentiation of *Ent. gallinarum* from *Ent. casseliflavus* (Table 10.10). Absence of pigmentation and positive hippurate and acid production from D-cycloserine are most useful, although unpigmented strains of *Ent. casseliflavus* have also been described (Vincent *et al.*, 1991). Motility is preferably tested at 30°C, and pigmentation may take 2 days to develop fully.

10.7.4.1.6 Occurrence and habitat. The species was described and derives its name from poultry (Bridge and Sneath, 1982). Devriese *et al.* (1991d), however, found very few strains (0.5% of isolates) in chickens and, among various other animal hosts, only a few isolates in cats (Devriese *et al.*, 1992b). The actual habitat of *Ent. gallinarum* is unknown.

10.7.4.1.7 Pathogenicity. Ruoff *et al.* (1990) found 1% *Ent. gallinarum* in their series of clinical enterococcal isolates. The importance of this vancomycin resistant species increased probably because of antibiotic usage.

10.7.4.2 Enterococcus casseliflavus *(Collins* et al., *1984).* Formerly *Streptococcus faecium* subsp. *casseliflavus* or *Streptococcus casseliflavus* (Vaughn *et al.*, 1979). Type strain: Mundt MUTK 20 (LMG 10745; ATCC 25788; NCFB 2372). From plant material.

10.7.4.2.1 Genomic characterization. Reported data are similar to those of *Ent. gallinarum.*

10.7.4.2.2 Chemical composition. As described for *Ent. gallinarum*, except that the fatty acid *cis*-11,12-methylenoctadecanoic acid is either absent or present in only trace amounts, and that menaquinones are present. The major isoprenologues are MK-7 and MK-8 (Collins and Jones, 1979).

10.7.4.2.3 Morphology and growth characteristics. Enterococcus casseliflavus is normally pigmented and motile, but non-pigmented and non-motile strains may occur (Vincent *et al.*, 1991). The type strain ATCC 25788, notably, was shown to be non-motile by these authors.

10.7.4.2.4 Biochemical characteristics. Besides the common enterococcal characteristics (Table 10.1) *Ent. casseliflavus* shows the group characteristics listed in Tables 10.3 and 10.9 and the species specific traits shown in Table 10.10.

10.7.4.2.5 Identification. Pigmentation, motility and intrinsic vancomycin-low level (MIC 8–16 μg/ml) resistance are most useful identification characteristics. The differentiation from other yellow pigmented enterococci is shown in Table 10.6, and the characters distinguishing between this species and *Ent. gallinarum* in Table 10.10 (see also section 10.7.4.1 – *Ent. gallinarum*). It should be noted that unpigmented and also non-motile *Ent. casseliflavus* strains have been described (Vincent *et al.*, 1991).

10.7.4.2.6 Occurrence and habitat. Enterococcus casseliflavus is associated with plant material (Vaughn *et al.*, 1979).

10.7.4.2.7 Pathogenicity. The species is found infrequently in infections (1% of the clinical isolates of Ruoff *et al.*, 1990).

10.7.4.3 Enterococcus flavescens *(Pompei* et al., *1992a).* Type strain: Pompei CA 2 (LMG 13518; CCM 4239). From a human sepsis. Taxonomic status is uncertain (see Table 10.2).

10.7.4.3.1 Genomic characterization. The G+C content of the DNA was found to range from 42 to 43 mol%. *Enterococcus casseliflavus* appeared to be the closest relative in DNA homology tests.

10.7.4.3.2 Morphology and growth characteristics. As described for *Ent. casseliflavus.*

10.7.4.3.3 Chemical composition. The major fatty acids are hexadecanoic acid and monounsaturated octadecenoic acid.

10.7.4.3.4 Biochemical characteristics and identification. As described for *Ent. casseliflavus*, except for the ribose reaction which is negative, and absence of haemolysis on sheep blood agar. Growth at 10 and 45°C is variable.

10.7.4.3.5 Occurrence, habitat and pathogenicity. Enterococcus flavescens has been isolated from human clinical specimens. Its habitat and pathogenic significance are unknown.

10.7.5 The Enterococcus cecorum *species group*

This group is composed of two species *Ent. cecorum* and *Ent. columbae*, both isolated from animal intestines, which are less closely related than are the species within other species groups. Phenotypically they appear to be more related to each other than to other species. A list of characteristics common to these two species is shown in Tables 10.3 and 10.11. They are unique among the enterococci in being carboxyphylic and in producing

Table 10.11 Common characteristics of the tentative *Enterococcus cecorum* species group (*Ent. cecorum* and *Ent. columbae*), other than, or differing from the general and differentiating test results listed in Tables 10.1, 10.2 and 10.3

Carboxyphylic
Poor growth on enterococcal selective media
No growth at 10°C
Growth at 6.5% NaCl broths often poor or negative
Hippurate usually negative
N-Acetyl-β-glucosaminidase usually positive
Glycyl-tryptophane arylamidase usually positive
β-Mannosidase negative*
Acid produced from
D-raffinose
Sucrose
Some strains produce acid from
Amidon
Glycogen
5-Ketogluconate
Rhamnose
D-Turanose
D-Tagatose
Acid not produced from α-methyl-D-mannoside

*Results with *Ent. columbae* are not available.

usually alkaline phosphatase (Tables 10.2 and 10.3). Other useful tests differentiating this group from others are shown in Table 10.3.

10.7.5.1 Enterococcus cecorum *comb. nov. (Williams* et al., *1989).* Formerly *Streptococcus cecorum* (Devriese *et al.*, 1983). Type strain: Devriese A60 (LMG 12902; ATCC 43196; DSM 20682; NCFB 2674). From chicken intestines.

10.7.5.1.1 Genomic characterization. The DNA base composition of the type strain is 37 mol% G+C. Originally described as *Streptococcus cecorum* by Devriese *et al.* (1983), it was reclassified by Williams *et al.* (1989) as an *Enterococcus* species, based on results of reverse transcriptase sequencing of its 16S rRNA. With the exception of a certain affinity to *Ent. columbae*, *Ent. cecorum* was not found to form a species group with other enterococci as studied by Williams *et al.* (1991).

10.7.5.1.2 Chemical composition. The peptidoglycan of *Ent. cecorum* is of the Lys-D-Asp type (Schleifer and Kilpper-Bälz, 1987). The glycerol teichoic acid group D antigen is absent. Main long-chain fatty acids (only examined in *Ent. cecorum*) are straight-chain saturated and mono-unsaturated hexadecanoic and octadecanoic acids. No *cis*-11,12-methylen-octadecanoic acid as in *Ent. faecalis* and in the *Ent. faecium* species group, and no tetradecanoic acid as in the *Ent. avium* species group.

10.7.5.1.3 Morphology and growth characteristics. Cells are coccal 1–1.3 μm in diameter and occur in pairs, short chains or small groups; within the chains and groups the cells are arranged in pairs. They are non-motile and colonies are circular with entire edges, low convex, partially translucent and non-pigmented. They reach 1–2 mm in diameter after 1 day and 2 to 3 mm after 2 days in 3–10% CO_2 in air or anaerobically in a $H_2 + CO_2$ atmosphere. *Enterococcus cecorum* shows less abundant growth in air. Uniform turbidity is produced in brain heart infusion broth.

10.7.5.1.4 Biochemical characteristics. Enterococcus cecorum differs from most enterococci in several characters: it does not grow in media containing 0.04% NaN_3 and grows poorly or not at all in 6.5% NaCl; it grows at 45°C but not at 10°C. Arginine is not hydrolysed and pyrrolidonyl-arylamidase is negative but alkaline phosphatase is positive. Most strains produce *β*-glucuronidase. Other characteristics are listed in Tables 10.3 and 10.11. The differentiation from *Ent. columbae* is shown in Table 10.12. Minimal inhibitory concentrations of antibiotics and inhibitory substances used in selective media are listed by Devriese *et al.* (1983). Certain characteristics differ among strains isolated from different host species (Devriese *et al.*, 1991a, 1992b). The identification of strains from

Table 10.12 Characteristics differentiating between *Enterococcus cecorum* and *Ent. columbae*, two carboxyphilic species from animal intestines

Characteristic	*Ent. cecorum*	*Ent. columbae*
β-Galactosidase (API substrate 1)	D−*	+
β-Glucuronidase	D+	−
Alkaline phosphatase	D+	+
Hippurate hydrolysis	D−	−
Acid from		
L-Arabinose	−	D+
D-Arabitol	D−	D
5-Keto-gluconate	D	−
α-Methyl-D-glucoside	−	−
Mannitol	D	+
Melezitose	D	D−
Rhamnose	D−	D
D-Tagatose	−	D
Sorbitol	D	D+
Trehalose	+	D+
D-Turanose	−	D−
D-Xylose	−	+
Xylitol	−	D−

*D, different; D+, usually positive; D−, usually negative.

canaries which differ from the others in arginine hydrolysis, VP, inulin and melibiose, as *Ent. cecorum* is uncertain.

10.7.5.1.5 Identification. Two characteristics which distinguish most strains of *Ent. cecorum*, from all other presently known enterococci, are the production of alkaline phosphatase and β-glucuronidase. *Enterococcus columbae* is also alkaline phosphatase positive, but shows negative β-glucuronidase reactions. β-Glucuronidase negative *Ent. cecorum* are rare in poultry, the animal host from which it was originally described, but relatively frequent in other animals (Devriese *et al.*, 1991a). These strains are difficult to differentiate from *Ent. columbae*. Certain additional differential characters may be helpful (Table 10.12). *Streptococcus bovis* also may be confused with *Ent. cecorum*. This species is β-glucuronidase negative and most *Strep. bovis* strains grow on Rogosa medium when incubated anaerobically in the presence of CO_2.

10.7.5.1.6 Occurrence and habitat. Enterococcus cecorum is the most frequent enterococcal species isolated from the intestines of adult chickens (Devriese *et al.*, 1991d). It is much more rare in young chicks and is also rare in other bird species. It can also be found more frequently than other enterococci in young ruminating cattle (Devriese *et al.*, 1992a) and is associated also with dogs, cats, pigs and horses (Devriese *et al.*, 1991a, 1992b).

10.7.5.1.7 Pathogenicity. Enterococcus cecorum is probably not pathogenic to chickens or other animals.

10.7.5.2 Enterococcus columbae *(Devriese* et al., *1990).* Type strain: Devriese STR 345 (LMG 11740; NCIMB 13013). From pigeon intestines.

10.7.5.2.1 Genomic characterization. The DNA base composition of the type strain is 38.2% G+C. DNA homology values are higher with *Ent. cecorum* than with other enterococcal species. This was confirmed by 16S rRNA sequencing but the two species show distinct lines of descent within the genus *Enterococcus* (Williams *et al.*, 1991).

10.7.5.2.2 Chemical composition. Enterococcus columbae does not contain serologically demonstrable group D antigen.

10.7.5.2.3 Biochemical characteristics. Enterococcus columbae differs from most other enterococci in being unable to grow in media containing 0.04% NaN_3 and in growing poorly or not at all in 6.5% NaCl. Arginine is not hydrolysed and pyrolidonylarylamidase is negative but alkaline phosphatase is positive. Other characteristics in common with *Ent. cecorum* are listed in Tables 10.3 and 10.11 and its differentiation from the latter species in Table 10.12.

10.7.5.2.4 Morphology and growth characteristics. In broth appearing as short chains of two to four Gram-positive cocci. Growth on agar better when incubated at 35–37°C in 3–10% CO_2 than in air. This CO_2 requirement is not absolute. *Enterococcus columbae* ressembles in this aspect another intestinal inhabitant of animals, *Ent. cecorum.*

10.7.5.2.5 Identification. A remarkable characteristic which *Ent. columbae* shares with most strains of *Ent. cecorum*, is the production of alkaline phosphatase. It can usually be differentiated from the latter species by its negative β-glucuronidase. Additional differentiating characteristics which help to differentiate *Ent. columbae* from β-glucuronidase negative *Ent. cecorum* are listed in Table 10.12.

10.7.5.2.6 Occurrence and habitat. Enterococcus columbae has been found to date only in domestic pigeons (*Columba livia*). It appears to be the dominant bacterial species of the intestinal flora of healthy pigeons.

10.7.5.2.7 Pathogenicity. The species may be found in diverse lesions in pigeons, but probably it is not a primary pathogen in these animals.

10.7.6 Miscellaneous species

10.7.6.1 Enterococcus sulfureus *(Martinez-Murcia and Collins, 1991).* Type strain: Mundt MUTK 31 (LMG 13084; NCFB 2379). From plant material.

10.7.6.1.1 Genomic characterization. The DNA base composition was determined as 38 mol%. Forms a distinct line of descent within the genus.

10.7.6.1.2 Chemical composition. No data available.

10.7.6.1.3 Morphology and growth characteristics. Cells are as typical for the genus. They are non-motile and yellow pigmented. Grows at 10°C but not at 45°C.

10.7.6.1.4 Biochemical characteristics. As in Table 10.1. *Enterococcus sulfureus* differs from most enterococci in being unable to grow at 45°C. Other characteristics are as follows: arginine dihydrolase and hippurate are negative; α- and *β*-galactosidase are positive. Acid is produced from gluconate, 2-ketogluconate, melezitose and D-raffinose. Acid is not produced from adonitol, L-arabinose, D- and L-arabitol, dulcitol, glycerol, inulin, mannitol, D-lyxose, sorbitol, D-tagatose, D-xylose and xylitol.

10.7.6.1.5 Identification. The differentiation of *Ent. sulfureus* from other yellow pigmented species is given in Table 10.6.

10.7.6.1.6 Occurrence and habitat. Isolated from plants.

10.7.6.1.7 Pathogenicity. Unknown.

10.7.6.2 Enterococcus saccharolyticus *comb. nov. (Rodrigues and Collins, 1990).* Formerly *Streptococcus saccharolyticus* (Farrow *et al.*, 1984). Type strain: Kruze HF 62 (LMG 11427; ATCC 43076; DSM 20726; NCFB 2594). From straw bedding.

10.7.6.2.1 Genomic characterization. This species was originally described by Farrow *et al.* (1984) as *Streptococcus saccharolyticus* in a description of several groups of *Strep. bovis*-like strains. Rodrigues and Collins (1990) reclassified it as an enterococcal species, mainly based on reverse transcriptase sequencing of its 16S rRNA. The species exhibited 96–97% sequence homology with other enterococci, and 89–90% with lactococci and streptococci. To date no closer relationship with certain other enterococcal species has been described. The G+C content of its DNA ranges from 37.6 to 38.3 mol%.

10.7.6.2.2 Chemical composition. Enterococcus saccharolyticus does not exhibit the group D antigen. Other characteristics have not been described.

10.7.6.2.3 Morphology and growth characteristics. Gram-positive cocci, mostly in pairs or short chains. They are non-motile. Colonies are circular, smooth and entire, non-pigmented and non-haemolytic. Grows at 10 and 45°C but not at 50°C.

10.7.6.2.4 Biochemical characteristics. Many of the characteristics traditionally attributed to enterococci do not apply to *Ent. saccharolyticus*: the strains grow poorly in 6.5% NaCl broth; VP, pyrrolidonylarylamidase and arginine hydrolysis are negative. Other test results are as follows: α- and β-galactosidase are positive and hippurate is negative.

Enterococcus saccharolyticus derives its name from its acid-producing activity from a wide range of carbohydrates, a distinguo which was of more importance at the time this species was still classified with the streptococci. In addition to the compounds listed in the general Table 10.1, amidon, D-arabitol, 2-keto-gluconate, inulin, mannitol, α-methyl-D-glucoside, melibiose, melezitose, D-raffinose, sorbitol, trehalose and D-turanose are also fermented. Acid production from L-sorbose is variable. Acid is not produced from L-arabinose, L-arabitol, adonitol, 5-keto-gluconate, dulcitol, gluconate, glycerol, α-methyl-D-mannoside, D-lyxose, rhamnose, D-tagatose, D-xylose and xylitol, in addition to the reactions listed in Table 10.1.

10.7.6.2.5 Identification. No short identification procedure can be advised. Extensive identification testing is necessary, possibly by comparing test results obtained with commercial identification galleries with the characteristics listed above. The species can be separated from *Ent. cecorum*, which it resembles phenotypically most closely, by its negative VP, alkaline phosphatase and β-glucuronidase.

10.7.6.2.6 Occurrence and habitat. Enterococcus saccharolyticus strains have been isolated from straw bedding and skin of cattle (Farrow *et al.*, 1984), but the species is probably not a component of the intestinal flora of bovines or other domestic animals. It was not detected in faecal samples from cattle by Devriese *et al.* (1992a).

10.7.6.3 Enterococcus dispar *(Collins* et al., *1991).* Type strain: Facklam E18-1 (LMG 13521; NCFB 2821; NCIMB 13000). From a human.

10.7.6.3.1 Genomic characterization. G+C content of the DNA was determined to be 39 mol%. Forms a distinct line of descent within the genus *Enterococcus*.

10.7.6.3.2 Chemical composition. The murein type is Lys-D-Asp. D antigen is not present. Other characteristics have not been described.

10.7.6.3.3 Morphology and growth characteristics. Cells are as typical for the genus. They form non-pigmented colonies and are non-motile. Grows at 10°C but not at 45°C.

10.7.6.3.4 Biochemical characteristics. Enterococcus dispar shows the general characteristics of Table 10.1. It is unable to grow at 45°C, in 0.04% NaN_3 and in 6.5% NaCl (in our hands). APPA activity and acid production from D-cyclodextrin have not been reported. Other characteristics are as follows: arginine, α- and *β*-galactosidase are positive. Acid is produced from glycerol, α-methyl-D-glucoside, D-raffinose, D-turanose, D-tagatose and 2-ketogluconate. Acid is not produced from L-arabinose, D- and L-arabitol, D-xylose, adonitol, L-sorbose, rhamnose, dulcitol, mannitol, sorbitol, inulin, melezitose, amidon, xylitol, D-lyxose, gluconate and 5-ketogluconate. PYRA and melibiose reactions are variable.

10.7.6.3.5 Phenotypic identification. A full identification procedure is necessary. The carbohydrate fermentation pattern of *Ent. dispar* most closely resembles that of *Ent. faecalis* (Table 10.3). It can be differentiated primarily by its lower resistance to selective agents and growth conditions.

10.7.6.3.6 Occurrence and habitats. Enterococcus dispar has been isolated only rarely from humans.

10.7.6.3.7 Pathogenicity. Unknown.

10.7.7 Species of uncertain taxonomic status

Tetragenococcus solitarius was originally described as *Enterococcus solitarius* (Collins *et al.*, 1989) but later reclassified by workers of the same group in the new genus *Tetragenococcus*, which contains alo the former *Pediococcus halophilus* (Collins *et al.*, 1990).

Confusion exists about *Enterococcus seriolicida*, a fish pathogen (Kusuda *et al.*, 1991). The type strain probably does not belong to the genus *Enterococcus* (Collins, M.D., pers. comm.). *Ent. seriolicida* shows SDS-PAGE whole-cell protein patterns which are almost identical to the patterns obtained from strains of *Lactococcus garvieae* (Pot *et al.*, unpublished).

References

Aguirre, A. and Collins, M.D. (1992) Phylogenetic analysis of *Alloiococcus otitis* gen. nov., sp. nov., an organism from human middle ear fluid. *International Journal of Systematic Bacteriology*, **42**, 79–83.

Andrews, F.W. and Horder, T.J. (1906) A study of streptococci pathogenic for man. *Lancet*, **2**, 708–713.

Batish, V.K. and Ranganathan, B. (1984) Occurrence of enterococci in milk and milk products. II. Identification and characterization of prevalent types. *New Zealand Journal of Dairy Science and Technology*, **19**, 189–196.

Batish, V., Chandler, H. and Ranganathan, B. (1982) Characterization of deoxyribonuclease positive enterococci isolated from milk and milk products. *Journal of Food Protection*, **45**, 348–351.

Bernsmann, P., Alpert, C.A., Muss, P., Deutscher, J. and Hengstenberg, W. (1982) The bacterial PEP-dependent phosphotransferase system mechanism of gluconate phosphorylation in *Streptococcus faecalis*. *FEBS Letters*, **138**, 101–103.

Betzl, D., Ludwig, W. and Schleifer, K.H. (1990) Identification of lactococci and enterococci by colony hybridization with 23S rRNA-targeted oligonucleotide probes. *Applied and Environmental Microbiology*, **6**, 2927–2929.

Bridge, P.D. and Sneath, P.H.A. (1982) *Streptococcus gallinarum* sp. nov. and *Streptococcus oralis* sp. nov. *International Journal of Systematic Bacteriology*, **32**, 410–415.

Bridge, P.D. and Sneath, P.H.A. (1983) Numerical taxonomy of *Streptococcus*. *Journal of General Microbiology*, **129**, 565–597.

Buttiaux, R. (1958) Les streptocoques fécaux des intestins humains et animales. *Annales de l'Institut Pasteur*, **94**, 778–782.

Chander, H., Ranganathan, B. and Singh, J. (1979) Role of some fatty acids on the growth and lipase production by *Streptococcus faecalis*. *Journal of Food Science*, **44**, 1566–1567.

Chirurgi, V.A., Oster, S.E., Goldberg, A.A., Zervos, M.J. and McCabe, R.E. (1991) Ampicillin-resistant *Enterococcus raffinosus* in an acute-care hospital: case-control study and antimicrobial susceptibilities. *Journal of Clinical Microbiology*, **29**, 2663–2665.

Clewell, D.B. and Gawron-Burke, C. (1986) Conjugative transposons and the dissemination of antibiotic resistance in streptococci. *Annual Reviews in Microbiology*, **40**, 635–659.

Clewell, D.B. and Weaver, K.E. (1989) Sex pheromones and plasmid transfer in *Enterococcus faecalis*. *Plasmid*, **21**, 175–184.

Collins, J.E., Bergeland, M.E., Lindeman, C.J. and Duimstra, J.R. (1988) *Enterococcus* (*Streptococcus*) *durans* adherence in the intestine of a diarrheic pup. *Veterinary Pathology*, **25**, 396–398.

Collins, M.D. and Jones, D. (1979) The distribution of isoprenoid quinones in streptococci of serological groups D and N. *Journal of General Microbiology*, **114**, 27–33.

Collins, M.D., Jones, D., Farrow, J.A.E., Kilpper-Bälz R. and. Schleifer, K.H. (1984) *Enteroccus avium* nom. rev., comb. nov.; *E. casseliflavus* nom. rev., comb. nov.; *E. durans* nom. rev., comb. nov.; *E. gallinarum* comb. nov.; and *E. malodoratus* sp. nov. *International Journal of Systematic Bacteriology*, **34**, 220–223.

Collins, M.D., Farrow, J.A.E. and Jones, D. (1986) *Enterococcus mundtii* sp. nov. *International Journal of Systematic Bacteriology*, **36**, 8–12.

Collins, M.D., Ash, C., Farrow, J.A.E., Wallbanks, S. and Williams, A.M. (1989a) 16S Ribosomal ribonucleic acid sequence analyses of lactococci and related taxa. Description of *Vagococcus fluvialis* gen. nov., sp. nov. *Journal of Applied Bacteriology*, **67**, 453–460.

Collins, M.D., Facklam, R.R., Farrow, J.A.E. and Williamson, R. (1989b) *Enterococcus raffinosus* sp. nov., *Enterococcus solitarius* sp. nov. and *Enterococcus pseudoavium* sp. nov. *FEMS Microbiological Letters*, **57**, 283–288.

Collins, M.D., Williams, A.M. and Wallbanks, S. (1990) The phylogeny of *Aerococcus* and *Pediococcus* as determined by 16S rRNA sequence analysis: description of *Tetragenococcus* gen. nov. *FEMS Microbiology Letters*, **70**, 255–262.

Collins, M.D., Rodrigues, U.M., Pigott, N.E. and Facklam, R.R. (1991) *Enterococcus dispar* sp. nov. a new enterococcus species from human sources. *Letters in Applied Microbiology*, *12*, 95–98.

Cooper, K.E. and Ramadan, F.M. (1955) Studies on the differentiation between human and animal pollution by means of faecal streptococci. *Journal of General Microbiology*, **12**, 180–190.

Cruz Colque, I.J., Devriese, L.A. and Haesebrouck, F. (1993) Streptococci and enterococci associated with tonsils of cattle. *Letters in Applied Microbiology*, **16**, 72–74.

Defernando, G.D.G., Hernandez, P.E., Burgos, J., Sanz, B. and Ordonez, J.H. (1991) Extracellular proteinase from *Enterococcus faecalis* subsp. *liquefaciens*. 2. Partial purification and some technologically important problems. *Folia Microbiologica*, **36**, 429–436.

Deguchi, Y. and Morishita, T. (1992) Nutritional requirements in multiple auxotrophic lactic acid bacteria – Genetic lesions affecting amino acid biosynthetic pathways in *Lactococcus lactis*, *Enterococcus faecium*, and *Pediococcus acidilactici*. *Bioscience, Biotechnology and Biochemistry*, **56**, 913–918.

Deibel, R.H. (1964) The group D streptococci. *Bacteriological Reviews*, **28**, 330–366.

Devriese, L.A. and Haesebrouck, F. (1991) *Enterococcus hirae* in different animal species. *The Veterinary Record*, **129**, 391–392.

Devriese, L.A., Dutta, G.N., Farrow, J.A.E., Van de Kerckhove, A. and Phillips, B.A. (1983) *Streptococcus cecorum*, a new species isolated from chickens. *International Journal of Systematic Bacteriology*, **33**, 772–776.

Devriese, L.A., Van de Kerckhove, A., Kilpper-Bälz, R. and Schleifer, K.H. (1987) Characterization and identification of *Enterococcus* species isolated from animals. *International Journal of Systematic Bacteriology*, **37**, 257–259.

Devriese, L.A., Ceyssens, K., Rodrigues, U.M. and Collins, M.D. (1990) *Enterococcus columbae*, a species from pigeon intestines. *FEMS Microbiology Letters*, **71**, 247–252.

Devriese, L.A., Ceyssens, K. and Haesebrouck, F. (1991a). Characteristics of *Enterococcus cecorum* strains from different animal species. *Letters in Applied Microbiology*, **12**, 137–139.

Devriese, L.A., Collins, M.D. and Wirth, R. (1991b) The genus *Enterococcus*. In *The Prokaryotes* (eds Balows, A., *et al.*). Springer-Verlag, New York, USA, pp. 1465–1481.

Devriese, L.A., Ducatelle, R., Uyttebroek, E. and Haesebrouck, F. (1991c) *Enterococcus hirae* infection and focal necrosis of the brain of chicks. *The Veterinary Record*, **129**, 316.

Devriese, L.A., Hommez, J., Wyffels, R. and Haesebrouck, F. (1991d) Composition of the enterococcal and streptococcal intestinal flora of poultry. *Journal of Applied Bacteriology*, **71**, 46–50.

Devriese, L.A., Laurier, L., De Herdt, P. and Haesebrouck, F. (1992a) Enterococcal and streptococcal species isolated from faeces of calves, young cattle and dairy cows. *Journal of Applied Bacteriology*, **72**, 29–31.

Devriese, L.A., Cruz Colque, J.I., De Herdt, P. and Haesebrouck, F. (1992b). Identification and composition of the tonsillar and anal enterococcal and streptococcal flora of dogs and cats. *Journal of Applied Bacteriology*, **73**, 421–425.

Devriese, L.A., Cruz Colque, J.I., Haesebrouck, F., Desmidt, M., Uyttebroek, E. and Ducatelle, R. (1992c) *Enterococcus hirae* septicaemia of psittacine birds. *The Veterinary Record*, **130**, 558–559.

Drolet, R., Higgins, R. and Jacques, M. (1990) L'entéropathie associée à *Enterococcus (Streptococcus) durans* chez le procelet. *Le Médecin Vétérinaire du Quebec*, **20**, 114–118.

Dunny, G.M. (1990) Genetic functions and cell–cell interactions in the pheromone inducible plasmid transfer system of *Enterococcus faecalis*. *Molecular Microbiology*, **4**, 689–696.

Etheridge, M., Yolken, R.H. and Vonderfecht, S.L. (1988) *Enterococcus hirae* implicated as a cause of diarrhea in suckling rats. *Journal of Clinical Microbiology*, **26**, 1741–1744.

Facklam, R.R. and Collins, M.D. (1989) Identification of *Enterococcus* species isolated from human infections by a conventional test scheme. *Journal of Clinical Microbiology*, **27**, 731–734.

Facklam, R.R. and Moody, M.D. (1970) Presumptive identification of group D streptococci: the bile-esculin test. *Applied Microbiology*, **20**, 245–250.

Facklam, R.R., Hollis, D. and Collins, M.D. (1989) Identification of Gram-positive coccal and coccobacillary vancomycin-resistant bacteria. *Journal of Clinical Microbiology*, **27**, 724–730.

Farrow, J.A.E. and Collins, M.D. (1985) *Enteroccus hirae*, a new species that includes amino acid assay strain NCDO 1258 and strains causing growth depression in young chickens. *International Journal of Systematic Bacteriology*, **35**, 73–75.

Farrow, J.A.E., Kruze, J., Phillips, B.A., Bramley, A.J. and Collins, M.D. (1984) Taxonomic studies on *Streptococcus bovis* and *Streptococcus equinus*: Description of *Streptococcus alactolyticus* sp. nov. and *Streptococcus saccharolyticus* sp. nov. *Systematic and Applied Microbiology*, **5**, 467–482.

Finegold, S.M., Attebery, H.R. and Sutter, V. (1974) Effect of diet on human flora: comparison of Japanese and American diets. *American Journal of Clinical Nutrition*, **27**, 1465–1469.

Freney, J., Bland, S., Etienne, J., Boeufgras, J.M. and Fleurette, J. (1992) Description and evaluation of the semiautomated 4-hour Rapid ID 32 Strep method for identification of streptococci and members of related genera. *Journal of Clinical Microbiology*, **30**, 2657–2661.

Fuller, R., Coates, M.E. and Harrison, G.F. (1979) The influence of specific bacteria and a filterable agent on the growth of gnotobiotic chicks. *Journal of Applied Bacteriology*, **46**, 335–342.

Garg, S.K. and Mital, B.K. (1991) Enterococci in milk and milk products. *Critical Reviews in Microbiology*, **18**, 15–45.

Garvie, E.I. and Farrow, J.A.E. (1981) Sub-divisions within the genus *Streptococcus* using deoxyribonucleic acid/ribosomal ribonucleic acid hybridization. *Zentralblatt für Bakteriologie, Parasitenkunde, Infektionskrankheiten und Hygiene (1. Abteilung Originale)*, C **2**, 299–310.

George, R.C. and Uttley, A.H.C. (1989) Susceptibility of enterococci and epidemiology of enterococcal infections in the 1980s. *Epidemiology and Infection*, **103**, 403–413.

Grayson, M.L., Eliopoulos, G.M., Wennersten, C.B., Ruoff, K.L., Klimm, K., Sapico, F.L., Bayer, A.S. and Moellering, R.C. (1991) Comparison of *Enterococcus raffinosus* with *Enterococcus avium* on the basis of penicillin susceptibility, penicillin-binding protein analysis, and high-level aminoglycoside resistance. *Antimicrobial Agents and Chemotherapy*, **3**, 1408–1412.

Guthof, O. (1955) Über eine neue serologische Gruppe α-hemolytischer Streptokokken (serologische Gruppe Q). *Zentralblatt für Bakteriologie, Parasitenkunde, Infektionskrankheiten und Hygiene (1. Abteilung Originale)*, **164**, 60–69.

Hall, L.M.C., Duke, B., Guiney, M. and Williams, R. (1992) Typing of *Enterococcus* species by DNA restriction fragment analysis. *Journal of Clinical Microbiology*, **30**, 915–919.

Heller, K. and Röschenthaler, R. (1978) β-D-Phosphogalactosidase-galactohydrolase of *Streptococcus faecalis* and the inhibition of its synthesis by glucose. *Canadian Journal of Microbiology*, **24**, 512–515.

Ike, Y., Hashimoto, H. and Clewell, D.B. (1987) High incidence of hemolysin production by *Enterococcus (Streptococcus) faecalis* strains associated with human parenteral infections. *Journal of Clinical Microbiology*, **25**, 1524–1528.

Jacobs, N.J. and Van Demark, P.J. (1960) Comparison of the mechanism of glycerol oxidation in aerobically and anaerobically grown *Streptococcus faecalis*. *Journal of Bacteriology*, **79**, 532–538.

Jayaro, B.M., Dorie, J.J.E. and Oliver, S.P. (1992) Restriction length fragment polymorphism analysis of 16S ribosomal DNA of *Streptococcus* and *Enterococcus* of bovine origin. *Journal of Clinical Microbiology*, **30**, 2235–2240.

Johnson, D.D., Duimstra, J.R., Gates, C.E. and McAdaragh, J.P. (1983) Streptococcal colonization of the pig intestine. *Proceedings of the Veterinary Infectious Disease Organization*, University of Saskatchewan, pp. 279–289.

Jones, D. (1978) Composition and differentiation of the genus *Enterococcus*. In *Streptococci* (eds Skinner, F.A. and Quesnel, L.B.). Academic Press, London, UK, pp. 1–49.

Kalina, A.P. (1970) The taxonomy and nomenclature of enterococci. *International Journal of Systematic Bacteriology*, **20**, 185–189.

Kariyama, R. and Shockman, G.D. (1992) Extracellular and cellular distribution of muramidase-2 and muramidase-1 of *Enterococcus hirae* ATCC 9790. *Journal of Bacteriology*, **174**, 3236–3241.

Kaufhold, A. and Ferrieri, P. (1991) Isolation of *Enterococcus mundtii* from normally sterile body sites in two patients. *Journal of Clinical Microbiology*, **29**, 1075–1077.

Kielwein, G. (1978) Vorkommen und Bedeutung von Enterokokken in Milch und Milchprodukten. *Archiv für Lebensmittelhygiene*, **29**, 127–128.

Kilpper-Bälz, R. and Schleifer, K.H. (1981) DNA–rRNA hybridization studies among staphylococci and some other Gram-positive bacteria. *FEMS Microbiology Letters*, **10**, 357–362.

Kilpper-Bälz, R. and Schleifer, K.H. (1984) Nucleic acid hybridization and cell wall composition studies of pyogenic streptococci. *FEMS Microbiology Letters*, **24**, 355–364.

Kilpper-Bälz, R., Fischer, G. and Schleifer, K.H. (1982) Nucleic acid hybridization of group N and group D streptococci. *Current Microbiology*, **7**, 245–250.

Knight, R.G. and Shlaes, D.M. (1986) Deoxyribonucleic acid relatedness of *Enterococcus hirae* and "*Streptococcus mutans*" homology group II. *International Journal of Systematic Bacteriology*, **36**, 111–133.

Knight, R.G., Shlaes, D.M. and Messineo, L. (1984) Deoxyribonucleic acid relatedness among major human enterococci. *International Journal of Systematic Bacteriology*, **34**, 327–331.

Kusuda, R., Kawai, K., Salati, F., Banner, C.R. and Fryer, J.L. (1991) *Enterococcus seriolicida* sp. nov., a fish pathogen. *International Journal of Systematic Bacteriology*, **41**, 406–409.

Lacoux, P.A., Jordens, J.Z., Fenton, C.M., Guiney, M. and Pennington, T.H. (1992) Characterization of enterococcal isolates by restriction enzyme analysis of genomic DNA. *Epidemiology and Infection*, **109**, 69–80.

Lancefield, R.C. (1933) A serological differentiation of human and other groups of hemolytic streptococci. *Journal of Experimental Medicine*, **57**, 571–595.

London, J. and Meyer, E.Y. (1970) Malate utilization by group D streptococci. Regulation of malic enzyme synthesis by an inducible malate permease. *Journal of Bacteriology*, **102**, 130–135.

Ludwig, W., Seewaldt, E., Kilpper-Bälz, R., Schleifer, K.H., Magrum, L., Woese, C.R., Fox, G.E. and Stackebrandt, E. (1985) The phylogenetic position of *Streptococcus* and *Enterococcus*. *Journal of General Microbiology*, **131**, 543–551.

Martinez-Murcia, A.J. and Collins, M.D. (1991) *Enterococcus sulfureus*, a new yellow-pigmented *Enterococcus* species. *FEMS Microbiology Letters*, **80**, 69–74.

McCoy, T.A. and Wender, S.H. (1953) Some factors affecting the nutritional requirements of *Streptococcus faecalis*. *Journal of Bacteriology*, **65**, 660–665.

Mead, G.C. (1978) Streptococci in the intestinal flora of man and other non-ruminant animals. In *Streptococci* (eds. Skinner, F.A. and Quesnel, L.B.). Academic Press, London, UK, pp. 245–261.

Meyer, K. and Schonfeld, H. (1926) Über die Unterscheidung des Enterococcus von *Streptococcus viridans* und die Beziehungen beider zum *Streptococcus lactis*. *Zentralblatt für Bakteriologie Parasitenkunde, Infektionskrankheiten und Hygiene (1. Abteilung Originale)*, **90**, 402–416.

Miranda, A.G., Singh, K.V. and Murray, B.E. (1992) Determination of the chromosomal size of 3 different strains of *Enterococcus faecalis* and one strain of *Enterococcus faecium*. *DNA and Cell Biology*, **11**, 331–335.

Moellering, R. (1991) The *Enterococcus*: a classic example of the impact of antimicrobial resistance on therapeutic options. *Journal of Antimicrobial Chemotherapy*, **28**, 1–12.

Mundt, J.O. (1963) Occurrence of enterococci in animals in a wild environment. *Applied Microbiology*, **11**, 136–140.

Mundt, J.O. (1975) Unidentified streptococci from plants. *International Journal of Systematic Bacteriology*, **25**, 281–285.

Mundt, J.O. (1976) Streptococci in dried and in frozen foods. *Journal of Milk and Food Technology*, **36**, 364–367.

Niven, C.F. and Sherman, J.M. (1944) Nutrition of the enterococci. *Journal of Bacteriology*, **47**, 335–342.

Noble, C.J. (1978) Carriage of group D streptococci in the human bowel. *Journal of Clinical Pathology*, **4**, 1182–1186.

Nowlan, S.P. and Deibel, R.H. (1967) Group Q streptococci. I. Ecology, serology, physiology and relationship to established enterococci. *Journal of Bacteriology*, **94**, 291–296.

Orla-Jensen, S. (1919) The lactic acid bacteria. *Memoirs of the Royal Academy of Sciences in Denmark Section of Sciences Series 8*, **5**, 81–197.

Payne, J. (1978) Damage and recovery in streptococci. In *Streptococci* (eds. Skinner, F.A. and Quesnel, L.B.). Academic Press, London, UK, pp. 349–369.

Pette, J.W. (1955) De vorming van zwavelwaterstof in Goudse kaas, veroorzaakt door melkzuurbacteriën. *Netherlands Milk and Dairy Journal*, **10**, 291–302.

Piraz, G., El Kharroubi, A., Van Beeumen, J., Coeme, E., Coyette, J. and Ghuysen, J.M. (1990) Characterization of an *Enterococcus hirae* penicillin-binding Protein 3 with low penicillin affinity. *Journal of Bacteriology*, **172**, 6856–6862.

Pompei, R., Berlutti, F., Thaller, M.C., Ingianni, A., Cortis, G. and Dainelli, B. (1992a) *Enterococcus flavescens* sp. nov., a new species of enterococci of clinical origin. *International Journal of Systematic Bacteriology*, **42**, 365–369.

Pompei, R., Thaller, M.C., Pittaluga, F., Flore, O. and Satta, G. (1992b) Analysis of bacteriolytic activity patterns, a novel approach to the taxonomy of enterococci. *International Journal of Systematic Bacteriology*, **42**, 37–43.

Rao, C.U.M., Shankar, P.A. and Laxminarayana, H. (1988) A study of enterococci occurring in milk and dairy products. *Indian Journal of Dairy Science*, **39**, 281–285.

Reizer, J., Reizer, A. and Saier, M.H. (1992) The putative Na^+/H^+ antiporter (NapA) of *Enterococcus hirae* is homologous to the putative K^+/H^+ antiporter (KefC) of *Escherichia coli. FEMS Microbiology Letters*, **94,** 161–164.

Reuter, G. (1985) Selective media for group D streptococci. *International Journal of Food Microbiology*, **2**, 103–114.

Rodrigues, U. and Collins, M.D. (1990) Phylogenetic analysis of *Streptococcus saccharolyticus* based on 16S rRNA sequencing. *FEMS Microbiology Letters*, **71**, 231–234.

Rosan, B. and William, N.B. (1966) Serology of strains of *Streptococcus faecalis* which produce hyaluronidase. *Nature*, **212**, 1275.

Ruoff, K.L., De La Maza, L., Murtagh, M.J., Spargo, J.D. and Ferraro, M.J. (1990) Species identities of enterococci isolated from clinical specimens. *Journal of Clinical Microbiology*, **28**, 435–437.

Schleifer, K.H. and Kilpper-Bälz, R. (1984) Transfer of *Streptococcus faecalis* and *Streptococcus faecium* to the genus *Enterococcus* nom. rev. as *Enterococcus faecalis* comb. nov. and *Enterococcus faecium* comb. nov. *International Journal of Systematic Bacteriology*, **34**, 31–34.

Schleifer, K.H. and Kilpper-Bälz, R. (1987) Molecular and chemotaxonomic approaches to the classification of streptococci, enterococci and lactococci: a review. *Systematic and Applied Microbiology*, **10**, 1–19.

Schleifer, K.H., Kraus, J., Dvorak, C., Kilpper-Bälz, R., Collins, M.D. and Fischer, W. (1985) Transfer of *Streptococcus lactis* and related streptococci to the genus *Lactococcus* gen. nov. *Systematic and Applied Microbiology*, **6**, 183–195.

Sherman, J.M. (1937) The streptococci. *Bacteriological Reviews*, **1**, 3–97.

Sherman, J.M. and Wing, H.U. (1937) *Streptococcus durans. Journal of Dairy Science*, **28**, 165–167.

Sijpesteijn, A. (1970) Induction of cytochrome formation and stimulation of oxidative dissimilation by haemin in *Streptococcus faecalis* strain 10C1. *Journal of Bacteriology*, **96**, 1595–1600.

Skerman, V.D.B., McGowan, V. and Sneath, P.H.A. (1980) Approved lists of bacterial names. *International Journal of Systematic Bacteriology*, **30**, 225–420.

Slanetz, L.W. and Bartley, C.H. (1957). Numbers of enterococci in water, sewage and faeces determined by the membrane filter technique with an improved medium. *Journal of Bacteriology*, **74**, 591–595.

Stiles, M.E., Ramji, N.W., Ng, L.K. and Paradis, D.C. (1978) Incidence of group D streptococci with other indicator organisms in meats. *Canadian Journal of Microbiology*, **24**, 1502–1508.

Thiercelin, E. (1899) Sur un diplocoque saprofyte de l'intestin susceptible à devenir pathogène. *Comptes Rendues des Séances de la Société de Biologie*, **51**, 269–271.

Thiercelin, E. and Jouhaud, L. (1903). Reproduction de l'entérocoque; taches centrales; granulations peripheriques et microblastes. *Comptes Rendus des Séances de la Société de Biologie Paris*, **55**, 686–688.

Tzipori, S., Hayes, J., Sims, L. and Witters, M. (1984) *Streptococcus durans*: an unexpected enteropathogen of foals. *Journal of Infectious Diseases*, **150**, 589–593.

Vaughn, D.H., Riggsby, W.S. and Mundt, J.O. (1979) Deoxyribonucleic acid relatedness of strains of yellow-pigmented group D streptococci. *International Journal of Systematic Bacteriology*, **29**, 204–212.

Vincent, S., Knight, R.G., Green, M., Sahm, D.F. and Shlaes, D.M. (1991) Vancomycin susceptibility and identification of motile enterococci. *Journal of Clinical Microbiology*, **29**, 2333–2337.

Watanabe, T., Shimohashi, H., Kawai, Y. and Mutai, K. (1981) Studies on streptococci. I. Distribution of faecal streptococci in man. *Microbiology and Immunology*, **25**, 257–269.

Watanabe, T., Kumata, H., Sasamoto, M. and Shimizu-kadota, M. (1992) The distribution of homologous enterococcal plasmid DNA sequences in human faecal isolates. *Journal of Applied Bacteriology*, **73**, 131–135.

Wessels, D., Jooste, P.J. and Mostert, J.F. (1988) Die voorkoms van *Enterococcus* spesies in melk en suiwelprodukte. *Suid Afrikaans Tydskrif vir Suiwelkunde*, **20**, 68–72.

Whittenbury, R. (1978) Biochemical characteristics of *Streptococcus* species. In *Streptococci* (eds Skinner, F.A. and Quesnel, L.B.). Academic Press, London, UK, pp. 51–69.

Williams, A.M., Farrow, J.A.E. and Collins, M.D. (1989) Reverse transcriptase sequencing of 16S ribosomal RNA from *Streptococcus cecorum*. *Letters in Applied Microbiology*, **8**, 185–189.

Williams, A.M., Rodrigues, U.M. and Collins, M.D. (1991) Intrageneric relationships of enterococci as determined by reverse transcriptase sequencing of small-subunit rRNA. *Research in Microbiology*, **142**, 67–74.

Williamson, R., Gutmann, L., Horaud, T., Delbos, F. and Acar, J.F. (1986) Use of penicillin-binding proteins for the identification of enterococci. *Journal of General Microbiology*, **132**, 1929–1937.

Zitzelsberger, W., Götz, F. and Schleifer, K.H. (1984) Distribution of superoxide dismutases, oxidases, and NADH peroxidase in various streptococci. *FEMS Microbiology Letters*, **21**, 243–246.

11 Spore-forming, lactic acid producing bacteria of the genera *Bacillus* and *Sporolactobacillus*

D. FRITZE and D. CLAUS

11.1 Introduction

Lactic acid production is traditionally considered to be associated with non-spore-forming bacteria known as lactic acid bacteria. In addition to these organisms, a number of lactic acid forming aerobic spore-formers have been described which are allocated to the genera *Bacillus* and *Sporolactobacillus*. In 1872, Ferdinand Cohn, a German botanist and bacteriologist, established the generic name *Bacillus* to include rod-shaped bacteria that grow in filaments (Cohn, 1872). Only 4 years later Cohn and Robert Koch independently detected that two species of the newly created genus *Bacillus* were able to form resting stages which were not killed easily by mere boiling (Cohn, 1876; Koch, 1876). These heat-resistant bodies were denoted as endospores. The detection of these bacterial resting stages resistant to various adverse environmental influences finally finished off the last advocates of the obstinately defended theory of 'spontaneous generation'. It was also demonstrated that a developmental cycle linked cells and spores: that a cell forms a resistant spore and that this spore grows out again to become a vegetative cell.

The unusual ability of endospore formation as a criterion for bacterial classification was first proposed by de Bary (1884) and by Hüppe (1886) but most bacteriologists continued to use the generic name *Bacillus* for various rod-shaped bacteria. Only from 1923, after the publication of the first edition of *Bergey's Manual of Determinative Bacteriology*, was it generally accepted that the genus *Bacillus* should be restricted to all rod-shaped bacteria forming endospores and growing aerobically. Today, five additional genera of aerobic bacteria are accepted for endospore-forming organisms. Members of the genera *Sporosarcina* (two species) and *Thermoactinomyces* (eight species) are distinct from *Bacillus* species mainly in morphological respect. Strains of the genus *Sporolactobacillus* (one species), *Amphibacillus* (one species) or *Alicyclobacillus* (three species which formerly belonged to *Bacillus*) are rod-shaped like *Bacillus*. *Sporolactobacillus* and *Amphibacillus* differ from strains of *Bacillus* species mainly in their response to oxygen, since optimal growth occurs under microaerophilic conditions, or in the case of *Amphibacillus*, spore-formation occurs not only under aerobic but also under anaerobic conditions. In addition both of them lack catalase. Since catalase is also absent in some *Bacillus* species or strains, it seems unlikely that lack of this property warrants separate genus position for these bacteria. The genus *Alicyclobacillus* has been formed to encompass 3 ω-alicyclic fatty acid producing thermoacidophilic *Bacillus* species.

11.2 General properties of the genus *Bacillus* and other spore-forming organisms

The phenotypic differentiation of the genus *Bacillus* from other genera is mainly through three properties: rod-like shape, aerobic spore formation and Gram-positive cell wall. However, it is often not easy to provide evidence for these characteristics and allocation of strains to the genus *Bacillus* might be complicated.

It is generally accepted that the genus *Bacillus* comprises Gram-positive organisms, although strains or species reported to stain Gram-negative or Gram-variable are not uncommon. It is known that in many species Gram-positive staining occurs only in very young cultures. When examining *Bacillus* strains staining Gram-negative, Wiegel (1981) found a difference between Gram reaction and Gram type, which refers to the structure of the cell wall as seen in thin sections in the electron microscope. He could prove, and it was confirmed later in other cases, that the cell wall of *Bacillus* strains staining Gram-negative may be a typical Gram-positively structured cell wall.

Bacillus species are potentially able to form resting stages after the end of exponential cell growth. Because these resting stages are intracellular they are designated as endospores. They differ from vegetative cells in ultrastructure, chemical composition and refraction and can be detected by phase contrast microscopy due to their optical brightness (Figure 11.1). From a practical point of view the most important difference is, however, the resistance of endospores to chemical or physical stresses. Compared with vegetative cells, spores are more resistant to heat by a factor of 10^4–10^5 or to certain disinfectants (including ethanol) by a factor of about 10^4. They survive long periods under dry conditions and show a remarkable resistance also to ionizing radiation and to UV light. The degree of resistance often depends on the environmental conditions under which endospores are formed (Russell, 1982).

It is well known that *Bacillus* strains do not form endospores under all cultural conditions, and the ability to form spores is often difficult to demonstrate. This is most important from a taxonomic point of view. There have been organisms described as non-spore-formers which had to

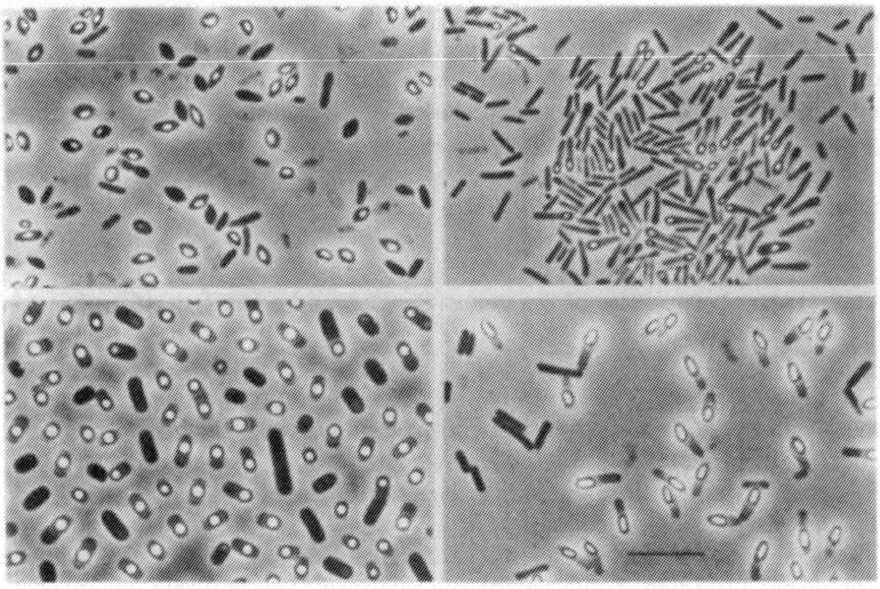

Figure 11.1 Sporulating and non-sporulating cells of some *Bacillus* species. Bar – 10 μm (refers to all photomicrographs).

be reclassified after the detection of spores, e.g. '*Lineola longa*' as '*Bacillus macroides*' (Bennet and Canale-Parola, 1965) or '*Lactobacillus cereale*' as *Bacillus coagulans* (Gordon *et al.*, 1973), and more may be found in the future. The detection of spore formation is also most important in the identification of bacterial isolates where non-sporulating *Bacillus* strains may be totally misidentified.

Certain *Bacillus* species or strains are known to easily 'lose' the ability to form endospores immediately after isolation or after only a few transfers on culture media. However, in most cases this is not a loss of genetic information like in spo^- mutants but a loss of the potency of expriming spore formation under the cultural conditions applied. With most strains often slight changes in cultural conditions are effective in restoring spore formation but with other strains it may be hardly achieved. Methods used by different authors vary from supplementation of media with trace elements to variations in the conditions of incubation and supply of nutrients. A survey of methods has been published by Kalakoutskii and Dobritsa (1984).

Some of the methods successfully used for the initiation of spore formation are the following:

- Influence of culture conditions:
  - Strains regularly transferred onto rich medium may readily lose

  their spore-forming property; too frequent transfers should be avoided.
  - Spore formation may be delayed for 1–2 weeks, therefore prolonged incubation is often necessary during which drying of plates must be avoided.
  - Growth temperature may be kept below optimum. In other cases growth temperature slightly under maximum proved successful.
- Influence of medium:
  - Spore formation is usually better on solid media.
  - Media may be prepared by substituting up to 50% of the distilled water by soil extract prepared from garden soil rich in organic matter (Gordon *et al.*, 1973).
  - For sporulation generally higher concentrations of trace elements, especially manganese, are needed than for vegetative growth (Charney *et al.*, 1951). Since peptone based media often contain very low concentrations of trace elements (Bovallius and Zacharias, 1971), media supplemented with manganous ions (as chloride or sulfate) in concentrations of 10–50 mg/litre may drastically improve spore formation. An additional supply of about 100 mg $CaCl_2$/litre may also be useful.
  - The concentration of nutrients may be reduced.

This list can be extended endlessly depending on who is working where with which strains. Each strain has to be treated individually and a standard procedure cannot be given.

11.3 *Bacillus* species forming lactic acid

A number of species allocated to the genus *Bacillus* are known to produce lactic acid. These are *B. coagulans*, *B. lentimorbus*, *B. popilliae*, *B. smithii*, and *B. stearothermophilus*. In addition, three not validly published species, namely '*B. laevolacticus*', '*B. racemilacticus*', and '*B. vesiculiferous*' belong to this group of organisms. As obviously not all species of the genus have been checked for possible production of lactic acid this property might be much wider spread than known today; for example, scarce indications on the occurrence of this end product can be found in older literature also for *B. subtilis*. Although metabolic end products of the different species of *Bacillus* depend on growth conditions, the typically excreted products are organic acids. Most of the fermenting *Bacillus* species exhibit a mixed acid fermentation. Strains of lactic acid forming *Bacillus* species have been isolated from soil (Allen, 1953), from spoiled food or milk (Hammer, 1915; Sarles and Hammer, 1932; Becker and Pederson, 1950), from diseased larvae of bees (Dutky, 1940), from the rhizosphere of various plants

(Nakayama and Yanoshi, 1967a,b) and from the intestine of crayfish (Trinkunaite *et al.*, 1987). Due to their ability of forming endospores resistant to air drying and to other stresses, and which enable them to survive long term under adverse conditions, most aerobic spore formers are ubiquitous and can be isolated from a wide variety of sources. Hence, it is not possible to judge ecology, which means the actual occurrence and growth of a given organism in a given place by the isolation of this organism from this place.

11.3.1 Bacillus coagulans

Hammer (1915) isolated several *Bacillus* strains from spoiled canned milk and described these as *Bacillus coagulans*. In a number of studies using these isolates considerable morphological and physiological inconsistencies were observed (Smith *et al.*, 1952; Bradley and Franklin, 1958; Seki *et al.*, 1978). This high variability had obviously led to the creation of a number of other species designations which later were recognized as being synonyms: '*Bacillus calidolactis*' (Hussong and Hammer, 1928), '*Bacillus thermoacidurans*' (Berry, 1933), '*Bacillus dextrolacticus*' (Anderson and Werkman, 1940), '*Bacillus thermoacidificans*' (Renco, 1942) or '*Lactobacillus cereale*' (Olsen, 1944).

Later, other workers observed that an organization of strains into clusters could be obtained when certain tests were applied (Wolf and Barker, 1968; Klaushofer and Hollaus, 1970; Logan and Berkeley, 1984). As a consequence, mol% G+C of the DNA and DNA/DNA homology studies were performed (Nakamura *et al.*, 1988) to find five DNA relatedness groups. Of those, DNA relatedness group 1 was identified to represent the species *Bacillus coagulans sensu stricto*. For these strains a homolactic fermentation was reported where L(+)-lactic acid predominates. DNA group 2 had to be described as a new species which was named *B. smithii*. The strains of the other groups of this study were identified as belonging to already known species of the genus *Bacillus*. Group 5, for example, was allocated to *B. stearothermophilus*.

11.3.2 Bacillus smithii

Bacillus smithii has been distinguished from *B. coagulans* and described as a separate species in 1988 (Nakamura *et al.*, 1988). Morphologically, these organisms are hardly distinguishable from most of the other lactic acid producing *Bacillus* species *B. coagulans*, *B. stearothermophilus*, '*B. laevolacticus*' and '*B. racemilacticus*'. The main properties of *B. smithii* in which it differs from its 'parent' species *B. coagulans* are its ability to grow at a higher maximum temperature, its larger diameter of cells, oxidase activity and its inability to grow in nutrient broth at pH 4.5 or 7.7 or to hydrolyse starch. L(+)-Lactic acid is predominantly produced.

11.3.3 Bacillus stearothermophilus

Thermophilic *Bacillus* strains with growth optima of 60°C and above have traditionally been assigned to the species *B. stearothermophilus* (and *B. coagulans*) and have been isolated from a variety of mesophilic and thermophilic environments (Allen, 1953; Bartholomew and Paik, 1966). The original description of *B. stearothermophilus* (Donk, 1920) was largely based on morphological and physiological characteristics and was later shown to comprise a variety of different groups of organisms (White *et al.*, 1994). In the 8th edition of *Bergey's Manual of Determinative Bacteriology* (Gibson and Gordon, 1974) it is noted that in glucose media most strains grow actively without oxygen until the pH level reaches 5.3–4.8. Other strains fail to grow anaerobically. Products of anaerobic fermentation are mainly L(+)-lactic acid plus small amounts of formic and acetic acid and ethanol in the ratio 2:1:1 (reported for 10 strains by McKray and Vaughn (1957)).

In a study on moderately acidophilic thermophilic *Bacillus* strains a number of strains of *B. stearothermophilus* including the type strain were shown to produce considerable amounts of lactic acid, mainly L(+)-lactate (Blumenstock, 1984). Some of the lactic acid producers which had been described as *B. stearothermophilus* because of their phenotype (Daron, 1967; Epstein and Grossowitz, 1969) but could not be allocated to the species *B. stearothermophilus* genetically (Sharp *et al.*, 1980) were considered to be an intermediate group of organisms between *B. stearothermophilus* and *B. coagulans*. In a recent study on 16S rRNA sequence homology (Rainey *et al.*, 1994) these strains were found to be located closely to *Bacillus pallidus*, the strains of which had been isolated from thermophilically treated waste water. The species is characterized through its thermophily (up to 70°C), its slightly alkaline requirements, and strictly aerobic growth. Whether the strains of *B. pallidus* are able to produce lactic acid is not known. Today nine thermophilic *Bacillus* species have been validly published as well as over six not validly described species.

11.3.4 Bacillus popilliae *and* Bacillus lentimorbus

Bacillus popilliae and *B. lentimorbus* both cause the 'milky disease' (named after the milky white appearance of the normally clear haemolymph of the diseased larvae due to very high numbers of spores) of the Japanese beetle larvae (*Popillia japonica*) and were first isolated and described by Dutky (1940). Unlike most other *Bacillus* species both organisms do not grow in nutrient broth. Both species are very fastidious and media for growth include casein digest, yeast extract and carbohydrates. They are much alike in their biochemical characteristics. They lack catalase which is present in all other members of the genus except *B. larvae*, *B. azotoformans*, *B. pulvifaciens* (very weak) and a number of strains of *B. stearothermophilus*.

Both species are described as facultative anaerobes but Pepper and Costilow (1964) failed to obtain significant growth in the complete absence of oxygen. Growth is definitely enhanced in the presence of oxygen. On the other hand there are some indications that during sporulation reduced oxygen concentrations are required.

While studying the utilization of carbohydrates Dutky (1947) and Steinkraus (1957) observed extensive acid production by a drop of pH in cultures and the need for high buffering capacity of media. Resting cells oxidize glucose with the production of CO_2, lactic acid, acetic acid, glycerol, ethanol, and trace amounts of acetoin and acetaldehyde. The first three products are the major ones, and their ratios may be varied by controlling the availability of oxygen. Practically no lactic acid is formed when oxygen is not limited but it may constitute up to 80% of the total acid when oxygen is greatly limited. In the absence of oxygen, however, glucose is not catabolized at all. An overview on the species has been published by Stahly *et al.* (1992).

11.3.5 Bacillus subtilis

Gary and Bard (1952) observed that the Marburg strain of *Bacillus subtilis*, when grown on complex media, fermented glucose to lactic acid but that it was not able to grow anaerobically. Obviously no further work has been concentrated by these or other authors on the species *B. subtilis* with respect to lactic acid.

11.3.6 'Bacillus laevolacticus'

Nakayama and Yanoshi (1967a) isolated from the rhizosphere of various plants catalase positive, acid tolerant, facultatively anaerobic, mesophilic *Bacillus* strains producing lactic acid. These were described as '*B. laevolacticus*' and '*B. racemilacticus*'. They were distinguishable from each other only through the ability of '*B. racemilacticus*' to grow in the presence of 3.5% NaCl and the production of DL-lactic acid instead of D(−)-lactic acid as was shown for '*B. laevolacticus*'. Differentiation of the strains from *B. coagulans* was on the basis of lower growth temperature, stronger acid tolerance, requirement of carbohydrates for growth and stereospecificity of produced lactate, namely DL-lactic acid from '*B. racemilacticus*', D(−)-lactic acid from '*B. laevolacticus*', and L(+)-lactic acid from *B. coagulans*. Collins and Jones stated in 1981 that, according to chemotaxonomical data, the acid-tolerant strains of '*B. laevolacticus*', '*B. racemilacticus*', '*B. myxolacticus*' (a strain which was later allocated to '*B. laevolacticus*'), '*B.*

dextrolacticus' (a strain which was later allocated to *B. coagulans*), and *Sporolactobacillus* could well be grouped as a common separate taxon close to the genus *Bacillus*. Yanagida *et al.* (1987a,b) included a large number of '*B. laevolacticus* and '*B. racemilacticus*' into their studies together with a strain of *Lactobacillus plantarum*. Clustering of certain strains was accomplished but a clear separation from strains of *Sporolactobacillus* was not possible whereas differentiation from the *Lactobacillus* strain was obvious.

In recent studies Pianka (1993) examined 22 catalase-positive, acid-tolerant, facultatively anaerobic lactic acid producing *Bacillus* strains in comparison with a number of strains of phenetically similar *Bacillus* species (*B. coagulans*, *B. smithii*), and *Sporolactobacillus*. According to mol% G+C (43–45%), DNA–DNA homology (72–98%) and phenetic similarity it was revealed that the '*B. laevolacticus*' group was highly homogeneous in phenotypic and genotypic characteristics. Through a number of phenotypic characteristics '*B. laevolacticus*' can clearly be differentiated from other lactic acid producing *Bacillus* and *Sporolactobacillus* species which are *B. coagulans*, *B. smithii*, '*B. racemilacticus*', '*B. vesiculiferous*' and *Sporolactobacillus inulinus*. Two of the strains formerly named '*B. racemilacticus*' had to be reclassified as belonging to the species '*B. laevolacticus*' thereby making the borderline between the two species more clear. Glucose is fermented to D(–)-lactic acid equivalent to 94–99% of sugar consumed; the final pH is 3.8–3.2. No growth occurs in carbohydrate free media and in the presence of glucose also active anaerobic growth is supported. These studies finally proved that the as yet not validly published species '*B. laevolacticus*' indeed merits species rank.

11.3.7 'Bacillus racemilacticus'

In their original publication Nakayama and Yanoshi (1967a) described two isolates (M5, M14) which they assigned to their newly established species '*B. racemilacticus*'. These two and two further strains of these authors (M39, M64) have been examined by Yanagida *et al.* (1987a) and assigned to the clusters '*B. laevolacticus*' (M14, M64), *Sporolactobacillus* (M5) or *B. coagulans* (M39). Subsequent studies (Pianka, 1993) suggest, that today only one strain remains (M5) which definitely deserves the name of '*B. racemilacticus*'. The other strains previously assigned to this species were shown to belong to *B. coagulans* (M39) and '*B. laevolacticus*' (M14, M64). Strain M5 corresponds to the original description given by Nakayama and Yanoshi (1967a) for '*B. racemilacticus*'. It produces DL-lactic acid in a homolactic fermentation of glucose. In most other respects this organism is similar in properties to '*B. laevolacticus*' with the exceptions of its higher growth temperature of 45°C, its less acidic pH in VP broth of 4.3, and its inability to produce acid from mannitol and starch.

11.3.8 'Bacillus vesiculiferous'

The spore forming, Gram-positive organism '*B. vesiculiferous*' was isolated under anaerobic conditions (Trinkunaite *et al.*, 1987). Culturing of the organism under a pO_2 of less than 0.005 atm resulted in rather slow growth (7–25 days to form colonies). Under microaerophilic conditions at a pO_2 of higher than 0.005 atm and up to 0.1 atm culture growth is more rapid (3–5 days until colony formation) but no growth is observed under air. Spore formation seems to be sharply inhibited under anaerobic conditions. With this organism the authors observed the unusual property of formation of extracellular brightly glistening globules which were later identified as gas containing structures.

'Gas-balloon' formation together with microaerophily and absence of catalase and superoxide dismutase led the authors to the proposal of a new species although it has not yet been validly published to date. The organism was clearly distinguishable from other catalase-negative *Bacillus* species, namely *B. larvae*, *B. popilliae*, and *B. lentimorbus*. They described the organism to exhibit a heterolactic fermentation and optimal conditions for growth were pO_2 = 0.03–0.08 atm, pH = 5.0–6.0, temperature = 26–37°C. Although this isolate most closely resembles *Sporolactobacillus* it was placed into the genus *Bacillus* because of presence of cytochromes (cytochrome *b* and *c*), respiratory type of energy-yielding metabolism, and heterofermentative lactic acid production. In the laboratories of the DSM good growth was obtained on MRS medium (Merck) under anaerobic conditions (anaerobic jar, Oxoid).

11.4 *Sporolactobacillus inulinus*

In 1963 Kitahara and Suzuki described a single endospore-forming and catalase-negative strain, which grew best under microaerophilic conditions and showed a homolactic fermentation producing D(–)-lactic acid. They created the new subgenus *Sporolactobacillus* within the Lactobacillaceae. Later, *Sporolactobacillus* was transferred as an independent genus to the family Bacillaceae, comprising only the single species *Sporolactobacillus inulinus* (Kitahara and Lai, 1967).

In the meantime a large number of other sporolactobacilli have been isolated. They have been grouped with *Spl. inulinus* or were described as strains of the new species '*Spl. laevas*', '*Spl. laevas* var. *intermedius*' and '*Spl. racemicus*' (Nakayama and Yanoshi, 1967b; Amemiya and Nakayama, 1980; Yanagida *et al.*, 1987a). The taxonomic position of these strains is still uncertain so that the species names have not been approved and are considered as not validly published. *Sporolactobacillus inulinus* may ferment broths of up to 20 or 30% glucose completely to D(–)-lactic acid

(Gibson and Gordon, 1974; French Patent 1.356.647, 1964). The other strains form DL- and/or D(−)-lactic acid (Nakayama and Yanoshi, 1967b; Yanagida *et al.*, 1987a).

The main criterion to establish the genus *Sporolactobacillus* and its separation from the genus *Bacillus* was the lack of catalase in the former genus. It has been pointed out earlier that several *Bacillus* species also lack this enzyme so that there seems to be no need to separate the two genera. Studies on the phylogenetic relationships, however, support the view of separate positions.

The genus *Sporolactobacillus* was treated in detail in Vol. 1 of this series (Sharma, 1992) and was also reviewed by Norris *et al.* (1981), Kandler and Weiss (1986) and Claus *et al.* (1992).

11.5 Phylogenetic relationships

Phylogenetic studies using the comparison of the oligonucleotide sequences of the 16S rRNA have shown that taxa might group together which, according to classical taxonomy, would not have been found to be closely related. The gram positive bacteria can, phylogenetically, be divided into two major groups, with either a low or a high proportion of mol% G+C in the DNA. The low-G+C group includes the genera *Bacillus*, *Clostridium*, *Staphylococcus*, *Streptococcus*, *Lactococcus*, *Lactobacillus*, *Thermoactinomyces*, and *Mycoplasma* (a wall-less variant of this grouping). The high-G+C group is formed by organisms which can be summed up under the term 'actinomycetes'.

A possible hierarchical structure for the Gram-positive bacteria is given in Woese (1987) and discussed by Priest (1993) with respect to relationship to lactic acid bacteria:

> Assuming the low G+C-content Gram-positive bacteria to be an evolutionary branch of significant depth, it is apparent that endospore forming bacteria encompass all the lactic acid bacteria, the staphylococci, and various other genera. Aspects of endospore formation and structure in *Bacillus* and *Clostridium* are highly conserved, making it likely that sporulation evolved once only. Given the greater phylogenetic depth of the genus *Clostridium* than of the genus *Bacillus*, the progenitor of the low G+C-content Gram-positive line must have been an anaerobic spore former. It follows that genera *Kurthia*, *Pediococcus*, and *Lactobacillus* and other members of the group must have lost the ability to differentiate into spores. Alternatively, but on current evidence more unlikely, sporulation may not have been a single evolutionary event but may have arisen in two phylogenetic lines (*Bacillus* and *Clostridium*).

Bacteria sharing characteristics of two of the named genera, *Lactobacillus* and *Bacillus*, were isolated by Kitahara and Suzuki (1963). The lack of catalase, microaerophilic growth, and lactic acid fermentation accommodated these strains with *Lactobacillus*, but the production of typical endospores and cell walls containing diaminopimelic acid were consistent with the description of *Bacillus*. Accordingly, they were placed in a new genus, *Sporolactobacillus*.

Early results of Fox *et al.* (1977) had shown that *Sporolactobacillus* may well be included into the genus *Bacillus* as it is currently defined. Both genera are definitely included into the same family according to 16S rRNA cataloguing (Stackebrandt and Woese, 1981). According to the latter study it seemed as if *Sporolactobacillus* does not necessarily need a genus of its own, since it grouped closely with *Bacillus coagulans*, which it also resembles physiologically. However, in a more recent phylogenetic study (Ash *et al.*, 1991) *Sporolactobacillus* seems to form a single line of descent far from rRNA cluster 1 of *Bacillus* where *B. coagulans* is found. This is confirmed and extended in the studies of Suzuki and Yamasato (1994) on 16S rRNA sequences of a set of lactic acid producing organisms assigned to the genera *Bacillus* or *Sporolactobacillus*. They revealed that most of the strains grouped more or less around the type strain of *Sporolactobacillus inulinus* forming several subclusters while the only remaining strain truly representing '*B. racemilacticus*' (see above) formed a separate branch and one strain was found within rRNA group 1 (Ash *et al.*, 1991) of *Bacillus*.

Among others, Stackebrandt *et al.* (1987) and Farrow *et al.* (1994) found that by 16S rRNA analyses a considerable number of organisms described as non-spore-formers clustered closely with certain members of the genus *Bacillus*. Another non-spore-forming organism that has recently been found to be phylogenetically closely related to the genus *Bacillus* is the genus *Saccharococcus* (Rainey and Stackebrandt, 1993). Although being a lactic acid producer with which spore formation was not observed, these organisms were not allocated to known taxa of lactic acid bacteria because of their ability to produce catalase, and were put into a separate genus. Its single thermophilic species *Saccharococcus thermophilus* exhibits relatively high sequence homology with the organisms of cluster 5 of Ash *et al.* (1991) where *Bacillus stearothermophilus*, *B. thermoglucosidasius* and *B. kaustophilus* are found.

The genus *Bacillus*, in the definitions as accepted today, is known to be heterogeneous. To date it comprises about 75 validly described species, the mol% G+C contents of which vary from about 34–65%. For the definition of a procaryotic genus, however, it has been proposed that the genetic diversity among the species of a genus should span not more than 10–12 mol% G+C in chromosomal base composition. Its physiological abilities range from acidophilic (growth down to pH 1–2), neutrophilic, and alkaliphilic (growth at up to pH 11) to psychrophilic (growth around zero and not over 25°C) and thermophilic (not at 30°C, well over 70°C), comprising heterotrophs and autotrophs (living with carbon monoxide or dioxide and O_2 and H_2 ('Knallgasbakterien')), N_2-fixing organisms, etc. Being a group of bacteria where mixed acid fermentation is common the property of lactic acid formation seems not to be unusual among the spore-forming aerobic or facultatively anaerobic organisms.

It has been proposed for some time that the genus be broken up to constitute several new taxonomic units. In 1974 Gibson and Gordon discussed a number of names which should be given to groups of *Bacillus* species with specific properties in common. According to numerical taxonomy Priest and co-workers (1988) found that at least five new genera would probably have to be formed to fulfil the requirements of a 'good' genus. By sequence analysis Rössler *et al.* (1991) grouped the genus *Bacillus* into four major clusters (*B. subtilis* cluster, *B. brevis* cluster, *B. alvei* cluster and a *B. cycloheptanicus* branch) which was confirmed and amended by Ash *et al.* (1991) who defined five major groupings and a number of single lineages of species. It has since been discussed that these groupings could be the basis for future splitting of the genus into at least six new genera.

According to these phylogenetic studies the metabolic end product of lactic acid (D(−)-, L(+)- or DL-lactic acid) can at least be found with organisms belonging to three major phylogenetic groups as defined today, namely rRNA group 1 and group 5 of Ash *et al.* (1991) and the once single lineage of *Sporolactobacillus inulinus* which today has shown to be surrounded by a number of strains and species.

11.6 Ecology and habitats of lactic acid producing spore-formers

For most spore-forming bacteria soil is considered to be their natural habitat where they may survive also under a number of adverse conditions due to the resistance of their endospores. From soil, they may contaminate virtually everything – spread about by dust or by any other means – and this is one important reason why the distribution of spore-formers is ubiquitous. Isolation of spore-forming organisms from air-dried or heated samples as is usually done, therefore, may reflect only the presence of metabolically inactive endospores, which had been deposited in this place randomly and survived for shorter or longer time. This is why the presence of spore-formers in a certain habitat may have no ecological background. An example may be the repeated isolation of the obligate thermophile *B. stearothermophilus* from samples of deep sea sediments where it is found in rather high numbers and where the conditions for its growth never prevail. For such reasons, the role of *Bacillus* species in ecosystems is generally poorly understood.

Like most other spore-forming bacteria, strains of all lactic acid spore-formers have been isolated from different soils as well as from a diversity of other samples. The thermophilic species are especially involved in the decomposition of plant litter accumulated in masses or in composting of any organic material. After mesophilic microorganisms have raised the temperature in the piles by their metabolic activities, the development of facultative or obligate thermophiles usually can be observed. Temperature in the piles finally increases to about 75°C where *B. stearothermophilus* has its growth maximum. After activities of the latter bacterium have ceased, temperature in the pile will decrease again. A similar process can be observed during fermentation of tobacco or cocoa beans, where *B. coagulans* and *B. stearothermophilus* may become a significant percentage of the total bacterial population. Up to now, it is difficult to ascribe a role to the thermophilic organisms in the latter fermentation (Ostovar and Keeney, 1973). Another well-known example of the activities of thermophiles is the phenomenon of spontaneous heating of hay which bears a certain residual water content and which is stored in barns. As in compost the biological heating process starts with the development of mesophilic microorganisms, followed by the growth of thermophiles. Shortly above 75°C biological heating within the hay is stopped but may be continued by chemical processes which finally may lead to spontaneous self ignition of the hay. In the first half of this century in Europe this was a relatively frequent reason why barns burnt down and considerable financial losses were counted.

Bacillus coagulans has been originally isolated from spoiled canned milk. Additional isolates were from fresh or dried soil, hot springs, rhizosphere, decomposing plant litter, compost, canned food, tomato juice, fresh or spoiled milk, cheese, silage, sugar and sugar beet extraction juice, and diverse other samples like medicated cream (Nakamura *et al.*, 1988).

Bacillus smithii, which was formerly considered to belong to *B. coagulans*, can be found in soil and in all habitats in which *B. coagulans* has been found. *Bacillus smithii* has a maximum growth temperature 5°C higher than *B. coagulans* and is therefore often detected in extraction juice of sugar beets together with *B. stearothermophilus* (Klaushofer and Hollaus, 1970).

The first named and described strains of the obligate thermophilic species *B. stearothermophilus* were isolated from cans of spoiled corn and string beans. The species is widely distributed and can be isolated from soil, hot springs, extraction fluids in sugar factories or sugar, composts, canned or dried foods or milk, and from a number of habitats with permanent low temperatures.

Bacillus popilliae and *B. lentimorbus* are found only in association with a specific disease of insects of the order Coleoptera and have been first isolated from the hemolymph of larvae of the Japanese beetle. Strains of the species can be found also in soil surrounding diseased larvae. Both species have been effectively used for the biological control of these beetles (Stahly *et al.*, 1992).

'*Bacillus laevolacticus*' and '*B. racemilacticus*' have been isolated only from the rhizosphere of different plants in Japan and from soil samples from Japan and Southeast Asia (Nakayama and Yanoshi, 1967a; Amemiya and Nakayama, 1980). They may be present also in soils of other regions.

The single strain of '*B. vesiculiferous*' had been isolated under anaerobic conditions from the intestines of the crayfish *Pacifastacus leniusculus* (Trinkunaite *et al.*, 1987).

Whereas the original strain of *Spl. inulinus* has been isolated from a sample of chicken feed (Kitahara and Suzuki, 1963), additional strains of the species are from soil or sea mud (Amemiya and Nakayama, 1980). The incidence of *Sporolactobacillus* in the environment seems to be very low. From a total of 700 samples of soil, foods and feed Doores and Westhoff (1983) were able to isolate only two strains. '*Sporolactobacillus laevas*', '*Spl. leavas* var. *intermedius*', and '*Spl. racemicus*' have been isolated up to now only from the rhizosphere of wild plants (Nakayama and Yanoshi, 1967b).

11.6.1 Lactic acid spore-formers and foods

According to their acidity, foods are often classified into three broad groups: (1) 'high acid' foods with a pH lower than 3.7, (2) 'acid' foods showing a pH of 3.7 to 4.6, and (3) 'low acid' foods with a pH above 4.6. Because heat-resistant spores may germinate freely when these foods have been canned, temperatures well above 100°C are employed to ensure sterility.

Whereas high acid foods generally prevent the growth of spore forming bacteria, *B. coagulans* is known to multiply in acid and low acid canned foods. Since only souring but no gas is produced by the bacterium the type of spoilage is known as 'flat sour' spoilage. As mentioned earlier, strains of *B. smithii* were classified until recently as *B. coagulans*. They differ mainly in their slightly higher maximum growth temperature.

Bacillus stearothermophilus is less aciduric than *B. coagulans* and therefore may cause flat sour spoilage only in low acid foods. Due to their extreme heat resistance spores of *B. stearothermophilus* may survive the sterilization process. Spores of this thermophile, however, will not develop except at temperatures above 30°C (in pure culture the minimum growth temperature of different strains is 30–45°C, the maximum temperature 65–75°C). Spoilage of canned food by *B. stearothermophilus* therefore is, at least in temperate climates, not often observed. Other lactic acid spore-formers are not known as food spoiling organisms.

11.6.2 Lactic acid spore-formers in sugar factories

In diffusion water used for the extraction of sugar beets thermophilic bacteria may reach numbers up to 10^7/ml within a few hours at 70°C. They grow in the hot pressed juices, in hot wells, in pipes and in sediment or storage tanks. At 65°C also other thermophilic spore-formers are present which were considered earlier as highly thermophilic variants of *B. coagulans* (Klaushofer and Hollaus, 1970; Klaushofer *et al.*, 1971), but are now known to belong to the new species *B. smithii* (Nakamura *et al.*, 1988). In the extraction fluid the thermophilic bacteria ferment sugar to lactic acid. The amount of sugar destroyed may be up to 1% of sucrose extracted (Stark *et al.*, 1953). Spores of thermophilic *Bacillus* species present in the final sugar may be of major concern to the canning industry because they are incorporated in canned vegetables if sugar is added to the brines.

11.7 Applications and products

Most *Bacillus* species are known to degrade a series of natural polymers and may themselves be rich sources of different extracellular enzymes. Strains of *B. coagulans* and *B. stearothermophilus* have been used to study thermostable enzymes like amylase, pullulanase, xylanase, lipase, proteinase, α-glucosidase, cyclodextrin glycosyltransferase, glucose isomerase, 1,4-β-galactosidase as well as proteases, DNA polymerase and restriction endonucleases. A number of these thermostable enzymes are produced commercially (Kristjansson, 1989; Priest, 1989; Sharp *et al.*, 1992; Zukowski, 1992).

Strains of the groups '*B. racemilacticus*' and '*B. laevolacticus*' produce mainly D(−)-lactic acid, which may be used in the food industry as well as in cosmetic or pharmaceutical products (de Boer *et al.*, 1990). With species which produce L(+)-lactic acid a racemase may be induced through accumulation of this isomer so that the L(+)-form is converted into the D(−)-form until equilibrium is reached. D(−)-Lactic acid formers produce this enantiomer exclusively. Besides this metabolic end-product these organisms produce polysaccharides which have been examined for their possible antitumour activity (Amemiya and Nakayama, 1980).

11.7.1 Strains for antibiotic testing and quality control

For the detection of penicillins and other inhibitory substances in milk two strains of *B. stearothermophilus* (ATCC 10149/DSM 6790 and DSM 1550) are in use (Galesloot and Hassing, 1962; DIN, 1982; IDF, 1970). Spore preparations of two other strains of this species are used as control strains for steam sterilization (Greene, 1992; US Pharmacopeia XXI, 1985).

11.7.2 Genetic engineering

The observation that a small (2178 bp) cryptic plasmid, pWVO1, from *Lactococcus lactis* subsp. *cremoris* Wg2, was able to replicate in *B. subtilis* (Vosman and Venema, 1983) has led to the construction of a series of small versatile vectors from this plasmid (Kok *et al.*, 1984; Van Der Vossen *et al.*, 1985) and from the highly related plamid pSH71, isolated from *Lactococcus lactis* subsp. *lactis* NCDO 712 (de Vos, 1987). The basic pWVO1-derived vector is pGK 12 (Kok *et al.*, 1984).

An important property of pWVO1 and pSH71, and one which they share with pMV158 (and its deletion derivative pLS1) from *Streptococcus agalactiae* (Lacks *et al.*, 1986), is their promiscuous nature (Kok *et al.*, 1984; de Vos, 1987). pGK12 has been shown to replicate in a large variety of Gram-positive and Gram-negative bacteria, including several *Bacillus* species (*B. subtilis*, *B. cereus*, *B. thuringiensis*) all species of lactic acid bacteria, various streptococci, *Enterococcus faecalis*, *Staphylococcus aureus*, *Clostridium*, *Pseudomonas*, several phototrophic bacteria, *Listeria* and *E. coli*.

Rabinovitch *et al.* (1985) have used the streptococcal plasmid pSM19035 for efficient cloning in *B. subtilis*. This large plasmid (27 kb) can transform competent *B. subtilis* as a monomer.

11.8 Descriptions of the spore-forming, lactic acid producing bacteria

11.8.1 The genus Sporolactobacillus

Members of this genus are Gram-positive, rods, sparse-peritrichously flagellated, have ellipsoidal endospores, swelling the sporangium, are aerobic to microaerophilic and catalase-negative.

11.8.1.1 Sporolactobacillus inulinus. This has been thoroughly described in the first volume of this series by Sharma (1992).

11.8.1.2 Others. A number of described but not validly published species are '*Spl. laevas*', '*Spl. laevas* var. *intermedius*', and '*Spl. racemicus*'.

11.8.2 The genus Bacillus

These are mostly Gram-positive rods, peritrichously flagellated, endospores ellipsoidal or round, swelling or not swelling the sporangium, aerobic to facultatively anaerobic, mostly catalase positive.

The genus now comprises 75 recognized species, of which three have recently been transferred to a newly formed genus *Alicyclobacillus* (Wisotzkey *et al.*, 1992) to harbour the thermoacidophilic species with ω-alicyclic fatty acids in their cell membranes. Another spore forming genus recently described is *Amphibacillus* (Niimura *et al.*, 1990). The only species of this genus is alkaliphilic and is reported to combine properties described for *Bacillus* as well as those described for *Clostridium* and therefore is allocated an intermediate position between these two genera.

Species of the genus *Bacillus* of which it is definitely reported that they produce lactic acid are five recognized species (*B. coagulans*, *B. smithii*, *B. stearothermophilus*, *B. popilliae* and *B. lentimorbus*) and three not validly published species ('*Bacillus laevolacticus*', '*Bacillus racemilacticus*' and '*Bacillus vesiculiferous*'). Recent studies (Pianka, 1993) indicate that species rank should be conferred to '*Bacillus laevolacticus*'.

11.8.2.1 Bacillus coagulans sensu stricto. co.a'gu.lans. L.Part.adj. *coagulans*: curdling, coagulating. Cell width 0.4–0.8 μm, peritrichously flagellated. Terminally or subterminally located oval or cylindrical spores are produced that measure 0.6–0.8 by 1.3–1.7 μm. Sporangia slightly or not swollen. Young cells Gram-positive, with increasing age the cells become Gram-variable and finally Gram-negative. Potassium hydroxide and aminopeptidase tests are negative. Vegetative cells contain diaminopimelic acid in their cell walls. MK7 is the main menaquinone. The G+C content of the DNA ranges from 45 to 47% (T_m) (type strain 46%). *Bacillus coagulans* is chemo-organotrophic, facultatively anaerobic, facultative thermophile and cells grow variably at 60°C but not at 65°C. Growth at pH 5.7 (Sabouraud), and 4.5 and 7.7 in NB, no growth in the presence of 3% NaCl or lysozyme. Growth in the presence of 0.02% azide. Cells produce acetylmethylcarbinol, pH values in Voges–Proskauer broth are 4.0–4.4. No production of dihydroxyacetone, H_2S, or indole. Catalase is formed, but oxidase is not formed. Starch, pullulan, DNA and hippurate are hydrolysed. No hydrolysis of casein, chitin, egg yolk lecithin, or gelatin. No decomposition of arginine, lysine, ornithine, phenylalanien, tyrosine, or urea. Milk is acidified and coagulated. Fermentation of D-fructose, D-glucose, and trehalose. No fermentation of adonitol, L-arabitol, dulcitol, D- and L-fucose, inulin, 2- and 5-ketogluconate, D-lyxose, melezitose, α-methyl-D-mannoside, *β*-methyl-xyloside, L-sorbose, D-tagatose, and L-xylose according to API 50CHB system. Products of fermentation of glucose are mainly L(+)-lactic acid plus small amounts of 2,3-butanediol, acetoin, acetic acid and ethanol. Final pH is 4.0–5.0, varying with the strain, medium and conditions. Fermentable sugars increase the amount of aerobic growth on agar, and they support anaerobic growth. Aciduric; for initiation of growth optimum pH level is close to 6; minimum 4.0–5.0 in different strains. Type strain: ATCC 7050, CCM 2013, DSM 1, NCIB 9365, NCTC 10334.

11.8.2.2 Bacillus smithii (*Nakamura* et al., *1988*). smi'thi.i. L.gen.n. *smithii* named after Nathan R. Smith, an American bacteriologist. Width of vegetative rods is 0.8–1.0 by 5.0–6.0 μm. Cells are motile by peritrichous flagellation. Oval or cylindrical endospores (0.6–0.8 by 1.3–1.5 μm) are produced terminally or subterminally in non-swollen sporangia or in some cases slightly swollen sporangia. Agar colonies are non-pigmented, translucent, thin, smooth, circular, and entire and measure about 2 mm in diameter. The cell walls of vegetative cells contain diaminopimelic acid. The main menaquinone is MK7; MK6 is also present. The mol% G+C ranges from 38.1 to 40.4 (Bd) for 21 strains and from 38.7 to 39.7 (T_m) for five strains. *Bacillus smithii* is chemo-organotrophic and facultatively thermophilic. Growth is at 25–60°C; most strains grow at 65°C. Facultatively anaerobic. Growth at pH 5.7 in Sabouraud, no growth at pH 4.5 or 7.7 in nutrient broth. No growth in the presence of 3% NaCl, 0.001% lysozyme, or 0.02% azide. Acetylmethylcarbinol, H_2S, and indole are not produced. In Voges–Proskauer broth, the pH ranges between 4.3 and 4.7. Catalase and oxidase are produced. Nitrate is not reduced to nitrite. DNA and hippurate are hydrolysed. Casein, chitin, egg yolk lecithin, and gelatin are not hydrolysed. Starch is weakly hydrolysed. *o*-Nitrophenyl-*ß*-galactosidase is not produced. Acetate, fumarate, malate, and succinate are utilized. Arginine dihydrolase, lysine and ornithine decarboyxlases, phenylalanine and tryptophan deaminases, and urease are not produced. Tyrosine is not decomposed. Litmus milk is not coagulated or acidified, but is usually alkalized. Acid but no gas is produced from D-fructose, D-glucose, and trehalose. When lactic acid is produced, mainly the D(+)-lactic acid can be detected. Type strain: NRRL NRS-173, DSM 4216, IFO 15311.

11.8.2.3 Bacillus stearothermophilus. The species *B. stearothermophilus* as described today is evidently a mixture of several thermophilic species. As shown in the numerical analysis of White *et al.* (1994) a preliminary description of the species *sensu stricto* may be as follows. Catalase- and oxidase-negative, acid from dextrin, fructose, glucose, glycerol, maltose, mannose, no acid from adonitol, arabinose, dulcitol, erythritol, inositol, inulin, lactose, rhamnose, hydrolysis of pullulan, starch and tributyrin, no hydrolysis of carboxymethylcellulose, hippurate, pectin, pustulan, RNA, tyrosine (and no pigment from tyrosine), xylan, no growth in the presence of 0.02% sodium azide, mol% G+C 52–59 (one strain exhibiting high DNA–DNA homology to the type strain has been reported to have a G+C base ratio of 43%). Type strain: ATCC 12980, CCM 2062, DSM 22, IAM 11062, NCIB 8923, NCTC 10339.

11.8.2.4 'Bacillus laevolacticus'. lae.vol.lac.'ti.cus. M.L.adj. *laevolacticus* referring to D(−)-lactic acid; the organisms produce mainly D(−)-lactic acid. Young cells are gram positive. Cell width 0.4–0.7 μm. Motile through means of few polarly and laterally inserted flagella. Spores ellipsoid measuring 0.6–0.8 μm in width to 0.8–1.2 μm and swelling the sporangium. On glucose–$CaCO_3$ agar most of the strains appear as typical white pin point colonies, about 2 mm in diameter, with a clear halo around the colony produced by acid production. The peptidoglycan side chains are directly linked via *m*-diaminopimelic acid. MK7 is the main (>90%) menaquinone. Ubiquinones not present. The mol% G+C of the DNA spans from 43 to 45% (T_m). The cellular fatty acids are predominantly composed of branched-chain anteiso C15:0 and C17:0 fatty acids. The DNA base composition of the type strain is 43%. *Bacillus laevolacticus* is chemo-organotrophic and does not grow in NB or NA. Glucose or other carbohydrates are needed for growth. It is facultatively anaerobic, catalase-positive, oxidase-negative and mesophilic. Maximum temperature for growth 40°C. Aciduric, growth at pH 4.5. Voges–Proskauer positive; pH in VP medium is 3.8–4.0. No growth in the presence of lysozyme or 5% NaCl. Hydrolysis of starch and pullulan. Citrate and propionate not utilized. No hydrolysis of gelatin, DNA, tyrosine, or casein. No production of indole. Nitrate not reduced to nitrite. Egg-yolk lecithinase negative. Deamination of phenylalanine negative. Acid production from glucose and mannitol but not from arabinose or xylose. No gas from glucose. From glucose predominantly D(−)-lactic acid is produced. Type strain: M8, deposited as ATCC 23492, DSM 442, IAM 12321, NCIB 10269.

11.8.2.5 'Bacillus racemilacticus'. Description largely as for '*B. laevolacticus*' with the following exceptions (data based on only one strain). Vegetative cells produce 'banana'-shaped spores (best seen in young stages) being 0.8 μm wide and 1.2–1.3 μm long. Maximum temperature for growth is 45°C. Inability to produce acid from mannitol and starch. In VP broth the pH is 4.3. Mol% G+C of the DNA is 37%. Relevant amounts of DL-lactic acid are produced. Reference strain: M5, ATCC 23497, DSM 2309, JCM 2518, NCIMB 10275.

11.8.2.6 'Bacillus vesiculiferous' *(Trinkunaite* et al., *1987).* ve.si.cu.li.'fe.rous, a bacillus carrying vesicles. Morphology: Gram-positive, straight or curved rods with rounded ends, 0.8–1.2 × 3–6 μm, rarely in chains, motile with lateral flagella. Spores oval (0.6–0.8 × 0.8–1.0 μm) central or paracentral. Globules of poly-*β*-hydroxybutyrate are present in the cytoplasm. The cell wall is Gram-positive in structure and consists of one layer. Glistening gas balloons are formed on cells during their growth at low pO_2 on solid media containing thioglycolate and casein hydrolysate, but no carbohydrates. At $pO_2 < 0.005$ atm colonies are round, snow white, fluffy, convex. At pO_2 optimal for the growth, colonies are round convex, yellowish-brown, with a glossy-matt surface. Microaerophilic, optimal pO_2 for growth on solid media is 0.03–0.08 atm. Cytochromes *b* and *c* present. Catalase and superoxide dismutase are not formed. Gelatin, casein and starch are not hydrolysed. On media containing peptone (or casein hydrolysate) plus carbohydrate growth is possible also under anaerobic conditions. The products of glucose fermentation are lactate, acetate, and propionate. Acetoin is not formed. Glucose, fructose, galactose, maltose, arabinose, and trehalose are fermented (with formation of acids but no gases). Nitrate is reduced to nitrite. Hippurate is not hydrolysed. Tyrosine is not decomposed. Phenylalanine is not deaminated. Indole is formed on media containing tryptophan. VP broth and citrate–salt medium are not alkalified. Growth is inhibited by 0.001% lysozyme, 10 μg/ml penicillin or 7% NaCl. Growth occurs within the pH range of 4.0–7.0 (optimum pH 5.0–6.0) and the temperature range of 18–40°C (optimum 26–37°C). The G+C content (T_m) in the DNA is 39 mol%. Reference strain: GB–1, DSM 5538.

References

Allen, M.B. (1953) The thermophilic aerobic sporeforming bacteria. *Bacteriology Reviews*, **17**, 125–173.

Amemiya, Y. and Nakayama, O. (1980) Polysaccharide formation by spore-bearing lactic acid bacteria. *Journal of General and Applied Microbiology*, **26**, 159–166.

Anderson, A.A. and Werkman, C.H. (1940) Description of a dextrolactic acid forming organism of the genus *Bacillus*. *Iowa State College Journal of Science*, **14**, 187–194.

Ash, C., Farrow, J.A.E., Wallbanks, S. and Collins, M.D. (1991) Phylogenetic heterogeneity of the genus *Bacillus* revealed by comparative analysis of small-subunit-ribosomal RNA sequences. *Letters in Applied Microbiology*, **13**, 202–206.

Bartholomew, J.W. and Paik, G. (1966) Isolation and identification of obligate thermophilic sporeforming bacilli from ocean basin cores. *Journal of Bacteriology*, **92**, 635–638.

Becker, M.E. and Pederson, C.S. (1950) The physiological characters of *Bacillus coagulans* (*Bacillus thermoacidurans*). *Journal of Bacteriology*, **59**, 717–725.

Bennet, J.F. and Canale-Parola, E. (1965) The taxonomic status of *Lineola longa*. *Archiv für Mikrobiologie*, **52**, 197–205.

Berry, R.N. (1933) Some new heat resistant acid tolerant organisms causing spoilage in tomato juice. *Journal of Bacteriology*, **25**, 72–73.

Blumenstock, I. (1984) *Bacillus coagulans* HAMMER 1915 und andere thermophile oder mesophile, säuretolerante *Bacillus*–Arten–eine taxonomische Untersuchung. PhD thesis, Universität Göttingen, Germany.

Bovallius, A. and Zacharias, B. (1971) Variations in the metal content of some commercial media and their effect on microbial growth. *Applied Microbiology*, **22**, 260–262.

Bradley, D.E. and Franklin, J.G. (1958) Electron microscopy survey of the surface configuration of spores of the genus *Bacillus*. *Journal of Bacteriology*, **76**, 618–630.

Charney, J.C., Fischer, W.P. and Hegarty, C.P. (1951) Manganese as an essential element for sporulation in the genus *Bacillus*. *Journal of Bacteriology*, **62**, 145–148.

Claus, D., Fritze, D. and Kocur, M. (1992) Genera related to the genus *Bacillus*. In *The Prokaryotes*, Vol. II, 2nd edn (eds Balows, A., Trüper, H.G., Dworkin, M., Harder, W. and Schleifer, K.-H.) Springer-Verlag, New York, USA, pp. 1769–1791.

Cohn, F. (1872) Untersuchungen über Bakterien. *Beiträge zur Biologie der Pflanzen*, **1**, 127–224.

Cohn, F. (1876) Untersuchungen über Bakterien, IV. Beiträge zur Biologie der Bacillen. *Beiträge zur Biologie der Pflanzen*, **2**, 249–277.

Collins, M.D. and Jones, D. (1981) Distribution of isoprenoid quinone structural types in bacteria and their taxonomic implications. *Microbiological Reviews*, **45**, 316–354.

Daron, H.H. (1967) Occurrence of isocitrate lyase in a thermophilic *Bacillus* species. *Journal of Bacteriology*, **93**, 703–710.

de Bary, A. (1884) *Vergleichende Morphologie und Biologie der Pilze, Mycetozoen und Bakterien*. Wilhelm Engelmann, Leipzig, Germany.

de Boer, J., Teixeira de Mattos, M.J. and Neijssel, O.M. (1990) D(−)-Lactic acid production by suspended and aggregated continuous cultures of *Bacillus laevolacticus*. *Applied Microbiological Biotechnology*, **34**, 149–153.

de Vos, W.M. (1987) Gene cloning and expression in lactic streptococci. *FEMS Microbiological Reviews*, **46**, 281–95.

DIN (1982) Nachweis von Hemmstoffen in Milch. DIN 10182 Teil 3. Beuth Verlag, Berlin, Köln, Germany.

Donk, P.J. (1920) A highly resistant thermophilic organism. *Journal of Bacteriology*, **5**, 373–374.

Doores, S. and Westhoff, D.C. (1983) Selective method for the isolation of *Sporolactobacillus* from food and environmental sources. *Journal of Applied Bacteriology*, **54**, 273–280.

Dutky, S.R. (1940) Two new spore-forming bacteria causing milky diseases of Japanese beetle larvae. *Journal of Agricultural Research*, **61**, 57–68.

Dutky, S.R. (1947) Preliminary observations on the growth requirements of *Bacillus popilliae* Dutky and *Bacillus lentimorbus* Dutky. *Journal of Bacteriology*, **54**, 267.

Epstein, I. and Grossowitz, N. (1969) Prototrophic thermophilic bacillus: isolation, properties, and kinetics of growth. *Journal of Bacteriology*, **99**, 414–417.

Farrow, J.A.E., Wallbanks, S. and Collins, M.D. (1994) Phylogenetic interrelationships of round-spore-forming bacilli containing cell walls based on lysine and the non-spore-forming genera *Caryophanon*, *Exiguobacterium*, *Kurthia*, and *Planococcus*. *International Journal of Systematic Bacteriology*, **44**, 74–82.

Fox, G.E. Pechmann, K.J. and Woese, C.R. (1977) Comparative cataloguing of 16S ribosomal ribonucleic acid: molecular approach to procaryotic systematics. *International Journal of Systematic Bacteriology*, **27**, 44–57.

Galesloot, Th.E. and Hassing, F. (1962) Een snelle en gevoelige methode om met papierschijfjes penicilline in melk aan te tonen. *Netherlands Milk Dairy Journal*, **16**, 89–95.

Gary, N.D. and Bard, R.C. (1952) Effect of nutrition on the growth and metabolism of *Bacillus subtilis*. *Journal of Bacteriology*, **64**, 501–512.

Gibson, T. and Gordon, R.E. (1974) *Bacillus*. In *Bergey's Manual of Determinative Bacteriology*, 8th edn (eds Buchanan, R.E. and Gibbons, N.E.). Williams and Wilkins, Baltimore, MD, USA, pp. 529–550.

Gordon, R.E., Haynes, W.C. and Pang C.H.-N. (1973) *The Genus* Bacillus. US Department of Agriculture, Washington, DC, USA.

Greene, V.W. (1992) Sterility assurance concepts, methods and problems. In *Disinfection, Preservation and Sterilization* (eds Russell, A.D., Hugo, W.B. and Ayliffe, G.A.J.). Blackwell Scientific Publications, Oxford, UK, pp. 605–624.

Hammer, B.W. (1915) Bacteriological studies on the coagulation of evaporated milk. Iowa Agricultural Experimental Station Research Bulletin, **19**, 119–131.

Hussong, R.V. and Hammer, B.W. (1928) A thermophile coagulating milk under practical conditions. *Journal of Bacteriology*, **15**, 179–188.

Hüppe, F. (1886) *Die Formen der Bakterien und ihre Beziehungen zu den Gattungen und Arten, Wiesbaden*. Verlag C.W. Kreidel VIII, Germany.

IDF (1970) *Detection of Penicillin in Milk by a Disk Assay Technique*. (International Standard FIL-IDF 57:1970). International Dairy Federation, Brussels, Belgium.

Kalakoutskii, L.V. and Dobritsa, S.V. (1984) Effect of nutrition on cellular differentiation in prokaryotic microorganisms and fungi. In *CRC Handbook of Microbiology*, Vol. VI (eds Laskin, A.I. and Lechevalier, H.A.). CRC Press, Boca Raton, Fl, USA, pp. 17–121.

Kandler, O. and Weiss, N. (1986) Genus *Sporolactobacillus* Kitahara and Suzuki 1963. In *Bergey's Manual of Systematic Bacteriology*, Vol. 2 (eds Sneath, P.H.A., Mair, N.S., Sharpe, M.E. and Holt, J.G.). Williams and Wilkins, Baltimore, MD, USA, pp. 1139–1141.

Kitahara, K. and Lai, C.-L. (1967) On the spore formation of *Sporolactobacillus inulinus*. *Journal of General and Applied Microbiology*, **13**, 197–203.

Kitahara, K. and Suzuki, J. (1963) *Sporolactobacillus* nov. subgen. *Journal of General and Applied Microbiology*, **9**, 59–71.

Klaushofer, H. and Hollaus, F. (1970) Zur Taxonomie der hochthermophilen, in Zuckerfabriksäften vorkommenden aeroben Sporenbildner. *Zeitschrift der Zuckerindustrie*, **9**, 465–470.

Klaushofer, H., Hollaus, F. and Pollack, G. (1971) Microbiology of beet sugar manufacture. *Process Biochemistry*, **6**,(6), 39–41.

Koch, R. (1876) Die Aetiologie der Milzbrandkrankheit, *Beiträge zur Biologie der Pflanzen*, **2**, 277–310.

Kok, J., Van der Vossen, J.M.B.M. and Venema, G. (1984). Construction of plasmid cloning vectors for lactic streptococci which also replicate in *Bacillus subtilis* and *Escherichia coli. Applied and Environmental Microbiology*, **48**, 726–731.

Kristjansson, J.K. (1989) Thermophilic organisms as source of thermostable enzymes. *Trends in Biotechnology*, **7**, 349–353.

Lacks, S.A., Lopez, P., Greenberg, B. and Espinosa, M. (1986) Identification and analysis of genes for tetracycline resistance and replication region. *Journal of Bacteriology*, **157**, 445–453.

Logan, N.A. and Berkeley, R.C.M. (1984) Identification of *Bacillus* strains using the API system. *Journal of General and Microbiology*, **130**, 1871–1882.

McKray, G.A. and Vaughn, R.H. (1957) The fermentation of glucose by *Bacillus stearothermophilus. Food Research*, **22**, 494–500.

Nakamura, L.K., Blumenstock, I. and Claus, D. (1988) Taxonomic study of *Bacillus coagulans* Hammer 1915 with a proposal for *Bacillus smithii* sp. nov. *International Journal of Systematic Bacteriology*, **38**, 63–73.

Nakayama, O. and Yanoshi, M. (1967a). Spore-bearing lactic acid bacteria isolated from rhizosphere. I. Taxonomic studies on *Bacillus laevolacticus* nov. sp. and *Bacillus racemilacticus* nov. sp. *Journal of General and Applied Microbiology*, **13**, 139–153.

Nakayama, O. and Yanoshi, M. (1967b) Spore-bearing lactic acid bacteria isolated from rhizosphere. II. Taxonomic studies on the catalase negative strains. *Journal of General and Applied Microbiology*, **13**, 155–165.

Niimura, Y., Koh, E., Yanagida, F., Suzuki, K.-I., Komagata, K. and Kozaki, M. (1990) *Amphibacillus xylanus* gen. nov., sp. nov., a facultatively anaerobic sporeforming xylan-digesting bacterium which lacks cytochrome, quinone, and catalase. *International Journal of Systematic Bacteriology*, **40**, 297–301.

Norris, J.R., Berkeley, R.C.W., Logan, N.A. and O'Donnell, A.G. (1981) The genera *Bacillus* and *Sporolactobacillus*. In *The Prokaryotes* (eds Starr, M.P., Stolp, H., Trüper, H.G., Balows, A. and Schlegel, H.G.). Springer-Verlag, Berlin, Germany, pp. 1711–1742.

Olsen, E. (1944) En sporedannende maelkesyrebakterie *Lactobacillus cereale* (nov. sp.). *Kemisk Maandesblad, den Nordisk Handelsblad Kemisk Industri*, **25**, 125–130.

Ostovar, K. and Keeney, P.G. (1973) Isolation and characterization of microorganisms involved in the fermentation of cocoa beans. *Journal of Food Science*, **38**, 611–617.

Pepper, R.E. and Costilow, R.N. (1964) Glucose catabolism by *Bacillus popilliae* and *Bacillus lentimorbus. Journal of Bacteriology*, **87**, 303–310.

Pianka, S. (1993) Taxonomische Untersuchung an zwei neuen *Bacillus*-Arten: *Bacillus aminovorans* und *Bacillus laevolacticus*, PhD thesis, Technische Universität Braunschweig, Brannschweig, Germany.

Priest, F.G. (1989) Products and applications. In *Bacillus* (ed. Harwood, C.R.). Plenum Press, New York, USA, pp. 293–320.

Priest, F.G. (1993) Systematics and ecology of *Bacillus*. In Bacillus subtilis *and other Gram-Positive Bacteria* (eds Sonenshein, A.L., Hoch, J.A. and Losick, R.). American Society for Microbiology, Washington, DC, USA, pp. 3–16.

Priest, F.G., Goodfellow, M. and Todd, C. (1988) A numerical classification of the genus *Bacillus. Journal of General Microbiology*, **134**, 1847–1882.

Rabinovich, P.M., Haykinson, M.Y., Arutyunova, L.S., Yomantas, Y.V. and Stepanov, A.I. (1985) The structure and source of plasmid DNA determine the cloning properties of vectors for *Bacillus subtilis. Basic Life Sciences*, **30**, 635–56.

Rainey, F.A. and Stackebrandt, E. (1993) Phylogenetic analysis for the relationship of

Saccharococcus thermophilus to *Bacillus stearothermophilus*. *Systematic Applied Microbiology*, **16**, 224–226.

Rainey, F.A., Fritze, D. and Stackebrandt, E. (1994) The phylogenetic diversity of thermophilic members of the genus *Bacillus* as revealed by 16S rDNA analysis. *FEMS Microbiology Letters*, **115**, 205–212.

Renco, P. (1942) Richerce su un fermento lattico sporingo (*Bacillus thermoacidificans*). *Annals of Microbiology*, **2**, 109–114.

Rössler, D., Ludwig, W., Schleifer, K.H., Lin, C., McGill, T.J., Wisotzkey, J.D., Jurtshuk Jr, P. and Fox, G.E. (1991) Phylogenetic diversity in the genus *Bacillus* as seen by 16S rRNA sequencing studies. *Systematic Applied Microbiology*, **14**, 266–269.

Russell, A.D. (1982) *The Destruction of Bacterial Spores*. Academic Press, London, UK.

Sarles, W.B. and Hammer, B.W. (1932) Observations on *Bacillus coagulans*. *Journal of Bacteriology*, **23**, 301–314.

Seki, T., Chung, C.-K., Mikami, H. and Oshima, V. (1978) Deoxyribonucleic acid homology and taxonomy of the genus *Bacillus*. *International Journal of Systematic Bacteriology*, **28**, 182–189.

Sharma, V.K. (1992) Sporolactobacilli. In *The Lactic Acid Bacteria*, Vol. 1 (ed. Wood, B.J.B.). Elsevier Applied Science, London, UK, pp. 431–446.

Sharp, R.J. Brown, K.J. and Atkinson, A. (1980) Phenotypic and genotypic characterization of some thermophilic species of *Bacillus*. *Journal of General Microbiology*, **117**, 201–210.

Sharp, R.J., Riley, P.W. and White, D. (1992) Heterotrophic thermophilic bacilli. In *Thermophilic Bacteria* (ed. Kristjansson, J.K.). CRC Press, Boca Raton, FL, USA, pp. 19–50.

Smith, N.R., Gordon, R.E. and Clark, F.E. (1952) *Aerobic Sporeforming Bacteria* (Agriculture monograph no. 16). US Department of Agriculture, Washington, DC, USA.

Stackebrandt, E. and Woese, C.R. (1981) The evolution of prokaryotes. *Symposium of the Society of General Microbiology*, **32**, 1–32.

Stackebrandt, E., Ludwig, W., Weizenegger, M., Dorn, S., McGill, T.J., Fox, G.E., Woese, C.R., Schubert, W. and Schleifer, K.-H. (1987) Comparative 16S rRNA oligonucleotide analyses and murein types of round-spore-forming bacilli and non-spore-forming relatives. *Journal of General Microbiology*, **133**, 2523–2529.

Stahly, D.P., Andrews, R.E. and Yousten, A.A. (1992) The genus *Bacillus*-insect pathogens. In *The Prokaryotes*, Vol. II, 2nd edn (eds Balows, A., Trüper, H.G., Dworkin, M., Harder, W. and Schleifer, K.-H.). Springer-Verlag, New York, USA, pp. 1697–1745.

Stark, J.B. Goodban, A.E. and Owens, H.J. (1953) Beet sugar liquors. Determination and concentration of lactic acid in processing liquor. *Journal of Agricultural and Food Chemistry*, **1**, 564–566.

Steinkraus, K.H. (1957) Study on the milky disease organism. II. Saprophytic growth of *Bacillus popilliae*. *Journal of Bacteriology*, **74**, 625–632.

Suzuki, T. and Yamasato, K. (1994) Phylogeny of spore-forming lactic acid bacteria based on 16S rRNA gene sequences. *FEMS Microbiology Letters*, **115**, 13–18.

Trinkunaite, L.L., Duda, V.I., Mityushina, L.M., Lebedinskii, A.V. and Krivenko, V.V. (1987) A new spore-forming bacterium *Bacillus vesiculiferous* sp. nov. *Microbiologiya*, **56**, 108–113.

Van der Vossen, J.M.B.M., Kok, J. and Venema, G. (1985) Construction of cloning, promoter-screening, and terminator-screening shuttle vectors for *Bacillus subtilis* and *Streptococcus*. *Applied and Environmental Microbiology*, **50**, 540–542.

Vosman, B. and Venema, G. (1983) Introduction of a *Streptococcus cremoris* plasmid in *Bacillus subtilis*. *Journal of Bacteriology*, **156**, 920–921.

White, D., Sharp, R.J. and Priest, F.G. (1994) A polyphasic taxonomic study of thermophilic bacilli from a wide geographical area. *Antonie van Leeuwenhoek*, **64**, 357–386.

Wiegel, J. (1981) Distinction between the Gram reaction and the Gram type of bacteria. *International Journal of Systematic Bacteriology*, **31**, 88.

Wisotzkey, J.D., Jurtshuk Jr, P., Fox, G.E., Deinhard, G. and Poralla, K. (1992) Comparative sequence analyses on the 16S rRNA (rDNA) of *Bacillus acidocaldarius, Bacillus acidoterrestris*, and *Bacillus cycloheptanicus* and proposal for creation of a new genus, *Alicyclobacillus* gen. nov. *International Journal of Systematic Bacteriology*, **42**, 263–269.

Woese, C. (1987) Bacterial evolution. *Microbiology Reviews*, **51**, 221–271.
Wolf, J. and Barker, A.N. (1968) The genus *Bacillus*: aids to the identification of its species. In *Identification Methods for Microbiologists*, part B (eds Gibbs, B.M. and Skinner, F.A.). Academic Press, New York, USA, pp. 93–109.
Yanagida, F., Suzuki, K.-I., Kaneko, T., Kozaki, M. and Komagata, K. (1987a) Morphological, biochemical, and physiological characteristics of sporeforming lactic acid bacteria. *Journal of General and Applied Microbiology*, **33**, 33–45.
Yanagida, F., Suzuki, K.-I., Kaneko, T., Kozaki, M. and Komagata, K. (1987b) Deoxyribonucleic acid relatedness among some spore-forming lactic acid bacteria. *Journal of General and Applied Microbiology*, **33**, 47–55.
Zukowski, M.M. (1992) Production of commercially valuable products. In *Biology of Bacilli: Applications to Industry* (eds Doi, R.H. and McGloughlin, M.). Butterworth-Heinemann, Boston, MS, USA, pp. 311–337.

Index

Note: Systematic names enclosed in double quotation marks are obsolete, or do not have official status at present.

MIX
Papier aus verantwortungsvollen Quellen
Paper from responsible sources
FSC® C105338

FSC
www.fsc.org

If you have any concerns about our products,
you can contact us on
ProductSafety@springernature.com

In case Publisher is established outside the EU,
the EU authorized representative is:
Springer Nature Customer Service Center GmbH
Europaplatz 3, 69115 Heidelberg, Germany

Printed by Libri Plureos GmbH
in Hamburg, Germany